**International
Association
of Fire Chiefs**

**National
Fire Protection
Association**

Fire Officer

Principles and Practice

SECOND EDITION

D0721420

JONES AND BARTLETT PUBLISHERS

Sudbury, Massachusetts

BOSTON   TORONTO   LONDON   SINGAPORE

Jones and Bartlett Publishers
World Headquarters
40 Tall Pine Drive
Sudbury, MA 01776
978-443-5000
info@jbpub.com
www.jbpub.com

National Fire Protection Association
1 Batterymarch Park
Quincy, MA 02169-7471
www.NFPA.org

International Association of Fire Chiefs
4025 Fair Ridge Drive
Fairfax, VA 22033
www.IAFC.org

Jones and Bartlett's books and products are available through most bookstores and online booksellers. To contact Jones and Bartlett Publishers directly, call 800-832-0034, fax 978-443-8000, or visit our website, www.jbpub.com.

Substantial discounts on bulk quantities of Jones and Bartlett's publications are available to corporations, professional associations, and other qualified organizations. For details and specific discount information, contact the special sales department at Jones and Bartlett via the above contact information or send an email to specialsales@jbpub.com.

The procedures and protocols in this book are based on the most current recommendations of responsible sources. The International Association of Fire Chiefs (IAFC), National Fire Protection Association (NFPA®), and the publisher, however, make no guarantee as to, and assume no responsibility for, the correctness, sufficiency, or completeness of such information or recommendations. Other or additional safety measures may be required under particular circumstances.

Some images in this book feature models. These models do not necessarily endorse, represent, or participate in the activities represented in the images.

Notice: The individuals described in "You are the Fire Officer" and "Fire Officer in Action" throughout the text are fictitious.

Additional illustration and photographic credits appear on page 393, which constitutes a continuation of the copyright page.

Production Credits

Chief Executive Officer: Clayton E. Jones
Chief Operating Officer: Donald W. Jones, Jr.
President, Higher Education and Professional Publishing:
  Robert W. Holland, Jr.
V.P., Sales and Marketing: William J. Kane
V.P., Design and Production: Anne Spencer
V.P., Manufacturing and Inventory Control: Therese Connell
Publisher, Public Safety Group: Kimberly Brophy
Senior Acquisitions Editor, Fire: William Larkin
Associate Editor: Amanda Brandt
Associate Production Editor: Sarah Bayle

Associate Photo Researcher: Jessica Elias
Director of Sales: Matthew Maniscalco
Director of Marketing: Alisha Weisman
Marketing Manager, Fire: Brian Rooney
Cover Image: © Mark C. Ide
Text Design: Anne Spencer
Cover Design: Kristin E. Parker
Composition: International Typesetting and Composition
Text Printing and Binding: Courier Corporation
Cover Printing: Courier Corporation

To order this product, use ISBN: 978-1-4496-0162-1

Library of Congress Cataloging-in-Publication Data
Fire officer : principles and practice. — 2nd ed.
    p. cm.
  Includes index.
  At head of title: International Association of Fire Chiefs, National Fire Protection Association.
  ISBN-13: 978-0-7637-5835-6
  ISBN-10: 0-7637-5835-3
  1. Fire departments—Management—Vocational guidance. 2. Fire extinction—Vocational guidance. 3. Fire chiefs—Certification. I. International Association of Fire Chiefs.
  TH9119.F568 2010
  363.37068—dc22
                        2009018595
6048

Printed in the United States of America
17 16 15 14   10 9 8

Brief Contents

Contents

Fire Officer: Principles and Practice, Second Edition

The National Fire Protection Association (NFPA) and the International Association of Fire Chiefs (IAFC) are pleased to bring you the *Second Edition* of *Fire Officer: Principles and Practice*, a modern integrated teaching and learning system for the Fire Officer I and II levels. Fire officers need to know how to make the transition from fire fighter to leader. *Fire Officer: Principles and Practice, Second Edition* is designed to help fire fighters make a smooth transition to fire officer.

Covering the entire scope of *NFPA 1021, Standard for Fire Officer Professional Qualifications*, 2009 Edition, *Fire Officer* combines current content with dynamic features and interactive technology to better support instructors and help prepare future fire officers for any situation that may arise.

The *Second Edition* also reflects the latest developments in:

- Building a personal development plan through education, training, self-development, and experience, including a description of the Fire and Emergency Services Higher Education program.
- Introducing organizational change into your department.
- The impact of blogs, video sharing, and social networks.
- How to budget for a grant.
- Changes in the National Response Framework and National Incident Management System. Additional items related to fire fighter safety and health are included.

Safety is Principle

The *Second Edition* features a laser-like focus on fire fighter safety. Reducing fire fighter injuries and deaths requires the dedicated efforts of every fire fighter, fire officer, fire department, and the entire fire community working together. It is with this goal in mind that we have integrated the 16 Firefighter Life Safety Initiatives developed by the National Fallen Firefighters Foundation into the text. Likewise, in each of the chapters, actual National Fire Fighter Near-Miss Reporting System cases are discussed to drive home safety and the lessons learned from those incidents. Some of the guiding principles added to the new edition include:

- Description of the "Everybody Goes Home" and the National Fire Fighter Near-Miss Reporting System, including over a dozen company officer near-miss examples throughout the text.
- Description of the IAFC/IAFF Firefighter Safety and Deployment Study.
- The latest fire fighter death and injury issues as reported by the NFPA® National Fallen Firefighters Foundation, IAFC, and IAFF, including results of a thirty-year retrospective study.
- Changes in fire-ground accountability and rapid intervention practices.
- Results of National Institute of Standards and Technology research on wind-driven fires, thermal imaging cameras, and fire dynamics as related to fire fighter survival.
- The latest developments in crew resource management.

Chapter Resources

Fire Officer: Principles and Practice, Second Edition serves as the core of a highly effective teaching and learning system. The features reinforce and expand on essential information and make information retrieval a snap. These features include:

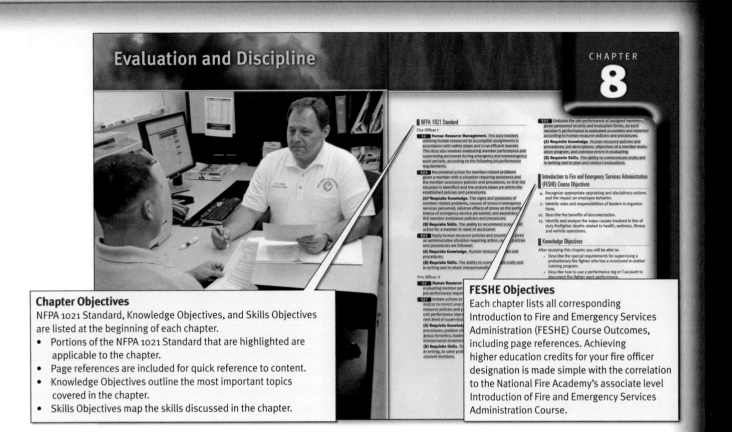

Evaluation and Discipline

Chapter Objectives

NFPA 1021 Standard, Knowledge Objectives, and Skills Objectives
are listed at the beginning of each chapter.

- Portions of the NFPA 1021 Standard that are highlighted are
  applicable to the chapter.
- Page references are included for quick reference to content.
- Knowledge Objectives outline the most important topics
  covered in the chapter.
- Skills Objectives map the skills discussed in the chapter.

FESHE Objectives

Each chapter lists all corresponding
Introduction to Fire and Emergency Services
Administration (FESHE) Course Outcomes,
including page references. Achieving
higher education credits for your fire officer
designation is made simple with the correlation
to the National Fire Academy's associate level
Introduction of Fire and Emergency Services
Administration Course.

Fire Officer II Bar

A bar is used to flag content that applies only to
Fire Officer II students. Students preparing for
Fire Officer I level should skip over the content
marked by the Fire Officer II icon, or incorporate
it at the instructor's discretion for further study.

You Are the Fire Officer

For years, fire fighters at The Nickel spent the first part of the workday wearing physical training clothes and coveralls from morning line-up through grocery shopping. It was an effective use of time. After lunch the crew is showered, neat, and clean in their station uniforms, ready to perform community service, safety inspections, or any other interaction with the public.

This morning's line-up includes the following document from the fire chief's office with the banner message "Immediately Read and Implement." The fire chief has issued a directive to make sure personnel are complying with the standard operating procedure (SOP) on uniforms. That SOP requires members to be in station uniforms any time they are out in public. Although past practice has allowed crews to wear physical training clothes and coveralls when out in public, the SOP provides no exception.

This makes for an uncomfortable morning meeting with the crew. It is clear that two of the fire fighters are very angry and plan to continue to wear coveralls in the morning. One is going to call the shop steward. You think that this order is unreasonable.

1. How do you respond to an order that you do not agree with?
2. How do you implement an unpopular directive?
3. What can you do to maintain the effectiveness of the work group?

You Are the Fire Officer

Each chapter opens with a case study intended to stimulate classroom discussion, capture students' attention, and provide an overview for the chapter. At the end of each chapter, the You Are the Fire Officer: Conclusion summarizes possible solutions for engaging classroom discussion and activities

Introduction to the Fire Officer's Job

The fire officer is responsible for managing a work unit within the fire department. When we think about the duties of a fire officer, the most prominent vision is leading a team of fire fighters into a challenging emergency situation. However, much of what a fire officer does during a normal workday is routine administrative activities related to the work group, the assigned equipment, and the physical facility. The officer is expected to ensure that the work unit will be prepared to function effectively and efficiently when it is needed. That means managing personnel, resources, and programs. This chapter will look at a few overarching principles and examples of the day-to-day administration within a work group.

Supervising and managing fire officers usually report to higher-ranking chief officers. In a large fire department, the direct supervisor is usually an administrative fire officer, usually at the rank level of battalion or district chief. In a smaller organization, a fire officer might report directly to the fire chief or to a deputy or assistant chief. In most cases, this supervisor is an individual who has spent time coming up through the ranks, working as a supervising and managing fire officer.

The International Association of Fire Chiefs (IAFC) defines an administrative fire officer as a person who worked as a managing fire officer for 3 to 5 years, is certified at the NFPA Fire Officer III level, and has accomplished formal education equivalent to a bachelor's degree.

The Fire Officer's Tasks

Administrative fire officers working in cities and large counties responded to an email request to "list the most important tasks you want a new fire officer to do well." This resulted in a list of four basic tasks that they consider vital: beginning of shift report, notifications, decision making, and problem solving. They believed that a new fire officer who meets these task expectations is on the right track.

The Beginning of Shift Report
The administrative fire officers emphasized the importance for a new supervising fire officer to provide a prompt and accurate report at the start of the workday Figure 2-1. This report is provided from each work location to the battalion or district chief within the first quarter hour of the reporting time. The format of the report may be electronic, paper, or verbal. Some departments use

Voices of Experience

One of the circumstances of my promotion to Captain was a shift reassignment. Being scheduled for an overtime shift, I would return to my old shift not as just another one of the guys on the crew, but as an appointed leader.

After roll-call we began our day as usual by conducting apparatus and equipment checks. About 20 minutes had gone by and I noticed that one of the fire fighters on my crew had not checked his personal equipment or the apparatus he was assigned to check. As I began to leave the truck to go look for the fire fighter, he came through the truck bay carrying brushes and soap, and stated that he was going to wash the command vehicle. I told the fire fighter that he needed to check his equipment immediately. The fire fighter mumbled an acknowledgment and kept going. At this point I wasn't sure if this reply was discontent for the actual order or from the direction being given by me as an appointed person of authority.

I gave the fire fighter a few minutes to do what I had asked. Seeing that the fire fighter had not yet checked out his personal equipment I was angered. I found the fire fighter outside beginning to check out his assigned truck. In a rather firm tone I told the fire fighter, "You need to go check out your gear and air-pak....now!" The fire fighter replied, "I don't feel like dealing with this today; my Dad died one year ago today." At that time I realized that this was a pivotal point in handling this problem and though still angry myself, I needed to take a couple minutes to address the problem in a calmer state of mind. I told the fire fighter once again to do what I had asked and I returned to the office.

After a few minutes, I went back outside and asked the fire fighter to come to the office. I asked the acting station chief to come in as well. Understanding that the situation needed to be brought to a resolve with a positive outcome, I began by expressing my sympathy for the fire fighter's situation. At this point, the fire fighter was quite emotional and not wanting to resolve the conflict. The fire fighter stated that I was not his direct supervisor and that we were not going to resolve the issue. He then told the acting station chief that he wanted to go home. Though a bit upset by the fire fighter's venting, I understood the importance of my next reaction. I asked the fire fighter to calm down and hear me out. I told the fire fighter that this was not a personal attack or lack of compassion for his situation. I advised the fire fighter that in my position, I am held accountable for all of my crew members as well as myself and that he was quite simply being asked to fulfill his responsibilities. I re-expressed my sympathy for his personal situation and informed him that though we sometime have difficult times in life, we still have a job to do here as fire fighters. The fire fighter agreed and immediately went to complete the original task.

> " I advised the fire fighter that in my position, I am held accountable for all of my crew members as well as myself. "

Brandon Cunningham
Captain
Fort Gordon Fire and Emergency Services
Ft. Gordon, Georgia

Voices of Experience

In the Voices of Experience essays, veteran fire officers share their accounts of memorable incidents while offering advice and encouragement. These essays highlight what it is truly like to be a fire officer.

x68   Fire Officer: Principles and Practice

A may-day requires a complex response from the company operating within a hot zone or burning structure. The first obligation is to maintain radio discipline so that command can determine the may-day location and situation. The second obligation is to maintain company or sector integrity. Incident control evaporates if every fire fighter rushes to assist the fire fighter in trouble. The fire officer needs to maintain company integrity to facilitate the may-day rescue and continue to fight the fire or control the hazard. Changes in assignment come from the sector incident commander.

The fire officer may directly encounter a dangerous situation, such as a collapsing building, a person with a gun, or a

careening vehicle. The officer must make an immediate, clear, and autocratic command to protect the fire fighters under his or her supervision by removing them from the danger area. Then, the fire officer needs to inform command.

After every incident, the fire officer should briefly review the event while still at the scene or as soon as possible after returning to quarters. This is an opportunity to clarify issues and answer questions. The fire officer can reinforce good practices and immediately identify any unacceptable performance.

Near Miss REPORT

Report Number: 07-0001110

Event Description: We were dispatched to a reported structure fire. The first arriving officer had firsthand knowledge of the dwelling from previous runs there. Companies engaged in a primary search due to the possibility of squatters. After forcing the front door, two members with a 1 3/4" handline and thermal imaging camera (TIC) made their way into the front living area. They encountered a "Collyer Mansion Syndrome."

There was wall-to-wall debris approximately 3' high. The fire was noted to our right and to Side "C" of the dwelling. Advancing the line was slowed due to the debris. About 10' into the room the floor felt like it was sagging. Command was notified of the conditions. After advancing approximately 3' or 4' more the floor dropped even more significantly. Command was notified of the more significant sagging by the handline crew. Recommendations were made to Command to evacuate and go defensive.

Crews immediately backed out and a tower ladder went into operation with numerous handlines. The building was a 1 1/2 story Cape Cod packed with everything from books, monitors, furniture and even a cache of weapons. . . a total of 24 personnel to the scene. All members are equipped with portable radios. All companies have TICs. Attack line experience: Nozzle-13 years; Back-up less than 6 months. Full PPE was in place.

All companies were advised of the building's condition by the first arriving unit. Having live fire training and experience in reading a building's conditions is key. The floor which we were advancing our attack line on had its joists burned completely away. We were prevented from falling into the basement by the carpet being held in place by all of the wall-to-wall debris. This was not discovered until overhaul from the exterior was accomplished. ICS was used and is used on all incidents by this F.D. Weather was cold and dry.

Lesson Learned: Know building construction past and present. Example: a spongy roof 15–20 years ago means something different than a spongy roof today. All firefighters should focus on building construction during training, as it assists us with every aspect of a job (i.e., investigative work, inspections, fire attack, overhaul and salvage, forcible entry, and ventilation). Fires are burning hotter and faster than ever before. This creates fire conditions in which you cannot advance a line fast enough or the 1 3/4" line will be ineffective. The fire may overrun you causing confusion and separation of your crew leading to other problems.

Experience or inexperience can certainly make a difference. Keep experienced fire fighters with inexperienced firefighters when possible. Preferably keep inexperienced fire fighters with an experienced officer. This will ensure that fire fighters learn from experience, not from trial and error. The error part is unforgiving. The command decision to go defensive after hearing the condition report was a key. Having a company back out prematurely is a philosophy of some commanders, fortunately not on this night. Command had complete confidence in his crews and was able to trust in their decisions. Salty aggressive commanders need to learn from this. Technology is not the end all. Falling back on basic fire fighter training should always be considered. Crawling into this fire using a TIC for guidance might have been a big mistake. A thermal imager does not provide you depth perception nor does it feel for you. Using all of your senses, touch, feel and hearing, are senses which are available during a fire attack. Don't forget them.

Near-Miss Report
Utilizing incident data, National Fire Fighter Near-Miss Reporting System cases are discussed to highlight important points about safety and the lessons learned from real-life incidents.

Chapter 3   Fire Fighters and the Fire Officer   **55**

Assessment Center Tips

Know Your Organization's Procedures
Even if the normal practice for your department is to have a specialized or designated person to conduct an investigation of a harassment or hostile workplace complaint, the promotional candidate must know both local and federal procedures that must be followed when an employee files a complaint. Harassment and hostile workplace complaints require a specific and detailed response by the first-line supervisor.

An assessment center scenario may involve a fire fighter coming to the candidate for advice. During the interview process, the fire fighter reveals that he or she may be a victim of harassment or a hostile workplace. The candidate would be expected to explain the options available to the fire fighter in filing either a federal or a local complaint.

a drastic change from the concept of a fire station as a home away from home for a group of fire fighters, but it is necessary to ensure that the fire station maintains a professional work environment. In the eyes of the law and in the opinion of administrators and elected officials, the same rules of behavior apply to an office in city hall at 2 P.M. and a fire station at 2 A.M.

The fire officer can help maintain an appropriate work environment by encouraging and enforcing acceptable behavior whenever fire fighters are on duty. The fire officer accomplishes this by:

- *Educating the employees on the workplace rules and regulations that define expected behavior.* Start with the local government's "Code of Conduct" or other documents that outline the chief administrative officer's expectations of all municipal employees. This can usually be found in the municipality's mission statement, core values, or personnel regulations.
- *Promoting the use of "on duty speech."* The goal is not to change the thoughts or feelings of individual fire fighters, but to establish a workplace environment where certain behaviors and words are not used. Fire fighters can think what they want. However, while they are on duty, in the fire station, or in uniform, they cannot use certain words or phrases or act out certain behaviors.

Figure 3-15 The company officer must identify and correct unacceptable behavior.

- *Being the designated adult.* This requires the fire officer to model appropriate behavior as well as encourage and enforce the same behavior by the fire fighters. The fire officer must identify and correct unacceptable workplace behavior whenever it is observed. Ignoring a problem is, in reality, permitting it to continue. The fire officer who is a candidate for promotion is expected to identify, explain, and enforce the limits of unprofessional behavior  **Figure 3-15**.

A company-level officer should make it a practice to walk around the fire station at various times during the workday to observe what is going on. This walk-around is more important when the officer is in a big station with multiple companies and in combination career–volunteer departments, where there is a constant flow of people coming into and out of the fire station. This practice is not designed to catch someone doing something wrong; rather, it is to make sure that everything is functioning properly. The officer should routinely determine what the crew members are doing and check on the safety of the facility and equipment. At the same time, the officer should look out for unexpected situations and surprises. Having a reputation that you know what is going on in the station and will react to inappropriate situations goes a long way toward encouraging appropriate workplace behaviors **Figure 3-16**.

Assessment Center Tips
Assessment Center Tips offer hints on how to successfully complete promotional examinations.

Getting It Done

The Fire Company Defines Its Nonhostile Environment
A fire officer who unilaterally imposes new and severe requirements that affect the language fire fighters use in the workplace would encounter much resistance. One method of fostering the concept of "on duty speech" or appropriate workplace behavior is to have the fire fighters develop their own set of rules or behaviors. Provide the fire fighters with the source documents (federal and municipal codes and regulations), and ask them to consider what words or behaviors are inappropriate while at work.

Getting it Done
Getting it Done boxes offer suggestions and tips for applying theories, knowledge, and skills to complete tasks.

Company Tips
Company Tips offer practical advice and information on teamwork and communication.

Fire Marks
Fire Marks offer history, lore, legends, and other interesting information.

Safety Zone
Safety Zone offers suggestions and tips for keeping you and your crew safe, no matter what the task.

Fire Marks

Are You a RAT?
Departments use different titles or terms to identify the resources that will be committed to meet the two-in, two-out requirement. A FAST truck (firefighter assist search team), a RIT (rapid intervention team), RICO (rapid intervention company operations), RAT (rapid assist team), and other identifications cover the same assignment.

The two-in, two-out rule is directly related to the concept of a **rapid intervention crew (RIC)**. NFPA 1561, *Standard on Emergency Services Incident Management System*, defines an RIC as a minimum of two fully equipped personnel, standing by on site, in a ready state for immediate rescue of injured or trapped firefighting personnel. Many departments dispatch an additional fire company to working fire incidents with the specific assignment of establishing a rapid intervention crew.

In order to ensure that the RIC can recognize in a timely manner that a fire fighter is missing, a **personnel accountability system** is needed to track the identity, assignment, and location of all fire fighters operating at the incident scene **Figure 6-5**. An accountability system typically provides the following:

- A method to identify all personnel that are on the scene
- A method to identify personnel that are in the hazard area
- A process to quickly account for personnel when an evacuation order or unusual event occurs (e.g., a flashover, backdraft, or collapse)
- A process to ensure that personnel do not go unaccounted for an extended period of time
- A standardized method of notification and reaction to a may-day event

An accountability system must function at multiple levels at an incident scene. Although every fire officer must be accountable for all assigned company members, command officers are accountable for the location and function of full companies. An officer who is assigned to manage a sector for a division at an incident must know where every assigned company is operating and what they are doing at all times. The incident commander must know which companies are assigned to each sector or division.

Figure 6-5 A personnel accountability system.

Company Tips

Overriding the Two-In, Two-Out Rule
The OSHA rule includes an exception that would allow a first-arriving fire company to override the two-in, two-out rule and initiate operations in an IDLH environment if there is a chance that a victim will be rescued alive. The company officer must communicate this action to the others who are responding to the incident. The mere fact that there might be someone in danger is not sufficient justification to ignore the two-in, two-out rule; the exception exists only to avoid the situation where a procedural rule could stand in the way of saving a life.

For an effective accountability system, crews must routinely receive training on and follow the procedures. During an emergency situation, fire fighters will revert to habit. The proper use of an accountability system must be fully integrated as one of those habits.

A personnel accountability system is required by NFPA 1500, *Standard on Fire Department Occupational Safety and Health Program*. Some departments require a personnel accountability report (PAR) roll-call every 20 minutes during fire attack or other high-hazard operations. Each sector chief or company officer must account for the personnel operating under his or her command. It is also used after a sudden change in operating conditions, after a flashover or collapse, or when moving from offensive to defensive fire attack.

Air Management
Self-contained breathing apparatus (SCBA) provide a safe and reliable air supply that allows a fire fighter to operate in an IDLH atmosphere for a finite length of time. The length of time depends on the rate at which the individual consumes the available air supply, which varies considerably, depending on the user and the nature of the work being performed. The rated time is based on a standard consumption rate for an individual at rest and does not reflect typical experience in firefighting operations. Running out of air in an IDLH atmosphere can result in an unconscious fire fighter who will die of asphyxiation.

Some fire departments have re-evaluated how they track SCBA use during incidents. Some departments have replaced their 30-minute rated SCBA air tanks with 45- or 60-minute air tanks to increase the time a fire fighter can operate in IDLH atmospheres. Other departments have established or strengthened incident management procedures to more closely follow individual fire fighter air use.

Low-pressure warning devices, which provide an indication when the remaining air supply reaches a set point, are effective only if the fire fighter is able to exit to a safe atmosphere in the time that is provided. The low-pressure alarm might not provide sufficient exit time for a fire fighter who is deep inside a burning building or unable to exit immediately.

Many departments monitor the air levels of fire fighters on entry and exit. If a fire fighter fails to exit within the expected time period, an accountability roll-call is taken.

Teams and Tools
The minimum size of an interior work team is two fire fighters. Every team member must have full personal protective equipment,

---

... near-miss report. ... often as significant ... level examines resource management (staffing, training, budget resources, and equipment/facility resources) and organizational climate (chain of command, delegation of authority, risk management programs, and safety programs).

Reducing Deaths from Sudden Cardiac Arrest

Heart attacks are the leading cause of death for fire fighters. Although most line-of-duty fatalities for fire fighters under the age of 35 years are from traumatic injuries, fire fighters above the age of 35 years are more likely to die from a medical cause. The NFFF Life Safety Summit noted that a disproportionate number of fire fighters over the age of 49 die of cardiac arrest while on duty.

Before authorization to begin training as a fire fighter, every candidate should be examined by a qualified physician. Regular physical examinations should be scheduled for as long as the fire fighter is engaged in performing emergency duties, with an emphasis on identifying risk factors that could lead to a heart attack under stressful conditions. Fire officers should always look for indications that a member is in poor health or is unfit for duty and, if necessary, arrange for a special evaluation by a fire department–approved physician.

Physical fitness activities should be considered an essential component of every fire fighter's training regimen. Changes in lifestyle can often reduce the risk of a fatal heart attack and can include stopping smoking, lowering high blood pressure, reducing high blood cholesterol level, maintaining a healthy weight, and managing diabetes. Fire officers must understand that these changes are as important to the body as wearing full protective equipment.

The NFPA, the IAFC, the NFFE, the National Volunteer Fire Council (NVFC), and the IAFF have developed resources to help the fire officer encourage healthy living. Fire officers should advocate methods of making positive lifestyle changes to increase fire fighter safety. NFPA 1583, *Standard on Health-Related Fitness Programs for Fire Department Members*, provides a structure and resources for a fire officer to develop a health-related fitness program. The International Association of Fire Chiefs partnered with the IAFF to develop the Fire Service Joint Labor Management

Figure 6-3 The fire officer should strive to be physically fit.

Wellness Fitness Initiative. This initiative produced the Candidate Physical Aptitude Test, as well as a peer fitness training certificate program with the American Council on Exercise.

Regardless of what the fire department mandates, every fire officer should strive to be physically fit **Figure 6-3**. Fitness should be a personal priority and an expression of leadership.

Reducing Deaths from Motor Vehicle Collisions

Collisions account for the largest percentage of traumatic fire fighter deaths. The NFPA data from 1977 to 2007 show that three-quarters of the 406 fatalities from collisions were volunteer fire fighters. Almost 40 percent of the fire fighters died in their personal vehicles. More than three-quarters of the fatalities were not wearing a seat belt. Excessive speed for road conditions and operator error are frequently cited causes of these fatal collisions. Obeying traffic laws, using seat belts, driving sober, and controlling speed would prevent most of the fire fighter fatalities in road collisions.

Placing an untrained driver behind the wheel of an emergency vehicle places the occupants and the general public in immediate danger. The driver of an emergency vehicle is legally authorized to ignore certain restrictions that apply to other vehicles, but only when operating the emergency vehicle in a manner that provides for the safety of everyone using the roadways. Specific procedures for safe emergency response must be practiced. The fire officer is responsible for ensuring that the driver consistently follows the rules of the road for emergency response.

Safety Zone

Physical Fitness Tips
The NFFF summarizes the importance of physical fitness with the following items that support fire fighter maintenance:

- **Regular medical check-ups:** Yes, they can be a pain, but if you do not do it for you, do it for those who need you.
- **Regular exercise:** Even walking makes a *big difference!* Walk a mile a day and watch the changes.
- **Eat healthy:** Think about what you are eating, and then picture operating interior at a working fire 30 minutes later. Now, what do you want to eat?

Fire Officer in Action

This feature promotes critical thinking through the use of case studies and provides instructors with discussion points for the classroom presentation.

Wrap-Up

End-of-chapter activities reinforce important concepts and improve students' comprehension. Additional instructor support and answers for all questions are available on the Instructor's ToolKit CD-ROM.

Chief Concepts

Chief Concepts highlight critical information from the chapter in a bulleted format to help students prepare for exams.

Hot Terms

Hot Terms are easily identifiable within the chapter and define key terms that the student must know. A comprehensive glossary of Hot Terms also appears in the Wrap-Up.

Fire Officer *in Action*

One outcome of the fumigation investigation was discovering that there was an inappropriate procedure to ensure that the local fire companies knew when and where fumigations were occurring. The administrative fire officer who conducted the investigation has tasked your company to recommend a policy change to update the notification procedure. In addition, your company will prepare a field training drill on responding to fumigation incidents.

Within the company you have assigned the rookie to make a list of the address, location, and brand of every rapid entry safe location in the fire station's response area. This information will be added to the property information contained in the computer-aided dispatch program.

The fire officer is the first supervisor or manager to deal with external and internal problems and complaints. Documenting and properly processing each issue is a required activity that frequently is a component in a promotional exam. Beyond complying with the internal process, a fire officer encounters unique opportunities to learn from each problem and complaint as part of a personal professional development journey.

1. Allowing a person with a complaint to completely express grief, anger, pain, or resentment is an example of:
   A. active listening.
   B. paraphrase and feedback.
   C. emphatic listening.
   D. draining the emotional bubble.

2. It has been 6 weeks since implementing a new procedure to reduce the time from dispatch to wheels rolling. The new policy has increased the time it takes to get wheels rolling by 30 percent. You should:
   A. continue the new procedure for another 6 weeks.
   B. re-interview the fire fighters to make them feel understood.
   C. change the plan.
   D. return to the original procedure.

3. While taking notes during a citizen's complaint you realize that the resolution requires action by an executive fire officer. You should:
   A. complete active listening and explain that you will be forwarding the complaint to the appropriate administrative fire officer for investigation.

   B. stop active listening and provide the complainant with the contact information of the appropriate administrative fire officer.
   C. stop the process and ask to reschedule this session when the appropriate administrative fire officer is available.
   D. recommend that the complaint be filled out on a fire department form and sent to the fire chief's office.

4. Which of the following is NOT a response after completing an investigation?
   A. Take no further action.
   B. Recommend additional study.
   C. Suggest an alternative solution.
   D. Refer the issue to the administrative fire officer that can provide a remedy.

Wrap-Up

Chief Concepts

- Promotional examinations were a product of the 1883 Pendleton Civil Service Reform Act.
- Every civil service job has a narrative classified job description and a knowledge, skills, and abilities (KSA) worksheet.
- Promotional examinations assess the important KSAs and measure performance dimensions.
- Multiple-choice examinations concentrate on facts from the reading list.
- A local test committee, composed of existing officers with technical subject matter expertise, validates promotional examinations.
- Assessment centers provide a variety of role-playing exercises.
- During an in-basket assessment, you should remember to review; prioritize; identify resources, options, and alternatives; follow up; and notify.
- Emergency incident simulations fall into four formats: the "data dump" question, a mock emergency incident, an interactive emergency scene simulation, and a full-scale incident simulation.
- During an interpersonal assessment, remember to remain in control, exactly state the desired behavior, give the employee a deadline, arrange follow-up meetings, get the employee to buy into or take personal responsibility for the improvement plan, be empathetic but focused, explain consequences, and finish the session on a positive note.
- Candidates may be required to deliver a short presentation or write a memo or report as part of the assessment center process.
- Technical skills may be evaluated during promotional tests for highly specialized positions.
- The candidate needs to develop a personal study plan to master the content for a promotional examination. Study techniques include keeping a study journal and participating in role-playing activities.

Hot Terms

Assessment centers A series of simulation exercises to identify a candidate's competency to perform the job that is offered in the promotional examination.

Class specification A technical worksheet that quantifies the knowledge, skills, and abilities (KSAs) by frequency and importance for every classified job within the local civil service agency.

"Data dump" question A promotional question that asks the candidate to write or describe all of the factors or issues covering a technical issue, such as suppression of a basement fire in a commercial property.

Dimensions Attributes or qualities that can be described and measured during a promotional examination. Between 5 and 15 dimensions would be measured on a typical promotional examination. The six most common are oral communication, written communication, problem analysis, judgment, organizational sensitivity, and planning/organizing.

In-basket exercise A promotional examination component in which the candidate deals with correspondence and related items accumulated in a fire officer's in-basket.

Job description A narrative summary of the scope of a job. Provides examples of the typical tasks.

Knowledge, skills, and abilities (KSAs) The traits required for every classified position within the municipality. Defined by a narrative job description and a technical class specification.

Metropolitan (metro-sized) fire department A department with more than 400 fully paid fire fighters. The Metropolitan Fire Chiefs is a special interest group organized by the International Association of Fire Chiefs in 1965 to address the needs of large fire departments. Metro Chiefs is also a section of the National Fire Protection Association.

Personal study journal A personal notebook to aid in scheduling and tracking a candidate's promotional preparation progress.

Spoils system Also known as the patronage system. The practice of making appointments to public office based on a personal relationship or affiliation rather than because of merit. The spoils system scandals of the New York City "Tweed Ring" and the Tammany Hall political machine (1865–1871) resulted in Congress passing the Pendleton Civil Service Reform Act of 1883.

Instructor Resources

A complete teaching and learning system developed by educators with an intimate knowledge of the obstacles that instructors face each day supports *Fire Officer: Principles and Practice, Second Edition.* These resources provide practical, hands-on, time-saving tools such as PowerPoint presentations, customizable lecture outlines, test banks, and image/table banks to better support instructors and students. In addition, complete Fire Officer I, Fire Officer II, and Fire Officer I & II curriculum packages have been created to meet every instructor's needs.

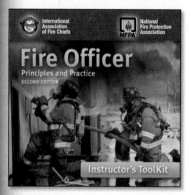

Instructor's ToolKit CD-ROM
ISBN: 978-0-7637-8365-5

Preparing for class is easy with the resources on this CD-ROM, and they have been formatted so that you can seamlessly integrate them into popular course administration tools. Instructors can access resources for Fire Officer I, Fire Officer II, or Fire Officer I & II combined.

The CD-ROM includes the following resources:

- **Adaptable PowerPoint Presentations.** Provide instructors with a powerful way to create presentations that are educational and engaging to their students. These slides can be modified and edited to meet instructors' specific needs.
- **Detailed Lesson Plans.** The lesson plans are keyed to the PowerPoint presentations with sample lectures, lesson quizzes, answers to all end-of-chapter student questions found in the text, and teaching strategies. Complete, ready-to-use lecture outlines include all of the topics covered in the text. The lecture outlines can be modified and customized to fit any course.
- **Image and Table Bank.** Instructors can use these graphics to incorporate more images into the PowerPoint presentations, make handouts, or enlarge a specific image for further discussion.

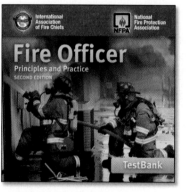

Instructor's Test Bank CD-ROM
ISBN: 978-0-7637-8462-1

The Instructor's Test Bank CD-ROM contains over 1,500 multiple-choice questions, and allows instructors to creat tailor-made classroom tests and quizzes quickly and easily by selecting, editing, organizing, and printing a test along with an answer key, including page references to the text. Each multiple choice question is referenced back to the corresponding NFPA 1021, *Standard for Fire Officer Professional qualifications,* 2009 Edition objective.

The questions found in the Instructor's Test Bank CD-ROM are separated by level so you can creat exams for Fire Officer I, Fire Officer II, or fire Officer I and II combined.

Student Resources

To help students retain the most important information and assist them in preparing for exams, Jones and Bartlett Publishers has developed a complete set of student resources.

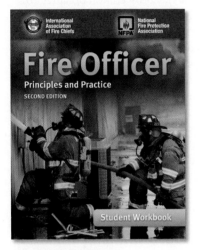

Student Workbook
ISBN-13: 978-0-7637-8367-9

This resource is designed to encourage critical thinking and aid comprehension of the material through the use of the following activities:

- Matching, fill-in-the-blank, short answer, and multiple-choice questions
- In-Basket Exercises
- Case Studies with corresponding questions

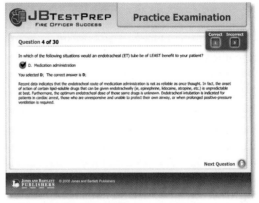

JBTest Prep: Fire Officer Success
ISBN-13: 978-0-7637-8369-3

JBTest Prep: Fire Officer Success is a dynamic program designed to prepare students to sit for Fire Officer I and II certification examinations by including the same type of questions they will likely see on the actual examination.

It provides a series of self-study modules, organized by chapter and level, offering practice examinations and simulated certification examinations using multiple-choice questions. All questions are page referenced to *Fire Officer: Principles and Practice, Second Edition* for remediation to help students hone their knowledge of the subject matter.

Students can begin the task of studying for certification examinations by concentrating on those subject areas where they need the most help. Upon completion, students will feel confident and prepared to complete the final step in the certification process—passing the examination.

Technology Resources

Digital Curriculum Solution Packages

Digital Curriculum Solution Packages allow educators to offer their students cutting-edge digital resources based on world class fire content. The innovative, multifaceted online tools facilitate absolute understanding for company officers. Gold standard content joins sound instructional design in a user-friendly online interface to give students a truly interactive, engaging learning experience with:

Navigate Course Manager:

Combining our robust teaching and learning materials with an intuitive and customizable learning platform, Navigate Course Manager enables instructors to create an online course quickly and easily. The system allows instructors to readily complete the following tasks:

- Customize preloaded content or easily import new content
- Provide online testing
- Offer discussion forums, real-time chat, group projects, and assignments
- Organize course curricula and schedules
- Track student progress, generate reports, and manage training and compliance activities

Navigate Test Prep:

Navigate TestPrep: Fire Officer I & II is a dynamic program designed to prepare students to sit for Fire Officer Certification examinations by including the same type of questions they will likely see on the actual examination.

It provides a series of self-study modules, organized by chapter and level, offering practice examinations and simulated certification examinations using multiple-choice questions. All questions are page referenced to *Fire Officer: Principles and Practice, Second Edition* for remediation to help students hone their knowledge of the subject matter.

Students can begin the task of studying for Fire Officer Certification examinations by concentrating on those subject areas where they need the most help. Upon completion, students will feel confident and prepared to complete the final step in the certification process--passing the examination.

Fire Officer Interactive

Fire Officer Interactive is a series of cutting-edge narrated lectures that are designed to reinforce the knowledge students learn in the classroom and to help them pinpoint any areas of weakness. This interactive resource engages today's students with an exciting and captivating learning experience that is unattainable through traditional printed workbooks or study guides.

Based on the *Fire Officer: Principles and Practice, Second Edition* textbook, Fire Officer Interactive helps to develop the competencies needed to keep students safe on the job

and covers the full scope of the 2009 edition of NFPA 1021, *Standard for Fire Officer Professional Qualifications.* These narrated lectures include:

- In-depth reporting and documentation to enable instructors to track students' progress and achievements.
- A tracking tool that holds students accountable for out-of-the classroom learning. Students are unable to "fast-forward" through the online lecture.
- Both text and audio narration are offered in order to address students' diverse learning styles. The text and audio narration can be enabled or disabled at any time.
- Anytime, anywhere access that provides total flexibility for certification or refresher training.
- A simple, user-friendly interface to ensure that students do not waste time learning how to navigate through the lecture.
- A safe and self-paced training environment that allows students to learn at their own convenience and pace.

Web Tools:

Unlock additional interactive educational resources such as chapter pretests, case studies and skill drill activities at www.fire.jbpub.com.

The Fire Officer Digital Curriculum Solution Packages have been created in three distinct levels: Level I only, Level II only or Level I & II combined. The packages are:

Fire Officer Advantage Package
ISBN: 978-1-4496-9765-5
- Printed Textbook
- Student Workbook
- Navigate Course Manager
- Web Tools

Fire Officer Preferred Package
ISBN: 978-1-4496-9766-2
- Printed Textbook
- Student Workbook
- Navigate Course Manager
- Navigate TestPrep
- Web Tools

Fire Officer Premier Package
ISBN: 978-1-4496-9767-9
- Printed Textbook
- Student Workbook
- Navigate Course Manager
- Navigate TestPrep
- Fire Officer Interactive
- Web Tools

Additional Fire Officer Resources

The following resources have been developed to help support future and current Fire Officers.

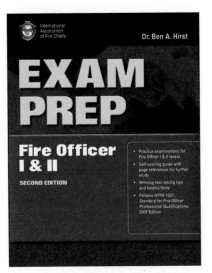

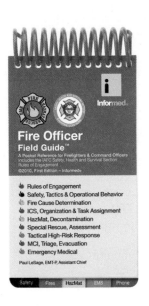

Exam Prep: Fire Officer I & II, Second Edition
ISBN 13: 9780763785970

The Second Edition of **Exam Prep: Fire Officer I & II** is designed to thoroughly prepare you for a Fire Officer I or II certification, promotion, or training examination by including the same type of multiple-choice questions you are likely to encounter on the actual exam.

To help improve examination scores, this preparation guide follows Performance Training Systems, Inc.'s Systematic Approach to Examination Preparation. **Exam Prep: Fire Officer I & II, Second Edition** is written by fire personnel explicitly for fire personnel, and all content has been verified with the latest reference materials and by a technical review committee.

Benefits of the Systematic Approach to Examination Preparation include:

- Emphasizing areas of weakness
- Providing immediate feedback
- Learning material through context and association

Exam Prep: Fire Officer I & II, Second Edition includes:

- Practice examinations for Fire Officer I & II levels
- Self-scoring guide with page references for further study
- Winning test-taking tips and helpful hints
- Coverage of *NFPA 1021, Standard for Fire Officer Professional Qualifications, 2009 Edition*

Your exam performance will improve after using this system!

Fire Officer Field Guide
ISBN 13: 9781890495435

Whether you are a command officer or a beginning fire fighter, the Fire Officer Field Guide™ will prove to be an essential tool throughout your career. This guide includes rapid-use checklists for safe operational behaviors, situational awareness tips, RIT team activation, mayday and rescue procedures, incident command with the latest NIMS developments, HazMat section with updated chemical characteristics, MCI, tactical tips, an updated emergency medical section, and more.

The Fire Officer Field Guide™ provides instant access to critical information in an easy-to-use checklist format. At only 3"x5", this guide fits in your pocket, is waterproof, alcohol-fast, and street tough. This guide covers:

- Rules of Engagement
- National ICS Profiles
- Updated Fire Strategy / Tactics
- Improved Operations Center Guidelines
- Mayday and Fire Fighter Rescue
- Evacuation and Abandon Structure
- Enhanced WMD and HazMat Response
- Special Rescue Situations
- Incident Command System
- Updated EMS
- Section Water Supply and Friction Loss

Also available in the following mobile app formats:

- iPhone
- Nook
- Android

Acknowledgments

Jones and Bartlett Publishers, the National Fire Protection Association and the International Association of Fire Chiefs would like to thank past editors, contributors, and reviewers of *Fire Officer: Principles and Practice.*

Editorial Board

Shawn Stokes
International Association of Fire
  Chiefs
Fairfax, Virginia

Kendall Holland
National Fire Protection
  Association
Quincy, Massachusetts

Mike Ward
Assistant Professor
The George Washington University
Washington, D.C.

Contributors

Robert H. Brown, EFO
  Coordinator, Fire Science Technology
    Management
  Red Rocks Community College
  Lakewood, Colorado

Bill Carey
  Hyattsville Volunteer Fire Department
  Hyattsville, Maryland

Brandon Cunningham
  Captain
  Fort Gordon Fire & Emergency
    Service
  Fort Gordon, Georgia

Mike Dalton
  Fire Investigator
  Knox County Fire Investigation Unit
  Knox County, Tennessee

Leonard C. Edge Jr.
  Instructor
  Fire Protection Technology
  Central Piedmont Community
    College
  Charlotte, North Carolina

David Hall
  Assistant Fire Chief
  Springfield, Missouri Fire Department
  Springfield, Missouri

Rudy Horist
  Assistant Chief
  Elgin Fire Department
  Elgin, Illinois

Randy T. Keirn, MPA, BSN, RN, EMT-P
  District Chief
  Lealman Fire District
  St. Petersburg, Florida

Richard Kosmoski
  Ex-Chief
  New Jersey State Certified Fire
    Service Instructor
  Middlesex County Fire Academy
  Sayreville, New Jersey

Peter J. Lamb
  Fire Chief
  North Attleboro Fire Department
  North Attleboro, Massachusetts

Paul LeSage
  Assistant Chief
  Tualatin Valley Fire and Rescue
  Aloha, Oregon

David Polikoff
  Captain
  Montgomery County Fire and
    Rescue
  Montgomery County, Maryland

Paul E. Ricci, EFO
  Assistant Chief
  Sandusky Fire Department
  Sandusky, Ohio

Keith D. Stahl
  Training Officer
  Calgary Fire Department
  Calgary, Alberta,
  Canada

Adam K. Thiel
  Fire Chief
  Alexandria Fire Department
  Alexandria, Virginia

Chris Watson
  Battalion Chief
  Austin Fire Department
  Austin, Texas

Steve Willis
  Fire Science Faculty member
  Southern Maine Community College
  South Portland, Maine

Dean S. Wrobbel, AAS
  Deputy Chief of Operations
  Saint Cloud Fire Department
  Saint Cloud, Minnesota

Reviewers

Christopher Albro
Warwick Fire Department
Warwick, Rhode Island

John P. Binaski, MS, EFO
College of the Sequoias
City of Tulare Fire Department
Visalia, California

Robert H. Brown, EFO
Coordinator, Fire Science Technology
 Management
Red Rocks Community College
Lakewood, Colorado

Josh Buckman
Indiana Firefighter Training System
Indianapolis, Indiana

Melvin Byrne, MS, EFO
Division Chief
Ashburn Volunteer Fire and Rescue
Virginia Department of Fire Programs
Leesburg, Virginia

Paul E. Calderwood
Deputy Fire Chief
Everett Fire Department
Everett, Massachusetts

Brandon Cunningham
Captain
Fort Gordon Fire and Emergency
 Service
Fort Gordon, Georgia

Kevin C. Easton
Division Chief
Sarasota County Fire Department
Sarasota, Florida

Leonard C. Edge Jr.
Instructor
Fire Protection Technology
Central Piedmont Community College
Charlotte, North Carolina

Todd A. Farley, CFO
Deputy Chief
Central Jackson County Fire
 Protection District
Blue Springs, Missouri

Lt. J. David Feichtner
Farmington Hills Fire Department
Farmington Hills, Michigan

Neil R. Fulton
Norwich Fire Department
Norwich, Vermont

Thomas Gaddie
Battalion Chief (ret.)
Lanier Technical College
Oakwood, Georgia

Mike Gagliano
Captain
Seattle Fire Department
Seattle, Washington

Michael A. Gorsuch
Station Captain
Columbia River Fire and Rescue
St. Helens, Oregon

Captain Ian M. Harmon
Fire Fighter/NREMT-P
Lexington, South Carolina

Rudy Horist
Assistant Chief
Elgin Fire Department
Elgin, Illinois

Mike Jones
Lieutenant, Fire Coordinator
Burleson, Texas Fire Department
TrainingDivision.com Fire Academy
Crowley, Texas

Randy T. Keirn, MPA, BSN, RN, EMT-P
District Chief
Lealman Fire District
St. Petersburg, Florida

Barb Kidd, MEd
Program Coordinator, Accreditation
 and Curriculum Services
Fire and Safety Division, Justice
 Institute of BC
New Westminster, British Columbia,
 Canada

Edward King
Deputy Chief
Vincennes Township Fire
 Department
Vincennes, Indiana

Michael Kinkade
Fire Chief, Forest Grove Fire and
 Rescue
President, Oregon Fire Instructors
 Association
Forest Grove, Oregon

**Kurt Klunder, Fire Medic Journeyman/
 NREMT-P, CCEMT-P**
Rapid City Fire Department
Rapid City, South Dakota

Richard Kosmoski
Ex-Chief
New Jersey State Certified
 Fire Service Instructor
Middlesex County Fire Academy
Sayreville, New Jersey

Jeff Lambert, MA
Fire Chief, Port Moody Fire Rescue
Fire and Safety Division, Justice
 Institute of BC
New Westminster, British Columbia,
 Canada

David B. Mattingly, MPA, CFO
Assistant Chief
Lexington Division of Fire and
 Emergency Services
Lexington, Kentucky

Michael A. Naglieri, AAS, BS
Associate Professor, Director,
 Fire Science
Lone Star College Cy-Fair
Cypress, Texas

Shawn Neat
City of Portland Maine
 Fire Department
Portland, Maine

Marion B. "Ollie" Oliver
Assistant Professor, Program Director
Fire Science Technology,
 Midland College
Driver/Operator and Paramedic (ret.)
Midland Fire Department
Midland, Texas

Bill Pfeifer
Training Specialist
Nebraska State Fire Marshal,
 Training Division
Madison, Nebraska

Jason Powell, BA, NREMT-P
Lieutenant
Johnson City Fire Department
Tennessee Fire Service and
 Codes Academy
Johnson City, Tennessee

Paul E. Ricci, EFO
Assistant Chief
Sandusky Fire Department
Sandusky, Ohio

Reviewers, *continued*

Derrick D. Richardson
Captain
Alton Fire Department
Alton, Illinois

Mike Richardson
Captain/Training Officer
St. Matthews Fire Department
Louisville, Kentucky

Douglas Rohn
Lieutenant
Madison Fire Department
Madison, Wisconsin

Reyes Romero
Deputy State Fire Marshal
New Mexico Firefighter
 Training Academy
Socorro, New Mexico

Trevor Sommerfeld
Training Officer
Calgary Fire Department
Calgary, Alberta,
Canada

Keith D. Stahl
Training Officer
Calgary Fire Department
Calgary, Alberta,
Canada

Robert Swiger
Retired Field Service Supervisor
North Carolina Department of
 Insurance Office State Fire Marshal
Fire Chief (ret.)
Carrboro Fire Department
Carrboro, North Carolina

Dean Williams
Battalion Chief
West Valley City Fire Department
West Valley City, Utah

Steve Willis
Fire Science Faculty member
Southern Maine Community College
South Portland, Maine

Jim Wishlove
Captain
Richmond Fire Rescue
Fire and Safety Division, Justice
 Institute of BC
New Westminster, British Columbia,
 Canada

Dean S. Wrobbel, AAS
Deputy Chief of Operations
St. Cloud Fire Department
St. Cloud, Minnesota

Gray Young
Chief of Fire Training, South
 Bossier Fire
Adjunct, LSU Fire and Emergency
 Training Institute
Minden, Louisiana

Photographic Contributors

We would like to extend a huge "thank you" to Glen E. Ellman, the photographer for this project. Glen is a commercial photographer and fire fighter based in Fort Worth, Texas. His expertise and professionalism are unmatched!

Thank you to the following organizations that opened up their facilities for this photo shoot:

Fort Worth Fire Department
Fort Worth, Texas

Portland Fire Department
Portland, Maine

Introduction to the Fire Officer

NFPA 1021 Standard

Fire Officer I

4.1* **General.** For qualification at Fire Officer Level I, the candidate shall meet the requirements of Fire Fighter II as defined in NFPA 1001, Fire Instructor I as defined in NFPA 1041, and the job performance requirements defined in Sections 4.2 through 4.7 of this standard. [p 5]

4.1.1 **General Prerequisite Knowledge.** The organizational structure of the department; geographical configuration and characteristics of response districts; departmental operating procedures for administration, emergency operations, incident management system and safety; departmental budget process; information management and recordkeeping; the fire prevention and building safety codes and ordinances applicable to the jurisdiction; current trends, technologies, and socioeconomic and political factors that affect the fire service; cultural diversity; methods used by supervisors to obtain cooperation within a group of subordinates; the rights of management and members; agreements in force between the organization and members; generally accepted ethical practices, including a professional code of ethics; and policies and procedures regarding the operation of the department as they involve supervisors and members. [p 4–5, 10–12, 15, 17]

4.1.2 **General Prerequisite Skills.** The ability to effectively communicate in writing utilizing technology provided by the AHJ; write reports, letters, and memos utilizing word processing and spreadsheet programs; operate in an information management system; and effectively operate at all levels in the incident management system utilized by the AHJ. [p 6, 9, 15]

Fire Officer II

5.1 **General.** For qualification at Level II, the Fire Officer I shall meet the requirements of Fire Instructor I as defined in NFPA 1041 and the job performance requirements defined in Sections 5.2 through 5.7 of this standard. [p 5]

5.1.1 **General Prerequisite Knowledge.** The organization of local government; enabling and regulatory legislation and the law-making process at the local, state/provincial, and federal levels; and the functions of other bureaus, divisions, agencies, and organizations and their roles and responsibilities that relate to the fire service. [p 4, 9–11, 15–16]

5.1.2 **General Prerequisite Skills.** Intergovernmental and interagency cooperation. [p 10, 15–16]

Additional NFPA Standards

NFPA 1001 *Standard for Fire Fighter Professional Qualifications*

NFPA 1041 *Standard for Fire Service Instructor Professional Qualifications*

Introduction to Fire and Emergency Services Administration (FESHE) Course Outcomes

3. Articulate the concepts of span of control, effective delegation and division of labor. [p 13]

5. Examine the history and development of management and supervision. [p 11–12, 15–16]

7. Identify roles and responsibilities of leaders in organizations. [p 4–6, 11]

10. Identify the importance of ethics as they apply to supervisors. [p 17]

Knowledge Objectives

After studying this chapter, you will be able to:

- Describe the roles and responsibilities of the Fire Officer I.
- Describe the roles and responsibilities of the Fire Officer II.
- Describe the fire service in the United States.
- Describe fire department organization.
- Describe the functions of management.
- Describe rules and regulations, policies, and standard operating procedures.
- Describe working with other organizations.
- Describe the challenges for the 21st-century fire officer.

Skills Objectives

There are no skills objectives for this chapter.

You Are the Fire Officer

*I*t is 40 minutes before the oncoming group of fire fighters will start to arrive, but you are already at the fire station. Today is your first day as a new supervising fire officer, assigned to Tower Ladder 5. Unable to sleep, you wanted to get in before the rest of the crew arrived. While the first pot of firehouse coffee is brewing, you look at the pictures and mementos displayed throughout the station. They document the people and incidents that made the rich history that defines the station known as "The Nickel."

You cannot miss the contrast of the pristine and shiny company officer helmet sitting on top of your weathered protective gear. What stories will you add to The Nickel?

1. What are the responsibilities of a supervising fire officer?
2. How does a new supervising officer work within the department's organizational structure?
3. How can you convert the theories and concepts memorized for a promotional exam into effective unit officer activities?

Introduction

This book provides information to meet the standards of the National Fire Protection Association (NFPA) 1021, *Standard for Fire Officer Professional Qualifications*, at the Fire Officer I and Fire Officer II levels. The professional qualification standards for fire officers are documented in NFPA 1021.

The NFPA 1021 standard defines four levels of fire officer. The technical committee that developed the current edition of NFPA 1021 performed a task analysis to validate the existence of four distinct fire officer levels and the specific requirements that should apply at each level.

The Fire Officer I level is the first step in a progressive sequence and is generally associated with an officer supervising a single fire company or apparatus. A Fire Officer I could also be assigned to supervise a small administrative or technical group. The next step, Fire Officer II, generally refers to the senior non–chief officer level in a larger fire department. An officer at this level could be the overall supervisor of a multiple-unit fire station. A Fire Officer II could also be in charge of a larger group performing a specialized service or a significant administrative section within the fire department.

Fire Officer III and IV generally refer to chief officer positions. An individual who is qualified at the Fire Officer III level might be qualified to work as a battalion or district chief in a large department and possibly as a deputy or assistant chief in a smaller organization. Fire Officer IVs tend to be fire chiefs or

hold senior positions in charge of a major component of the fire department.

An officer is responsible for being a leader and supervisor to a crew of fire fighters, managing a budget for the station, understanding the response district, knowing departmental operational procedures, and being able to manage an incident. The officer must also understand fire prevention methods, fire and building codes and applicable ordinances (laws enacted by a local government or municipality), and the department's records management system. At each higher level, there are increasing requirements for knowledge and management skills.

Fire Officer I

The Fire Officer I classification is generally bestowed upon an individual who supervises a single fire suppression unit or a small administrative group within a fire department. At this level, the emphasis is placed on accomplishing the department's goals and objectives by working through subordinates to achieve desired results. The Fire Officer I must be able to prioritize multiple demands on the time of the company or work group members and to delegate tasks to subordinates. These demands may be related to emergency operations, nonemergency tasks, or administrative functions.

The Fire Officer I performs administrative duties and supervisory functions that are related to a small group of fire department members. Typical administrative duties include record keeping, managing projects, preparing budget requests, initiating and completing station maintenance requisitions, and conducting preliminary accident investigations. Supervisory duties include making work assignments and ensuring that health and safety procedures are followed. Nonemergency duties could include developing pre-incident plans, providing company-level training, delivering public education programs, and responding to community inquiries.

Emergency duties include the ability to supervise a group of fire fighters performing company-level tasks, functioning as the initial arriving officer at an emergency scene, performing a size-up, establishing the incident management system, developing and implementing an incident action plan, deploying resources, and maintaining personnel accountability. Once the emergency incident has been mitigated, the Fire Officer I is expected to conduct a preliminary investigation to determine the cause, secure the scene to preserve evidence, and conduct a postincident analysis. Fire Officer I candidates are also required to meet all of the requirements of Fire Fighter II as defined in NFPA 1001, *Standard for Fire Fighter Professional Qualifications* and Fire Instructor I as defined in NFPA 1041, *Standard for Fire Service Instructor Professional Qualifications*.

The International Association of Fire Chiefs (IAFC) uses another term to distinguish the different company officers. The IAFC calls the Fire Officer I level a **Supervising Fire Officer** within its *Officer Development Handbook*.

Fire Officer II

The requirements for Fire Officer II begin with meeting all of the requirements of Fire Officer I as defined in NFPA 1021. As with Fire Officer I, the duties of Fire Officer II can be divided into administrative, nonemergency, and emergency activities.

Administrative duties include evaluating a subordinate's job performance, correcting unacceptable performance, and completing formal performance appraisals on each member. Other duties include developing a project or divisional budget, including the related activities of purchasing, soliciting, and rewarding bids, and preparing news releases and other reports to supervisors.

Nonemergency duties include conducting inspections to identify hazards and address fire code violations; reviewing accident, injury, and exposure reports to identify unsafe work environments or behaviors; and taking approved action to prevent reoccurrence of an accident, injury, or exposure. Other duties could include developing a pre-incident plan for a large complex or property; developing policies and procedures appropriate for this level of supervision; analyzing reports and data to identify problems, trends, or conditions that require corrective action; and then developing and implementing the required actions.

Emergency duties include supervising a multiunit emergency operation using an **incident command system (ICS)** and developing an operational plan to safely deploy resources to mitigate the incident. ICS defines the roles and responsibilities to be assumed by personnel and the operating procedures to be used in the management and direction of emergency operations. The Fire Officer II is also expected to determine the area of origin and preliminary cause of a fire and to develop and perform a postincident analysis of a multicompany operation.

The IAFC identifies the Fire Officer II level as a **Managing Fire Officer**. The goal of the IAFC *Officer Development Handbook* is to encourage company officers to acquire the appropriate levels of training, experience, self-development, and education throughout their professional journey to prepare for the Chief Fire Officer designation as the pinnacle of professional development.

Roles and Responsibilities of a Fire Officer

The roles and responsibilities of a fire officer differ from those of a fire fighter. Understanding the new roles is essential for the new fire officer to succeed.

Roles and Responsibilities for Fire Officer I

A Fire Officer I has the following roles and responsibilities:

- Supervises and directs the activities of a single unit
- Instructs members of the company regarding operating procedures, including duty assignments and giving special instructions when fighting fires
- Responds to alarms for fires, vehicle extrications, hazardous materials incidents, emergency medical incidents, and other emergencies as required
- Assumes command of emergency scenes, per the incident command system; analyzes situations; and determines proper procedures until being relieved by a higher ranking officer
- Administers emergency medical first aid and cardiopulmonary resuscitation (CPR), and attends to victims until primary medical personnel arrive
- Oversees routine and preventive maintenance and makes periodic inspections of their assigned apparatus
- Receives direction and instruction from the fire captain and battalion chief regarding station operations, grounds and building maintenance, and overall fire scene action
- Provides training to crew members regarding the apparatus operations, including leading practical training exercises; participates in departmental in-service training and drills
- Evaluates employee performance and conducts performance reviews
- Reads, studies, interprets, and applies departmental procedures, technical manuals, building plans, and so on
- Completes and maintains manual or computer records and prepares necessary reports on incidents, accidents, and personnel training
- Performs prefire planning activities, including touring and studying businesses for physical layout, possible hazards, location of water sources, exposure problems, potential life loss, and so on
- Conducts occupancy inspections
- Determines cause and a preliminary origin of the fire
- Participates, prepares, and delivers various public education programs regarding fire prevention and safety and conducts tours of the fire station as required

Near Miss REPORT

Report Number: 07-0000649

Event Description: We were responding to an automatic alarm. It was evening, clear conditions, dry roads. I was the officer on the ladder truck, following behind the engine. We entered an on ramp to a state road when I noticed the passenger door on the officer's side of the engine in front of us come open, then close. Both the engine and truck were placed in service while responding and returned to quarters. Once back at the station, I asked the firefighter who was riding in the seat what happened. He said as the truck rounded the curve of the on ramp, the door popped open and then slammed shut on its own. He stated it startled him but he considered himself fortunate because he was wearing his seat belt.

Lesson Learned: Wear your seat belt at all times.

- Assists in fire safety inspections of public and private buildings or property
- Participates in and oversees the periodic inspection and testing of equipment, such as hoses, ladders, and engines
- Works directly in firefighting activities; utilizes tools, equipment, portable extinguishers, hoses, ladders, and so on
- Takes appropriate action on the maintenance needs of equipment, buildings, and grounds
- Supervises and performs maintenance and cleaning work on fire equipment, buildings, and grounds

Additional Roles and Responsibilities for Fire Officer II

A Fire Officer II has the same roles and responsibilities as for Fire Officer I, with the following additional items:

- Supervises and directs activities of a multiunit station
- Completes employee performance appraisals
- Creates a professional development plan for members of the organization
- Leads water rescue, hazardous materials, or other special teams as assigned
- Ensures the safe and proper use of equipment, clothing, and protective gear and enforces departmental policies
- Participates in the formulation or evaluation of departmental or agency policies as assigned, implements new or revised policies, and encourages team efforts of fire personnel
- Participates in the formulation of the departmental budget and makes purchases within it
- Develops emergency incident operational plans requiring multiunit operations
- Prepares written report so major causes for local service demand are identified for various planning areas within the service area of the organization

This book covers only the roles and responsibilities of Fire Officers I and II according to NFPA 1021. Fire Officer III, IAFC's Administrative Fire Officer, and Fire Officer IV, IAFC's Executive Fire Officer, have more training and responsibilities.

The transition to fire officer is a big step in a career. It involves not only increased responsibility, but also a different role from that of a fire fighter. The officer is a part of management and is responsible for the conduct of others. The officer has to apply policies, procedures, and rules to subordinates and to different situations. This means being consistent and fair and not playing favorites. These changes often require difficult adjustments for the new officer. As an officer, you will be required to take actions that might not make you happy or popular, but they are your responsibility. It is like the role of a parent—often difficult, but ultimately rewarding.

Fire Service in the United States

The U.S. fire service originated as communities of citizens who responded when a fire broke out. Fighting fires was considered a civic duty, and no compensation was provided. Citizens volunteered their time to answer the call of public service, and each fire fighter had a regular occupation that provided a living. Over time, the fire service has evolved into many different methods of providing personnel when the alarm sounds.

Many fully volunteer departments are composed of members who are notified when there is an alarm. These men and women drop whatever they are doing to respond to the emergency. Some volunteer departments have a sufficient number of personnel and volume of calls that members are scheduled to be on standby or present at the fire station for specific shifts, according to a duty roster. In both forms of fully volunteer departments, the personnel are still unpaid for their services.

Some departments have moved away from the purely volunteer method of staffing due to an increasing number of alarms and a decreasing availability of volunteers. Some departments provide an incentive for fire fighters by paying them for each response to an alarm. These departments are termed "paid on call" or use part-time paid personnel.

Recruitment and retention are a challenge in volunteer fire departments that respond to several calls each day, particularly in areas where few individuals are available to serve as volunteers. The demands often exceed the amount of time volunteers are able to commit to the fire department, even if compensation is provided. A combination department uses full-time career personnel along with volunteer or paid-on-call personnel. This system usually provides faster response times because some personnel are on duty at the stations, ready to respond immediately. Frequently, the full-time staff consists of a minimum number of fire fighters, allowing for apparatus to respond and handle routine emergencies, such as medical assists, motor vehicle collisions, and incipient fires. The volunteer or paid-on-call staff are dispatched as a backup force when an incident exceeds the capabilities of the full-time personnel, such as a working structure fire.

A career department is staffed by full-time, paid personnel whose regular job is working for the fire department. These departments are typically found where the level of risk and call volumes require personnel to be on duty at the station at all times.

Although there are four common forms of staffing fire department organizations, most discussions divide fire fighters into two categories: career and volunteer. NFPA says there are 1.2 million fire fighters in the United States. Of this total, approximately 28% are full-time, career fire fighters and 72% are volunteers, which includes part-time and paid-on-call fire fighters. Three out of four career fire fighters work in communities with populations of 25,000 or more. Most volunteer fire fighters work in fire departments that protect small, rural communities with populations of 2500 or less.

There are 30,185 fire departments in the United States. Combination departments have varying proportions of career and volunteer members and can be mostly career or mostly volunteer.

Private industry and nongovernmental organizations operate fire brigades or plant emergency response teams to protect factories, processing plants, large private facilities, and isolated communities. Although many are established to handle incipient fire situations, some are organized along the lines of a municipal fire department, including fire officers.

Safety Zone

United States Fire Statistics
- There were 530,500 residential fires in the United States in 2007.
- A total of 3000 people died in residential fires in 2007.
- The total number of people killed in fires in 2007 was 3430.
- Every 20 seconds, a fire department responds to a fire somewhere in the United States.
- An estimated 32,500 intentionally set structure fires occurred in 2007.
- There were 840,500 outside fires in 2007.

History of the Fire Service

Since prehistoric times, controlled fire has been a source of comfort and warmth, but uncontrolled fire has brought death and destruction. Accounts from the Roman Empire describe community efforts to suppress uncontrolled fires. In 24 BC, the Roman emperor Augustus Caesar created what was probably the first fire department. Called the *Familia Publica*, it was composed of about 600 slaves who were stationed around the city to watch for and fight fires. But because the *Familia Publica* were slaves, they had little interest in preserving the homes of their masters and little desire to take risks, so fires continued to be a problem. By about AD 60, under the emperor Nero, the *Corps of Vigiles* had been established as the fire protectors. This group of 7000 free men was responsible for firefighting, fire prevention, and building inspections. The Corps adopted the formal rank structure of the Roman military, which is still used by most fire departments.

The first documented fire in North America was in Jamestown, Virginia, in 1607. The fire started in the community blockhouse and almost burned down the entire settlement. At that time, most structures were built entirely of combustible materials, such as straw and wood. Local ordinances soon required the use of less flammable building materials and mandated that fires be "banked," or covered over, throughout the night. In 1630, Boston, Massachusetts, established the first fire regulations in North America when it banned wood chimneys and thatched roofs. In the Dutch colony of New York in 1647, Governor Peter Stuyvesant not only banned wood chimneys and thatched roofs, but also required that chimneys be swept out regularly. Fire wardens imposed fines on those who did not obey the regulations; the money collected was used to pay for firefighting equipment.

Colonial fire fighters had only buckets, ladders, and fire hooks (tools used to pull down burning structures). Homeowners were required to keep buckets filled with water outside their doors and to bring them to the scene of the fire. Some towns also required that ladders be available so fire fighters could access the roof to extinguish small fires. If all else failed, the fire hook was used to pull down a burning building and prevent the fire from spreading to nearby structures. The "hook-and-ladder truck" evolved from this early equipment.

The first organized volunteer fire company was established in Philadelphia. The Union Fire Company was formed in 1735, under the leadership of Benjamin Franklin. Franklin recognized the many dangers of fire and continually sought ways to prevent it. For example, he developed the lightning rod to help draw lightning strikes, a common cause of fires, away from homes. Another early volunteer fire fighter, George Washington, imported one of the first fire engines from England, which he donated to the Alexandria (Virginia) Fire Department in 1765.

In 1871, two major fires significantly affected the development of both the fire service and fire codes. At the time, the city of Chicago was a boom town with 60,000 buildings—40,000 of which were constructed wholly of wood, with roofs of tar and felt or wooden shingles. With the lax construction regulations and no rain for 3 weeks, the city was extremely dry. On October 8, 1871, a fire started in a barn on the west side of the city. The fire department was already exhausted from fighting a four-block fire earlier in the day. Errors in judging signaling the alarm resulted in a delayed department response. The Great Chicago Fire burned

Figure 1-1 The Great Chicago Fire of 1871 caused the deaths of 300 people and led to changes in firefighting operations in the United States.

through the city for 3 days **Figure 1-1 ▲**. When it was over, more than 2000 acres and 17,000 homes had been destroyed, the city had suffered over $200 million in damage, 300 people were dead, and 90,000 were homeless.

At the same time, another fire was raging 262 miles north of Chicago in Peshtigo, Wisconsin. Throughout the summer, the north woods of Wisconsin had experienced drought-like conditions. Logging operations had left pine branches carpeting the forest floor. A flash forest fire created a "tornado of fire" more than 1000 feet high and 5 miles wide. More than 2400 square miles of forest land burned, several small communities were destroyed, and more than 2200 people lost their lives. The Peshtigo firestorm even jumped the 60-mile-wide Green Bay to destroy several hundred square miles of land and additional settlements on Wisconsin's northeast peninsula. Although not as publicized as the Chicago fire, Peshtigo would become the deadliest fire in United States history.

These events would forever change the U.S. fire service and the role of the fire officer. Communities began to enact strict building and fire codes. The development of water pumping systems, advances in firefighting equipment, and improvements in communications and alarm systems all helped ensure that such tragedies would not recur.

Safety Zone

Canadian Fire Statistics
In Canada, fire loss statistics are compiled in each province by Human Resources Development Canada. The results are forwarded to the Canadian Council of Fire Marshals and Fire Commissioners, which produces an annual fire loss report for Canada. In 2002, the latest year for published results, the statistics revealed:

- There were 22,186 residential fires, resulting in the loss of 250 lives.
- The total number of people killed in fires was 304.
- Every 2.5 minutes, a fire department responds to a fire somewhere in Canada.
- An estimated 9990 intentionally set fires occurred.

Fire Marks

An Ounce of Prevention
Along with the first U.S. fire company, Benjamin Franklin also organized the first fire insurance company in the United States and coined the phrase "an ounce of prevention is worth a pound of cure," an apt motto for fire safety. Early insurance companies marked the homes of their policy holders with a plaque, or **fire mark**, that showed the name or logo of the insurance company. The insurance company paid fire departments to respond to those buildings displaying their fire mark. Sadly, others were left to burn. There are still some communities in the United States that fund their fire departments through annual subscription fees.

■ Fire Equipment

Buckets gave way to hand-powered pumpers in 1720 when Richard Newsham developed the first such pumper in London, England. Dozens of individuals powered the pump, making it possible to propel a steady stream of water from a safe distance. By 1829, more powerful, steam-powered pumpers had been developed and began to replace the hand-powered pumpers. Many volunteer fire fighters felt threatened by the steam engines and fought against their use. Steam engines were heavy machines that were pulled to the fire by a trained team of horses. They required constant attention, which limited their use to larger cities that could meet the costs of maintaining the horses and the steamers.

The advent of the internal combustion engine in the early 1900s enabled most communities to have machine-powered pumpers. Today, both staffed and unstaffed firehouses have fire engines ready to respond at any hour of the day or night. Although they require regular maintenance, current equipment does not require the constant attention that was demanded by horses and steam engines. Modern fire apparatus carry water, a pumping mechanism, hoses, equipment, and personnel. One new fire suppression vehicle can outperform several older vehicles.

The progress in fire protection equipment extends beyond vehicles. Without an adequate water supply, apparatus would be helpless. The advent of municipal water systems provided large quantities of water to extinguish major fires.

The Romans developed the first municipal water systems, just as they had developed the first fire companies. It wasn't until the 1800s, however, that water distribution systems were developed to support fire suppression efforts. George Smith, a fire fighter in New York City, developed the first fire hydrants in 1817. He realized that using a valve to control access to the water in the pipes would enable fire fighters to tap into the system when there was a fire. These valves, or fireplugs, were used with both aboveground and belowground piping systems.

■ Communications

Because small fires are more easily controlled, the sooner a fire department is notified, the more likely it will be able to extinguish the fire and minimize losses. During the Colonial period, a fire warden or night watchman patrolled neighborhoods and sounded the alarm if a fire was discovered. Some towns, including Charleston, South Carolina, built a series of fire towers where wardens would watch for fires. In many towns, ringing the community fire bell or church bells alerted citizens to a fire.

The introduction of public call boxes in Washington, D.C., during the 1850s was a major advance. Call boxes, located around the city, enabled citizens to send a coded telegraph signal to the fire department dispatch center. The fire department would know the location of the fire alarm box by the number of bells in the coded signal. When fire fighters arrived at the alarm box, the caller could direct them to the exact location of the fire. Similar systems are still operating, but most have been replaced by more immediate and effective communications systems. Cellular telephones enable citizens to report an emergency from almost anywhere, anytime. The introduction of computer-aided dispatch and automatic vehicle locators have improved response times because the closest available fire units can be quickly sent to the emergency.

Communications are also vital for a fire officer to coordinate the firefighting efforts effectively. During the firefight, officers must be able to communicate with fire fighters or summon additional resources. Improvements in communication systems are tied to the history of the fire service. Today's two-way radios enable fire units and individual fire fighters to remain in contact with each other at all times. Before electronic amplification and two-way radios became available, the chief officer would shout commands through his trumpet **Figure 1-2 ▾**. The **chief's trumpet**, or bugles, eventually became a symbol of authority. Although chief officers no longer use trumpets for communicating, the

Figure 1-2   The chief's trumpet, used for amplification before electronic devices, eventually became a symbol of authority.

use of trumpets to symbolize the rank of chief also signifies the chief's need to communicate clearly.

Building Codes

Throughout history, fires have served as an impetus for communities to establish building codes. Although the first building codes, developed in ancient Egypt, focused on preventing building collapse, building codes were quickly recognized as an effective means of preventing, limiting, and containing fires.

Colonial communities had few building codes. The first settlers had a difficult time erecting even primitive shelters, which often were constructed of wood with straw-thatched roofs. The fireplaces used for cooking and heating may have had chimneys constructed of smaller logs. The all-wood construction and open fires meant that fires were a constant threat. As communities developed, they enacted codes restricting the hours during which open fires were permitted and the materials that could be used for roofs and chimneys. In 1678, Boston required that "tyle" or slate be used for all roofs. After the British burned Washington, D.C., in 1814, codes prohibiting the building of wooden houses were adopted. Building codes also began to require the construction of a fire-resistive wall, or firewall, of brick or mortar between two buildings.

Building codes not only govern construction materials, but also frequently require built-in fire prevention and safety measures. Required fire detection equipment notifies both building occupants and the fire department. Built-in fire suppression or sprinkler systems help to contain a fire to a small area and prevent small fires from becoming major fires. Fire escapes, stairways, doors that unlock when the alarm sounds, and doors that open outward enable occupants to escape a burning building safely. Without modern building code requirements, high-rise buildings and large shopping centers could not be built safely.

Fire officers are often on the front line of ensuring that the fire codes are obeyed. They must also understand built-in fire protection systems and how they affect firefighting operations.

Code Development

Building codes and fire codes have evolved over many years. The first codes were locally developed and were often influenced by local preferences and customs. The insurance industry played a major role in the development of the first model codes, which were offered to local jurisdictions as a proposed minimum standard. Local communities could make the code stricter as needed.

Today, model codes and standards are written by national organizations, such as the NFPA. Volunteer committees of citizens and representatives of businesses, insurance companies, and government agencies explore and develop proposals that are debated and reviewed by various groups. The final result, called the **consensus document**, is then presented to the public. Many states and municipalities adopt selected NFPA codes and standards into their administrative laws.

Paying for Fire Service

Many early volunteer fire departments were funded by donations or subscriptions, and many volunteer departments still rely on this source of revenue to purchase equipment and pay operating expenses. The first fire wardens were employed by communities and paid from community funds.

Fire insurance companies were established in England soon after the Great Fire of London in 1666, to help cope with the financial loss from fires. The companies would collect fees (premiums) from homeowners and businesses and pledged to repay the owner for any losses due to fire.

Because the insurance companies could save money if the fire was put out before much damage was done, they often agreed to pay a fire company for trying to extinguish a fire. As mentioned earlier, houses that had insurance were designated with a fire mark Figure 1-3 ▾ . Most fire companies were loosely governed, and more than one company might show up to fight a fire. If two fire companies arrived at a fire, however, a dispute might arise over which company would collect the money. This hastened many municipalities to begin assuming the responsibility for providing fire protection. Local tax revenues pay for most career fire departments today and support many volunteer organizations. Regardless of the form of the organization, good fire officers are vital to efficient and effective operations.

Training and Education

The first fire fighters required simply muscular strength and endurance to pass buckets or operate a hand pumper. As equip-

Figure 1-3 A fire mark indicated the homeowner had insurance that would pay the fire company for extinguishing the blaze.

ment became more complex, the importance of formalized training and good judgment increased.

Fire fighters operate sophisticated technical equipment, including large vehicles, radios, thermal imaging cameras, and self-contained breathing apparatus. These tools, as well as better fire-detection devices, increase the safety and effectiveness of modern fire fighters. The most important resources on the fire scene remain the knowledgeable, well-trained, physically capable fire fighters who have the ability and the determination to attack a fire.

Fire Department Organization

As soon as a firefighting exceeded one person and one bucket, there was a need for an organization to focus individual efforts and provide structure for the endeavor. The organization reflects the unique characteristics of the community and the conditions that resulted in the organization of a fire department.

Organizations will evolve as the department grows, the leadership changes, and the community determines what services and activities are needed by the fire department. This section will look at the formal conditions and practices found in most departments.

Source of Authority

Governments—whether municipal, county, state, provincial, or national—are charged with protecting the welfare of the public against common threats. Fire is one such peril; an uncontrolled fire threatens everyone in the community. Citizens accept certain restrictions on their behavior and pay taxes to protect themselves and support the common good. People charged with protecting the public are given certain authority to enable them to perform effectively. For example, fire departments can legally enter a locked home without permission to extinguish a fire and protect the public. Extinguishing a fire is considered to be an important measure to protect the community.

In most areas, the fire service draws its authority from the governing entity responsible for protecting the public from fire—whether it is a town, a city, a county, a township, or a special fire district. The head of the fire department, usually the **fire chief**, is accountable to the leader of the governing body, such as the city council, the county commission, the mayor, or the city manager. Because of the relationship between a fire department and a local government, fire fighters should consider themselves civil servants, working for the tax-paying citizens who fund the fire department.

Federal and state governments also grant authority to fire departments and operate their own fire departments and fire protection agencies to protect federal or state properties, particularly for military installations and wildland areas. Some private corporations have government contracts to provide fire protection services or offer subscription services to private property owners.

Most urban and suburban fire departments are organized by a municipal or county government. Typically, these agencies fall under the organizational umbrella as a department, just like the police department, the public works department, or a human resources department.

Another form of organization that is usually similar to that of a municipal or county department is a fire protection district. A fire protection district is a special political subdivision that can be established by a state or a county, with the single purpose of providing fire protection within a defined geographic area. It has oversight by a fire district board that is usually elected by the voters in the district. A fire district operates very much like a school district and has the ability to set a tax rate, collect taxes, and issue bonds.

In some states, a volunteer fire department can be established by charter and is independent of any local government body. The fire department may be a private association rather than a governmental entity. This type of department is not funded directly through taxes, although it may receive a grant or contract with the local government to provide services. Funding can also be obtained from fund-raising events, donations, subscriptions, and fees for service.

■ Chain of Command

The organizational structure of a fire department consists of a **chain of command**. The ranks may vary in different departments, but the basic concept is generally the same. The chain of command creates a structure for managing the department as well as for directing fire-ground operations. Fire fighters usually report to a supervising officer who is responsible for a single fire company (e.g., an engine company) on a single shift. Some fire departments have only one officer rank and some have two or more officer ranks (sergeants, lieutenants, and captains).

When there are two levels of officers in a fire department, the senior officer has more authority, functioning as a managing fire officer. A managing fire officer could be directly responsible for supervising a fire company on one shift and also responsible for coordinating all of the company's activities with other shifts. A managing fire officer could also be in charge of all of the companies on one shift in a multi-unit fire station.

Supervising and managing fire officers report directly to an administrative fire officer. In a large organization, there are often several levels of administrative fire officers, usually called chiefs. **Battalion chiefs**, or district chiefs, are responsible for managing the activities of several fire companies within a defined geographic area, usually in more than one fire station. A battalion chief is usually the officer in charge of a single-alarm working fire.

Above battalion chiefs are **division chiefs**, deputy chiefs, and/or **assistant chiefs**. Officers at these levels are usually in charge of major functional areas, such as training, emergency operations, support services, and fire prevention within the department. They can also have responsibility for relatively large geographic areas, including several battalions or districts. These officers report directly to the chief of the department.

The fire chief (or chief of the department) is the executive fire officer that has overall responsibility for the administration and operations of the department. The fire chief can delegate responsibilities to other members of the department but is still responsible for ensuring that these activities are properly carried out.

The chain of command is used to implement department rules, policies, and procedures. This organizational structure enables a fire department to determine the most efficient and effective way to fulfill its mission and to communicate this information to all members of the department **Figure 1-4 ▶**. Using the chain of command ensures that a given task is carried out in a uniform manner.

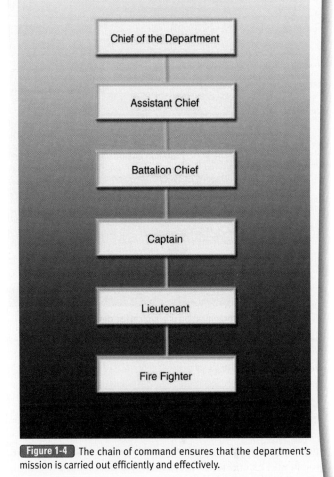

Fire Department Organizational Chart

- Chief of the Department
- Assistant Chief
- Battalion Chief
- Captain
- Lieutenant
- Fire Fighter

Figure 1-4 The chain of command ensures that the department's mission is carried out efficiently and effectively.

■ Basic Principles of Organization

The fire department uses a paramilitary style of leadership. Most fire departments are structured on the basis of four management principles:

1. Unity of command
2. Span of control
3. Division of labor
4. Discipline

Unity of Command

Unity of command is the theory that each fire fighter answers to only one supervisor, and each supervisor answers to only one boss **Figure 1-5 ▶**. In this way, the chain of command ensures that everyone is answerable to the fire chief and establishes a direct route of responsibility from the chief to the fire fighter.

At a fire ground, all functions are assigned according to incident priorities. A fire fighter with more than one supervising officer during an emergency could be overwhelmed with various conflicting assignments. The incident priorities may not get accomplished in a timely and efficient manner.

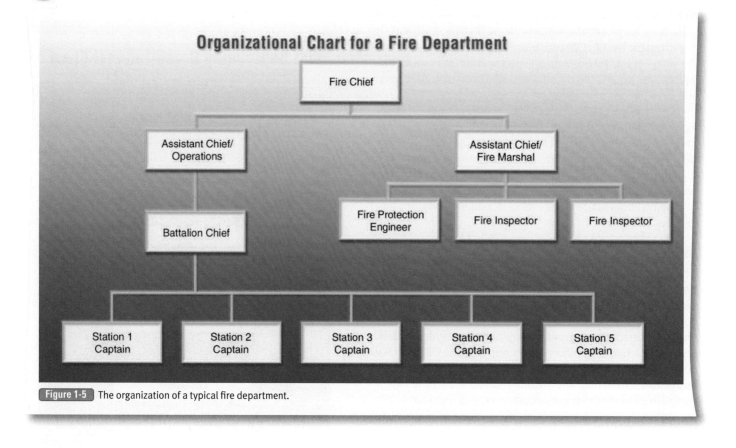

Figure 1-5 The organization of a typical fire department.

Span of Control

Span of control refers to the maximum number of personnel or activities that can be effectively controlled by one individual (usually three to seven). Most experts believe that span of control should extend to no more than five people, but this number can change, depending on the assignment or the task to be completed. A fire officer must recognize his or her own span of control in order to be effective.

Division of Labor

Division of labor is a way of organizing an incident by breaking down the overall strategy into smaller tasks. Some fire departments are divided into units based on function. For example, the functions of engine companies are to establish water supplies and flow water; truck companies perform forcible entry and rescue functions. Each of these functions can be divided into multiple assignments, which can then be assigned to individual fire fighters. With division of labor, the specific assignment of a task to an individual makes that person responsible for completing the task and prevents duplication of job assignments.

Discipline

Discipline is the set of guidelines that a department establishes for fire fighters. Discipline encompasses behavioral requirements, such as always following orders from superior officers and performing up to expectations. Standard operating procedures, suggested operating guidelines, policies, and procedures are all forms of discipline because they outline how things are to be done, and usually how far a person can

go without requesting further guidance. Firefighting demands strong discipline to operate safely and effectively. Discipline can also be positive, when it defines appropriate action, or it can be corrective, when it responds to inappropriate actions or behaviors.

■ Other Views of Organization

There are several different ways to look at the organization of a fire department.

Function

Fire departments can be organized along functional lines. For example, the training division is responsible for leading and coordinating department-wide training activities. Engine companies and truck companies have certain defined functional responsibilities at a fire. Hazardous material squads have different functional responsibilities that support the overall mission of the fire department.

Geography

Each fire department is responsible for a specific geographic area. Fire stations are located throughout a community to ensure a rapid response time to every area, and each station is responsible for its own specific geographic area. This also enables the fire department to distribute and use specialized equipment efficiently throughout the community.

Staffing

A fire department must have sufficient trained personnel available to respond to a fire at any hour of the day, every day of the year. Staffing issues affect all fire departments—career departments, combination departments, and volunteer departments.

In volunteer departments, it is important to ensure that there are enough responders available at all times, particularly during the day. In the past, when many people worked in or near the community where they lived, volunteer response time was not an issue. People often now have longer commutes to work and work longer hours, however, so the number of people available to respond can be severely limited.

The volunteers are also challenged with ever-increasing time demands to obtain additional emergency service certifications, licenses, and continuing education. Although some of this training can be accomplished while on duty at the fire station, some of it will require additional time at an academy, specialized training, or a testing facility.

In combination departments the challenge is to make sure the appropriate mix of qualifications, certifications, and capabilities is present to deliver prompt and safe response. Within the mix of volunteer and career members there needs to be a credentialed fire company commander, a qualified apparatus operator, and structural or wildland fire fighters. In many communities the career employee will have multiple certifications, such as apparatus operator and EMT-Paramedic, and be expected to fill any position that is not covered by a volunteer member.

In career departments the challenge is to ensure all of the assigned positions are covered by appropriately qualified personnel. Although each career position has a home fire station and fire company, it is common for up to a third of the on-duty workforce to be assigned to a different fire company or different fire station. The company officer needs to make work assignments based on individual qualifications, experience, and capability.

In all three types of department, the company officer functions as the staffing coordinator and gatekeeper, implementing the department's rules, regulations, and procedures to ensure proper fire company staffing.

The Functions of Management

Fire officers are managers, and like all other managers, they have specific functions that they must perform, regardless of the size or type of organization. The four functions were originally identified by Henri Fayol and published in the *Bulletin de la Société de l'Industrie Minérale* in 1916. Constance Storrs provided the

English translation in the 1949 textbook *General and Industrial Management*. The four functions are:

- Planning
- Organizing
- Leading
- Controlling

Planning means developing a scheme, program, or method that is worked out beforehand to accomplish an objective. The fire officer is responsible for developing a work plan for a fire company or an administrative work group. The fire officer develops plans to achieve departmental, work unit, and individual objectives. Short-range planning covers developing a plan that extends up to a year. Medium-range planning covers planning that is 1–3 years in advance. Long-range planning covers events longer than 3 years in advance.

Planning includes establishing goals and objectives and then developing a plan to meet and evaluate those goals and objectives. This could be as simple as planning the daily activities for the fire company or as complex as developing an annual budget. Planning can also include emergency activities, such as developing strategies and tactics and an incident action plan.

Organizing means putting together into an orderly, functional, structured whole. The fire officer takes the available people, equipment, structure, and time and develops them into an orderly, functional, and structural unit to implement the plan and deliver the expected services. Fire officers decide which companies will perform certain duties when they arrive on an emergency scene. They also decide what station duties each member of the crew will perform.

Leading means guiding or directing in a course of action. The act of **leadership** is a complex process of influencing others to accomplish a task. When most people think of managing, they think of the leading function of management. Leading is the human side of managing. It includes motivating, training, guiding, and directing the employees.

Controlling means restraining, regulating, governing, counteracting, or overpowering. Fire officers are in the controlling function when they consider the impact on the budget before making purchases, when they conduct employee performance appraisals, or when they ensure compliance with departmental policies.

Fire officers use the functions of management to get work accomplished by and through others. The four functions are a continuous cycle. Although fire officers at all levels use all four functions, each level may use them to different degrees.

Voices of Experience

For many years the fire service has done a good job of training our new fire fighters to handle the job. We send them through a rookie school and teach them the basic principles of becoming fire fighters. Once on the job we place them under capable officers to guide them and continue to train them.

We do much the same for our driver/operators. We send them through extensive training and help them to earn certifications in driving practices, pump operation, and aerial operation. Once on the job we again place them with officers who nurture them and make sure they understand the job they are encountering.

With that said, the question is, "what do we do for our officers?" In most cases in the past, we just promote them and send them on their way to be officers. The department I retired from believed in training and educating its new officers. With good training manuals there is no reason not to have our officers prepared.

Leonard C. Edge Jr.
Instructor
Central Piedmont Community College
Charlotte, North Carolina

"What do we do for our officers?"

Rules and Regulations, Policies, and Standard Operating Procedures

Fire officers must thoroughly know the department's regulations, policies, and standard operating procedures. This is essential to ensure a safe and harmonious working environment. Although a fire fighter is required to follow all regulations, policies, and procedures, the fire officer must not only follow them, but also ensure compliance by subordinates. The transition to fire officer is often a difficult one because of the new role the officer is now required to fill. For officers to enforce the organizational rules, they must understand the differences among rules and regulations, policies, and standard operating procedures.

Rules and regulations are developed by various government or government-authorized organizations to implement a law that has been passed by a government body. Rules may also be established by the local jurisdiction that sets the conditions of employment or internally within a fire department. For example, an organization may have a rule that states that employees with more than 15 years of service will receive 10 shifts of vacation. A fire department may have a rule that requires all members to wear their seat belts when riding in vehicles. Rules and regulations do not leave any room for latitude or discretion.

Policies are developed to provide definite guidelines for present and future actions. Fire department policies outline what is expected in stated conditions. Policies often require personnel to make judgments and to determine the best course of action within the stated policy. Policies governing parts of a fire department's operations may be enacted by other government agencies, such as personnel policies that cover all employees of a city or county. An example of a policy is one that states that the fire officer shall ensure that station sidewalks are maintained to provide safety from slips and falls during snow and ice accumulation. Because it gives the officer latitude in determining how to ensure the safety of pedestrians, this is a policy.

Standard operating procedures (SOPs) are written organizational directives that establish or prescribe specific operational or administrative methods to be followed routinely for the performance of designated operations or actions. SOPs are developed within the fire department, are approved by the chief of the department, and ensure that all members of the department approach a situation or perform a given task in the same manner. SOPs provide a uniform way to deal with emergency situations, enabling different stations or companies to work together smoothly, even if they have never worked together before. They are vital because they enable everyone in the department to function properly and know what is expected for each task. Fire officers must learn and frequently review departmental SOPs. An example of an SOP is a statement of the step-by-step process (procedure) to be used whenever vertical ventilation is required.

Some fire departments prefer the term suggested operating guidelines (SOGs) instead of SOPs because conditions often require the fire officer to use personal judgment in determining the most appropriate action for a given situation. The term SOG suggests that there is a specific step-by-step procedure that should be used, but it allows the officer to deviate from the step-by-step procedure if the conditions warrant. The distinction between an SOP and an SOG is very subjective.

Working with Other Organizations

Fire departments are part of the structure of the community. To fulfill its mission, the fire department must often interact with other organizations. A motor vehicle crash provides a good example of this need. An area with a centralized 9-1-1 call center could dispatch the fire department, a separate emergency medical service provider, law enforcement officials, and tow truck operators to the same incident. At the scene, they all must work together to solve the problem. Fire officers frequently have to request and then interact with other agencies.

21st-Century Fire Officer Challenges

The fire service is responding to challenges that will require changes in the structure, task, and mission of the 21st-century fire department. The paramilitary structure of the fire department was established in the 1860s, when cities were replacing independent volunteer companies with municipal fire departments based on the Civil War military deployment model. A rigid command and control process remains essential when operating at emergency scenes. Away from emergencies, however, departments are using the concepts of employee empowerment, decentralized decision making, and delegation to fully engage fire fighters in the required tasks to prepare and maintain readiness for a wide range of community needs. This "all-hazards" approach is especially important when considering nonfire incidents, a crumbling built environment, homeland security, cultural diversity, and ethics.

Explosion of Nonfire Incidents

Battling a structure fire is a common perception of a "typical day at work." Fire Fighter I and II focus on this task. The NFPA notes, however, that firefighting is one of the least frequent activities performed by fire companies. It accounts for less than 10% of the response workload based on data submitted to the National Fire Incident Reporting System (NFIRS). Of the runs requiring fire suppression tasks, almost half are for structure fires.

Emergency medical service (EMS) calls have become the most frequent activity, accounting for 55% of fire company responses. The last decade shows an increase in EMS workload that exceeds demographics, with some cities noticing the number of EMS calls increasing while the population is declining. In some fire departments, EMS calls represent up to 80% of a fire company's response workload.

Activated fire protection system alarms are the second most common response, representing 20% of the responses. The majority of these activations are due to faulty alarm systems, good intent, or false calls. The company officer must work to remain vigilant in events that result in no service in most of the responses, with an occasional incipient fire or an inferno in less than 1 out of 100 activated fire alarm responses.

Investigating an odor, hazardous condition, or other service call is the third most common response, representing 13% of the responses. Fire fighters encounter many opportunities to be creative problem solvers and deliver outstanding customer service on these events.

Deterioration, or Crumbling, of the Built Environment

Although the number of structure fires continues to decline, the rate of fire fighters being killed or seriously injured while operating in burning structures continues to climb. Flashover and structural collapse are the primary causes of death within a burning structure.

Fire fighters operating within burning structure fires in the 1990s died mostly from smoke inhalation. They were lost or trapped and ran out of air. Fire fighters who have died in structure fires in the 2000s were mainly killed by a flashover or structural collapse. They would fall through the floor or quickly be overwhelmed by a flashover. Most of these events occurred in Type V lightweight wood component structures; these buildings represent most of the structures built since 1985. Fire officers should note that many of the deaths in the early 21st century were suffered by members of the first crew that entered the burning building.

Century-old buildings, while robust, have many worn or rusted out building components. Fire escapes are pulling out of the walls and structural components are crumbling. Most of the renovations are made using lightweight wood elements, some without the benefit of a fire code inspection after changes are made to the occupancy or use of the building.

Protecting the Homeland

Since the September 11, 2001, terrorist attacks, fire officers have become part of the first line of homeland protectors. This responsibility continues to evolve. Fire officers have received additional training in responding to chemical, biological, and nuclear weapons of mass destruction. The federal government reorganized emergency services resources into the U.S. Department of Homeland Security. Local fire department incident management is affected by the National Response Framework, which must be implemented if the jurisdiction plans to receive federal assistance after a major natural or man-made disaster.

These new challenges are changing the federal and, in some cases, regional or local organizational structures. The fire officer is also observing changes in the community's expectations from the fire department. For example:

- Hundreds of thousands of responses to investigate "white powder" and "suspicious package" incidents after anthrax-contaminated letters were discovered in late 2001

- Close work with law enforcement on responses to "active shooter" events, where members of the fire department are deployed with law enforcement entry teams

Other public safety agencies are getting into areas that used to be the domain of the fire department. Hundreds of police departments have created hazardous materials response teams, training police officers to the hazardous materials technician level and procuring equipment that may exceed the fire department resources.

The combining of the roles covers all agencies. The Washington, D.C., area had a massive response of multiple public safety agencies during the 2002 sniper attacks. Every agency had a role to play after a shooting, many performing non-traditional activities. Every activity included minute-by-minute media coverage from the ground and the air. These events disrupt the normal fire department activities and emergency responses. The public expectation is that the public safety services function as a seamless operation, with total cooperation and coordination among the police, fire, and emergency medical service agencies and at the local, state, and federal levels.

To manage this effectively, a fire officer must understand the other agencies and how they interact with the department. Federal, state, and local response plans identify which organizations are responsible for each area of the incident.

For a local incident, the local fire service is often given primary responsibility for search, rescue, and fire extinguishment. Law enforcement is given responsibility for criminal investigation and scene security. Emergency management is responsible for evacuation notification. The American Red Cross may be responsible for establishing evacuation shelters.

As incidents grow, the Federal Bureau of Investigation (FBI) may take a lead role in an investigation. The Federal Emergency Management Agency (FEMA) may become the primary agency responsible for coordination of the incident, which could include the use of urban search and rescue (USAR) teams. Fire officers must understand their role within the local, state, and federal response plans.

Fire officers must also be aware that there are individuals and organizations that wish to create chaos and harm emergency service workers. They research fire department activities to exploit weaknesses and identify opportunities. The fire officer must be vigilant for threats to fire fighters and the department.

Cultural Diversity

Like the world in general, the fire service is experiencing a change in the diversity of the organization. Fifty years ago, the fire service was composed of virtually all white males. However, this began to change in the 1960s. At first, the integration of women and minorities was one of assimilation. The fire service desired to make those who were different fit into the mold of the traditional fire service. Although the fire service began including a few women and minorities, they either assimilated or they left.

This view is beginning to change. It is becoming more commonplace to have a blend of men, women, Caucasians, Hispanics, African Americans, Asian Americans, Native Americans, and others, each bringing their strengths and unique perspectives to the fire service. This integration is far from complete, and many departments struggle with bridging these relationships. The fire officer of the future must look beyond the physical attributes of individuals and match the individual's strengths with the organization's needs in order for the organization, the individual, and the officer to be successful.

Ethics

Inappropriate behavior by individuals in power or holding the public trust gets frequent attention by the media. The fire service is not exempt. Most of the time, fire officers make ethical decisions, choosing to do the "right" thing; however, some officers make unethical choices (which may then influence fire fighters to make unethical choices). When they do, these choices often appear in the newspaper and have very negative consequences for the individual and the organization.

Ethical choices are based on a value system. The officer has to consider each situation, often subconsciously, and make a decision based on his or her values. If the organization has clear values that are part of a strong organizational culture, the officer uses the organization's value system. If the values are not clear, the individual substitutes his or her own value system.

The key to improving ethical choices is to have clear organizational values. This can be accomplished by:

- Having a code of ethics that is well known throughout the organization
- Selecting employees who share the values of the organization
- Ensuring that top management exhibit ethical behavior
- Having clear job goals
- Having performance appraisals that reward ethical behavior
- Implementing an ethics training program

Even at the company level, these values can be implemented to help prevent undesirable ethical choices. One way to help judge a decision is to ask yourself three questions:

- What would my parents and friends say if they knew?
- Would I mind if the paper ran it as a headline story?
- How does it make me feel about myself?

Asking these questions can help prevent an event that could devastate the department's and the fire officer's reputation for years to come.

Summary

From the first time citizens gathered together to confront a hostile fire, someone needed to organize, plan, control, and lead the firefighting effort. Formation of volunteer fire companies included a hierarchy that included a foreman, who was appointed to run the fire company. The company officer structure expanded when cities used Civil War generals to organize the individual fire companies into a citywide fire department with battalions and divisions.

The foundation of company officer practice came from World War II combat experience. These elements were codified when the National Fire Protection Association established the Professional Qualification standards, adopting NFPA Standard 1021 for fire officers in 1976. The International Association of Fire Chiefs expanded company officer development in 2003, providing an *Officer Development Handbook* to encourage company officers to acquire the appropriate levels of training, experience, self-development, and education throughout their professional journey to prepare for the Chief Fire Officer designation, the pinnacle of professional development.

Moving into a fire officer role represents a change in your roles and responsibilities. You will need to work with the people under your supervision to accomplish emergency tasks and routine assignments. The impact of your efforts is amplified when you accomplish tasks and assignments as a supervising fire officer, just as the fire-ground orders by the foreman of the Union Hose Company were amplified when he used the chief's trumpet.

You Are the Fire Officer: Conclusion

The next person to arrive at The Nickel is Captain Jean Torres. Torres is the senior fire officer on your shift and, as a Fire Officer II, operates as a Managing Fire Officer as defined by the IAFC. A Managing Fire Officer will supervise a larger work group that may include more than one Supervising Fire Officer.

Torres greets you with a large smile, "Welcome to The Nickel. I try to arrive before the crew so I can plan the work day." While walking to the office, Torres outlines the expectations for a new supervising fire officer: "You are responsible for the personnel assigned to Tower Ladder 5. Make sure they are properly trained in all the tools and equipment on the tower. You will help them stay physically and mentally prepared to work at any emergency incident. Because of retirements, most of the fire fighters have less than 5 years on the job.

"Learn the district. That includes the buildings, target hazards, and unique communities in the neighborhood.

"Practice your craft as a supervising fire officer. The promotional process means that you have assembled a large toolbox of management theories, concepts, and practices. Now you need to find the proper supervisory tool based on the situation, the individuals involved, and your personal style. I will provide feedback and suggestions."

Reaching the Captain's desk, Torres pointed to the daily planner. "My responsibility is to command the engine company and oversee all Station 5 operations. That means no surprises from the tower company. If there is an injury, property damage accident, or incident, I need to hear about it from you and not from the battalion chief, headquarters, or the News at Noon. We will work these issues out together."

Pointing to the "Everyone Goes Home" poster from the National Fallen Firefighter Foundation, Torres concludes this orientation with two directives:

1. Everyone has their seat belt on before the tower starts to respond.
2. The tower comes to a complete stop at every stop sign and red light intersection.

Wrap-Up

Chief Concepts

- Major Fire Officer I responsibilities:
  - Supervising a single unit
  - Conducting crew training
  - Conducting prefire plans
  - Presenting public education presentations
- Major Fire Officer II responsibilities:
  - Supervising multiple units
  - Conducting occupancy inspections
  - Conducting employee performance appraisals
- There are more than 1.2 million fire fighters in the United States, of which 28% are career and 72% are volunteer.
- Benjamin Franklin established the first organized volunteer fire company in 1735 in Philadelphia.
- Speaking trumpets were originally acquired for company officers to shout orders at the scene of a fire.
- Historical events of the past have shaped the fire service of today.
- A chain of command is a continuous line of authority that exists from the top of the organizational structure to the lowest level.
- Military organizational concepts that still serve the fire officer are:
  - Unity of command
  - Division of labor
  - Span of control
  - Discipline
- Fire officers use leadership and delegation to accomplish departmental objectives and goals.
- The four-step management process includes:
  - Planning
  - Organizing
  - Leading
  - Controlling
- Rules are statements that do not allow for latitude.
- Policies give officers decision-making abilities within specified parameters.
- Standard operating procedures establish standardized step-by-step methods for handling a given situation.
- The 21st-century fire officer challenges include:
  - Increase in nonfire incidents
  - Deterioration of the built environment
  - Protecting the homeland
  - Cultural diversity
  - Ethics

Hot Terms

Assistant or division chief Midlevel chief who often has a functional area of responsibility, such as training, and answers directly to the fire chief.

Battalion chief Usually the first level of fire chief; also called district chief. These chiefs are often in charge of running calls and supervising multiple stations or districts within a city. A battalion chief is usually the officer in charge of a single-alarm working fire.

Chain of command The superior–subordinate authority relationship that starts at the top of the organization hierarchy and extends to the lowest levels.

Chief's trumpet An obsolete amplification device that enabled a chief officer to give orders to fire fighters during an emergency; precursor to a bullhorn and portable radios.

Consensus document A code or standard developed through agreement between people representing different organizations and interests. NFPA codes and standards are consensus documents.

Controlling Restraining, regulating, governing, counteracting, or overpowering.

Decision making The process of identifying problems and opportunities and resolving them.

Discipline A moral, mental, and physical state in which all ranks respond to the will of the leader. Also, the guidelines that a department sets for fire fighters to work within.

Division of labor The production process in which each worker repeats one step over and over, achieving greater efficiencies in the use of time and knowledge; also, the formal assignment of authority and responsibility to job holders.

Fire chief The highest ranking officer in charge of a fire department. The individual assigned the responsibility for management and control of all matters and concerns pertaining to the fire service organization.

Fire mark Historically, an identifying symbol on a building to let fire fighters know that the building was insured by a company that would pay them for extinguishing the fire.

Incident command system (ICS) A system that defines the roles and responsibilities to be assumed by personnel and the operating procedures to be used in the management and direction of emergency operations; the system is also referred to as an incident management system (IMS).

Wrap-Up

Leadership A complex process by which a person influences others to accomplish a mission, task, or objective and directs the organization in a way that makes it more cohesive and coherent.

Leading Guiding or directing in a course of action.

Managing Fire Officer Description from the IAFC *Officer Development Handbook* on the tasks and expectations for a Fire Officer II. Encourages the company officer to acquire the appropriate levels of training, experience, self-development, and education to prepare for the Chief Fire Officer designation.

Organizing Putting together into an orderly, functional, structured whole.

Planning Developing a scheme, program, or method that is worked out beforehand to accomplish an objective.

Policies Formal statements that provide guidelines for present and future actions; policies often require personnel to make judgments.

Rules and regulations Developed by various government or government-authorized organizations to implement a law that has been passed by a government body.

Span of control The maximum number of personnel or activities that can be effectively controlled by one individual (usually three to seven).

Standard operating procedures (SOPs) A written organizational directive that establishes or prescribes specific operational or administrative methods to be followed routinely for the performance of designated operations or actions.

Supervising Fire Officer Description from the IAFC *Officer Development Handbook* on the tasks and expectations for a Fire Officer I. Encourages the company officer to acquire the appropriate levels of training, experience, self-development, and education to prepare for the Chief Fire Officer designation.

Unity of command The management concept that a subordinate should have only one direct supervisor, and a decision can be traced back through subordinates to the manager who originated it.

Fire Officer *in Action*

Prior to your arrival at The Nickel, a senior Tower Ladder 5 fire fighter was assigned to be the acting officer. On emergency incidents, he was expected to fully function as a supervising fire officer. The department expects acting officers to function in any of the incident management roles assumed by a lieutenant.

Back at the station, the acting officer functions as a substitute teacher. The only administrative or supervisory tasks accomplished were to handle issues that affected immediate readiness and resources. After you meet the Tower Ladder 5 crew, you plan to spend some one-on-one time with the acting officer. You want to be sure that all the long-term readiness and administrative issues are identified and handled.

1. What is unity of command?
   A. Fire fighters answer to only one supervisor.
   B. Number of people one fire officer can supervise effectively
   C. Method of directing an incident
   D. A set of guidelines established for fire fighters to follow

2. _____ means "Putting together into an orderly, functional, structured whole."
   A. Standard operating procedures
   B. Span of control
   C. Organizing
   D. Controlling

3. The location and year of the deadliest fire in United States history is _____.
   A. Boston, 1919
   B. San Francisco, 1906
   C. Baltimore, 1904
   D. Peshtigo, 1871

4. "Selecting employees who share the values of the organization" describes _____.
   A. cultural diversity
   B. ethics
   C. policies
   D. planning

Preparing for Promotion

NFPA 1021 Standard

Fire Officer I

NFPA 1021 contains no Fire Officer I Job Performance Requirements for this chapter.

Fire Officer II

5.2.3 Create a professional development plan for a member of the organization, given the requirements for promotion, so that the individual acquires the necessary knowledge, skills, and abilities to be eligible for the examination for the position. [p 30]

(A) Requisite Knowledge. Development of a professional development guide and job shadowing.

(B) Requisite Skills. The ability to communicate orally and in writing.

Introduction to Fire and Emergency Services Administration (FESHE) Course Outcomes

1. Identify career development opportunities and strategies for success. [p 25–36]

Knowledge Objectives

After studying this chapter, you will be able to:

- Discuss the origin of civil service promotional examinations.
- Describe how a promotional examination is prepared.
- Identify the elements of a promotional examination.
- Identify the components of an assessment center.
- List techniques for studying for a promotional examination.

Skills Objectives

There are no skills objectives for this chapter.

he department issued a "Notice of Promotional Opportunity" for Fire Lieutenant. The test will be held in 7 months, utilizing a 100-question multiple-choice exam as part one.

Candidates who score a 70 percent or higher on the exam will proceed to part two, the assessment center. This will contain four components: an in-basket exercise, an emergency incident simulation, an interpersonal interaction, and a presentation to an interview board. The announcement includes a list of 15 references.

This is the first promotional test for which you are eligible. You ask the veterans for advice.

A fire fighter with decades on the job says that others started studying months ago. You should not expect to do well on your first promotional exam. The competition is so tight that the difference between the first- and fifteenth-ranked candidates in the last lieutenant exam was 1.47 points. "Promotional exams are like the Olympics: years of preparation for minutes of performance."

The captain disagrees, pointing out that he placed third on his first promotional exam after 4 months of preparation. The captain encourages you to order the books tonight. He states that studying for a couple hours every day, on and off duty, will pay off.

The lieutenant says it is a waste of time to study obsessively for the test. He guesses that the department will need to make six promotions a year to cover anticipated retirements. The promotion list is good for 4 years, so you need to place in the top 25 of the list to get promoted. It takes a lot more time and effort to score in the top 5 versus the top 25. The lieutenant says to spend about an hour a day in test preparation while working in the fire station. Use off-duty time for other activities: spend time with your family, take a college class, run a part-time business, or work fire department overtime. Those activities provide a better return on your investment.

You are confused and frustrated.
1. How much time do you really need to prepare for the exam?
2. If you pass the test, are you ready to be a supervising fire officer?
3. What is wrong with concentrating on being a good fire fighter for another 4 years?

Introduction to Preparing for Promotion

The purpose of this chapter is to provide a general description of the civil service promotional examination process that is used by most fire departments. There are many variations of the testing procedures and promotional processes. You need to understand the specific process that is used in your organization. Many organizations change elements or assessment tools from examination to examination.

The Origin of Promotional Examinations

Prior to the Civil War, most government jobs were awarded according to the patronage or **spoils system**; those in power could appoint people to public office based on a personal relationship or political affiliation, rather than on merit. The best jobs went to political supporters and often required a payment to the individual who had the power to make appointments. The system was ripe for corruption.

Congress enacted the Pendleton Civil Service Reform Act in 1883 in response to the abuses at Tammany Hall in New York City and many other publicized abuses of the patronage system. The Pendleton Act established the civil service system in the federal government and provided a model for the civil service systems that were developed by many states and cities in subsequent years. The spoils system was gradually replaced by merit selection and promotion for most government employees. The process of developing promotional examinations for fire officer positions, based on testing for specific knowledge and skills, was derived directly from this act.

Fire Marks

Metropolitan Fire Department

When New York City created the paid Metropolitan Fire Department in 1865, Tammany Hall was the seat of political power. William March "Boss" Tweed was the chairman of the Democratic Party and the Grand Sachem (leader) of the Society of St. Tammany, which had been founded in 1789 as a club for patriotic and fraternal purposes. Between 1865 and 1871, Boss Tweed and Tammany Hall exploited the spoils process in New York City, swindling an estimated $75 million to $200 million from the city. The public began to demand political reform.

Getting It Done

Variations in Examinations

The fire officer promotional process remains an inexact science. Many different variations of examination and testing procedures are in use, with differences in emphasis and on the weights assigned to particular **dimensions** (attributes or qualities that can be described and measured during a promotional examination). The rank (sergeant, lieutenant, or captain) and the classified job description determine the technical, theoretical, or behavioral emphasis of the examination.

Sizing Up Promotion Opportunities

The opportunities for promotion depend on factors such as the size and organizational structure of your fire department. In a very small department, there could be only a few officer positions, and promotional openings may not be available unless someone retires. If you are working in a **metropolitan (metro-sized) fire department**, an agency with more than 400 fully paid fire fighters, your opportunities for promotion are higher than average simply because of the number of openings that can be anticipated.

A combination of growing communities and retiring baby boomers is creating the need for more fire fighters and more fire officers. The fire fighters and fire officers who began their careers in the 1970s are reaching retirement age. Some departments will see more than half of their existing workforce retire by 2015.

Many suburban and rural communities are adding career fire fighters to handle increasing workloads, often to supplement volunteer coverage for weekday emergencies. Loudoun County, Virginia, located 35 miles northwest of Washington, D.C., is a rural agricultural and outer suburb bedroom community. In the 1980s, only a dozen career fire fighters worked for the county, and the fire stations were operated entirely by volunteers. Loudoun County became the third fastest growing county in the nation in the 1990s, fueled by affordable housing close to Washington, D.C., and a booming information technology industry. As large residential communities replaced farms, the number of career fire fighters increased from 14 in 1991 to 148 in 2003 to nearly 500 in 2008.

Of course, not all fire departments are expanding, and fire officers do not automatically retire at the first opportunity. Firefighting is a lifestyle, like the military, medicine, and other public safety careers. Rotating shift work, the unique team-based workforce, and hours of tedium mixed with moments of intense life-saving activity create a special work environment. The rich tradition of service and sacrifice extends beyond the on-duty hours and becomes a major component of an individual's life. Some fire fighters represent the third or fourth generation of a firefighting family. Many fire fighters and officers enjoy their work so much that it is common to find them working a decade beyond their retirement eligibility dates or fighting to eliminate the mandatory retirement age Figure 2-1 ▶.

Many other factors can influence the number of promotional opportunities that are available in a fire department. Sometimes the numbers decrease, particularly when the organization is downsized or restructured.

During the Clinton administration, Vice President Al Gore directed an effort to reinvent the federal government. Prominent in local government reinvention was to reduce the number of middle managers. Reinvention resulted in the elimination or civilianization of headquarters and staff positions that previously had been held by fire officers. The first decade of the 21st century shows a reduction in the size of local government.

In most fire departments, completion of a promotional examination process creates an eligibility list that lasts from 2 to 6 years. Depending on local practice, the list may be either rank ordered or banded. On a rank-ordered list, the highest-scoring candidate is ranked #1, the second-highest-scoring candidate is

Figure 2-1 Many fire fighters and officers enjoy their work so much that it is common to find them working beyond their retirement eligibility.

ranked #2, and so forth. Procedure usually is to promote individuals in the order in which they placed on the eligibility list.

Other departments use a banded list, in which the candidates are placed into bands, or groups, of promotional candidates. The bands are usually identified as "Highly Qualified," "Qualified," and "Not Qualified." In this case, all of the candidates on the Highly Qualified list are considered equally qualified for promotional consideration, followed by all of the Qualified candidates.

Postexamination Promotional Considerations

Regardless of the eligibility-ranking scheme, the jurisdiction will make promotions to meet departmental and community needs. A basic requirement is that the candidate must be medically qualified to assume the new job. Most promotional job announcements require successful candidates to have an appropriate medical or physical performance rating. A candidate who is medically restricted or cannot complete the required physical ability assessment may not be considered for promotion until these issues are resolved Figure 2-2 ▾ .

Figure 2-2 Most promotional job announcements require that the candidate have an appropriate physical performance rating.

In addition, the candidate must be free of active formal discipline. Departments often identify a required time period to pass between a formal disciplinary action and consideration for an officer promotion. The period varies from a month to a full year. That time period is described in the jurisdiction's personnel regulations, the department's administrative procedures, or the current labor contract. For example, a candidate who receives a 2-day suspension without pay for repeated tardiness in reporting for duty may not be considered for promotion for 365 days, even after ranking #1 on a lieutenant eligibility list.

The increasing complexity of the fire department mission adds variables to the decision of who should be promoted. Some specialty supervisory positions require additional certifications or work experience. For example, to be promoted into the position of arson squad captain, the candidate also needs to have certification as an NFPA 1033 Fire Investigator and 2 years of experience as a fire investigator. Other assignments may include preferred qualifications. For example, the department may prefer promoting bilingual candidates in fire companies that serve ethnic communities.

The promotional process is constantly changing to meet community needs and respond to legal and political challenges. Each jurisdiction has a promotional process that evolved as a result of community needs, consent decrees, lawsuit settlements, arbitration decisions, grievance settlements, labor contracts, and memoranda of understanding. It is important to learn what rules, goals, or practices affect the department's promotional practice.

Preparing a Promotional Examination

The preparation of a promotional examination usually involves a combined effort between the fire department and the municipality's human resources section. Some jurisdictions may contract with a testing organization or a consultant to organize, develop, or deliver a promotional exam. Public safety promotional examinations are the most complex and detailed examinations conducted by a municipality.

The fire department usually establishes a test preparation committee consisting of three or more officers, usually chaired by a chief officer. These individuals are the subject matter experts

Figure 2-3 If the examination is developed within the agency, test committee members establish its content.

and are responsible for validating the technical content of the promotional examination. If the promotional examination is developed within the agency, the test committee members write the questions and establish its content **Figure 2-3 ▲** .

Company Tips

Administrative Positions

Uniformed officers staffed most of the 1970s-era fire department administrative supervisory positions. The personnel director was a deputy chief, a battalion chief ran the fire academy, the boss of the apparatus shop was a captain, and a lieutenant was in charge of the warehouse. In addition to reducing the number of middle-management and administrative positions in local government during the 1990s, many of the remaining middle manager jobs were converted to civilian positions. Here is an example of what happened to those four uniformed supervisory positions in one fire department.

The city consolidated the fire and police department personnel directors, replacing two uniformed chief officer positions with a single public safety personnel director. The public safety personnel director is a 27-year-old with a master's degree in human resources. The department privatized the fire academy, and the battalion chief position at the academy was eliminated. A community college now runs the recruit school under a service contract. A civilian project manager is in charge of contract compliance and makes sure the college delivers the required training products.

The city also civilianized the fire apparatus shop supervisor, replacing the fire captain with an employee from the city's fleet management division. The new boss of the apparatus shop is a 50-year-old master mechanic with decades of experience running a fleet of heavy equipment rigs.

The fire department warehouse is gone, absorbed into the citywide resource management system. A logistics specialist and an accountant coordinate the fire department supply requests.

The combination of civilianization and reduction of middle-management positions has reduced the number of uniformed fire officer positions in many fire departments, thereby reducing the number of potential officer promotions.

Charting the Required Knowledge, Skills, and Abilities

The municipality's personnel or human resources department uses two documents to define the **knowledge, skills, and abilities (KSAs)** that are required for every classified position within the municipality: a narrative job description and a technical class specification. Fire officers need to be familiar with both of these documents to prepare for a promotional examination.

A narrative **job description** summarizes the scope of the job and provides examples of the typical tasks a person holding that position would be expected to perform. The job description also lists the KSAs needed at the time of appointment and any special requirements that apply. Examples of special requirements could include possession of a valid driver's license and an Emergency Medical Technician certificate. Other requirements might include a Class "A" medical rating, passing a work performance test, or possession of technical or professional certifications such as NFPA Fire Officer I. When the human resources department posts a promotional announcement, a full or partial narrative job description is usually included.

Human resources departments also prepare a technical **class specification** worksheet to quantify the KSA components of every classified municipal job **Figure 2-4 ▶** . The classification system is a core component of the civil service system and is used to determine the compensation level for a position. The classification details would show why the base salary for a battalion chief is set 35% higher than for a lieutenant, based on the KSAs.

The worksheet also provides a map for the promotional examination in order to identify the factors that need to be evaluated. A promotional examination should focus on the unique, high-importance tasks that distinguish one position from another.

Periodically, a classification specialist from the human resources department surveys the individuals within a classification to rank the frequency and importance of a wide range of job tasks. This survey is performed to validate and update the job description and the KSA technical worksheet.

Promotional Examination Components

There is no perfect promotional exam. Through trial and error, research, grievances, and court decisions, the following components are frequently utilized components of a promotional exam.

The decision of which components to use is influenced by time requirements, expense, staff expertise, and past experience. These components can be locally developed or provided by a vendor, consultant, or assessment specialist.

Multiple-Choice Written Examination

Multiple-choice written examinations are widely used in the promotional process because they can be structured to focus on very specific subjects and factual information. There is no element of style or creativity in answering a multiple-choice question; the candidate only has to select the appropriate answer. The scoring is equally straightforward because an answer is simply right or wrong.

NEW HANOVER COUNTY, NC
CLASS SPECIFICATION

CLASS TITLE: Fire Lieutenant

CLASS CODE:		
DEPARTMENT:	**ACCOUNTABLE TO:**	**FLSA STATUS:**
Fire Services	Fire Captain	Non-exempt

CLASS SUMMARY:

Incumbents are responsible for shift supervision of County firefighters. Duties include: supervising and evaluating staff; overseeing responses to reports of fires; supervising rescue operations; preparing work schedules; writing reports of firefighting and rescue activities; training personnel; coordinating equipment and facility maintenance; presenting fire prevention programs; and, representing the department at special events.

DISTINGUISHING CHARACTERISTICS:

The Fire Lieutenant is the second level in a five level firefighter series. The Fire Lieutenant is distinguished from the Firefighter/Apparatus Operator in that it has shift supervisory responsibilities. The Fire Lieutenant is distinguished from the Fire Captain which has full supervisory authority.

DUTY NO.	**TYPICAL CLASS ESSENTIAL DUTIES:** (These duties are a representative sample; position assignments may vary.)	FRE-QUENCY	
1.	Supervises two or more full-time staff to include: prioritizing and assigning work; conducting performance evaluations; ensuring staff are trained; and, making hiring, termination, and disciplinary recommendations.	Daily 25%	
2.	Oversees response to reports of fires which includes: preparing pre-incident surveys; securing the scene; extinguishing the fire; salvaging structures and their contents; and providing emergency medical services to injured parties.	Daily 15%	
3.	Supervises rescue operations by overseeing extrication activities and providing emergency medical services to injured parties.	Daily 15%	
4.	Prepares daily and weekly work schedules for firefighters.	Daily 10%	
5.	Prepares written reports of fire and rescue activities.	Daily 10%	
6.	Provides training to shift personnel and students at the County fire academy on firefighting, rescue, and emergency medical topics.	Daily 10%	
7.	Coordinates firefighter maintenance of vehicles, fire and rescue equipment, and facilities.	Daily 5%	
8.	Presents fire prevention and education programs to businesses, schools, and community groups.	Weekly 5%	
9.	Represents the department at special events including parades and open houses.	Occasion-ally 5%	
10.	Performs other duties of a similar nature or level.	As Required	
11.	Performs work during emergency/disaster situations.	As Required	

Figure 2-4 Sample technical class specification worksheet. Courtesy of New Hanover Fire Rescue, Wilmington, NC.

NEW HANOVER COUNTY, NC
CLASS SPECIFICATION

CLASS TITLE: Fire Lieutenant

POSITION SPECIFIC RESPONSIBILITIES MIGHT INCLUDE:
• Does not apply.

Knowledge (position requirements at entry):
Knowledge of:
• General principles of fire science;
• Emergency management techniques;
• Basic principles of rescue;
• Hazardous materials management techniques;
• Emergency medical practices;
• Departmental policies and practices;
• Local and state fire ordinances.

Skills (position requirements at entry):
Skill in:
• Performing fire suppression and rescue operations;
• Driving a vehicle;
• Preparing and making presentations;
• Preparing written incident reports;
• Using a computer and related software applications;
• Supervising and evaluating employees;
• Communication, interpersonal skills as applied to interaction with coworkers, supervisor, the general public, etc. sufficient to exchange or convey information and to receive work direction.

Training and Experience (positions in this class typically require):
High School Diploma or General Equivalency Diploma (GED) and five years of related firefighting experience, including two years of progressively responsible supervisory experience; or, an equivalent combination of education and experience sufficient to successfully perform the essential duties of the job such as those listed above; Associate's Degree in Fire Science preferred.

Licensing/Certification Requirements (positions in this class typically require):
• Class B Driver's License;
• Firefighter II Certification;
• Must be able to obtain EMT and Fire Instructor Level II certifications within one year and Level I Fire Inspector and ERT Certifications within two years.

Physical Requirements/Working Conditions:
Positions in this class typically require: climbing, balancing, stooping, kneeling, crouching, crawling, reaching, standing, walking, pushing, pulling, lifting, fingering, grasping, feeling, talking, hearing, seeing, and repetitive motions.

Heavy Work: Exerting up to 100 pounds of force occasionally, and/or up to 50 pounds of force frequently, and/or up to 20 pounds of forces constantly to move objects.

Incumbents may be subjected to moving mechanical parts, electrical currents, vibrations, fumes, odors, dusts, gases, poor ventilation, chemicals, oils, extreme temperatures, inadequate lighting, work space restrictions, intense noises, and travel.

NOTE:
The above job description is intended to represent only the key areas of responsibilities; specific position assignments will vary depending on the business needs of the department.

Classification History:
Draft prepared by Fox Lawson and Associates LLC (CC).
Date: 10/99

Figure 2-4   *(Continued)*

The multiple-choice written examination concentrates on facts that can be found within the materials on a reading list. The reading list generally includes textbooks, reference books, standard operating procedures manuals, department rules and regulations, and other locally developed reference materials.

Using the KSA technical worksheet, the test committee determines the number of questions to include from each knowledge area.

The supervising fire officer examination usually includes many technical questions covering engine, truck, and rescue company operations. These questions are directed toward a candidate's ability to demonstrate the basic knowledge that would be important for a fire officer. In general, the first-level supervisory examination has the longest and most diverse reading list. Up to 70% of this test focuses on the technical aspects of a supervising fire officer.

Typical technical questions cover building construction, incident management, hydraulics, emergency medical care, and firefighting tactics. Many examinations also test the candidate's knowledge in specialized company tasks, such as technical/heavy rescue, truck company operations, and rapid intervention teams.

Supervisory questions tend to focus on the immediate requirements affecting fire company preparedness, such as a subordinate unfit for duty or injured on the job. A supervising fire officer examination might include a question about how to handle a fire fighter reporting late for duty. The hierarchical nature of the fire department requires that the first-level supervisor identify the problem, stop the behavior, and report to or consult with a senior officer or chief before taking any significant supervisory action.

A managing fire officer examination usually includes fewer technical questions and more management and administration questions because the higher-level position involves more management responsibilities. The process would assume that the candidate for a higher-level position has already met the qualifications for the lower-level position. The second-level supervisor might also be expected to develop a fire fighter's work improvement plan, prepare a budget proposal, evaluate fire company performance, or develop a department-wide training plan. For example, a managing fire officer candidate could be asked questions that relate to evaluating a fire fighter's reporting-time performance over a year. The supervisor should be able to identify patterns and determine the underlying causes for tardy reporting.

There are three options for constructing a multiple-choice written examination. Sometimes, the local exam committee develops the test and the committee members write the questions. There are also private companies that develop generic examinations used by many fire departments. The department has the third option of hiring a consultant to write a more specific examination that is directed toward local priorities.

Larger fire departments are more likely to develop an examination internally because they have the necessary resources. For example, the fire chief in a large city can probably assign a command officer and five to seven company officers to write the questions for a supervising fire officer examination.

The committee first determines how many questions are needed to assess the candidate's knowledge within each particular area. For example, the committee might decide that nine building construction questions should be included in an examination. Two of the committee members are assigned to work together and develop 15 multiple-choice questions on building construction. The whole committee would then meet to select the nine best questions for the examination.

A smaller department would be more likely to go to an external source for the examination questions, simply because of the time and effort that would be required to write and validate them internally. Writing good examination questions is difficult work. Each question has to be fully researched and carefully worded. A question can be challenged and invalidated if the wording is not clear, if there is no correct answer, or if more than one of the answers that are provided could be correct.

In many fire departments, multiple-choice written examinations are the only assessment tools used for promotional examinations. The simplicity of a multiple-choice examination can also be viewed as a weakness. A candidate who knows all of the facts can do very well on this type of examination, even if the individual is unable to apply the information in a real-life situation. The description "Book smart, street dumb" has often been used in complaints about a process that relies solely on written multiple-choice questions to promote officers.

■ Assessment Centers

Some public safety agencies began to use **assessment centers** in promotional examinations in the late 1970s. An assessment center is a series of simulation exercises that are used to evaluate a candidate's competence in performing the actual tasks associated with a job. The assessment center process was developed in the officer corps of the German army in the 1920s. The goal of the program was to select future officers based on their predictive performance in a 2- to 3-day assessment procedure. The same type of process can be developed for fire officers.

In-Basket Exercises

One of the most common assessment center events is the **in-basket exercise** Figure 2-5 ▶ . The candidate has to deal with a stack of correspondence and related items that have accumulated in a fire officer's in-basket. This timed exercise measures the candidate's ability to organize, prioritize, delegate, and follow up on administrative tasks. Those who do well on in-basket exercises tend to demonstrate successful managerial performance in real life.

A typical in-basket contains the following items:

- Instructions for the exercise
- A calendar
- An organizational chart or list of personnel
- Ten to 30 exercise items

An in-basket assessment generally begins by positioning the candidate as a newly promoted officer reporting to the fire station for the first time. The former officer suddenly retired or left the fire department, leaving a stack of issues that require

Figure 2-5 The in-basket exercise is a common assessment center event.

attention in the in-basket. The candidate has a specified amount of time to go through the officer's in-basket and decide what to do with each item.

The situation is contrived so that the candidate is unable to directly call or contact anyone else for advice or assistance. The candidate might be provided with copies of the department's standard operating procedures and reference manuals. Hidden within the in-basket are surprises and time-critical issues, such as an important report that is overdue or an activity that has to be performed immediately. Frequently, the candidate encounters a scheduling conflict, such as an order from the operations chief to have the engine company report to a multiple-company drill at the same time the apparatus is scheduled for an annual pump test at the shop. Many in-baskets also include a writing exercise that could require the candidate to prepare a letter for the chief's signature or complete an incident report using correct formatting, spelling, and grammar.

From the test administrator's viewpoint, in-baskets are generally easy to organize and grade. This type of test can be difficult to develop for the fire officer level, however, because many of the typical job tasks in a fire station do not readily lend themselves to an in-basket assessment.

The following is a suggested method for handling promotional in-baskets:

- **Review.** Look at every item and determine which items are important (critical) and urgent (time constraints), which are important but not urgent, and which are unimportant.
- **Prioritize.** Handle the important and urgent items first, followed by the important and nonurgent items. When these are finished, you will have handled all items or will have only unimportant items left.
- **Identify resources/options/alternatives.** Determine who can handle the items, what options you may have in how they can be handled, and possible alternative courses of action.
- **Follow-up.** Provide some form of follow-up on all delegated items.
- **Make notifications.** Ensure that appropriate persons are notified of necessary and critical information.

Candidates can prepare for an in-basket exercise by practicing on sample exercises that can be obtained from study materials publishers. Both lieutenant and captain in-basket exercises are available; these include answers with explanations and descriptions of the dimensions assessed.

■ Emergency Incident Simulations

Emergency incident simulations are often included in a promotional examination to test the candidate's ability to perform in the role of an officer at a fire or some other type of situation. Emergency incident management usually occupies a small amount of an officer's actual time, yet the required skill sets can be the most critical for evaluation in a promotional examination. An officer must be able to lead, supervise, and perform a set of essential skills at an emergency incident scene. Emergency incident simulations fall into one of four formats.

In the first format, the candidate is provided information concerning an emergency situation, usually in a written format that includes pictures or pre-incident plan information. The candidate has to explain the actions that would be taken in the situation that is described and what factors would be considered in the making of those decisions. This is known as a **"data dump" question** because it provides an opportunity for the candidate to demonstrate a depth of knowledge about a particular subject. For example, the evaluators might find out how much expertise the candidate can demonstrate regarding basement fires in mercantile properties.

In the second format, the candidate is provided with a set of basic information concerning an emergency incident to begin the exercise. As the simulation progresses, the candidate is provided with additional updates of the situation, such as, "Immediately after you arrive on Engine 1, the store manager reports an acrid, unknown odor coming from a leaking package in the rear of the store. It has made a dozen people sick. What actions will you take? What will you report to dispatch?" The candidate has to react to the unfolding situation, which can change based on the answers to the previous questions.

In the third format, the candidate participates in an interactive emergency scene simulation in a classroom or incident simulation trainer. The candidate is presented with a dynamic emergency incident through a multimedia format that typically includes full-color pictures depicting the incident. The graphics often include simulated smoke and flames superimposed on photos of an actual building in the community. Using a handheld radio, the candidate is expected to command the event using the local jurisdiction's standard operating procedures and incident command tools (accountability boards, command post clipboards, incident management system worksheets) that are appropriate to what a person at that rank would use on the street **Figure 2-6 ▶**.

This type of exercise is complex and can be expensive because the test-site administrator might need to arrange for role players and technical support staff to run the simulation event for every candidate.

The fourth simulation format attempts to make the conditions as realistic as possible by actually having the candidate don protective clothing, climb into the officer's seat on an apparatus,

Figure 2-6 The candidate is expected to command the simulated event.

and respond to a realistic scenario with a crew of fire fighters. This type of exercise is usually performed at a large urban or regional training academy, where there are buildings and props that allow for a consistent simulation for all of the candidates. The real-life format allows a candidate to fully function in the role of an officer; however, it requires extensive preparations and a large supporting cast.

Interpersonal Interaction

An interpersonal interaction exercise is designed to test a candidate's ability to perform effectively as a supervisor. In a typical

Getting It Done

Training Centers

Montgomery County, Maryland, Public Safety Training Academy unveiled the Command Development Center in 2007. This is one of a few virtual reality simulation training centers in the United States. It also includes what is believed to be the largest fire and rescue tactical "table-top" venue in the nation. This sophisticated center provides real-time scenarios using viewpoints from many incident management positions.

Warren Township Division Chief Don Abbott was training the Marion County, Indiana, HazMat Task Force. Chief Abbott developed a 1:87 scale (HO scale) model city capable of handling 139 different training sequences. The disaster city grew to a 12' (3.65-meter) by 18' (5.49-meter) diorama featuring 234 buildings and 160 vehicles, with telephone and radio capabilities.

Retiring from the fire department in 1994, Don and Bev Abbott started using "Abbottville" for hazardous material, disaster, and incident management training. In 2002, Don Abbott accepted a project manager position with the Phoenix Fire Department and created the Command Training Center (CTC) in the former quarters of Fire Station 30. Up to 14 incident management positions can be staffed, with dynamic scenes delivered in real-time evolutions.

The experience in developing the Phoenix CTC and training thousands of company and command officers influenced the subsequent virtual reality simulation centers. Some promotional candidates will be tested in a virtual simulation center.

Near Miss REPORT

Report Number: 09-0000268

Event Description: While operating at a second alarm fire in a multiple-family dwelling with commercial occupancies on the first floor, units on the first floor interior setback of a Laundromat reported weakened floor joists above their heads. A second report from members attempting to ventilate the roof of the setback reported being unable to vent due to poured concrete on the roof. A third report came in from the second floor interior reporting a refrigerator that had just shifted and was tilting. An immediate order was given to withdraw from the interior of the building. Shortly thereafter, a collapse of the second floor occurred into the rear of the Laundromat. Fortunately, no members sustained injury.

Lesson Learned: The importance of accurate reporting of all information to the incident commander is demonstrated here. All reports collectively led the incident commander to withdraw members. If any of these pieces of information had been left out, the decision process may have been different, potentially leading to tragedy. All reports must be acknowledged by the recipient. If not, the sender must make continued attempts. The information you have may seem unimportant but, when pieced together with other transmissions, may save a life.

Figure 2-7 An interpersonal interaction exercise tests a candidate's ability to perform effectively as a supervisor.

interpersonal interaction scenario, the candidate has to deal with a role player who has a problem, complaint, or dilemma. The candidate is usually provided with background information to prepare for the situation, then has 10 to 20 minutes of face-to-face interaction with the role player **Figure 2-7 ▲**.

Many of these assessments require the candidate to deal with a poorly performing or troubled employee. The background information could include the employee's last evaluation report, a sick leave use printout, or recent disciplinary action. In many cases, the role player has important information related to the performance issue that is revealed only if the candidate asks the right questions during the interview phase. After the interview, the candidate may have to write up an employee improvement plan or a discipline letter **Figure 2-8 ▾**.

The key points of an interpersonal interaction assessment are to:

1. Maintain control of the interview.
2. Tell the employee the exact behavior you expect (show up on time, wear your seat belt, etc.).

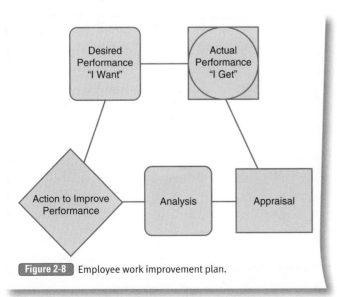

Figure 2-8 Employee work improvement plan.

3. Give the employee a deadline to demonstrate a consistent behavioral change.
4. Specifically arrange for follow-up meetings. ("We will meet once a week to review your progress in meeting this goal.")
5. Attempt to get the employee to buy into or take personal responsibility for the improvement plan.
6. Be an empathetic listener, but remain focused on the reason for the interview.
7. Clearly explain the consequences if the employee's behavior does not change or improve.
8. Try to finish the session on a positive note ("You are a valuable member of this company and I am counting on you . . .").

A successful approach for an interpersonal interaction exercise is for the candidate to demonstrate the qualities of an "extreme supervisor." This means that any issue is handled in strict accordance to the rules and regulations of the organization.

An alternative scenario for this type of assessment is to have the role player confront the candidate as an angry or frustrated citizen. This exercise would assess how well the candidate can handle a customer service problem or a citizen complaint. This type of exercise is usually based on some situation that has actually occurred.

Writing or Speaking Exercise

A candidate should expect to deliver a short oral presentation or write a memo or report as part of the assessment center process. The oral presentation assesses oral communication, planning and organizing skills, and persuasiveness. Sometimes, the candidate is required to write a report or a letter for a superior officer's signature. The written report assesses the candidate's written communication, problem analysis, and decision-making skills.

The oral presentation could be incorporated into one of the other exercises. For example, a candidate might be required to make a brief oral presentation to explain the size-up considerations of the emergency incident simulation.

Many assessment centers include an oral interview session in which the candidate is asked questions by an interview board **Figure 2-9 ▸**. One of the most popular questions for this type of interview is to ask the candidate why he or she should be selected for promotion. This provides an opportunity for the candidate to demonstrate skills or knowledge in a particular area or even to correct an error in a previous exercise. In some cases, a candidate's ability to recognize an error and correct a response could be viewed as equal to having the right answer to every question.

Videos are available from emergency service publishers and individual subject matter experts to assist candidates in preparing to make a great presentation in an oral interview. Chapter 14, Fire Officer Communications, discusses fire officer communication skills in more detail.

Technical Skills Demonstration

Fire officers are expected to be working supervisors and to be skilled in both task- and tactical-level activities. For some supervising fire officer examinations, the candidates might also be required

Figure 2-9 Many assessment centers include an oral interview session.

Assessment Center Tips

Preventing Injury and Death
NIOSH Publication 99–146, *Preventing Injuries and Deaths of Firefighters Due to Structural Collapse*, is a good example of a special interest bulletin that could provide valuable information for a promotional examination. This particular publication makes 10 recommendations that every fire officer should know:

1. Ensure that the incident commander conducts an initial size-up and risk assessment at the incident scene before beginning interior firefighting.
2. Ensure that the incident commander always maintains close accountability for all personnel at the fire scene—both by location and by function.
3. Ensure that at least four fire fighters are on the scene before entering a structure and beginning interior firefighting at a structural fire (two fire fighters inside the structure and two fire fighters outside).
4. Establish rapid intervention crews and make sure they are positioned to respond to emergencies immediately.
5. Equip fire fighters who enter hazardous areas (e.g., burning structures or structures suspected of being unsafe) to maintain two-way communications with the incident commander.
6. Ensure that standard operating procedures and equipment are adequate and efficient to support radio traffic at multiple-responder fires.
7. Provide all fire fighters with Personal Alert Safety System (PASS) devices and make sure they wear and activate them when they are involved in firefighting, rescue, or other hazardous duties.
8. Conduct prefire planning and inspections that cover all building materials and components of a structure.
9. Transmit an audible tone or alert immediately when conditions become unsafe for fire fighters.
10. Establish a collapse zone around buildings with parapet walls.

to demonstrate very specific technical skills. This type of exercise is often included in promotional tests for officers who would command highly specialized teams, such as urban search and rescue, hazardous materials, fire investigations, and paramedic teams.

Preparing for a Promotional Examination

To prepare for an examination, the candidate must master both the content of the examination (technical and reference information) and the process (answering multiple-choice questions or role playing as an officer). Most of the content comes from the materials on the reference reading list; however, you should also be prepared for questions that relate to current issues and recent local history. For example, if your department recently had a problem with on-duty fire fighters taking emergency incident scene videos and posting them on a website, do not be surprised to find that issue somewhere in the examination. You should stay current by reading fire service publications and periodicals to keep up with trends and issues.

Mastering the content for an examination requires a personal study plan. Many tests require the candidates to absorb a tremendous quantity of factual information, which can require a major investment of time and effort. It is not unusual for candidates to spend more than a year preparing for an examination, setting aside a period of time every day for reading and answering practice questions.

Fire fighters who participate in structured programs and study groups perform better than fire fighters who study on their own. Many fire departments have officially organized programs to help candidates prepare for examinations. In other cases, more informal study groups or preparatory programs are available for candidates who want to be well prepared. If your fire department does not have a promotional preparation program, you can set up your own study group. The individuals support each other, share materials, and even develop their own multiple-choice questions **Figure 2-10 ▼**.

Building a Personal Study Journal

A valuable tool in studying for a promotional examination is to create a **personal study journal**. This journal can be used to set up your personal study schedule, keep track of your progress, and make notes about confusing, interesting, or important facts you will discover.

Figure 2-10 Fire fighters who establish study groups often perform better than fire fighters who study on their own.

Voices of Experience

Like many of you, when I started as a volunteer in the fire service over 20 years ago, the last thing I was thinking about was being promoted to an officer position. I was caught up in the glamour of saving lives and fighting fire. As I started testing for full-time fire department jobs I soon learned that education was important and I started pursuing educational opportunities. Once I was successful in landing a full-time fire department job, I was once again living the dream and it seemed like the five years before I was eligible to test for lieutenant was an eternity, so I focused on being a fire fighter.

As most experienced fire fighters will tell you, the time in the fire service will fly by and you'd better be preparing for your next position along the way. As a young know-it-all fire fighter I thought that I had plenty of time, but all of a sudden it was time to test. I thought I wasn't ready to be promoted so I wasn't going to test, but an assistant chief of mine told me "if you don't test and try to be part of the solution, don't complain if things are done differently than you think is right." I decided that I would test, without much preparation, and my test scores showed. I was very frustrated with my ranking because I thought I deserved better.

> "As a young know-it-all fire fighter I thought that I had plenty of time, but all of a sudden it was time to test."

As luck would have it, my assistant chief was there to set me straight and he asked, "How long did you prepare for this lieutenants' test?" I had to admit that all I did was to read one or two of the five books, and he said, "It wasn't enough, was it?" I was then faced with a dilemma: did I really want to get promoted or not? I decided that I did want to be promoted and started preparing. Over the next two years I studied and asked questions and I improved from the bottom of the list to third on the list. I still didn't feel I did the best I could have and it drove me to study harder and take more courses. I studied throughout the next two years without waiting for the announcement to come out. This time my studying paid off and I was on top of the list and was promoted. I have never stopped preparing for my next promotion; I take classes, study how other officers lead, and try to do my best every day.

As an aspiring officer you need to have the right attitude. I had a rough period of time during my first five years as a full-time fire fighter and hated everything that the department and administration did. At one point I decided that this was no way to go through my career and I became committed to getting promoted. I am living proof that it is never too late to change your life and your attitude and you can overcome large odds and have a successful career. No matter what happens, always try your hardest. You will experience setbacks in your career, but find a mentor you can talk to and work through your feelings and you will be successful.

Dean S. Wrobbel
Deputy Chief
St. Cloud Fire Department
St. Cloud, Minnesota

You could use a three-ring binder and divide it into sections. The first section could include all of the information about the promotional examination—a copy of the official announcement and copies of all of the documents the candidate was required to submit. Any test-related documents from the department can also be placed in this section.

The second section would contain a calendar, usually in the one-month-per-page format. The calendar covers the period from the start of the preparation process through the last promotional test activity. Important deadlines are highlighted, and the candidate can create a study plan, working backward from the examination date.

The third section would contain the written reference materials. Subdivided by reference, these sections would cover the items used to master the written material. Some candidates outline each chapter; others make up practice tests. You should make a note of any questions you miss in practice tests to be sure that you go back and obtain the correct answer. This information can be extremely valuable during the final review before the examination date.

The fourth section would cover information about the announced or anticipated components of the promotional examination. Some departments provide information to the candidates about topics that may be emphasized or areas of the references that will be excluded. For example, the department may clarify that candidates are expected to conform to the may-day procedure that was in effect before July 1. There may be a new procedure in effect on the day of the promotional test, but the exam will test the candidate on the earlier procedure. This is also where you would store information about the exam structure and process.

Preparing for Role Playing

The most effective candidates in role-playing exercises are the ones who act naturally during the examination, while making supervisory decisions that are strictly based on policies and regulations. This is sometimes described as performing in the "extreme supervisor" role. Many successful candidates model their performance in the examination, as well as afterward, on the behavior of successful and respected officers. It is a good idea to pay attention to the supervisors and officers who are known for being efficient and effective.

One of the best ways to prepare for role playing is to experience working in one of the busier or larger fire stations within your department. More supervisory issues come up in a fire station with 14 fire fighters and 3 officers on each shift than in a station with 3 fire fighters and 1 officer. Similarly, a fire company that runs 15 calls a day encounters more problem-solving opportunities than a company that runs 3 calls a day. Working under one of the widely respected officers is also a good way to develop supervisory skills. Candidates who have worked for a supervisor who demonstrates leadership and provides a positive role model have a significant advantage over candidates who have never had that experience.

Summary

Taking a promotional exam is quite an undertaking. You invest a huge amount of time preparing for a few minutes of performance. An effective promotional process allows the department to identify which candidates possess the knowledge, skills, and abilities to function as a supervisory or managing fire officer. A perfect promotional process will provide feedback to the candidate on strengths, weaknesses, and appropriate job-fit for a fire officer assignment.

Promotional processes are not perfect, so the candidate needs to identify areas of weakness or poor job-fit well before the promotional exam. Like the construction of the promotional exam, this personal assessment is based on the unique features of the department and the individual. The best way to start is to obtain the narrative job description and the class specification sheet from the personnel office or human resources. These two documents create the map in developing the promotional exam.

Getting It Done

Textbook Study Guides

There are publishers that produce study guides with hundreds of multiple-choice questions taken from fire service textbooks and references. The answer to each question is keyed to a page number in the textbook or source material. Some book publishers release their own study guides, whereas others specialize in producing promotional study guides for specific examination levels.

In addition to printed study guides, some publishers produce computer-based practice test programs accessible either on disk or online. A version of computer-based assessment, a student can have the software assemble test questions based on level of difficulty and subject area. Once the student has completed the test, the program generates the results, with links back to textbook references.

You Are the Fire Officer: Conclusion

You have been talking with your friends about the upcoming exam. Some are creating a study group, including a website. A recruit school colleague says, "You can't be promoted if you don't take the exam." She suggests that you join the study group and give it your best shot.

The next day, you ask the captain for study plan suggestions. He shows you the personal study journal that he has made for the battalion chief's exam. He offers to spend some time every on-duty evening to assist the members who are studying for promotional exams.

Wrap-Up

Chief Concepts

- Promotional examinations were a product of the 1883 Pendleton Civil Service Reform Act.
- Every civil service job has a narrative classified job description and a knowledge, skills, and abilities (KSA) worksheet.
- Promotional examinations assess the important KSAs and measure performance dimensions.
- Multiple-choice examinations concentrate on facts from the reading list.
- A local test committee, composed of existing officers with technical subject matter expertise, validates promotional examinations.
- Assessment centers provide a variety of role-playing exercises.
- During an in-basket assessment, you should remember to review; prioritize; identify resources, options, and alternatives; follow up; and notify.
- Emergency incident simulations fall into four formats: the "data dump" question, a mock emergency incident, an interactive emergency scene simulation, and a full-scale incident simulation.
- During an interpersonal assessment, remember to remain in control, exactly state the desired behavior, give the employee a deadline, arrange follow-up meetings, get the employee to buy into or take personal responsibility for the improvement plan, be empathetic but focused, explain consequences, and finish the session on a positive note.
- Candidates may be required to deliver a short presentation or write a memo or report as part of the assessment center process.
- Technical skills may be evaluated during promotional tests for highly specialized positions.
- The candidate needs to develop a personal study plan to master the content for a promotional examination. Study techniques include keeping a study journal and participating in role-playing activities.

Hot Terms

Assessment centers A series of simulation exercises to identify a candidate's competency to perform the job that is offered in the promotional examination.

Class specification A technical worksheet that quantifies the knowledge, skills, and abilities (KSAs) by frequency and importance for every classified job within the local civil service agency.

"Data dump" question A promotional question that asks the candidate to write or describe all of the factors or issues covering a technical issue, such as suppression of a basement fire in a commercial property.

Dimensions Attributes or qualities that can be described and measured during a promotional examination. Between 5 and 15 dimensions would be measured on a typical promotional examination. The six most common are oral communication, written communication, problem analysis, judgment, organizational sensitivity, and planning/organizing.

In-basket exercise A promotional examination component in which the candidate deals with correspondence and related items accumulated in a fire officer's in-basket.

Job description A narrative summary of the scope of a job. Provides examples of the typical tasks.

Knowledge, skills, and abilities (KSAs) The traits required for every classified position within the municipality. Defined by a narrative job description and a technical class specification.

Metropolitan (metro-sized) fire department A department with more than 400 fully paid fire fighters. The Metropolitan Fire Chiefs is a special interest group organized by the International Association of Fire Chiefs in 1965 to address the needs of large fire departments. Metro Chiefs is also a section of the National Fire Protection Association.

Personal study journal A personal notebook to aid in scheduling and tracking a candidate's promotional preparation progress.

Spoils system Also known as the patronage system. The practice of making appointments to public office based on a personal relationship or affiliation rather than because of merit. The spoils system scandals of the New York City "Tweed Ring" and the Tammany Hall political machine (1865–1871) resulted in Congress passing the Pendleton Civil Service Reform Act of 1883.

Fire Officer *in Action*

You have joined the study group started by fellow recruit school graduates. Once every 2 weeks the study group meets to review the material and take practice exams. Each member outlines a portion of a reference book and writes multiple-choice questions. The password-protected website includes a blog and an area to post the outlines, test questions, and other study material.

1. Which of the following actions is most valuable in the early phase of preparing for an announced promotional exam?
   A. Read more books on promotional exams.
   B. Ask individuals who scored high in the last promotional exam for preparation suggestions.
   C. Ask senior officers what they think will be on the promotional exam.
   D. Obtain old promotional exams from larger fire departments.

2. A key component in completing an in-basket exercise is to:
   A. provide a "data-dump" response to each item.
   B. demonstrate your mastery of a technical topic.
   C. clearly show what items you consider "urgent."
   D. properly document the follow-up activities.

3. A feature of an interpersonal interaction exercise is that the role player:
   A. has important information that is revealed only after detailed questioning by the candidate.
   B. reveals all of the essential information in the first 3 minutes of interaction.
   C. is expected to surprise or confuse the candidate through words or actions.
   D. does not know what the "correct" candidate response should be.

4. While preparing for the promotional exam, you learn about a firefighting technique that is superior to the method your department uses. You should:
   A. incorporate that technique into your response to the emergency management incident simulation.
   B. recommend to the test committee that they integrate this technique into the promotion process.
   C. identify locations within the promotional references that support the superior technique.
   D. make sure your exam responses reflect the current formalized practices of your department.

Fire Fighters and the Fire Officer

NFPA 1021 Standard

Fire Officer I

4.1.1 **General Prerequisite Knowledge.** The organizational structure of the department; geographical configuration and characteristics of response districts; departmental operating procedures for administration, emergency operations, incident management system and safety; departmental budget process; information management and recordkeeping; the fire prevention and building safety codes and ordinances applicable to the jurisdiction; current trends, technologies, and socioeconomic and political factors that affect the fire service; cultural diversity; methods used by supervisors to obtain cooperation within a group of subordinates; the rights of management and members; agreements in force between the organization and members; generally accepted ethical practices, including a professional code of ethics; and policies and procedures regarding the operation of the department as they involve supervisors and members. [p 45–53]

4.2.5* Apply human resource policies and procedures, given an administrative situation requiring action, so that policies and procedures are followed. [p 50–55]

(A)* Requisite Knowledge. Human resource policies and procedures.

(B) Requisite Skills. The ability to communicate orally and in writing and to relate interpersonally.

4.4 **Administration.** This duty involves general administrative functions and the implementation of departmental policies and procedures at the unit level, according to the following job performance requirements. [p 42–45]

4.4.2 Execute routine unit-level administrative functions, given forms and record-management systems, so that the reports and logs are complete and files are maintained in accordance with policies and procedures. [p 42–44]

(A) Requisite Knowledge. Administrative policies and procedures and records management.

(B) Requisite Skills. The ability to communicate orally and in writing.

Fire Officer II

NFPA 1021 contains no Fire Officer II Job Performance Requirements for this chapter.

Introduction to Fire and Emergency Services Administration (FESHE) Course Outcomes

4. Recognize appropriate appraising and disciplinary actions and the impact on employee behavior. [p 49–50]

7. Identify roles and responsibilities of leaders in organizations. [p 42, 44–47, 49, 52–55]

10. Identify the importance of ethics as they apply to supervisors. [p 50, 52]

11. Identify the role of a company officer in Incident Command System. [p 46–47]

Knowledge Objectives

After studying this chapter, you will be able to:

- Describe the fire officer's vital tasks.
- Describe a typical fire station workday.
- Describe the transition from a fire fighter to a fire officer.
- Describe the activities a fire officer performs to maintain an effective working relationship with his or her supervisor.
- Describe integrity and ethical behavior.
- Describe how to maintain workplace diversity.
- Describe the concept of the fire station as a business work location.

Skills Objectives

After studying this chapter, you will be able to:

- Function as a newly assigned fire officer, and with a description of a fire station, work group, and schedule, prepare a beginning of shift report or activity plan.
- Function as a fire officer and demonstrate the effective issuing of an unpopular order to a fire company.
- Demonstrate making a decision consistent with the department's core values, mission statement, and value statements given an ethical dilemma.
- Conduct an initial interview and notifications consistent with the department's policy, rules, and regulations given a harassment or hostile workplace complaint.

For years, fire fighters at The Nickel spent the first part of the workday wearing physical training clothes and coveralls from morning line-up through grocery shopping. It was an effective use of time. After lunch the crew is showered, neat, and clean in their station uniforms, ready to perform community service, safety inspections, or any other interaction with the public.

This morning's line-up includes the following document from the fire chief's office with the banner message "Immediately Read and Implement." The fire chief has issued a directive to make sure personnel are complying with the standard operating procedure (SOP) on uniforms. That SOP requires members to be in station uniforms any time they are out in public. Although past practice has allowed crews to wear physical training clothes and coveralls when out in public, the SOP provides no exception.

This makes for an uncomfortable morning meeting with the crew. It is clear that two of the fire fighters are very angry and plan to continue to wear coveralls in the morning. One is going to call the shop steward. You think that this order is unreasonable.

1. How do you respond to an order that you do not agree with?
2. How do you implement an unpopular directive?
3. What can you do to maintain the effectiveness of the work group?

Introduction to the Fire Officer's Job

The fire officer is responsible for managing a work unit within the fire department. When we think about the duties of a fire officer, the most prominent vision is leading a team of fire fighters into a challenging emergency situation. However, much of what a fire officer does during a normal workday is routine administrative activities related to the work group, the assigned equipment, and the physical facility. The officer is expected to ensure that the work unit will be prepared to function effectively and efficiently when it is needed. That means managing personnel, resources, and programs. This chapter will look at a few overarching principles and examples of the day-to-day administration within a work group.

Supervising and managing fire officers usually report to higher-ranking chief officers. In a large fire department, the direct supervisor is usually an administrative fire officer, usually at the rank level of battalion or district chief. In a smaller organization, a fire officer might report directly to the fire chief or to a deputy or assistant chief. In most cases, this supervisor is an individual who has spent time coming up through the ranks, working as a supervising and managing fire officer.

The International Association of Fire Chiefs (IAFC) defines an **administrative fire officer** as a person who worked as a managing fire officer for 3 to 5 years, is certified at the NFPA Fire Officer III level, and has accomplished formal education equivalent to a bachelor's degree.

The Fire Officer's Tasks

Administrative fire officers working in cities and large counties responded to an email request to "list the most important tasks you want a new fire officer to do well." This resulted in a list of four basic tasks that they consider vital: beginning of shift report, notifications, decision making, and problem solving. They believed that a new fire officer who meets these task expectations is on the right track.

The Beginning of Shift Report

The administrative fire officers emphasized the importance for a new supervising fire officer to provide a prompt and accurate report at the start of the workday **Figure 3-1**. This report is provided from each work location to the battalion or district chief within the first quarter hour of the reporting time. The format of the report may be electronic, paper, or verbal. Some departments use sophisticated online staffing systems, like TeleStaff, that provide real-time scheduling that conforms to departmental operational requirements. The chiefs rely on this information to make staffing adjustments at the beginning of the shift. An accurate report is needed to ensure that adequate staffing and equipment are in place and ready for the balance of the shift Figure 3-2 ▾.

The first part of the report provides the on-duty staffing information and sick leave list and identifies any positions that need to be filled for that day. The positions are a priority because someone who worked the previous shift has to remain on duty until a relief person shows up to fill any position that remains vacant. The battalion chief moves available staff to cover the vacancies noted on the beginning of shift reports, and any vacancies that remain have to be covered by fire fighters working overtime. It takes time to reassign on-duty personnel, call the

Figure 3-1 The beginning of shift report is a vital company officer task.

Today's Date is: May 29

	Total	Paramedics	Fire Officer	EMS Officer	Prearranged Callback	Annual Leave	Vacancies	Detail Out-of-Operations	Injury Lv or Light Duty	LWOP	Fire OIC	EMS OIC
Minimums	10	3	2	1							**Fire OIC**	**EMS OIC**
Today's staffing	9	3	1	1	1	1	0	1	0	0	Smyth	Willow
Next day staffing	8	2	1	0							Smyth	????

Today's shortage	Why?	PM Surplus	Next Day's Shortage	Why?
Engine Officer	O/R	none	Engine Officer	Off Rep
			Medic Officer	Leave

Sick Leave	Detailed Out of Ops	Next Day APPROVED Leave
None	Capt. Johnson	FF Tolliver
		Lt. Willow

Injury Leave/Light Duty
None

	Vehicle Status	
	Engine 7746	Eng 46
Messages for the Chief:	Rescue 7099	Res 46
Vehicle 7234 overdue for preventative maintenance	Medic 6322	Med 46
Rescue 46 thermal imager broken	Reserve Engine 7234	Eng 35
Furnace malfunctioning	Reserve Medic 4276	Med 11
Fire Chief at 46 for dinner @ 1830	Battalion 9 5040	shop
	Reserve Suburban 5107	BC 09

Figure 3-2 Example of a municipal beginning of shift report.

overtime personnel, and get everyone to their work locations so the holdover personnel can be released.

It is maddening when a new fire officer sheepishly calls the administrative fire officer 2 hours into a work shift to report that a fire fighter from the previous shift is still on involuntary holdover because someone failed to note the problem on the beginning of shift report. By the time the position is covered, the fire fighter on involuntary holdover may have remained on duty for several more hours.

The report also projects the staffing for the next day. This allows the chief to make assignments in anticipation of the known absences for the next workday, instead of trying to solve all of the problems during the first 45 minutes of the next workday.

In addition, the report notes the location and condition of all of the apparatus or rolling stock, such as a reserve pumper that has been loaned to another station or an ambulance at the shop. Finally, the report provides the chief with any "must know" information that will require immediate attention.

Notifications

The second most important issue noted by the administrative fire officers was that the new supervising fire officer must make prompt notifications. Some types of information have to be passed up the chain of command quickly. For example, all injury and infectious disease exposure reports have to be processed without delay. If a fire fighter is exposed to a possible blood-borne pathogen early Saturday morning, the exposure report cannot sit on the fire officer's desk until Monday morning before it gets to the battalion chief. The designated infection control officer has to be informed immediately to get patient information while it still can be easily obtained. The same priority applies to any information the chief needs to know about when it is current, particularly before someone at a higher level calls to ask about it. Many chiefs call this the "no surprises" rule.

Decision Making and Problem Solving

The administrative fire officers rated the third and fourth issues as equal in importance. Some chiefs complained that new supervising fire officers were hesitant to make decisions. They seemed to want the chief to make the hard decisions and to enforce unpopular rules. The chiefs wanted the new officers to run their companies and make the decisions that are within their scope of responsibility. The chiefs were available for consultation, but they expected their officers to run the fire stations.

Some of the chiefs also noted that they spent a lot of time listening to fire officers complain about problems without proposing any solutions. The chiefs preferred the officers who would think through the problem and propose a solution. The most valuable proposals would consider the larger picture of how a possible solution would impact the rest of the department **Figure 3-3 ▸**.

■ Example of a Typical Fire Station Workday

A supervising fire officer is responsible for accomplishing the fire department mission through the efforts of the fire fighters under his or her command. At the company level, this requires a balance of management and leadership skills. The officer has to organize the work and provide the leadership to ensure it gets done safely and effectively.

Figure 3-3 Discuss possible solutions to a problem with the chief.

The fire department has an agency-wide mission that is translated into annual goals. These goals are used to develop annual, quarterly, and monthly objectives for each fire company. The administrative fire officer and the supervising or managing fire officer meet regularly to set the objectives and to review progress. The monthly goals show up as the planned activities on the fire company daily planner.

The following is an example of a 24-hour shift in a fire station. The starting point is a schedule that ensures all of the required tasks are completed in a logical order. The fire officer has to anticipate that emergency incidents will alter the workday and will require adjustments in the schedule.

0700 Line-up and equipment check. Send beginning of shift report to chief. Clean quarters, empty trash, clean dishes.

0800 Dust and vacuum all carpeted areas. Sweep all tile floors.

0830 Physical training and outside skill drill. Go to store to pick up groceries.

1100 Heavy cleaning (while still in physical training clothes).
- Monday: Air out bunkroom and rotate mattresses; clean all windows.
- Tuesday: Clean utility rooms and shop area.
- Wednesday: Clean and inventory emergency medical service (EMS) equipment, self-contained breathing apparatus (SCBA), and decontamination areas.
- Thursday: Move recyclables outside for pickup, then clean weight room and lockers.
- Friday: Scrub kitchen and clean out refrigerators.

1130 Scrub bathrooms after fire fighters clean up from physical training.

Noon Lunch

1330 Scheduled productivity activity (e.g., fire safety inspections, school visits, inside or outside training).

1800 Dinner, followed by kitchen clean-up. Run dishwasher.

1930 Individual study time, occasional fire safety inspections (nightclubs) or drills.

Near Miss REPORT

Report Number: 07-0000886

Event Description: At the beginning of the shift the four-person engine crew was doing their routine truck check of the engine. The driver started the engine and attempted to place the engine in pump. When the driver shifted the engine from neutral to drive the engine drove forward because the parking brake apparently had not been engaged from the previous shift. The driver had not completely exited the engine and was able to stop the truck abruptly with his foot on the brake before striking the closed bay door. The other three personnel were on the truck at the time of the incident. Two of the personnel were half in the truck on steps leading into the cab with the cab doors open checking various equipment. There were no injuries or damage from this incident. The incident could have been much worse if there had been any personnel walking or checking equipment at the front of the engine or if the engine had driven into the bay door causing the door to fall. The worst outcome could have been personnel falling off the truck.

Lesson Learned: Do not take for granted that when the truck is turned off in the bay that the parking brake has been engaged. Driver shall be seated in the vehicle, make sure the parking brake is engaged and foot covering the brake. Make sure the engine is set correctly in the pump mode of the cab portion before exiting the engine.

2130 Remove all trash, tidy up day room, and make final pass through the kitchen.

Special station activity

- First and third Thursdays of the month: Scrub apparatus bay floor.
- Second and fourth Fridays of the month: Wax kitchen floor.
- February: Safety officer inspection of facility, apparatus, and personal protective equipment (PPE).
- March and September: Steam clean carpeted areas.
- April: Fire chief's annual inspection of the fire station.
- May: EMS Week open house.
- October: Fire Prevention open house.

Typical Volunteer Duty Night

The following example of a volunteer duty night is less detailed than the previous schedule, but has the same essential tasks as a 24-hour shift in a municipal fire department, including equipment maintenance, training, and station maintenance. Many departments use a day book or monthly calendar to plan the duty crew activities.

1700 Evening duty crew starts. Equipment check.

1800 Dinner, followed by kitchen clean-up. Run dishwasher.

1930 Classroom session, skill drill, or community outreach activity.

2130 Remove all trash, tidy up, and make final pass through the kitchen. Take clean dishes and cups out of dishwasher and put away.

General station and apparatus cleaning are conducted on weekends. One section of the fire station gets a major cleaning two or three times a year (e.g., wax floors, scrub apparatus floors). Specialized heavy cleaning, such as steam cleaning carpets and waxing apparatus, is scheduled throughout the year.

The Transition from Fire Fighter to Fire Officer

There are four times in a fire fighter's career when a major change occurs in how the individual relates to the formal fire department organization. The first change occurs when the fire fighter completes the probationary training period. Completing the initial training and probationary period is a major milestone. It is marked in many departments by a change in helmet color or shield. The second change takes place when the fire fighter successfully completes a promotional process and starts working as a fire company commander. The third event is when the fire officer completes another level of training, advances through the promotional process, and starts working as a chief officer. The fourth event is when a fire fighter retires.

In all four situations, a significant change occurs in the individual's relationship to the organization and to the other members of the fire department. Part of this change is the individual's sphere of responsibility within the formal organization. As a fire fighter, there is a mutual responsibility among the crew members on the rig. They work together to accomplish fire-ground tasks and look out for each other.

A promotion from fire fighter or driver/operator to company-level officer is a large step. The company-level officer is directly responsible for the supervision, performance, and safety of a crew of fire fighters. There is a sacred duty to ensure that all of the fire company team members remain safe when operating in a hostile or hazardous work environment. The company-level officer functions as a working supervisor, sharing the hazards and work conditions with the fire company team **Figure 3-4 ▶**.

This change in the sphere of responsibility often requires the new officer to change some on-duty behaviors or practices. The formal fire department organization considers a fire officer to be the fire chief's representative at the work location.

Figure 3-4 The company-level officer is responsible for the supervision, performance, and safety of a crew of fire fighters.

This creates an expectation that the fire officer will behave in a way that is appropriate for a first-line supervisor. Behavior that was acceptable for a fire fighter may be unacceptable for a fire officer. For example, consider a fire fighter who is known for developing elaborate practical jokes within the fire station environment. As a new fire officer, this individual would have to consider the impact that being seen as a practical joker would have on his or her ability to function as an effective supervising fire officer.

Conversely, the new fire officer needs to consider how to respond to pranks and verbal jabs from fire fighters. What may have been an appropriate response from another fire fighter may no longer be the best response from an officer. Wearing the fire officer badge enhances the effect and consequences of any action or response.

Promotion to a chief officer rank changes the individual's relationship to the organization and the members to an even greater degree. A command officer has less of a hands-on role than an officer of a company. The command officer is typically directly responsible for several fire companies and must depend on company officers to provide direct supervision over crews performing fire-ground tasks, often in hazardous conditions. In many cases, the command officer works outside the hazardous area but is still responsible for whatever happens inside.

Fire Marks

James O. Page
In July 1971, James O. Page was promoted to battalion chief in the Los Angeles County Fire Department. In the same year he was admitted to the California Bar and finished the manuscript for his fire company supervision textbook, *Effective Company Command*. He became widely known as an author and publisher, the "father" of fire-based emergency medical services, and a practicing attorney.

Fire Officer as Supervisor-Commander-Trainer

In his book, *Effective Company Command*, James O. Page divided the company officer's duties into three distinct roles: supervisor, commander, and trainer.

Supervisor

In this role, the fire officer functions as the official representative of the fire chief. That means the fire chief expects that every fire officer will issue orders and directives and conduct business in a way that meets the chief's objectives. The fire officer will supervise the fire company in a manner consistent with the rules and regulations of the fire department. For example, if all fire companies are expected to spend Tuesday afternoon conducting fire safety inspections in commercial properties, the fire officers are responsible for ensuring that their individual fire stations complete this task.

Unpopular Orders and Directives

On occasion, a fire officer may be required to issue and enforce unpopular orders Figure 3-5 ▼ . Even if the officer disagrees with a particular directive, the formal organization requires and expects the officer to carry out that directive to the best of his or her ability. A fire officer can improve his or her effectiveness in handling an unpopular order by determining the story or history behind the order, which would enable the fire officer to put the directive in perspective.

When faced with an unpopular order, the fire officer should express any concerns and objections with his or her supervisor in private. This is the time for the fire officer to discuss suggestions for modifying or reversing the order. Occasionally, special circumstances may make the order difficult to implement, and the supervisor might be able to authorize adjustments. Once the meeting with the officer's supervisor is over, the formal organization expects the officer to enforce the order as issued or amended.

Telling the fire fighters that their officer does not agree with an order undermines the officer's authority and supervisory ability. The fire chief expects an officer to perform the required supervisory tasks, and the fire fighters have to understand that the fire officer does not make all the rules or have a choice about which ones to enforce. Enforcing unpopular orders is part of the job.

Commander

When operating at the scene of an emergency incident, the fire officer is expected to function as a commander and exercise

Figure 3-5 An officer sometimes has to issue and enforce unpopular orders.

Figure 3-6   At the scene of an emergency incident, the fire officer may function as the initial incident commander.

strong direct supervision over the company members. In some cases, the fire officer could be responsible for directing the actions of additional resources, or he or she might be functioning as the initial incident commander **Figure 3-6**.

Functioning as the initial incident commander on a major emergency is one of the higher-profile roles of a fire officer. The ability to bring order out of the chaos of an emergency incident is an art that requires a well-developed skill set. The fire officer needs to be clear, calm, and concise in the initial radio transmissions. The communication of incident size-ups must be consistent with the organization's requirements and incident management system.

Developing a command presence is a key part of mastering the art of incident command. Command presence is the ability of an officer to project an image of being in control of the situation. In order to be a successful leader, the officer must convince others to follow by demonstrating the ability to take charge and make the right things happen. A fire officer who is going to establish command upon arriving at an emergency incident should have a detailed knowledge of the responding companies, a mastery of the local procedures, and the ability to issue clear direct orders. Fire fighters are aggressive, action-oriented people who also can be compulsive. It is important that a new fire officer develop a command presence in order to focus the efforts of this action-oriented team **Figure 3-7**.

Trainer

The fire officer has the responsibility of making sure the fire fighters under his or her command are confident and competent in their skills. The company-level officer is responsible for the performance level of the fire company and has to establish a set of expectations that the company will perform at the highest level possible **Figure 3-8**. In a large department, one fire company may need to have a higher level of specialized skill in one area than in another. For example, a fire company may have a specialty assignment or a co-located specialized unit that the fire company staffs when it is requested. Technical rescue

Figure 3-7   It is important for the fire officer to develop a command presence.

Figure 3-8   The officer must ensure that fire fighters are confident and competent in their skills.

rigs, decontamination units, foam pumpers, mass casualty units, and mobile command posts are examples of co-located specialized units.

In addition, the fire company's response district may require a higher level of fire fighter skill or knowledge. Consider hose handling skills. A fire company in a high-rise district would require different hose handling skills and competencies than those required of a company in a mountainous wildland interface area. Although both companies need to demonstrate the basics of operating a hose line from a standpipe, the high-rise fire company should have a much higher level of expertise and competence in this skill. Conversely, the mountain fire company usually has a higher level of expertise and competence in rope rescue evolutions.

The company-level officer is the key to developing these competencies within the company. Page makes three specific recommendations to assist fire officers in this task: develop a personal training library, know the neighborhood, and use problem-solving scenarios.

Developing a Personal Library

James O. Page's personal library starts with a three-ring notebook with subject matter tabs. The subject tabs could be from the topic headings in NFPA 1001, from the recruit school curricula, or from a personal list of important topics. Every time the fire officer attends a training event, the notes from that session are placed into the three-ring binder **Figure 3-9 ▼** . Every related handout, product information sheet, and other related items are also placed into the three-ring binder. When the officer is preparing to present a class that covers that topic, the personal training library is the first stop.

Leo D. Stapleton, retired Boston fire commissioner, advocates that officers maintain a personal journal where they record the incidents they run and the issues they handle. For working incidents, that would include what was encountered, what they learned, and how they would handle it the next time. Stapleton encourages these entries as soon as possible after the incident, while the information is vivid in the officer's memory.

Today, a fire officer can make an electronic version of Page's three-ring binder and Stapleton's journal with the use of a laptop

Thirty Years on the Line

Thirty Years on the Line was written by Boston Deputy Chief Leo D. Stapleton in 1982 to teach the firefighting lessons he learned; it contains a series of short stories involving actual incidents and real characters. By 2007, Stapleton had published nine fiction and three nonfiction books about Boston fire fighters. This includes a six-novel, 10-year epic following a group that select firefighting as their profession; they start as Ffops, fire fighters on probation, in 1996, and conclude in 2006 with one of them, Donald Holden, becoming a deputy chief.

computer, scanner, and digital camera. The fire officer can collect PowerPoint presentations and video clips, download files from the Internet, and convert various paper media into Adobe portable document format (.pdf) files. A digital projector allows all these materials to be used in classroom sessions.

In addition to the three-ring notebook or laptop computer, Page and Stapleton recommend that the fire officer obtain personal copies of the textbooks and references used in fire fighter training and promotional examinations. You can highlight, tab, and write in this copy of the book until it becomes your personal reference. You can write notes in the margins and even note disagreements with the author's statement or point of view to personalize the learning experience. This is especially valuable when you are reading a book in preparation for a promotional exam **Figure 3-10 ▼** .

Figure 3-9 Notes from training events should be placed in a three-ring binder.

Figure 3-10 Highlight, tab, and write in your personal copies of textbooks used for training.

Know the Neighborhood

Fire fighters should have a detailed knowledge of the environment they protect. That requires going out to walk through each nonresidential structure, from the roof through the subbasement; these walk-throughs reinforce the written preincident plan or diagrams **Figure 3-11 ▼**. During the walk-through, the fire officer should take pictures to capture significant details. A detailed overhead view of many communities can be obtained from aerial photographs, digital maps, or Google Earth. This aerial view is especially valuable for apartment complexes, business parks, and retail areas. The maps and photos can be used to locate access routes, plan the placement of first-alarm apparatus, and identify exposure problems.

Knowing the neighborhood may include working with the building owners and occupants to practice incident action plans and procedures. After the September 11, 2001, terrorist attacks, many commercial and office buildings revised their internal emergency plans, upgraded their security systems, and introduced access restrictions. Fire officers should strive to establish a working relationship with every building manager or emergency program manager in their response district.

Use Problem-Solving Scenarios

The fire officer can help the company members become more skilled and knowledgeable by providing opportunities to use their problem-solving skills. Instead of reading the code regulations to provide training for his company members, Page would present fact-based situations and require them to use the code to solve the problem. This technique forced the fire fighters to identify the occupancy use group, identify the issues, look up the applicable regulations, and make decisions. This is an excellent way for adults to learn concepts, regulations, and decision-making skills. The same technique can be used for fire-ground hydraulics, technical rescue scenarios, and incident management procedures.

The same problem-solving approach can be used in reviews of pre-incident action plans. The fire officer can construct various emergency incident scenarios for training sessions, based on actual buildings and potential situations. There are computer-based fire incident simulators that can accept digital pictures taken during walk-through visits to create customized emergency incident scenarios.

▌ The Fire Officer's Supervisor

Every fire officer has a supervisor. In municipal fire departments, a fire officer's supervisor is usually a command-level officer (a battalion chief, a district chief, or a battalion commander) who supervises numerous fire companies within a geographical area. The battalion chief's supervisor could be a deputy chief, who reports directly to the fire chief. In the same manner, the fire chief would report to the mayor, the city manager, the commissioner of public safety, or an individual at an equivalent level. Under the chain of command, the fire chief's orders and directives are passed down through the deputy chief to the battalion or district chief, who then ensures that the fire officers enact the orders or directives.

Regardless of the organization structure, every fire officer has an obligation to work effectively with a supervisor. Three activities are necessary for this:

- Keep your supervisor informed.
- Make appropriate decisions at your level of responsibility.
- Consult with your supervisor before making major disciplinary actions or policy changes.

No supervisor likes surprises. Part of the supervisor–subordinate relationship is to make sure the supervisor is not surprised or blindsided **Figure 3-12 ▼**. For example, if a fire company

Figure 3-11 Walk-throughs help fire fighters get to know the properties in the neighborhood.

Figure 3-12 Always keep your supervisor informed.

has a vehicle crash at 10:00 P.M., the chief needs to know about it within the hour. It would be poor form for the chief to find out about the accident from a story on the 11 o'clock news.

Supervising and managing fire officers should not hesitate to make decisions appropriate for their level of responsibility. This means that problems should be addressed and situations resolved where and when they occur. If a fire officer has the authority to solve a problem, he or she should not be waiting for the supervisor to arrive to solve it.

This usually covers fire station–level activities such as maintenance, training, and community outreach. At fire stations with rotating career work shifts, the expectation is that most issues between the shifts are resolved at the supervising fire officer level. A fire officer who is working in a staff position should make the decisions appropriate for that position. Volunteer officers should make decisions appropriate for their level within the organization.

Some issues require the fire officer to consult with a supervisor before making a decision or taking action. Policy changes cannot be performed in an administrative vacuum. If a decision is going to have an impact that goes beyond the fire officer's scope of authority, it is time to talk to the supervisor.

This policy also applies before major disciplinary actions are taken. Consultation with a supervisor is required by the municipal personnel regulations to ensure that all major discipline is delivered in a consistent and impartial manner. This is also a recommended practice in most volunteer fire departments.

Getting It Done

Do Not Hold Your Breath
A large western fire department uses the term "holding your breath" to describe probationary fire fighters and new fire officers who attempt to demonstrate a behavior that is desired by the department but is alien to the individual. At some point, the person will dramatically revert to the original behavior. For example, it is unreasonable to expect a fire fighter who struggles to meet the minimum physical capability to suddenly start Olympic-level training because of a promotion to company officer. It is reasonable that the new fire officer acknowledges the gap between the current and the preferred capability and make an effort to improve personal performance. The public sharing of the need to improve performance could result in an on-duty group effort that improves the company-level physical capability.

Integrity and Ethical Behavior

The formal organization provides the new supervising fire officer with the symbols of power and authority that are associated with the badge, insignia, and distinctive markings on the officer's helmet. The individual needs to provide the core values of integrity and ethical behavior that, combined with the formal symbols, create an effective fire officer. An unethical fire officer is ineffective and damages the department's reputation. Left uncorrected, unethical and corrupt fire officers corrode the department's ability to deliver services and maintain the public trust.

Integrity

Integrity refers to the complex system of inherent attributes that determine a person's moral and ethical actions and reactions, including the quality of being honest.

The fire officer should "walk the talk" and demonstrate the behaviors that he or she says are important **Figure 3-13 ▶**. If the company officer says that physical fitness is important, then the fire fighters should see their officer performing physical fitness training during the workday.

Integrity can be demonstrated by a steadfast adherence to a moral code. This is a combination of a fire officer's internal value system and the fire department's official organizational value system. Formal organizations publish their expectations as a code of ethics, a code of conduct, or a list of value statements.

Ethical Behavior

The fire officer position provides a wide range of opportunities to demonstrate **ethical behavior**. The fire officer demonstrating ethical behavior makes decisions and models behavior consistent with the department's core values, mission statement, and value statements.

Figure 3-13 The fire officer should demonstrate the behaviors that he or she says are important.

Voices of Experience

A key responsibility of the fire officer is to ensure the company is properly trained. Over the years, I have found many new fire officers want to be liked by their crew even at the expense of doing their job well. Often this translates into not making the crew do things they don't want to do. It is easy to sit around and have an extra cup of coffee and tell another war story rather than taking the crew out to the bay to hone their skills. Before you know it, the morning or evening is nearly gone and it's "not worth" starting anything at that point. The new fire officer who falls into this trap is falling into a vicious cycle. The more you don't get up and train the more the crew doesn't want to get up and train.

As a newly promoted company officer, I fell into the trap. I really wanted to fit in, so I thought I would ease into breaking the pattern of being in a "holiday routine," everyday. The crew liked it, but soon I found basic skills lacking at fire scenes. I had a very difficult time getting even a limited amount of cooperation. With all future crews, I made the expectations clear from the first day and rarely had an issue.

> **"It is more important to have the respect and trust of your personnel than to be liked."**

To make sure you don't fall victim to the same trap I did, you should understand a basic principle of supervision: It is more important to have the respect and trust of your personnel than to be liked. While being liked is an asset, new fire officers tend to make it more important than any other trait. Crew members want their supervisor to have integrity, which means doing the right thing even when it isn't what they prefer to do. Additionally, fire fighters care about their own safety, and training directly impacts that. Fire fighters trust officers who care about them and their safety.

To gain the respect and trust of your crew, first provide them with an understanding of why you are committed to training. Meet with your crew during the first tour together. Discuss your belief in training and how you want everyone to go home safely. Knowing your motives will help dispel the belief that you are only trying to make yourself look good. Second, set clear expectations about crew training. Failing to set clear expectations from the beginning almost always leads to problems down the road as you try to make up for it later. Third, set high expectations. If you don't expect much from your fire fighters, you won't get much. Lastly, develop quality training. I've rarely heard a fire fighter complain about good training. It is your responsibility to figure out how to make it "good" in the fire fighter's eyes.

David Hall
Assistant Fire Chief
Springfield, Missouri Fire Department
Springfield, Missouri

Paragraph A.1.3 in the "Annex A: Explanatory Material" section of NFPA 1021, *Standard for Fire Officer Professional Qualifications*, makes the following statement:

> Fire officers are expected to be ethical in their conduct. Ethical conduct includes being honest, doing "what's right," and performing to the best of one's ability. For public safety personnel, ethical responsibility extends beyond one's individual performance. In serving the citizens, public safety personnel are charged with the responsibility of ensuring the provision of the best possible safety and service.
>
> Ethical conduct requires honesty on the part of all public safety personnel. Choices must be made on the basis of maximum benefit to the citizens and the community. The process of making these decisions must also be open to the public. The means of providing service, as well as the quality of the service provided, must be above question and must maximize the principles of fairness and equity as well as those of efficiency and effectiveness.

Fire department activity tends to be high profile, regardless of the task. A simple trip to the grocery store to pick up dinner creates public attention. The fire officer should act as if someone is always documenting his or her actions when out of the fire station.

Workplace Diversity

The civil rights of Americans are established by federal laws, which are enforced by the Equal Employment Opportunity Commission (EEOC). Title VII of the Civil Rights Act of 1964 covers state and local governments, schools, colleges, and unions. This law states that it is illegal for an employer to:

> (1) fail or refuse to hire or discharge any individual, or otherwise discriminate against any individual with respect to compensation, terms, conditions, or privileges of employment because of such individual's race, color, religion, sex, or national origin, or (2) to limit, segregate, or classify employees or applicants for employment in any way that would deprive any individual of employment opportunities or otherwise adversely affect status as an employee because of such individual's race, color, religion, sex or national origin.

The Equal Employment Opportunity Act of 1972 amended the Civil Rights Act of 1964 and expanded its coverage to include almost all public and private employers with 15 or more employees. In general, the 1972 act covers volunteer fire departments and other nonprofit emergency service organizations.

The Civil Rights Act of 1991 provides additional compensatory and punitive damages in cases of intentional discrimination under Title VII and the Americans with Disabilities Act of 1990. The changes introduced in the 1991 act increased the number of discrimination lawsuits filed against organizations.

Many fire departments have made changes to their recruitment, hiring, and promotion practices in order to comply with the civil rights laws. Some departments took action to diversify their workforces without the prompting of the courts. Where this did not occur, the legal remedies have ranged from consent decrees, in which the fire department agrees to accomplish specific diversity goals within a specific time, to court orders outlining specific hiring practices.

__Diversity__ as applied to fire departments means the workforce should reflect the community it serves. In all cases, the overarching goal is for the fire department to reflect the diversity of the community. Consider the example of a fire department that is 90% Caucasian in a community where the population is 40% African American, 20% Latino, and 20% Asian American. The fire department population is not reflecting the community.

If the department has agreed to an EEOC consent decree, it has promised the court that it will work to hire qualified individuals who reflect the community. A consent decree can require a variety of activities, including community outreach, job fairs, pre-employment preparation, and peer group coaching. The court formally meets with the fire department representative periodically to see how well the department is progressing. Some fire departments that worked under Justice Department consent decrees in the 1970s have successfully met the court's goals and are no longer under court supervision.

Some departments operate under a specific court-mandated hiring process. Using our earlier example, the court could require the department to hire two African Americans, one Latino, and one Asian American before it can hire one additional Caucasian. This requirement would remain in effect until the hiring diversity goal is accomplished.

Fire Marks

Ricci vs. DeStefano

In 2009, the Supreme Court struck down a New Haven decision to throw out a 2003 promotional test that the city felt promoted too few minorities. The 5-to-4 ruling in *Ricci vs. DeStefano* confronts a persistent issue raised under Title VII of the Civil Rights Act of 1964: how to meet a requirement that requires both equal access and equal results.

The Fire Officer's Role in Workplace Diversity

Jack W. Gravely is a lawyer, subject matter expert, and trainer on workplace diversity. He has been a frequent speaker at public safety agencies on issues of racial and cultural diversity. From 1985 to 1988, he was the Special Assistant to the County Manager for Equal Employment Opportunity/Affirmative Action (EEO/AA) in Arlington County, Virginia. In that position, Gravely developed the policies and staffing changes necessary to address the sudden influx of Central Americans, Asians, and Ethiopians who moved into the county. Gravely was appointed Director of Workplace Diversity for the Federal Communications Commission in 1995. He went on to work as a broadcaster and consultant for a diversity-based multimedia company.

Gravely points out that a fire officer today has the benefit of four decades of EEO/AA court decisions to guide decision making.

When he started diversity training classes, the emphasis was on the language of the regulations and their potential impact. The EEOC files about 400 lawsuits every year; there are hundreds of court decisions and a large body of case law that present a clear and generally consistent policy on how a supervisor should behave in the workplace. Gravely recommends that a fire officer focus on actionable items and hostile workplace.

Actionable Items

Actionable items are employee behaviors that require an immediate corrective action by the supervisor. Dozens of lawsuits have shown that failing to act when these situations occur is likely to create a liability and a loss for the department. The best example is the use of certain words in the workplace. An employee's use of derogatory or racist terms about people from other ethnicities, religions, or genders requires immediate corrective action by the supervisor. Regardless of the conditions, context, or situation, such words are inappropriate and represent a potential million-dollar liability to the organization.

The fire officer must act immediately in these situations. That means speaking with the offending fire fighter, in private, and counseling the fire fighter that the use of such words is unacceptable in the fire station or in any situation while the individual is representing the department (either on duty or off duty but in uniform). The fire officer should provide the fire fighter with the fire department's or municipality's EEO/AA policy statement and, if applicable, the code of conduct. The fire officer should also maintain a formal or informal record of the counseling session Figure 3-14 ▾. If there is any doubt that the message was understood, the fire officer should ensure that a higher-level supervisor is informed of the action that has been taken.

The same policies should apply to fire fighters regularly assigned to the fire company, fire fighters detailed in or visiting the fire station, and other uniformed or civilian members of the fire department. Case law has shown that use of unacceptable language requires an immediate response. An officer's failure to act has been interpreted as official condoning or encouragement of such behavior.

Although there is enough case law to provide general guidelines and some specific examples, the subject of what constitutes

harassment remains a dynamic aspect of the work environment. For example, in one city, a judge ruled that use of the term "boy" contributed to a hostile workplace environment in the fire department. A new recruit complained that the term was used to harass him. In this instance, the recruit was Caucasian and the two senior fire fighters accused of harassment were African American.

The fire officer needs to stay informed about the organization's EEO/AA and diversity policies. Most large organizations have a diversity or EEO/AA office that can provide up-to-date information and answer questions.

Hostile Workplace and Sexual Harassment

In 1993, the EEOC amended the guidelines on sexual harassment. The amended regulations broadened the types of harassment that are considered illegal and included a requirement that employers have a duty to maintain a harassment-free work environment. The standard for evaluating sexual harassment is what a "reasonable person" in that same or similar circumstances would find intimidating, hostile, or abusive. The 1993 guidelines clearly state that employers are liable for the acts of those who work for them if the organization knew or should have known about the conduct and took no immediate, appropriate, corrective action. That is why a fire officer must immediately respond to any utterances of offensive or derogatory language in the work environment.

Sexual harassment is unwanted, uninvited, and unwelcome attention and intimacy in a nonreciprocal relationship. The abuse of power is an essential component of sexual harassment. The 1993 EEOC guidelines state that verbal and physical conduct of a sexual nature is harassment when the following conditions are present:

- The employee is made to feel that he or she has to endure such treatment in order to remain employed.
- Whether or not the employee submits or rejects such treatment is used when making employment decisions.
- The employee's work performance is affected.
- An intimidating, hostile, or offensive work environment is present.

The term "hostile work environment" can be used to describe a broad range of situations in which an employee is subject to discrimination in the workplace. The generalized hostile workplace definition can apply to a variety of circumstances that do not necessarily include a specific abuse of supervisory power.

Gravely believes that hostile workplace complaints will shape workplace diversity in the 21st century. There are few quid pro quo sexual harassment cases—cases in which a supervisor specifically promises a work-related benefit in return for a sexual favor. The trend is toward more complaints about a hostile workplace, which can involve sexual issues.

The start of the 21st century is providing many examples of the impact of fire department hostile work environment complaints, most involving large court-directed settlements for the complainant. In one case, a Latino put dog food in the meal of an African American. During the handling of the complaint, two white supervisors were disciplined. It has cost the municipality over $5 million in settlements and legal fees.

Figure 3-14 Privately discuss inappropriate behaviors with fire fighters.

Handling a Harassment or Hostile Workplace Complaint

Fire fighters who want to initiate a harassment complaint have a choice of three methods for doing so. They can start with the federal government, they can start with the local government, or they can start within the fire department—it is their choice. If the process starts within the fire department, a fire officer may be the first formal point of contact. In this case, the fire officer would function as the first step of a multistep procedure.

The fire officer should know the department's procedure for handling a harassment or hostile workplace complaint. The fire officer's designated role in conducting an investigation of an EEO complaint depends on the procedures adopted by the jurisdiction or the fire department. In some cases, an officer's role is limited to starting the process by making appropriate notifications. Many local governments have specially trained EEO staff that conduct an investigation. In other cases, the fire officer might be required to perform the initial investigation and submit a report. Following are some general guidelines:

- *Keep an open mind.* Many fire officers have a hard time believing that discrimination or harassment could be happening right under their noses. Failure to investigate a complaint is the most common reason a local government is found liable. Every complaint must be investigated and documented. Do not come to any conclusions until your investigation is complete. Follow your organization's procedures and inform your supervisor.
- *Treat the person who files the complaint with respect and compassion.* Employees often find it extremely difficult to complain about discrimination or harassment. When an employee comes to you with concerns about discrimination or harassment, be professional, but also be understanding.
- *Do not blame the person filing the complaint.* Case law and current practice dictate that it is the complainant who determines whether the situation is hostile or is harassment. Blaming the person bringing the complaint to you is the second most common way that local governments lose harassment or hostile workplace complaints.
- *Do not retaliate against the person filing the complaint.* It is against the law to punish someone for complaining about discrimination or harassment. The most obvious forms of retaliation are termination, discipline, demotion, or threats to do any of these things. More subtle forms of retaliation could include changing the shift hours or work location of the accuser, even if the intent is to remove the alleged victim from the problem.
- *Follow established procedures.* Local government personnel regulations generally provide a detailed procedure on how to handle harassment or hostile workplace complaints. Follow the procedures.
- *Interview the people involved.* An initial investigation usually involves conducting interviews of the people who are involved in the situation. This usually starts with the person who made the complaint. The interviewer needs to find out exactly what the employee is concerned about. Get details: what was said or done, when and where, and the names of those present. Then talk to any employees who are being accused of discrimination or harassment. Get details from them as well. Be sure to interview any witnesses who may have seen or heard any problematic conduct. Take notes about your interviews and gather any relevant documents.
- *Look for corroboration or contradiction.* Discrimination and harassment complaints often involve "he said/she said" situations. The accuser and the accused offer different versions of an incident, leaving you with no way of knowing who is telling the truth. You may have to turn to other sources for clues. Witnesses may have seen part of an incident. In some cases, documents, such as emails and posted notes, prove one side right.
- *Keep it confidential.* A discrimination complaint can polarize a workplace. The unique team nature of 24-hour fire service work creates a close environment where it is difficult to maintain confidentiality. The fire officer must insist on and enforce confidentiality during the investigation.
- *Write it all down.* Take notes during all interviews. Before the interview is over, go back through your notes with the interviewee to ensure accuracy. Keep a journal of the investigation. Write down the steps you have taken to get at the truth, including dates and places of interviews. Keep a list of all documents that are reviewed. Document any action taken against the accused or the reasons for deciding not to take action. Anticipate that all of this written record will be used in any subsequent civil service or court actions.
- *Cooperate with government agencies.* If the fire fighter files a complaint with another government agency (either the federal EEOC or an equivalent state agency), that agency may investigate. Notify your supervisor as soon as you receive a call or visit from an EEOC investigator. You will probably be asked to provide certain documents, to give your side of the story, and to explain any efforts you made to deal with the complaint yourself. Be cautious, but cooperative.

Regardless of where the complaint is filed, the fire chief is required to take corrective action if the investigation confirms that the complaint has merit. If the department concludes that some form of discrimination or harassment occurred, formal corrective action could include mandatory training, work location transfer, or demotion. Termination may be proposed for the more egregious kinds of discrimination and harassment, such as threats, stalking, or repeated and unwanted physical contact.

The Fire Station as a "Business Work Location"

The fire officer needs to consider the fire station or other fire department facility as a business work location. This is a drastic change from the concept of a fire station as a home away from home for a group of fire fighters, but it is necessary to ensure that the fire station maintains a professional work environment. In the eyes of the law and in the opinion of administrators and elected officials, the same rules of behavior apply to an office in city hall at 2 P.M. and a fire station at 2 A.M.

The fire officer can help maintain an appropriate work environment by encouraging and enforcing acceptable behavior whenever fire fighters are on duty. The fire officer accomplishes this by:

- *Educating the employees on the workplace rules and regulations that define expected behavior.* Start with the local government's "Code of Conduct" or other documents that outline the chief administrative officer's expectations of all municipal employees. This can usually be found in the municipality's mission statement, core values, or personnel regulations.
- *Promoting the use of "on duty speech."* The goal is not to change the thoughts or feelings of individual fire fighters, but to establish a workplace environment where certain behaviors and words are not used. Fire fighters can think what they want. However, while they are on duty, in the fire station, or in uniform, they cannot use certain words or phrases or act out certain behaviors.

Figure 3-15 The company officer must identify and correct unacceptable behavior.

- *Being the designated adult.* This requires the fire officer to model appropriate behavior as well as encourage and enforce the same behavior by the fire fighters. The fire officer must identify and correct unacceptable workplace behavior whenever it is observed. Ignoring a problem is, in reality, permitting it to continue. The fire officer who is a candidate for promotion is expected to identify, explain, and enforce the limits of unprofessional behavior **Figure 3-15 ▲**.

A company-level officer should make it a practice to walk around the fire station at various times during the workday to observe what is going on. This walk-around is more important when the officer is in a big station with multiple companies and in combination career–volunteer departments, where there is a constant flow of people coming into and out of the fire station. This practice is not designed to catch someone doing something wrong; rather, it is to make sure that everything is functioning properly. The officer should routinely determine what the crew members are doing and check on the safety of the facility and equipment. At the same time, the officer should look out for unexpected situations and surprises. Having a reputation that you know what is going on in the station and will react to inappropriate situations goes a long way toward encouraging appropriate workplace behaviors **Figure 3-16 ▶**.

Figure 3-16 The officer should routinely check in with crew members.

Summary

Although it is far less exciting than moving an attack line down a hot and smoky hallway, keeping up on the administrative side of the day-to-day operations contributes to the readiness of the fire company. A work group with a clear and well-organized plan for the day accomplishes more than a group that is waiting to respond to the crisis of the day.

A new supervising fire officer has a different relationship with the administrative fire officer. Besides the boss–subordinate relationship, the new supervising fire officer is part of the management team. The administrative fire officer functions as a coach and mentor.

It will be difficult to sell the concept of the fire station as a business work location, but court settlements and high profile incidents are making it clear that the 21st-century fire station can no longer be a fire fighter's home away from home. It is considered a place of employment like the police station, city hall, or the hospital.

You Are the Fire Officer: Conclusion

You have three choices: enforce the SOP to the letter as outlined by the chief's directive, modify the enforcement after consulting with your supervisor, or ignore the directive and permit the wearing of physical training gear and coveralls in public. Ignoring the ban may make you popular with the fire fighters, but it could endanger your rank. Ignoring a directive from the fire chief is not a career-enhancing activity.

The most appropriate response is to enforce the uniform SOP as outlined by the fire chief. Sudden directives come from a situation several levels above the fire officer. The formal organization expects that the fire fighters will immediately comply.

If possible, consult with your supervisor to get the background on the directive. In this case, the chief informs you that the enforcement directive came in response to an issue that is about to hit the media. You explain the impact this enforcement will have on your crew. The chief says that he understands your issue, but it is imperative that every time the fire department is in public they will be in their station uniforms and not coveralls. There is no room for adjustment at this time. You need to enforce this SOP to the letter until the situation changes.

You have a follow-up meeting with the crew and explain that The Nickel will comply with the letter of the uniform SOP. You ask the members how to change the daily work routine to make this work.

Wrap-Up

Chief Concepts

- A fire officer is responsible for accounting for the people and resources at a fire station and work location. This may require a report at the beginning of duty, outlining staffing and equipment status.
- Transitioning from fire fighter to fire officer changes how the individual relates to the formal fire department organization and the role the fire officer plays with fellow fire fighters.
- A fire officer has a larger sphere of responsibility when supervising a work group than he or she had as a fire fighter.
- A fire officer should "walk the talk" and demonstrate integrity by behaving ethically.
- Fire officers have a supervisor. Keep your supervisor informed, make appropriate company-level decisions, and consult with your supervisor before making major disciplinary or policy changes.
- The Equal Employment Opportunity Commission is the federal agency empowered to enforce compliance with the Civil Rights Act of 1964, the Equal Employment Opportunity Act of 1972, and the Civil Rights Act of 1991. Current fire department recruitment, hiring, and promotion practices are guided by the EEOC to achieve workplace diversity.
- An employee can file an EEO complaint in one of three ways: with the federal EEOC office, with the municipality's EEOC/diversity office, or with the fire department.
- The 1993 amendment of the 1980 EEOC sexual harassment guidelines clarified the issue of hostile workplace complaints. The complainant determines whether conditions are creating a hostile workplace or sexual harassment.
- Follow the local procedures when encountering a harassment or hostile workplace complaint. Take notes, make appropriate notifications, and do not take sides.
- The trend in local government is to expect the fire station to comply with the same behavior rules as are applied to other offices. The fire station is a business work location.
- Fire officers should educate and encourage the concept of "on duty speech" to maintain a nonhostile workplace.

Hot Terms

Actionable items Employee behavior that requires an immediate corrective action by the supervisor because dozens of lawsuits have shown that failing to act will create a liability and a loss for the department.

Administrative fire officer IAFC description of a person who has worked as a managing fire officer for 3 to 5 years, is certified at the NFPA Fire Officer III level, and has accomplished formal education equivalent to a bachelor's degree.

Diversity A fire workforce that reflects differences in terms of age, cultural background, race, religion, sex, and sexual orientation.

Ethical behavior Decisions and behavior demonstrated by a fire officer that are consistent with the department's core values, mission statement, and value statements.

Fire Officer *in Action*

You get an email from a colleague who suggests you check out the unofficial website for the fire station. There is a picture of the fire station kitchen with dirty cookware in the sink and dirty dishes left on the counter. The caption: "Another mess left by A shift."

When you return to work, you ask the B shift managing fire officer if there is a problem with the condition of the station when B shift relieves A shift. She says that sometimes the kitchen is left messy, especially when there is a middle-of-the-night run and the crew decides to make a snack.

You thank her for the information and say that you will work to avoid a repeat of the situation. You ask that the B shifter who runs the unofficial website consider removing the photo because he has made his point. In today's morning meeting you will discuss the need to clean up after the midnight munchies.

The company officer has a wide range of responsibilities. Some require immediate individual action, others need coordination, and a few make the fire officer feel like a schoolteacher in the middle of a playground spat.

1. While operating at a 2 A.M. traffic accident, one of your fire fighters comes into contact with the blood and body fluids of one of the patients. You should:
   A. write a memo about the exposure.
   B. notify your supervisor as soon as you can.
   C. make a notation in the next workday beginning of shift report.
   D. finish clearing the incident, and then have the fire fighter proceed to the hospital to obtain treatment.

2. A contractor is in your fire station kitchen using derogatory and racist terms. You should:
   A. ignore him and move your crew out of the kitchen.
   B. notify that employee's supervisor of the statements.
   C. notify your supervisor and the fire department EEO/AA representative.
   D. immediately meet with this person in private and state that the use of such words is unacceptable at this work location.

3. You are on a public service call and encounter a situation that none of you have encountered before. The fire company devises a solution that falls into the gray area of authorized fire department activities. You should:
   A. do nothing that is out of the clear scope of fire department–approved activities.
   B. follow through with the solution even though it is contrary to your personal ethics.
   C. follow through with the solution if it is consistent with the department's and your ethics.
   D. ask your supervisor for permission to follow through with the solution.

NFPA 1021 Standard

Fire Officer I

4.1.1 General Prerequisite Knowledge. The organizational structure of the department; geographical configuration and characteristics of response districts; departmental operating procedures for administration, emergency operations, incident management system and safety; departmental budget process; information management and recordkeeping; the fire prevention and building safety codes and ordinances applicable to the jurisdiction; current trends, technologies, and socioeconomic and political factors that affect the fire service; cultural diversity; methods used by supervisors to obtain cooperation within a group of subordinates; the rights of management and members; agreements in force between the organization and members; generally accepted ethical practices, including a professional code of ethics; and policies and procedures regarding the operation of the department as they involve supervisors and members. [p 70–71]

4.2 Human Resources Management. This duty involves utilizing human resources to accomplish assignments in accordance with safety plans and in an effective manner. This duty also involves evaluating member performance and supervising personnel during emergency and nonemergency work periods, according to the following job performance requirements. [p 62–63, 68, 70]

4.2.5 Apply human resource policies and procedures, given an administrative situation requiring action, so that policies and procedures are followed. [p 68, 70]

(A) Requisite Knowledge. Human resource policies and procedures.

(B) Requisite Skills. The ability to communicate orally and in writing and to relate interpersonally.

4.2.6 Coordinate the completion of assigned tasks and projects by members, given a list of projects and tasks and the job requirements of subordinates, so that the assignments are prioritized, a plan for the completion of each assignment is developed, and members are assigned to specific tasks and both supervised during and held accountable for the completion of the assignments. [p 63–72]

(A) Requisite Knowledge. Principles of supervision and basic human resource management.

(B) Requisite Skills. The ability to plan and to set priorities.

Fire Officer II

5.2 Human Resource Management. This duty involves evaluating member performance, according to the following job performance requirements. [p 70]

5.2.1 Initiate actions to maximize member performance and/or correct unacceptable performance, given human resource policies and procedures, so that member and/or unit performance improves or the issue is referred to the next level of supervision. [p 70]

(A) Requisite Knowledge. Human resource policies and procedures, problem identification, organizational behavior, group dynamics, leadership styles, types of power, and interpersonal dynamics.

(B) Requisite Skills. The ability to communicate orally and in writing, to solve problems, to increase team work, and to counsel members.

Introduction to Fire and Emergency Services Administration (FESHE) Course Outcomes

3. Articulate the concepts of span and control, effective delegation, and division of labor. [p 71–72]

4. Recognize appropriate appraising and disciplinary actions and the impact on employee behavior. [p 70–71]

5. Examine the history and development of management and supervision. [p 63]

7. Identify roles and responsibilities of leaders in organizations. [p 62]

8. Compare and contrast the traits of effective versus ineffective supervision and management styles. [p 62–70]

Knowledge Objectives

After studying this chapter, you will be able to:

- Understand principles of supervision and basic human resource management.
- Coordinate the completion of assigned tasks and projects.

Skills Objectives

There are no skills objectives for this chapter.

*I*t has been 4 months since you started as a supervising fire officer at Tower 5. Captain Jean Torres, as the managing fire officer, is reviewing your work performance. Torres notes a decline in the completion of tasks by the tower company and expresses concern at the fewer number of skill drills and target hazard walk-throughs.

You explain that you are using the principles of Empowered Employee Knowledge (EEK). You conducted a Circle of Consensus and had the members of the tower company determine what training they needed. A primary assumption of EEK is that the employee will know what they need and will naturally seek out education and training opportunities. You mention that EEK is a business book bestseller and has a helpful website, called "Theory Y in the 21st Century."

Torres gets irritated and suggests that implementing the latest business best-selling concept may not be an appropriate task for a supervising fire officer. Torres asks you to consider the following Theory X directive: stop EEKing around and perform a departmental skill drill every day and a target hazard walk-through every week.

1. Why are there so many management concepts with contradictory information?
2. How do these theories assist you to execute supervisory tasks?
3. Is there a way for you to evaluate a new management concept or procedure before using it at the fire station?

Introduction to Management Concepts

Management is the science of using available resources to achieve desired results. A formal definition of management would probably refer to the systematic pursuit of practical results, using available human and knowledge resources in a concerted and reinforcing way.

A fire officer is a manager who has been given the responsibility to direct and supervise a group of fire fighters, as well as apparatus, equipment, facilities, and other resources, in order to achieve certain outcomes. The desirable outcomes begin with protecting people and property from a variety of undesirable situations. Additional desirable outcomes include ensuring that the work is performed safely, efficiently, promptly, and in accordance with a long list of rules, regulations, procedures, and additional concerns.

Managing People

Most fire officers will find that their greatest challenge has to do with managing people. It is the workers who get the job done. The manager is responsible for performing a set of functions that direct and coordinate their efforts, provide them with the necessary tools and resources, and ensure that the outcome meets the standards. Fire departments are labor intensive, with tasks completed by skilled workers **Figure 4-1 ▶**. In order to be effective as a manager, a fire officer has to develop skills that are directly related to managing human resources.

The concept of management came out of the Industrial Revolution. The introduction of steam power in the late 1700s led to the creation of large factories in Europe, which in turn created the need for management. Adam Smith noted that prior to the Industrial Revolution it took several hundred years for a country or region to develop a tradition of labor and the expertise

Figure 4-1   Managing people can be the fire officer's greatest challenge.

in manual and managerial skills that are needed to produce and market a given topic. That was far too long to meet the needs of the rapidly growing textile factories, railroads, and other forces of the Industrial Revolution. Management was essential to produce results quickly and on a large scale.

Human resource management is built from two generalized schools of management thought: scientific management and humanistic management. Each school developed a set of theories on how supervisors can manage people to accomplish tasks in the work environment. Think of these theories and the related practices as tools in the fire officer's management toolbox. The fire officer will not necessarily use all of the tools or the same tools for every situation. The selection of management tools must consider the situation, the task, and the individuals or the team that will be involved in producing a desired outcome. Some of these theories are directly applicable for use by a fire officer in assigning tasks or responsibilities to crew members.

Fire Marks

A Paramilitary Legacy

Organizational structure and development influence management concepts. Municipal fire department structure was created at the end of the Civil War. City leaders asked West Point–educated generals to organize the individual volunteer fire companies into a municipal fire department. The first formal fire company leaders were the individual fire company foremen. When the company foreman was off, an assistant or acting foreman was in charge of the fire company.

Individual fire companies were organized into battalions consisting of four to eight fire stations. Battalions were organized into divisions, reflecting geographic boundaries. The city created a department and appointed a chief officer who oversaw all of the fire companies in the municipality.

The city fire department supplemented, replaced, or evicted the volunteer fire companies that were established before 1860. Unlike other city employees, career fire fighters were on continuous duty, with a couple of hours off for meals and one or two days off a month. They lived in the fire station.

Scientific Management

Steam power and the creation of large European factories in the late 1700s started the Industrial Revolution and the need for management. By time New York City established the Metropolitan Fire Department in 1865, the Industrial Revolution was in full song. Engineers were the movers, shakers, and creators of the product-based business world.

Adam Smith noted that it took several hundred years for a country or region to develop a tradition of labor and the expertise in manual and managerial skills needed for the Industrial Revolution. That was far too long. Enter the engineering approach to management, education, and training.

Frederick Winslow Taylor

Frederick Winslow Taylor decided to forsake Harvard for a career in industry. In 1874, the skills he needed were not taught in universities; they were learned on the shop floor. Taylor was hired as a laborer in the machine shop of Midvale Steel.

At Midvale, Taylor developed and put into place the basic elements of what later came to be known as **scientific management**. Scientific management is based on the breaking down of work tasks into constituent elements; the timing of each element based on repeated stopwatch studies; the fixing of piece rate compensation based on those studies; the standardization of work tasks on detailed instruction cards; and generally, the systematic consolidation of the shop floor's brain work in a "planning department."

In 1911, Taylor published *The Principles of Scientific Management*, in which he described how the application of the scientific method to the management of workers could greatly improve productivity. Scientific management methods called for optimizing the ways tasks were performed and simplifying the jobs so workers could be trained to perform a specialized sequence of motions in the one "best" way.

Prior to scientific management, work was performed by skilled craftsmen who served lengthy apprenticeships to learn their trades. Individual workers made their own decisions about how their job was to be performed. Scientific management took away much of this autonomy and converted skilled crafts into a series of simplified jobs that could be performed by unskilled workers. To determine the optimal way to perform a job, Taylor performed experiments that he called time studies (also known as time and motion studies). These studies featured use of a stopwatch to measure a worker's sequence of motions, with the goal of determining the most efficient way to perform each task. That worker and others could then be trained to perform those same specific tasks in the same efficient manner.

Taylor's Four Principles of Scientific Management

After years of various experiments to determine optimal work methods, Taylor proposed the following four principles of scientific management:

- Replace "rule-of-thumb" work methods with methods based on a scientific study of the tasks.
- Scientifically select, train, and develop each worker, rather than passively leaving them to train themselves.

- Cooperate with the workers to ensure that the scientifically developed methods are being followed.
- Divide work nearly equally between managers and workers so that the managers apply scientific management principles to planning the work and the workers actually perform the tasks.

Factories that implemented these principles, including Henry Ford's Model T automobile factories, achieved impressive productivity. Versions of scientific management can still be found in the 21st-century workplace.

Firefighter Combat Challenge competitors use time and motion studies to improve performance. When the first Combat Challenge was held at the Maryland Fire and Rescue Institute in May 1991, the team from Prince William County, Virginia, won with a time of 10:08 minutes. Within a decade, the winning time to complete the same course had dropped to less than 2 minutes. By closely analyzing every movement, the competitors learned how to perfect their techniques and shave seconds from every step.

Humanistic Management

One of the problems with the scientific management school was that people were considered as carbon-based cogs in a production line. Taylor considered the workers as cheap, stupid, and interchangeable. Each worker was trained to perform just one task, a small portion of a production process, and to repeat that task at a mind-numbing rate. The **humanistic management** school shifted the focus to pay more attention to the workers and to working conditions that would make them more productive.

The humanistic management school started with Professor George Elton Mayo, a Harvard University industrial psychology professor. From 1924 until 1933, the Western Electric Hawthorne plant was the site of a series of experiments where changes in the working condition were evaluated against productivity in an electronic relay assembly line. Hawthorne effect means people may improve their performance or behavior not because of any specific condition being tested, but simply because of the attention they receive.

Humanistic management research continued through the work of Douglas McGregor and Abraham H. Maslow.

■ McGregor: Theory X and Theory Y

Douglas McGregor, a social psychologist and professor of management at the Massachusetts Institute of Technology, studied the work environment from the mid-1930s through the mid-1950s. The economic conditions during this time period were different from the situation at the start of the Industrial Revolution. McGregor's study period covered the Depression, World War II, and the booming prosperity of the 1950s that created the middle class. McGregor's 1960 publication, *The Human Side of Enterprise*, summarized the results of his research and added the Theory X and Theory Y concepts to the manager's motivational toolbox.

Whereas immigrants to the United States in the early 20th century worked to survive, most workers in the mid-20th century enjoyed much better retirement and financial resources. McGregor observed that in spite of these improvements, many mid-20th-century workers were dissatisfied with their jobs. He concluded that worker motivation is directly related to autonomy and responsibility; workers with greater autonomy are more likely to be motivated in their jobs. McGregor also felt that modern employment often stifled human creativity and impaired motivation.

McGregor developed the Theory X and Theory Y concepts to define the problem in terms of the manager's or the organization's view of the workers.

A Theory X manager believes that people do not like to work, so they need to be closely watched and controlled. Theory X makes sense when you look at the Industrial Age working conditions of the late 1800s. The assembly line factories were horrible. They had scant light, poor ventilation, and unsafe and unsanitary working conditions. Some workers had to relieve themselves at their workstations. Many of the factory workers were recently arrived immigrants, with no industrial work skills and limited ability to read or write English. Many were children. No wonder they had to be coerced and threatened!

A Theory Y manager has an entirely different view of employee creativity and motivation. A Theory Y manager believes that people do like to work and that they need to be encouraged, not controlled.

These two distinctly different philosophical models of worker motivation and supervisory strategies can be observed in many organizations today. The working world has changed significantly since 1960, especially with the increased participation of female managers, and Theory Y is the prevailing trend. Nevertheless, academic and business writers point out that job satisfaction and employee loyalty are far lower now than they were in 1960. Theory X remains a common managerial style.

Theory X versus Theory Y for the Fire Officer

McGregor's Theory X and Theory Y models can be valuable tools for a fire officer. Individuals seek to become fire fighters because they are attracted to the work. Few people are forced by economic conditions to become fire fighters. The fire service requires a strong personal commitment, beginning with the initial investment in physical and technical training. Many fire fighter activities are difficult, physically demanding, and unpleasant, yet most fire fighters love their work. They tend to operate from a different worldview than most workers in 21st-century jobs, such as those in information technology or service industries.

The fire officer is supervising a self-selected workforce that is dedicated and enthusiastic about their responsibilities as fire fighters. The officer's challenge is often to steer their efforts in the right direction and to create an effective team **Figure 4-2 ▾** . The Theory Y concepts can be used effectively in many situations to encourage fire fighter creativity.

Theory Y does not work for every situation, however. There are three specific situations where the fire officer must behave as a Theory X manager, at least temporarily. The first situation is when operating at a fire or other high-risk activity—the fire officer must provide close and autocratic supervision. The second situation is when the fire officer must take control of a workplace conflict and issue specific directions to defuse the situation. The final situation is when a fire officer is near the end of a series of negative disciplinary measures; for example, the Theory X concept and techniques are applicable in an involuntary work performance meeting with a subordinate.

■ Maslow: Hierarchy of Need

Abraham H. Maslow was a psychologist who researched mental health and human potential. He is most often associated with the concepts of a hierarchy of needs, self-actualizing persons, and

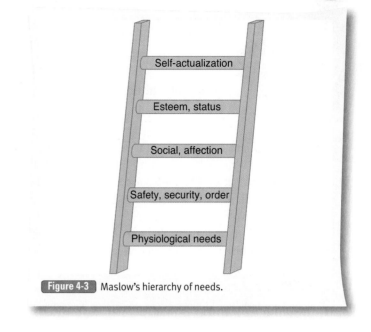

Figure 4-3 Maslow's hierarchy of needs.

peak experiences. Maslow's concept saw human needs arranged like a ladder or a pyramid **Figure 4-3 ▴** .

Level One: Physiological Needs

The most basic human needs, at the bottom of the ladder, are physical—air, water, food, and shelter. It is hard to be creative or productive when you are exhausted and hungry. For the fire officer, this means to stay alert for conditions where the fire fighters are hungry, exhausted, dehydrated, too hot, or too cold. Sometimes fire fighters have to work at the extremes of discomfort and personal inconvenience due to the nature of emergency activities, but these situations should be the exceptions and not a normal way of getting the job done. Making a fire company work without a rest and rehydration period will fail to meet their most basic needs **Figure 4-4 ▾** . Meals may be delayed, but food is needed to keep fire fighters working, just as fuel is needed to keep apparatus operating.

Figure 4-2 The fire officer must steer the fire fighter's efforts in the right direction to create an effective team.

Figure 4-4 Fire fighters need periods of rest and rehydration to meet their most basic physiological needs.

Near Miss REPORT

Report Number: 08-0000009

Event Description: Fire companies had been working in the cold for about 6 hours at a dwelling fire. The temperatures during the operation ranged from 19 degrees to a high of 26 with wind chills dropping those temperatures by about another 10 degrees. Winds were gusting to about 25 MPH. The water supply was provided via tanker shuttle and portable pond for the majority of the operation. Near the end of the operations, a tanker was hooked to the supply pumper directly via 5" supply hose to allow the breaking down of the ponds. The tanker ran out of water and was leaving the scene to re-fill. The driver did not complete a circle of safety and failed to disconnect the 5" supply line from the tanker. When pulling away from the scene, the driver dragged the supply line taking out three fire fighters in the process. None were seriously hurt, but some experienced minor knee pain. There was significant damage to the pump of the tanker and minor damage to the supply engine.

Lesson Learned: Driver fatigue after working long hours driving tankers for shuttle operation was a contributing factor. Changing from the portable drafting pond to a direct hose connection with the tanker and the changing of the "routine" of dumping water and returning to the fill site may have contributed to the near miss.

Level Two: Safety, Security, and Order

Safety, security, and order needs are next. Safety is obviously a primary concern in the fire service, although a fire fighter's perceptions of safety are probably quite different from those of a typical office or factory worker. If the fire fighters feel that their fire officer is leading them in an unsafe manner or a particular policy or practice in the fire department is exposing them to an avoidable risk, safety is likely to become a very significant issue.

Security is more closely associated with maintaining employment or status within the organization. It could be as subtle as a fear that a personality clash with a particular company officer will destroy or impede a fire fighter's career. An individual who feels threatened may do what is necessary to survive, but will not feel highly motivated to advance the fire officer's or the fire department's objectives.

Imagine the impact on your fire company if a notice from fire administration suddenly announced that all incumbent fire fighters would be required to achieve paramedic training, and that any fire fighter who does not pass the National Registry certification exam within 1 year will be terminated. Although this would be a great example of Theory X coercion, it would also create tremendous fire fighter stress.

The need for security and order could be a factor when a fire department is undergoing a significant reorganization, such as the appointment of a new fire chief or the combining of departments into a regional authority. Any type of change in the established order is likely to arouse insecurities. The anticipation of significant changes in the organization or budget-related cutbacks can easily impede fire fighter creativity and motivation.

Getting It Done

Consistency Counts

Given a choice, fire fighters prefer a company officer who is consistent. The brilliant officer who is inconsistent creates concern and doubt in the fire fighter.

Consider the dilemma of laying a supply line by the first arriving engine company in departments without a standard operating procedure. Instead of focusing on the upcoming fire problem, the fire fighter is concentrating on clues to see if the rig is slowing down to deploy a supply line. A consistent response reduces confusion and allows supervisors and fire fighters to concentrate on the emergency scene issues.

Level Three: Social Needs and Affection

These are the psychological or social needs that are related to belonging to a group and feeling acceptance by the group. Group acceptance and belonging are particularly strong forces within the fire service. The majority of fire fighters have some type of identification on their personal vehicles that visibly declares their local fire department affiliation and membership in the fire service at large. Fire fighters wear their company patches proudly and often engage in social and recreational activities with their co-workers Figure 4-5 ▶ . Transferring a fire fighter away from a "good" assignment can be a strong demotivator, just as suspending a volunteer from riding can have a significant impact at the social and affection levels.

Level Four: Esteem and Status

Promotions, gold badges, special awards, and take-home fire department vehicles are all symbols that apply to the esteem and status level Figure 4-6 ▶ . Membership in an elite fire department

Figure 4-5 Fire fighters often engage in social and recreational activities with their co-workers.

unit is often an indicator of special qualifications or achievements. Fire fighters will invest money and off-duty time in order to compete for a high-esteem or high-status position. For example, fire fighters will spend months preparing for a fire officer promotional exam or taking courses to qualify for a rescue company. Most elite fire units have more applicants than available positions.

Level Five: Self-Actualization

Peak experiences are profound moments of love, understanding, happiness, or rapture, when a person feels more whole, alive, self-sufficient and yet a part of the world; more aware of truth, justice, harmony, and goodness. Self-actualizing people have many such peak experiences. Such people tend to focus on problems outside of themselves and have a clear sense of what is true.

Maslow felt that unfulfilled needs at the lower levels on the ladder would inhibit a person from climbing to the next step. As he pointed out, a person who is dying of thirst quickly forgets the thirst when deprived of oxygen.

Figure 4-6 Promotions and special awards are symbols that apply to the esteem and status level.

Company Tips

Building Trust in the New Fire Officer

It is vital for a new supervisor to make an effort t[...] with the workers. The failure to establish trust fron[...] ning of the relationship can quickly undermine a supervisor's career. This is particularly true for a new fire officer. Fire fighters make or break a new officer, and they will make every effort to protect themselves from an officer who appears to be unsafe, unstable, or unprepared for the job.

How can the new fire officer build trust? Here are five suggestions:

1. Know the fire officer job, both administrative and tactical.
2. Be consistent. Strive to provide a measured response to any problem, emergency, or challenge.
3. Walk your talk. Actions speak much louder than words.
4. Support your fire fighters. Make sure you help meet their physiological, safety, security, and order needs.
5. Make fire fighters feel strong. Help them to become competent and confident in their emergency service skills. Show how they can control their destiny.

Maslow's thinking was original. Most psychology research before him, including Maslow's first textbook, had been concerned with the abnormal and the ill. He wanted to know what constituted positive mental health. Maslow observed that "Human nature is not nearly as bad as it has been thought to be."

■ Blake and Mouton's Managerial Grid

The Exxon Corporation hired behavioral scientists Robert Blake and Jane Mouton in the early 1960s to perform a series of experiments designed to increase leadership effectiveness. In the 1990s the grid theory was applied to crew management of aircraft and aerospace teams—a concept later adopted by the fire service as "crew resource management" to improve incident safety.

The grid theory assumes that every decision made and every action taken in the workplace is driven by people's values, attitudes, and beliefs. At the individual level, these values are based on two fundamental concerns that influence behavior—a concern for people and a concern for results. Blake and Mouton developed a survey document with 35 questions that measures a person's level of concern in each area. The two values are plotted on an X–Y chart. Blake and Mouton describe five behavioral models based on a person's position on this grid.

Indifferent: Evade and Elude

The indifferent style represents the lowest level of concern for both results and people. The key word for this style is *neutral*. This is the least visible person in a team; he or she is a follower who maintains distance from active involvement whenever possible. An indifferent person carefully goes through the motions of work, doing enough to get by, but rarely making a deliberate effort to do more.

The stereotypical image of this personality is the bureaucratic government agency where everyone is treated like a number. This sort of workplace allows the person to blend in without attracting attention. In fact, he or she often seeks work that can be done in isolation in order to carry on without being

...urbed or noticed. The indifferent manager relies heavily on instructions and process, depending on others to outline what needs to be done. Reliance on instructions avoids the need to take personal responsibility for results. If problems arise, the indifferent person is often content to ignore or overlook them, unless the instructions specify how to react to that particular problem. The indifferent person might point out the problem to someone else, but would be unlikely to offer a solution. With no instructions, he or she simply carries on with the attitude that "This is not my problem."

Controlling: Direct and Dominate
The controlling person demonstrates a high concern for results, along with a low concern for others. The high concern for results brings determination, focus, and drive for success. This person is usually highly trained, organized, experienced, and qualified to lead a team to success. The low concern for others prevents the controlling person from being aware of others involved in an activity, beyond what is expected of them in relation to results. The controlling person expects everyone else to "keep up" with his or her efforts, and so moves ahead, intensely focused on results, often leaving others lost in the wake of his or her forceful initiative.

Accommodating: Yield and Comply
The accommodating person demonstrates a low concern for results with a high concern for other people. This individual maintains a heightened awareness of the personal feelings, goals, and ambitions of others, and always considers how proposed actions will affect them. He or she is approachable, fun, friendly, and always ready to listen with sympathy and encouragement.

Some cornerstone phrases of the accommodating attitude are "Let's talk about it," "What can I do to help?", and "Let me know what you think." The main weakness in this behavior lies in the focus of the discussions. Discussions with an accommodating person tend to include an overwhelming emphasis on personal feelings and preferences, while avoiding concrete issues. The discussion itself becomes the goal, so conversations can meander in any direction instead of concentrating on solving the problem.

The controlling and accommodating styles are diametrically opposed in their perspectives. Each of these orientations leads in a narrow and singularly focused manner by ignoring the other primary concern in the workplace.

Status Quo: Balance and Compromise
The status quo person believes there is an inherent contradiction between the two concerns, but does not value one concern over the other. Instead, the status quo person sees a high level of concern for either people or results as too extreme and tries to moderate both in the workplace.

The objective of the status quo person is to play it safe and work toward acceptable solutions that follow proven methods. This is a politically motivated approach that seeks to avoid risk by maintaining the tried and true course, following popular opinion and norms without pushing too hard in any direction.

Another key aspect of the status quo approach is to maintain popular status within the team and organization. This type of manager must be intelligent and informed enough to persuade people and companies to settle for a compromise—often less than they want and less than they could achieve. This requires being well liked, keeping well informed, and effectively convincing people that the consequences are not worth the risk. On the surface, this might make the status quo appear unbiased and impartial, but more accurately, the status quo approach represents a narrow view that underestimates people, results, and the power of change.

Sound: Contribute and Commit
The sound person sees no contradiction in demonstrating a high concern for both people and results at the same time. He or she feels no need to restrain, control, or diminish the concerns for both people and results in a relationship. The consequence is a freedom to test the limits of success with enthusiasm and confidence. The sound attitude leads to more effective work relationships based on "what's right" rather than "who's right."

The sound behavioral model is preferred for a candidate to become a successful fire officer. The full integration of concerns for both people and results is in contrast with the other behavioral styles. Each of the other models represents a weakness in one or both critical areas. The controlling person feels that a high concern for results is more important than a high concern for people. The accommodating person feels the reverse, that a high concern for people is more important than results. The status quo feels that a high concern for either people or results is too risky, and prefers to maintain the status quo, holding the safe middle ground. The indifferent sees any major concern for people or results as unrealistic and too demanding.

Applying Human Resource Management Principles

Managing fire fighters requires utilizing physical, financial, human, and time resources. Human resource management focuses on the task of managing people. Much of a fire officer's time is spent dealing with subordinates, so the fire officer must understand human resource management. Human resource management includes a wide variety of activities. Some of these are at the company level whereas others are typically addressed at the departmental or organizational level. Although some areas affect career departments more than volunteer departments, all functions are performed by all departments regardless of their structure. The typical human resource management functions include:

- Human resource planning
- Employee (labor) relations
- Staffing
- Human resource development
- Performance management
- Compensation and benefits
- Employee health, safety, and security

Human resource planning is the process of having the right number of people in the right place at the right time who can accomplish a task efficiently and effectively. Most often, this includes forecasting future staffing needs and determining how those needs can be met. When a fire department projects the number of retirements for the following year and determines that a new recruit class should begin prior to those vacancies, that

Voices of Experience

The layout fire fighter is responsible for pulling the section of 3" supply line with a hydrant wrench attached off the engine and dropping it near the hydrant, or appropriate location, for the start of the engine's water supply. This evening, I had asked an experienced member if the probie had been taught how to layout and he told me she had been. Likewise I went to her as well to make certain and to see if she might have any questions. Since there were none, and there was nothing to give me any concern, I posted the assignments and members began doing their checks.

Not long after, we received an alarm for an odor of gas inside an apartment building in our first-due area. The location was relatively close to quarters and we arrived on scene quickly. The apartment complex is a dead-end complex, one way in and out, and has the closest hydrant across the street from the entrance. We would drop our supply line at the entrance, a split lay, and the second-due engine would make the hydrant connection. As we pulled up to the entrance, I yelled out "Layout!", our command for the layout fire fighter to exit the cab, run to the rear of the engine, and pull off the supply line. This whole process takes just a few seconds. The driver looks back to make sure the layout fire fighter has cleared the backstep before proceeding to the reported building. An unusually long amount of time went by, and as the first-due truck pulled up behind us the driver and I were wondering where the probie was. Finally he saw her and we made our way to the apartment building.

> "I went over what went wrong and learned that she had only been told, by word of mouth, how to layout."

Once back at the station she and I went over what went wrong and learned that she had only been told, by word of mouth, how to layout. She had never placed her hands on the hose until that moment. I went back and brought over the experienced fire fighter who had initially "taught" the probie how to layout and had them both practice laying out, hands on the hose. This correction of instruction accomplished many things. First, it took away the possible embarrassment the probie might have had after failing a task. Second, it reinforced to both members that simple tasks need to be taught, including any hands-on application, whenever and wherever possible. Third, it gave a different way of correction to each member that didn't single out anyone. Finally, this retraining left each one fully aware of my expectations as the engine officer, of training each other, and of how we are to operate as a team.

Bill Carey
Hyattsville Volunteer Fire Department
Hyattsville, Maryland

Getting It Done

Assigning the Rookie a Mentor

One effective method of orienting a rookie is to select a fire fighter who exemplifies good performance. The fire officer will ask that fire fighter to become the "big brother" or "big sister" to the recruit. If the fire fighter accepts the responsibility, the recruit is then brought into the discussion so he or she knows that the big brother or big sister fire fighter is there to help him or her with any questions he or she might have. This gives the recruit an experienced fire fighter as a resource for questions.

department is fulfilling the human resource planning function. Typically, this is not an activity that is conducted at the company level other than at emergency scenes.

Employee relations include all activities designed to maintain a rapport with the employees. Typically, this is associated with working with the employee's union. Fire officers must know and understand all agreements between the union and the fire department. The fire officer should also be aware of the vast number of laws that regulate the relationship between employees and their employer. Ignorance of the law is no excuse, and fire officers who violate the law may quickly find themselves and their department faced with a federal lawsuit. Chapter 5, Organized Labor and the Fire Officer, focuses on this human resource function.

Staffing is the process of attracting, selecting, and maintaining an adequate supply of labor. The fire service traditionally has been fortunate in its ability to attract applicants. However, with the increase in the services provided and the greater educational requirements, this is becoming more difficult. Although some fire departments have developed programs to actively seek qualified women and minorities, most fire departments have not.

This function also includes labor force reductions. There are many methods of shrinking the fire department's labor supply; most fire departments have opted for using attrition and layoffs rather than terminations, transfers, and reduced workweeks. The staffing function is typically accomplished at the organizational level.

Human resource development includes all activities to train and educate the employees. This function is heavily dependent on the fire officer at the company level. The development of employees begins when they first arrive at the department. The fire department orients the recruits to the fire department's methods of operation through a recruit class. The process continues once they arrive at their fire station assignment. The fire company officer will usually sit down with each recruit and orient him or her to the job and the ways things are done at the fire station.

Fire departments require regular drilling and training to all personnel. This is also part of the human resource development function. The fire service is unlike most other private and public organizations in its ability to spend vast amounts of time solely to develop its employees while at work. Ensuring daily training is one of the most basic responsibilities the fire company officer has to do.

Performance management is the process of setting performance standards and evaluating performance against those standards. Generally, the standards are set at the fire organizational or fire departmental level; however, the evaluation of the fire fighter's performance is a primary function of the fire company

officer. This issue is covered in great detail in Chapter 8, Evaluation and Discipline.

Human resource management also includes the setting of compensation and benefits. Although the fire officer may not set the compensation policy, the fire officer must understand how the system is designed and to what benefits the employees are entitled. Most fire departments operate on a step-and-grade pay system. The position of fire fighter is established on a particular pay grade level that is composed of a number of steps. If the fire fighter demonstrates satisfactory performance, he or she will progress from one pay step up to the next until he or she eventually reaches the top step of the pay grade. The fire fighter will remain at this pay step and grade until he or she is promoted. He or she will then move to a new pay grade and progress up through the steps until he or she again reaches the top step.

The other most common systems that are used include merit-based pay and skill-based pay. In merit-based pay systems, the fire fighter is typically paid a base amount and then receives additional compensation for good performance. For example, a fire fighter who receives an outstanding evaluation might receive a bonus of 5% of his or her annual pay. Skill-based pay systems typically pay a base amount and then give additional compensation for any skills the fire fighter can demonstrate. For example, a fire fighter who is also a paramedic and a hazardous materials technician would be paid an additional amount because of his or her additional skills.

Traditionally, fire departments have good benefits compared to the private sector. Benefits include defined benefit retirement systems where the employee contributes a set percentage of his or her pay and receives a guaranteed benefit amount. It also includes defined contribution retirement plans where the employee pays in a percentage of pay and receives a benefit amount that is dependent upon the performance of the account. Typically, these accounts are in the form of a 457 deferred compensation plan. Benefits also include paid holidays, vacation leave, sick leave, and health, dental, vision, and life insurance.

Health, safety, and security include all activities to provide and promote a safe environment for fire fighters. The fire officer is responsible for many of these activities, such as ensuring seat belts are utilized, floors are dry, and personal protective equipment is used correctly. The fire department or fire organization may provide other health activities such as an Employee Assistance Program, which is covered in detail in Chapter 8, Evaluation and Discipline. The fire department may also offer tobacco cessation classes and incentives or diet and exercise information.

Utilizing Human Resources

The fire officer's ability to utilize the human resources that are assigned is an essential function of the position. The fire officer must be able to accomplish the department's work with and through other people. The fire officer must ensure that duties are carried out in accordance with departmental policies. Normally, this is done through direct supervision.

Direct supervision requires that the fire officer directly observe the actions of the crew. If the crew makes an interior attack, the fire officer is present. When a hose is loaded, the fire officer oversees the reloading. Direct supervision allows the fire company officer to ensure that departmental safety policies are used. The more dangerous the activity, the closer the direct supervision that is required. For nonemergency activities, less direct supervision is needed.

The more direct supervision that is needed, the less efficient the crew. The goal of the fire officer is to reduce the need for direct supervision and increase the utilization of the fire company. Frequently this is accomplished through a series of policies and procedures. Human resource policies and procedures guide the fire officer's decisions as they relate to personnel issues.

The fire officer must be familiar with the location and topical areas that are covered in these policies and procedures. The organization may have federal laws, a union contract, city regulations, and departmental policy that must all be followed. When presented with an issue, the fire officer must first determine which laws apply. The most common are the Fair Labor Standards Act (FLSA), the Civil Rights Act, the Age Discrimination Employment Act (ADEA), the Americans with Disabilities Act (ADA), the Family Medical Leave Act (FMLA), the Uniformed Service Employment and Reemployment Rights Act (USERRA), and the Health Insurance Portability and Accountability Act (HIPAA). The Department of Labor provides a great deal of information on the laws that affect fire fighters. The fire officer may also want to discuss issues that are affected by these laws with legal counsel.

Once the determination is made of any laws affecting a decision, the fire officer should review the fire department's policies to ensure compliance. If a labor contract exists, the fire officer must determine if the activity is addressed by the contract. If the fire organization has a human resource department, it may prove to be a valuable asset in determining the proper action by the fire officer.

The fire officer must base his or her actions on fairness and equity to the parties involved. For example, if an employee desires to schedule leave time in short increments the fire officer must determine whether there are any laws, organizational policies, or departmental policies that address the issue. If not, the fire officer must consider the impact of the request on the achievement of the department's mission. Lastly, the fire officer must consider whether the action is fair to all employees.

■ Mission Statement

One basic principle is for the fire officer to know and understand the fire department's mission. Frequently, the fire department's mission is expressed through a written mission statement. The mission statement is a formal document that outlines the basic reason for the organization and how it sees itself. It is designed to guide the actions of all employees.

■ Getting Assignments Completed

One of the greatest demands on the fire officer is the effective use of time. Fire officers will find they have a great number of demands on the company's time. This includes conducting public education, inspections, and other fire prevention efforts. It also includes training and education of the crew members, and routine duties such as cleaning the stations, doing paperwork, and maintaining the apparatus. And of course it includes responding to calls.

Some of these activities are known months in advance. Others may require the immediate response of the crew. The daily schedule of some crews is in a constant state of interruption due to the call volume. Other crews may only occasionally be interrupted. No matter which situation, the fire officer can ensure maximum efficiency by using good time management skills.

The first duty is to determine what activities are to be completed, when they must be completed, and how long it will take to complete. The fire officer can then categorize what needs to be done during the shift, the week, the month, and the year. Items that must be completed during the shift are a higher priority than those that must be completed next week. For example, completing the daily log is more important than completing a weekly inspection.

Occasionally, there is not sufficient time to complete all the required tasks during the shift. When this occurs, the fire officer must determine the fire department's priorities. This allows the fire officer to determine which activity must be completed and which will have to wait. For example, meeting the deadline for the employee's timesheets may be a higher priority than getting the fire engine to the service center for an oil change.

Because many activities, such as an emergency call, are not known until they occur, the fire officer must plan ahead and build in the expected interruptions. The fire officer must not wait until the last minute to have inspections completed because he or she may have calls that prohibit this from happening. The sooner the scheduled activity is completed, the more flexibility the fire officer has.

Once the activities have been prioritized and it is determined when they must be accomplished, the fire officer must develop a plan that lays out how the activities will be accomplished. Activities might require the entire company to be involved in the activity or they might require only a part of the crew to complete.

An example of splitting the crew up to accomplish multiple goals simultaneously would be to send a crew member to the store while another cleans the fire station kitchen and a third member completes the weekly inventory of supplies.

Having many tasks to coordinate can easily become overwhelming. One method to assist in making sure that activities are accomplished is to place all scheduled events on a monthly calendar. Inspections, public education, and special training should be noted. Employee leave time should also be noted. The calendar provides a visual method of tracking upcoming events.

Assessment Center Tips

Practice Techniques

An effective way to ensure that all tasks are properly handled in an assessment center is to practice the techniques while working as a fire fighter or supervising fire officer. A common error is not documenting a follow-up date to check on progress or use as a milestone target. This is documented in the response to an in-basket exercise by establishing a due date on the item as well as on the calendar.

Delegating tasks will also require the candidate to explain in detail what is required of the person with the delegated task in sufficient detail so that an assessor from outside the department can understand the assignment. Avoid fire department jargon, abbreviations, and catch phrases.

Getting It Done

Company Officer Delegation

There are seven steps in effective delegation:

1. Define your desired results.
2. Select the appropriate fire fighter.
3. Determine the level of delegation.
4. Clarify expectations and set parameters.
5. Give authority to match level of responsibility.
6. Provide background information.
7. Arrange feedback during the process.

Determining the level of delegation in Step 3 relates to the amount of decision-making authority provided to the fire fighter. Five options for the company officer are:

1. Take action independently. No need to report back to the officer.
2. Take action and report back to officer when done (like a task assignment within the incident command system).
3. Recommend action that the company officer must approve.
4. Provide two or more recommended actions from which the company officer will choose.
5. Provide information about the positives and negatives of different recommendations.

Another method is to create a "daily" file. Within this file, there is a page that describes each activity and when it is to be completed. The file is organized from the soonest activity to complete to the farthest. This allows the fire officer to quickly determine what needs to be done without having to look through a pile of papers.

One of the best tools to improve time efficiency is through delegation. As discussed previously, delegation allows subordinates to complete tasks they are capable of performing. These duties should be ones that allow the subordinate to grow. An example is the fire officer delegating the responsibility for ordering supplies to a fire fighter. Delegation allows the fire officer to focus on duties that cannot be delegated, such as performance appraisals.

Once a task is assigned, the fire officer must provide the fire fighter with regular follow-up. This can take many forms. At the station level, this is most often a verbal progress report. On more formal projects, the progress report may be in writing to provide long-term documentation.

Summary

Consider a house burning in Colonial times, when citizens responded from home in response to a fire alarm. When the third person joined the bucket brigade, one of the three needed to be in charge. By the time a dozen people were on the fire ground, management practices were used to coordinate the people on the bucket brigade, the people using the salvage bags, and the people using tools to pull down the combustible roof and disassemble the bed. The same human resource and organizational needs exist when a fire officer responds to a structure fire today, even though the technology has changed.

This chapter provides the foundation of human resource concepts and procedures that should be in your fire officer management toolbox of ideas, concepts, and practices. You will add to your toolbox through experience, formal courses, and self-study.

Dozens of management books are published every year that provide new information. The fire officer must consider whether these concepts and practices would work in the fire department environment.

You Are the Fire Officer: Conclusion

It has been 2 weeks since Captain Torres directed you to "stop EEKing around." You have conducted a skill drill every day, done three target hazard walk-throughs, and more closely monitored the completion of in-station tasks by the tower crew.

Captain Torres has you come to the office. "You know how you feel when you first arrive at a working structure fire, with smoke showing from many windows and report of people trapped? You want to do everything at once and use every tool carried on the tower. There are not enough people or time to use all of the tools to complete critical fire-ground tasks. The smart company officer sizes up the situation and selects the best tool to accomplish the task."

Torres continues, "Management concepts are like the tools on the aerial. You have collected a large set of management ideas, concepts, and practices that you carry in your management toolbox. The smart supervisor selects the best idea, concept, or practice for the situation. You need to pick the idea, concept, or practice that accomplishes the objective without pummeling fire fighters or eroding your authority."

You will see new management tools during your company officer tenure. Like new firefighting tools, they will have high promise. Until you evaluate how each new management tool works, by understanding its history and development and trying it out in a training session, it remains an unknown resource.

Just because it is new and comes from a best-selling book by a management guru does not mean it will work for you as a supervising fire officer.

Wrap-Up

Chief Concepts

- Scientific management breaks down work tasks into constituent elements. The timing of each element is based on repeated stopwatch studies in order to standardize work tasks into simple, repeatable tasks.
- Frederick Winslow Taylor's four principles of scientific management are:
  - Replace rule-of-thumb work methods with scientific study.
  - Scientifically select, train, and develop each worker.
  - Cooperate with workers to ensure methods are being followed.
  - Division of work: managers think, workers work.
- Taylor considered workers cheap, stupid, and interchangeable. They were trained to perform one small portion of a task that they would repeat at a mind-numbing rate.
- McGregor Theory X: People do not want to work.
- McGregor Theory Y: People do want to work.
- Maslow's hierarchy of needs is a ladder of five need levels. The supervisor's role is to meet the employee needs in order to proceed to the next level.
  - Level One: Physiological
  - Level Two: Safety, security, and order
  - Level Three: Social needs and affection
  - Level Four: Esteem and status
  - Level Five: Self-actualization
- Managing fire fighters requires physical, financial, human, and time resources.
- Human resource planning is the process of having the right number of people in the right place at the right time.
- The fire officer's ability to utilize the human resources that are assigned is essential.
- Direct supervision requires the fire officer to directly observe the actions of the crew.
- One of the greatest demands on the fire officer is the effective use of time.

Hot Terms

Humanistic management Emphasis on human need and attitude, motivation comes from within the employee and not from authoritarian control. Lead to Maslow's hierarchy of needs.

Scientific management The breakdown of work tasks into constituent elements; the timing of each element is based on repeated stopwatch studies; the fixing of piece rate compensation based on those studies; standardization of work tasks on detailed instruction cards; and generally, the systematic consolidation of the shop floor's brain work.

Fire Officer *in Action*

You are supervising a fire company with a new driver and rookie fire fighter. The first couple of emergency responses were a flurry of mistakes that slowed the response from the station and initial actions at the emergency scene. You need to develop a work improvement plan for both the driver and the fire fighter. The driver seems to lack confidence when driving the fire truck and has no idea how to get from the fire station to locations within the district. The rookie fire fighter is enthusiastic, suffering from episodes of adrenalin-induced mistakes.

1. One of the rookie's problems is stumbling out of the rig when arriving at the incident scene. You have the rookie practice donning the gear, grabbing the tools, and exiting the rig while in the station. This is an example of:
   A. scientific management.
   B. Theory X management.
   C. Theory Y management.
   D. hierarchy of needs.

2. Your company is operating at a greater alarm structure fire, operating a 2.5" (6-cm) attack line within a commercial building. The best principle to use during incident scene operations is:
   A. scientific management.
   B. Theory X management.
   C. Theory Y management.
   D. hierarchy of needs.

3. The company has been performing difficult overhaul operations for a long period of time. You tell the safety officer that your crew will need relief soon. This is an example of:
   A. Theory Y management.
   B. Maslow's safety, security, and order level.
   C. Maslow's physiological needs level.
   D. Blake and Mouton's balance and compromise.

4. Special awards identifying individual or company officer performance are an example of:
   A. Blake and Mouton's direct and dominate.
   B. Blake and Mouton's contribute and commit.
   C. Maslow's esteem and status level.
   D. Maslow's self-actualization level.

Organized Labor and the Fire Officer

NFPA 1021 Standard

Fire Officer I

4.1.1 **General Prerequisite Knowledge.** The organizational structure of the department; geographical configuration and characteristics of response districts; departmental operating procedures for administration, emergency operations, incident management system and safety; departmental budget process; information management and recordkeeping; the fire prevention and building safety codes and ordinances applicable to the jurisdiction; current trends, technologies, and socioeconomic and political factors that affect the fire service; cultural diversity; methods used by supervisors to obtain cooperation within a group of subordinates; the rights of management and members; agreements in force between the organization and members; generally accepted ethical practices, including a professional code of ethics; and policies and procedures regarding the operation of the department as they involve supervisors and members. [p 79–82, 84, 87–88]

4.2.5* Apply human resource policies and procedures, given an administrative situation requiring action, so that policies and procedures are followed. [p 90–92]

(A) Requisite Knowledge. Human resource policies and procedures.

(B) Requisite Skills. The ability to communicate orally and in writing and to relate interpersonally.

Fire Officer II

5.1.1 **General Prerequisite Knowledge.** The organization of local government; enabling and regulatory legislation and the law-making process at the local, state/provincial, and federal levels; and the functions of other bureaus, divisions, agencies, and organizations and their roles and responsibilities that relate to the fire service. [p 79–84, 89–90]

Introduction to Fire and Emergency Services Administration (FESHE) Course Outcomes

4. Recognize appropriate appraising and disciplinary actions and the impact on employee behavior. [p 90–92]
5. Examine the history and development of management and supervision. [p 79–86]
7. Identify roles and responsibilities of leaders in organizations. [p 78–79, 87, 90]

Knowledge Objectives

After studying this chapter, you will be able to:

- Discuss the lasting impact organized labor has had on fire fighter safety, working conditions, and procedures.
- Understand the diminished benefits of fire fighter strikes.
- Identify the increased benefits of political activism.
- Describe the steps of the grievance process.

Skills Objectives

After studying this chapter, you will be able to:

- Demonstrate the initial handling of an employee grievance.

*I*t is Saturday afternoon and the battalion chief tells you there is a spot available in next week's pump operator certification course. This is an NFPA 1002–compliant week-long course at the regional fire academy. After checking your records, you call Fire Fighter Trammel into your office and instruct him to report to the academy Monday morning at 0800.

"Sorry, but I can't go," states Trammel. "Our shift is on scheduled days off Monday, Tuesday, and Wednesday next week and I have child care issues." You calmly repeat the instruction to report to the Training Academy at 0800 hours on Monday and attend the class while on day work from Monday through Friday.

"Are you really sure about this?" Trammel asks. "I think the new labor contract requires a 72-hour notice for any work schedule changes."

You get frustrated and tell Trammel that this is a direct order.

"Okay," says Fire Fighter Trammel, "this is an unfair labor practice and a violation of the labor agreement between the IAFF local and the city fire department. This is your notice of my step 1 grievance. I am officially requesting that you comply with the 72-hour advance notice of a work schedule change. I will be contacting my battalion representative."

1. Does a labor contract override personnel regulations?
2. What is the role of a supervising fire officer in a labor dispute?
3. How can this situation be resolved fairly and appropriately?

Introduction to Organized Labor and the Fire Officer

Within most municipal fire departments, wages, working conditions, and many other aspects of the work environment are directly influenced by labor–management relations. It is important for the fire officer, particularly in a career fire department, to understand some of the history of labor relations to better function as a first-line supervisor. The range, scope, and tasks of a fire officer's supervisory activities are defined by three primary components:

- The local labor contract
- The municipality's personnel regulations
- The fire department's rules, regulations, and procedures

In many cases, a fire officer is both a supervisor representing management and a member of the bargaining unit represented by the union. This is an unusual situation when compared with most work environments, in which there is a very clear distinction between labor and management. It is especially important for a fire officer to clearly understand how this applies to the specific organization and the position he or she is occupying.

The U.S. fire service has been significantly influenced by organized labor activities, including several laws and regulations that impact the working conditions of fire fighters. Some results benefit all fire departments, from the largest all-career department to the smallest all-volunteer fire company. A career fire officer who is working in a municipal fire department is much more likely to be involved in an organized labor situation than a volunteer officer; however, most aspects of the relationship between the workers and the organization are similar within the volunteer organization.

At the fire company level, most career fire fighters work under a labor contract or some form of written agreement, such as a memorandum of understanding (MOU), between labor and management. The contract or MOU covers various working conditions, promotion/assignment practices, and problem-solving procedures. A labor contract is a negotiated legal agreement between the labor organization and the local jurisdiction. An MOU is a less powerful form of written agreement that is often used in jurisdictions where government employees do not have formal collective bargaining rights. **Collective bargaining** is a method whereby representatives of employees (unions) and employers determine the conditions of employment through direct **negotiation**, normally resulting in a written contract setting forth the wages, hours, and other conditions to be observed for a stipulated period (e.g., 3 years).

Organized labor focuses on the working conditions and benefits of the union's members. In departments with a contract, collective agreement, or MOU, each fire station or work location will have a **shop steward**, who is a union member appointed or elected to be the first line of labor representation at the workplace. The shop steward is a member of the workforce who has received additional training in labor relations. The steward enforces the contract or labor agreement and represents the union members at that fire station or work location. When handling issues of discipline, policy, or procedures, a fire officer may deal with a shop steward as the initial labor representative.

The relationship between the employer and the labor organization is determined by a wide variety of labor laws and regulations at the federal, state, and local levels. The balance of power between labor (representing the employees) and management (representing the fire chief/local government) swung back and forth during the 20th century. This balance of power depends on legislation and policies enacted by the federal government and decisions made by the Supreme Court. In the early part of the 20th century, the labor movement had a political advantage and was able to make considerable progress. At the start of the 21st century, however, management appeared to enjoy an advantage.

Legislative Framework for Collective Bargaining

Collective bargaining is regulated by a complex system of federal and state legislation. Federal legislation establishes a basic framework that applies to all workers, and the states have discretionary powers to adopt labor laws and regulations that do not violate the federal requirements. The application of this basic model to governmental employees is much more complex. Some of the federal labor laws do not apply to federal government employees, state government employees, or employees of local government agencies within the states.

Four major pieces of federal legislation have established the groundwork for the rules and regulations of the present collective bargaining system. Before the adoption of these federal laws, each labor case was decided by a judge in a local court, who applied the broad concepts used in common-law decisions. The federal legislation provides a set of guidelines for how each state or commonwealth can regulate collective bargaining. Fire fighters employed by local government agencies are subject to

state law that can require, permit, or prohibit collective bargaining for local public employees.

The four federal laws that regulate the collective bargaining system are the Norris-LaGuardia Act of 1932, the Wagner-Connery Act of 1935, the Taft-Hartley Labor Act of 1947, and the Landrum-Griffin Act of 1959. These federal laws, together with the Railway Labor Act of 1926 and some antitrust legislation, created the legal foundation for collective bargaining in the United States. Like most federal legislation, each act was designed to address a specific aspect of collective bargaining or to correct a problem. Each subsequent act built on the earlier legislation. Like a pendulum, each subsequent act may change the direction of the federal government in relation to collective bargaining issues.

Norris-LaGuardia Act of 1932

The Norris-LaGuardia Act of 1932 specified that an employee could not be forced into a contract by an employer in order to obtain and keep a job. Prior to this act, employers required workers to sign a pledge that they would not join a union as long as the company employed them. These pledges were called **yellow dog contracts**.

The local courts generally sided with management in any labor dispute where a yellow dog contract was in effect. The judge would order an injunction that either prohibited striking or prohibited picketing during a strike. The police would enforce these injunctions.

In reaction to this practice, the Norris-LaGuardia Act of 1932 said that yellow dog contracts were not enforceable in any court of the United States. This act made it almost impossible for an employer to obtain an injunction to prevent a strike.

President Franklin Roosevelt took many steps to bolster economic growth during the Great Depression. One of his initiatives was the National Industrial Recovery Act (NIRA) of 1933. Section 7a of the NIRA guaranteed unions the right to collective bargaining in order to keep wages at a level that would maintain the purchasing power of the worker. After the NIRA passed, workers flocked to join both the American Federation of Labor (AFL) and the new Congress of Industrial Organizations (CIO). The Supreme Court struck down the NIRA as unconstitutional in 1935.

Wagner-Connery Act of 1935

When the NIRA was overturned, employers were free to use **unfair labor practices** in the management of their employees. This led to the Great Strike Wave of 1933–1934. During this period, labor organized against management and conducted citywide strikes and factory takeovers in numerous industrial sectors. In response, Senator Robert Wagner of New York introduced the Wagner-Connery Act, which was quickly passed in Congress in 1935 to mitigate the revolutionary labor climate and avert further economic disruption. A 1936 strike in the automobile industry quickly brought the Wagner-Connery Act before the Supreme Court, where it was upheld as constitutional. The Wagner-Connery Act established the procedures that are commonly called collective bargaining.

The Wagner-Connery Act forms the basis of formal labor relations in the United States. It grants workers the right to decide, by majority vote, which organization will represent them at the labor–management bargaining table. The act also requires management to bargain with duly elected union representatives and outlaws yellow dog contracts.

Provisions of the Wagner-Connery Act also established the National Labor Relations Board (NLRB), which has the power to hold hearings, investigate labor practices, and issue orders and decisions concerning unfair labor practices. The act defined five types of unfair labor practices and declared them illegal:

1. Interfering with employees in a union
2. Stopping a union from forming and collecting money
3. Not hiring union members
4. Firing union members
5. Refusing to bargain with the union

The Wagner-Connery Act and favorable court decisions resulted in some unions becoming extremely powerful. Union membership swelled from 4 million members in 1935 to 16 million members in 1948. The next two acts were adopted to balance the power between unions and management.

Taft-Hartley Labor Act of 1947

The Taft-Hartley Labor Act of 1947, which was passed over the veto of President Harry Truman, was formulated during the period that followed World War II. At that time, industry-wide strikes threatened to undermine a smooth return to civilian production. The Taft-Hartley Labor Act was designed to modify the Wagner-Connery Act and swing the pendulum back toward the middle by reducing the power of unions. It also spelled out specific penalties, including fines and imprisonment, for violation of the act.

Taft-Hartley gave workers the right to refrain from joining a union and applied the unfair labor practice provisions to unions as well as to employers. It specifically prohibited a union from forcing management to fire antiunion or nonunion workers. Unions were required to engage in **good faith bargaining**, and a 60-day "cooling off" period was created, when a labor agreement ends without a new contract. The act also regulated much of the union's internal activities.

The most significant provision of the Taft-Hartley Labor Act that affects fire fighters is referred to as "strikes during a national emergency." In the event that an imminent strike could affect a major part of an industry and imperil the health and safety of the nation, the president is granted certain powers to help settle the dispute. The president can order employees back to work, compel **arbitration**, or provide economic, judicial, or political pressure to achieve a resolution of the dispute.

Landrum-Griffin Act of 1959

After the AFL and CIO merged in 1955 to become the AFL-CIO, Senator John McClellan conducted hearings that revealed evidence of crime and corruption in some of the older local unions. The Labor-Management Reporting and Disclosure Act, otherwise known as the Landrum-Griffin Act, passed in 1959, at the height of the furor.

Landrum-Griffin established a bill of rights for members of labor organizations. It required that unions file an annual report with the government listing the assets of the organization as well as the names and assets of every officer and employee. Minimum election requirements were mandated, as were the duties and responsibilities of union officials and officers. This act also amended portions of the Taft-Hartley Labor Act.

Collective Bargaining for Federal Employees

Collective bargaining rights for public employees have traditionally lagged behind those in the private sector. Federal legislation passed in 1912 prohibits federal employees from striking. There were fewer than 1 million unionized government employees in 1956. Federal legislation during the 1950s and 1960s allowed unions to grow in the public sector. By 1970, the number of unionized government employees had grown to 4.5 million. Relations between management and unions had matured.

President Kennedy issued Executive Order #10988 in 1963, which granted federal employees the right to bargain collectively under restricted rules. This was a significant milestone for public sector unions. President Nixon further expanded the rights of employee unions within the federal government. Nixon also established a Federal Labor Relations Council that is similar to the National Labor Relations Board for private sector unions.

State Labor Laws

In addition to the federal laws, each state exercises legislative control over several aspects of collective bargaining. Each state determines whether it will engage in collective bargaining with state government employees and whether local government jurisdictions within the state may engage in collective bargaining with their employees. Some states require municipal governments to bargain collectively with fire fighters' unions, some permit collective bargaining, and others limit or prohibit collective bargaining.

Strikes by state employees are illegal in all but 10 states, and many states also prohibit local government employees from striking. Forty percent of local governments have forbidden employee strikes; however, numerous municipal strikes occurred between 1967 and 1980. After this period, several states and municipalities adopted legislation to prohibit strikes or other adverse labor actions.

Fire fighters have the right to bargain collectively in half of the states. Those states authorize municipalities to recognize a local fire fighter labor organization as the bargaining unit and enter into a binding contract that covers fire fighter pay, benefits, working conditions, conflict resolution, promotions, and department practices. The remaining states place restrictions on bargaining with fire fighters.

Labor organizations are supporting an effort to pass federal legislation that will provide collective bargaining rights for public safety officers employed by states or their political jurisdictions. This proposed federal legislation will supersede state regulations.

Federal Labor–Management Conflicts

Two notable strikes involving federal government employees mark the beginning and end of a significant era in the evolution of labor–management relations: the postal service workers strike and the Professional Air Traffic Controllers Organization (PATCO) air traffic controllers strike.

Postal Workers' Strike: 1970

U.S. postal workers went on strike illegally in 1970. At that time, the Postmaster General was legislatively restricted from negotiating with the striking workers, but nevertheless, negotiations were conducted and a settlement was reached. The striking postal workers were reinstated without penalty, and the negotiated wage increases were adopted. Subsequently, Congress recognized the postal workers union for the purposes of collective bargaining. This strike set the tone for future civil service strikes, including several that involved fire fighters.

PATCO Air Traffic Controllers' Strike: 1981

One reason that public employees are not striking frequently today could be in reaction to results of the 1981 strike of air traffic controllers. The PATCO went on strike after reaching an impasse in contract negotiations with the Federal Aviation Administration.

President Reagan used the example of the 1919 Boston, Massachusetts, police strike when he described the actions he would take to end the walkout. When the PATCO workers failed to report to work as ordered, President Reagan fired every striking member and decertified the union. Military air traffic controllers were brought in as temporary replacements. Air traffic volume was reduced for 18 months while the Federal Aviation Administration hired and trained replacement workers.

Unlike the 1970 postal workers' strike, none of the PATCO strikers was hired back, and many of the union officials were subjected to years of aggressive litigation by the federal government. This action set the tone for employer/employee relations during the remainder of the Reagan administration. The frequency of public sector strikes in the United States decreased very significantly over the following two decades.

Right to Work

The collective bargaining rights of public employees in different states are related to the right-to-work issue. Under the Taft-Hartley Labor Act, workers have the right to refrain from joining a union. In 2009, **right-to-work** laws were in effect in 22 states. In those states, a worker cannot be compelled, as a condition of employment, to join or to pay dues to a labor union. The same states tend to limit or restrict collective bargaining for local government employees.

The purpose of the original legislation in 1947 was to prohibit the practice of closed shops, in which a worker must be a member of a particular union in order to work for the company. Open shops provide the worker with the option of remaining outside the union. Proponents of the open shop statute believe that the practice eliminates union collusion and exclusionary practices, which are now deemed illegal, and protects an individual's right to refrain from joining an organization.

Opponents of open shops express the opinion that right-to-work statutes reduce a union's bargaining power and place an unfair burden on the union members. They state that, in essence, the union has to bargain for the entire workforce, even for nonmembers. A worker who chooses not to join the union is afforded the same benefits as union members, without paying dues to support union activities, pay for attorney fees, or conduct research studies. Under the federal regulations, however, an individual who chooses not to join a union can still be compelled to pay a share of the cost of the union's representation in collective bargaining.

The right-to-work states remain a source of controversy among labor organizations. Although unions support overturning existing right-to-work statutes, advocacy groups defend the existing statutes and promote their adoption in additional states. Workers entering the labor market should be aware of the prevailing labor laws that pertain to them.

■ The International Association of Fire Fighters

The largest fire service labor organization in the United States, the International Association of Fire Fighters (IAFF), represents 295,000 fire fighters and emergency medical service personnel in the United States and Canada Figure 5-1 ▾ . The IAFF has provided almost a hundred years of support and advocacy for professional career fire fighters, and its accomplishments influence many aspects of a fire fighter's job. It is a very powerful organization, both politically and within the fire service at the national level.

The International Association of Fire Fighters was established on February 28, 1918. The three principle objectives of the IAFF at that time were to obtain pay raises, establish the two-platoon or 12-hour workday schedule, and ensure that appointments and promotions were based on individual merit, not political affiliation.

Figure 5-1 The International Association of Fire Fighters (IAFF) is the largest fire service labor organization. Courtesy of IAFF.

Near Miss REPORT

Report Number: 06-0000482

Event Description: Dispatched to an automatic fire alarm to a county owned building. Upon arrival, fire department personnel were met by cleaning crew who stated that the basement was filling with smoke. Fire fighters and Incident Commander went to basement area to investigate. The investigation discovered a leaking HVAC unit. This unit had refrigerant escaping from it exposing fire department crew to Chlorodifluormethane (Forane 22). This crew evacuated to fresh air, and those with SCBA donned their masks. County maintenance personnel arrived on scene and shut down the air handler. Once the leak was stopped, fire department assisted with venting the basement area. The county maintenance personnel continued to do cleanup as fire department personnel made ourselves available. The fire fighters exposed to the Forane 22 received treatment on scene from EMS including oxygen and pulse ox readings. Fire fighters declined hospital treatment at this time.

Lessons Learned: Lessons learned are: personnel may never know what they are being exposed to whether it is a light smoke or a chemical release. Prevention should include review of Operating Procedures that include SCBA use at incidents. Finally, a combination effort between labor and management should be entered to review and implement a specific SCBA SOP/SOG.

The IAFF is unique among labor organizations in its dominance in representing a profession. Other public service professions, such as law enforcement and education, have two or more national labor organizations competing to represent the employees. Although the International Brotherhood of Teamsters and the American Federation of State, County, and Municipal Employees also represent fire fighters, these organizations represent few fire fighters and have little influence over the firefighting profession.

All fire fighters, including volunteers, have benefited from the efforts of the IAFF and its local and state affiliates. Labor advocacy has improved the quality of protective clothing, the safety of firefighting equipment, the content of training programs, and advanced techniques of emergency incident operations. The generalized process for handling a grievance, which is covered in this chapter, also comes from the labor–management relationship.

Organizing Fire Fighters into Labor Unions

Regardless of time, statutory environment, or work culture, labor–management relationships have been present in every work environment that includes an employer and employees. The first paid fire department in the United States was established in 1853 in Cincinnati.

The career fire fighter's work environment at the start of the 20th century was grim. New York City fire fighters worked "continuous duty," 151 hours a week, with just 3 hours off each day to go home for meals. San Francisco fire fighters got 1 day off after five consecutive 24-hour duty periods. Career fire fighters spent most of their time in cramped, unhealthy, and unsafe fire stations, sharing living accommodations with the horses.

As paid fire departments were established, fraternal or benevolent fire fighter support organizations gradually became political advocates for the workers and began to define the labor–management relationship. The genesis of these organizations occurred entirely at the local level. Antitrust protection, legislated by the 1890 Sherman Act, was interpreted by the federal courts to prohibit union representation of multiple labor groups. Consequently, individual local unions could not organize to form regional or national labor organizations. This prohibition lasted until the Clayton Act was passed in 1914.

Labor Actions in the Fire Service

The fire service has experienced many adverse labor actions in response to poor labor–management relations **Figure 5-2 ▶**. A **strike**—the act of withholding labor for the purposes of effecting a change in wages, hours, or working conditions—is one of the most drastic labor actions. Although a strike is precipitated by a conflict between labor and management, the impact of a strike is often felt beyond the parties that are directly involved in the relationship. In the private sector, the consumer is often the ultimate victim of a strike, through a combination of inconvenience and economic impact. In the public sector, the safety and the welfare of the general public are often directly affected by major labor actions.

The potential impact on public safety is so severe that many states prohibit fire fighters from walking out. In many of the cases in which fire service strikes have occurred, the public and media have questioned labor's right to strike as a matter of ethics. The IAFF started with the premise that fire fighter strikes are inadvisable, and from 1930 to 1968 the IAFF charter included a

Figure 5-2 Picketing is one of the most visible labor actions.

no-strike clause. However, there were three periods during the 20th century when municipal fire fighters did go on strike in different U.S. cities.

Striking for Better Working Conditions: 1918–1919

The 1918–1921 strikes occurred during a period of economic instability following World War I. Most of the fire fighter strikes were efforts to establish a two-platoon system, obtain more pay, or simply gain the right to form a labor organization that would be recognized by the municipality. Organized labor reached a turning point in 1919. First, there was an industry-wide strike in the steel industry. Then, the country witnessed the first "general strike" that shut down Seattle, Washington, on February 6. A similar strike occurred in Winnipeg, Manitoba, Canada, on May 15, 1919.

Fire fighters in Memphis, Tennessee, threatened to strike in 1918 and succeeded in obtaining a two-platoon work schedule, allowing fire fighters more time to be at home. The old system in Memphis had allowed just 1–3 days off each month. The new system allowed fire fighters to spend every other day away from the fire station.

Many IAFF local unions were able to gain a two-platoon schedule and/or a pay raise by threatening to strike. Fire fighters on the two-platoon system could finally have a home away from the fire station, although they were still required to work 84 hours a week. They could marry and raise a family. Today, most municipal fire fighters work between 42 and 56 hours a week. The workweek for most federal and military fire fighters still exceeds 60 hours.

Striking to Preserve Wages and Staffing: 1931–1933

The Great Depression began with the stock market crash of 1929. For the next several years, local governments reduced wages, initiated furloughs (unpaid leave), and reduced workforce size. In Seattle, fire fighters had to take two 25-day furloughs in 1933. In addition to the furloughs, the city closed seven fire stations and eliminated eight engine companies, two hose companies,

two squad companies, two fireboats, and one truck company to reduce expenditures.

Several fire fighter strikes between 1931 and 1933 occurred in reaction to wage reductions and elimination of fire fighter jobs. One local union went on strike because the fire fighters had not been paid by the city for 2 months.

Organized labor and fire fighter strikes during this time seem to have accomplished the IAFF mission of improving working conditions. Davis Ziskind's 1940 book, *One Thousand Strikes of Government Employees*, examined 39 fire fighter strikes and lockouts from 1900 to 1937. About half of the strikes resulted in partial or complete victory by labor. Unlike the unfortunate result of the Boston police riot, Ziskind could only find one large-loss fire that occurred during a labor action—a large industrial fire occurred in Pittsburgh, Pennsylvania, during the first IAFF organized strike in 1918.

Staffing, Wages, and Contracts: The 1973–1980 Fire Fighter Strikes

The most recent series of fire fighter strikes occurred between 1973 and 1980. The IAFF voted to eliminate the 50-year-old no-strike clause in its constitution at its 1968 convention, at a time when the United States was in political turmoil. Most of these strikes occurred after labor and management reached an impasse while negotiating a new contract. An **impasse** occurs when the parties have reached a deadlock in negotiations.

In the 1970s, local governments were facing another recession that hit hard. Several large cities were facing bankruptcy. In the depths of the recession, New York City laid off more than 40,000 city employees on July 2, 1975, including 1600 fire fighters. Although 700 of the fire fighters were hired back within 30 days, 900 others lost their permanent fire fighter jobs. It would take 2 years before the city could rehire all of the laid-off fire fighters.

Negative Impacts of Strikes

There have been fewer fire fighter strikes in the United States or Canada since the 1980s. None of the strikes in the 1970s resulted in a net gain for organized labor. The extreme negative public reaction, the political response of changing legislation, and the lasting legacy of lost trust have caused fire fighter strikes to be viewed as counterproductive. Measures such as picketing and alternative pressure tactics that maintain emergency services have been used to influence public opinion in several fire departments. Fire service strikes, or job actions, have occurred more recently among other industrialized nations throughout the world, particularly in Europe, where military personnel have been used to fill in for fire fighters.

Establishing a Strong Supervisor/Employee Relationship

The basis for a strong, positive, and effective supervisor/employee relationship is open, honest, and constant communications between the fire officer and the fire fighter. Once those communication paths are opened and trusted, then they prove very beneficial through good times and major issue times. And it is through that open, honest communication that some form of agreement, compromise, or answer can be found.

Key recommendations that form the foundation of any strong supervisor/employee relationship between a fire officer and a fire fighter include:

1. Schedule regular one-on-one meetings between you and each member of your company. This establishes a personal connection and trust between you and each fire fighter. During that meeting, discuss job performance and expectations from both people involved. Give guidance and coaching where necessary.

2. Schedule regular meetings with the company as a whole. Use this time to discuss new policies and procedures, any concerns about station procedures (e.g., checking out apparatus, housekeeping, kitchen duty, etc.) and upcoming personnel or policy changes. This is also a good opportunity to obtain feedback and input from the company members. Keeping an open line of communications is critical to an effective supervisor/employee relationship.

3. If a disagreement arises, work together to articulate the concern and to develop possible solutions. When both parties decide together what they want, they are usually very successful in attaining a mutually acceptable goal. When only one side decides what they want, the success ratio goes way down.

4. If the personal and professional relationship between you and a fire fighter is rocky from the beginning and you both have decided to work to improve it, start by listing the areas in which you can succeed together. Set goals and deadlines. Start with goals that are easily attainable ("low-hanging fruit") and build on those successes. Ask yourselves if the goal you are defining has a positive impact on the company and the other station personnel.

Having a good relationship does not mean that you will always agree on everything; it means that when you agree to disagree you have discussed the issue and cannot find middle ground. In some cases, a third party will need to be brought in to help mediate the discussions to arrive at an acceptable solution.

To trust each other's intentions, the company officer and the fire fighter must be honest and up front. Sometimes the supervisor/employee relationship can take a detour when either side holds back information, exaggerates, or deceives. Unfortunately, suspicion usually leads to acrimony and acrimony tends to lead to even more suspicion. In any long-term relationship, bluffing or threats can only last so long and they are counterproductive. Any time a company officer feels that communications are not going well, the officer should focus on bringing them back into alignment by discussing feelings and concerns in an open exchange.

The progression of the profession of firefighting is just that: professionals working together for the common good and common goals. A cooperative, collaborative supervisor/employee relationship is the profession at its best. With hard work and open, honest communication, fire officers and fire fighters will continue to make it better and raise the level of our professionalism, trust, and stature with those we serve.

Positive Labor–Management Relations

The value of a positive and productive labor–management relationship is widely recognized. A healthy labor–management relationship is essential to producing positive outcomes and avoiding the strife and consequences of a confrontational climate. Successful relationships are built on trust, respect, and open lines of communication. Each side must be willing and able to focus on the mutual benefits of a positive relationship or face the consequences of a negative outcome.

The root cause of almost every labor disturbance is a failure to properly manage the relationship between labor and management. The traditional way of thinking was based on the premise that either labor or management must score a victory over the other to settle every point. A philosophical shift in labor–management relationships is moving away from confrontational strategies to cooperative relationships, often through **mediation**. Mediation is the intervention of a neutral third party in an industrial dispute. As with many traditions in the fire service, the ability and the willingness to change were produced by necessity.

Tremendous amounts of time, energy, and money can be wasted in the process of two sides trying to overwhelm each other instead of working together. A poor labor–management relationship usually produces casualties on both sides. Fire chiefs and union presidents can lose their positions of power and influence. Positive relationships based on mutual respect and understanding are much more likely to produce positive results.

Some fire departments exist merely to meet an internal determination of what makes up the basic requirements for public safety. Others are committed to excellence and continual progress. Most observers agree that the most successful and progressive fire departments put significant effort into managing their labor–management relationships instead of handling continual confrontations and power struggles **Figure 5-3 ▾**.

Figure 5-3 Successful fire departments put effort into managing their relationships.

Voices of Experience

1. Leave your ego at the door.
2. Communicate about concerns before they become problems.
3. Through mutual respect we can solve any problem.

This was the template for the working relationship between me, as Chief of Department, and the local union Executive Board.

After 27 years in the fire service I was promoted to the rank of chief. The first meeting I held was with the district's command staff. The second meeting was with the Executive Board of our local union. In this gathering, I communicated my intent to lead in a collaborative style. This meant that my office was open to ideas, suggestions, questions, discussions, and shared leadership.

I expressed my expectations. The union Executive Board members and I would meet at least once each month and more often if necessary. We would leave our egos at the door. Problems would be anticipated and brought to each other's attention before the problem became a grievance or disciplinary issue. For every problem brought to the table at least one solid, well-thought-out solution must be offered. We agreed that through mutual respect we could solve almost any problem.

I spent a lot of time (but not as much as I would have liked) visiting the stations and talking with and listening to the fire fighters and officers, all members of the union. I had three restrictions on any subject that may come up. I would not allow anyone to be spoken of disparagingly, I would not make any decision that did not belong to me during the visit, and I would not speak about any personnel issues or discuss matters in the executive sessions of the Board of Directors. A lot of rumors were silenced and the reasoning behind controversial decisions made were discussed, and sometimes debated, but always respectfully.

One afternoon, a few years into our collaborative administration, the union president called. He said that he wasn't sure he wanted to continue our collaborative style of management due to the fact some union members were saying he was "too close to the chief." This was an interesting statement, I replied, because many of the chief officers I worked with and a few of the members of the District Board thought I was letting the union run the department. To me this meant we were doing things just right!

This agreement held for the five years that I was chief. During this time we confronted many common problems: budget deficits, discipline issues, fairness complaints, and conflicts between staff personnel and line personnel. We went through the preparations for Y2K and the emotions of 9/11. We worked with regional departments to develop an Urban Search and Rescue team. I shared ideas regarding change with the Executive Board and they provided valuable insight. Through all of this and more I was confronted by the labor union's attorney twice and only once did an issue go before Civil Service Board.

> "My office was open to ideas, suggestions, questions, discussions, and shared leadership."

Voices of Experience

I received many acknowledgments as I retired. Of all the awards and plaques received, I am most proud of the plaque given to me by the union local thanking me for my collaborative efforts. Leave your egos at the door, communicate openly as appropriate, listen actively, bring solutions and not just problems and complaints, and respect each other no matter how difficult the issue. I believe these are the components to successful labor/management relationships.

Robert H. Brown E.F.O.
Coordinator, Fire Science Technology and Management
Red Rocks Community College
Lakewood, Colorado

Fire Marks

The Fair Labor Standards Act and Fire-Based EMS

The **Fair Labor Standards Act (FLSA)**, originally passed in 1938 as part of Roosevelt's New Deal, is an example of federal legislation that significantly affects the fire service. Once passed, federal legislation remains in effect unless it is ruled unconstitutional (as with the NIRA), amended (as with Taft-Hartley), or repealed. The FLSA gradually evolved with the adoption of new provisions and regulations that expanded its application to include fire departments.

The primary purpose of FLSA was to establish minimum standards for wages and to spell out administrative procedures covering work time and compensation, including overtime entitlement. For most workers, FLSA requires overtime to be compensated at time-and-a-half pay—150% of an employee's regular hourly wages. This legislation was adopted during a period of profound underemployment and was designed to encourage employers to hire more workers instead of paying current employees to work additional hours.

At the time, FLSA did not cover government employees; however, additional provisions were adopted in 1974 to make FLSA applicable to federal employees. In 1985, the Supreme Court ruled in *Garcia v. San Antonio MTA*, 469 US 528 (1985) that local union public employers must also abide by the FLSA regulations, unless Congress specifically legislates otherwise.

Amended regulations, adopted in 1986, established a special overtime rule for fire fighters. Most employees are entitled to be paid overtime starting at the 41st hour of a 7-day workweek. The average workweek for fire fighters across the United States at that time was determined to be 53 hours. Under the new federal regulations, public safety agencies do not have to pay fire fighters overtime until after they have worked 212 hours in a maximum 28-day cycle—which averages out to a 53-hour workweek. This is known as the 207(k) exemption and covers individuals whose work involves "the prevention, control, or extinguishment of fires" for 80% of their work time.

Anne Arundel County, Maryland, fire fighter/paramedics filed suit in 1990 for improperly calculated overtime payments. The FLSA description of fire fighter activities included "housekeeping, equipment maintenance, lecturing, attending community fire drills, and inspecting homes and schools for fire hazards" as incidental functions but did not include emergency medical services (EMS) activity. Some of the employees were assigned to paramedic ambulances and did not routinely participate in firefighting activities. Other employees, who were assigned to fire suppression companies, were actually spending more time handling EMS first-responder calls than performing fire suppression activities.

The plaintiffs claimed that the 207(k) exemption did not apply to them because more than 20% of their time was spent on nonfirefighting activities. They argued that their overtime pay should start at the 41st hour and not the 54th hour of work in an average workweek. U.S. District Court Judge Walter Black ruled in favor of the employees, creating a $4 million back payment obligation for the county.

The Anne Arundel case was one of many similar lawsuits and union grievances that were going on throughout the nation during the same time period. Dozens of cities were involved in FLSA lawsuits and were becoming obligated for huge amounts of retroactive back pay. Some of the settlements went back to the 1986 adoption of the amended FLSA regulations. The situation became even more complicated when the eighth U.S. Court of Appeals made a ruling that was in conflict with the ruling of the fourth U.S. Court of Appeals in the Anne Arundel case. When the Supreme Court declined to review the Anne Arundel case, the conflicting legal opinions were left to stand within the regions subject to each circuit court.

U.S. Representative Robert Ehlrich, a Maryland Republican, introduced House Resolution 1693, the Fire and Emergency Services Definition Act. This legislation expanded the FLSA definition of fire protection activities to include paramedics, emergency medical technicians, rescue workers, ambulance personnel, and hazardous materials workers. The Ehrlich resolution, which had the support of both the IAFF and the International Association of Fire Chiefs (IAFC), was passed on November 4, 1999. The Senate passed the same resolution a week later, and President Clinton signed the Fire and Emergency Services Definition Act into law on December 9, 1999. This 1999 law amended a portion of the original act that was passed in 1938 and expanded by the Supreme Court decision in 1985 to cover local government employees. It took almost a full decade to resolve the issue of FLSA coverage for fire fighters performing EMS duties.

There remain active court cases that represent millions of dollars of unpaid overtime. Some of the cases include staffing practices that started after the 1999 revision of the FLSA legislation.

For example, the U.S. Supreme Court declined to consider whether Los Angeles City dual role paramedics should qualify for the 207(k) exemption, leaving intact an August 2005 decision by the Ninth Circuit, which found that dual role paramedics are not covered by the exemption because they are not primarily responsible for fire suppression. This left a $5 million verdict for 119 paramedics. The Ninth Circuit found that in order for the fire fighters/paramedics to have a responsibility to engage in fire suppression they must have a real obligation to do so (*City of Los Angeles v. Cleveland*).

The Los Angeles ruling was one of three cases cited when the Third Circuit Court of Appeals reversed a lower court ruling in 2008 and held that paramedics employed by the Philadelphia Fire Department were entitled to time-and-a-half pay for hours worked after the 40th (*Lawrence v. City of Philadelphia*).

In some instances, management uses the moral ethos of public service as leverage against labor and vice versa. Public support is usually viewed as vital by both sides because elected officials, who represent the public, have ultimate control over the economic and policy issues.

The Growth of IAFF as a Political Influence

The economic turmoil and wage reductions in the 1990s were as severe in the 1930s, but organized labor took a different path. Instead of striking, labor worked to get political candidates who were supportive of labor into local and national political office. The attempted recall of Memphis elected officials in 1979 may have been the beginning of this new era of labor influence.

Fire fighters are particularly respected in political circles for the efforts they can put behind a political candidate who supports the objectives of the IAFF or a particular local union. In addition to assisting with political campaigns, organized labor found that "money talks." The creation in 1978 of FIREPAC, the IAFF's political action committee, was a natural progression of the labor organization's activity as a major influence in improving the fire fighter's work environment.

FIREPAC is a **political action committee (PAC)**, a special interest group that can solicit funding and lobby local and national elected officials for its cause. Funded by donations from individual IAFF members, FIREPAC promotes the legislative and political interests of the IAFF. The money is used to educate members of Congress about issues important to fire fighters and emergency medical personnel and to elect candidates to office that support those issues. In the 2007–2008 election cycle, FIREPAC raised more than $4.6 million in voluntary donations, which was contributed to 367 congressional candidates, with just under 30% of the funds going to Republican candidates. The election results showed that 92% of the candidates supported by FIREPAC won.

Using money and off-duty fire fighters to assist political candidates or causes may appear unseemly. However, Hal Bruno, former director of political reporting for ABC television and radio networks, has repeatedly pointed out in his *Firehouse* magazine columns that the fire service must be fully engaged in the political process. That means being an active part of the local political process.

Fire Marks

Seattle Medic One
The Seattle Medic One paramedic ambulance service began as a university research project on March 7, 1970. The Seattle City Council declined to fund continuation of the Medic One project within the fire department budget in 1972. Members of IAFF Local 27 scrambled to fund the lifesaving project through a special countywide tax levy. Two years later, in 1974, CBS news magazine *60 Minutes* profiled Seattle as the Best Place to Have a Heart Attack. Organized labor had to scramble again in 1997 when the voters defeated the continuing levy, requiring a special referendum in February 1998 to keep the 22 paramedic ambulances running in Seattle and King County. Both the jurisdiction and labor were aggressive in promoting the levy, which passed in 2007.

Labor–Management Alliances

In addition to political action, the IAFF has developed strategic labor–management alliances with the International Association of Fire Chiefs (IAFC) to promote mutually agreeable goals. In some situations, both labor and management want the same goal and they work together to achieve those goals. This chapter examines four national examples.

Fire Fighter Safety and Deployment Study

Building on the work started by the IAFF/ IAFC EMS Systems Performance Measurement program, a Department of Homeland Security (DHS)–funded research effort started in 2008 to gather data and conduct time-on-task experiments to develop a prospective deployment model. Joining the IAFF and IAFC in this effort were three national groups: the Center for Public Safety Excellence Commission on Fire Accreditation, the Department of Fire Protection Engineering at Worchester Polytechnic Institute, and the National Institute of Standards and Testing.

Outcome data are being gathered from more than 400 fire departments, including the 53 largest in the United States. The data obtained will be used to document experience with fireground and EMS operations and outcomes. Crews from Montgomery County, Maryland, and Fairfax County, Virginia, participated in the associated field experiments to determine time-on-task baseline performance capability and compare different staffing schemes with outcomes. There were 40 fire suppression and more than 100 EMS time-on-task experiments conducted. These experiments are similar to the 1979 LAFD Measure of Effectiveness task studies using three to six fire fighters on engine companies.

EMS Systems Performance Measurement

During the 1997 IAFF EMS Conference, focus group meetings were held to identify indicators of EMS system quality. At that time, the IAFF EMS Committee was charged with generating an agreed-upon set of performance indicators and associated measures for use by local fire departments in measuring and reporting EMS system performance capability. Working with the IAFC, the EMS Systems Performance Measurement instrument was constructed, field tested, validated, and released in 2002. This instrument consists of 15 EMS quality indicators, their definitions, and performance measures that are useful at the local level in reporting the value of a fire-based EMS system to local decision-makers and the public.

The instrument also provides background information relating the indicator to quality in an EMS system, explains any existing standards, notes the absence of standards, proposes a system goal, and provides for data collection of information related to each main measure. The indicators include call processing time, turnout time, time to first shock (defibrillation), employee turnover, patient outcome, protocol compliance, deployment of mobile resources, staffing, and employee illness and injury. The measurement instrument will provide system leaders the best way to collect relevant data and to report on that data in the future.

The IAFF and IAFC joint efforts to promote and protect fire-based EMS systems continue. In 2007, the two organizations along with other fire service groups, released a whitepaper on fire-based EMS, *Prehospital 9-1-1 Emergency Medical Response: The Role of the United States Fire Service in Delivery and Coordination*, as well as an associated video presentation, *EMS: The Right Response*.

The IAFC/IAFF Labor–Management Initiative

The IAFC/IAFF Labor–Management Initiative (LMI), formerly known as the Fire Service Leadership Partnership, was developed in 1999 to assist today's fire chiefs and union presidents in fostering cooperative and collaborative labor–management relationships. The same fundamental concepts that were involved in developing the Wellness-Fitness Initiative (discussed in the following section) provided the foundation for the LMI.

Scores of fire chiefs and union presidents have attended LMI programs to learn how to enhance their labor–management relationships. Many issues that face the fire service require continued cooperation and problem solving from both labor and management. Without the combined talent of all members of the fire service, fire fighters and fire chiefs will always be behind in their reaction to emerging issues, such as fire-based EMS, urban–wildland interface, occupational safety, and perhaps the greatest challenge to the fire service, response to terrorism. It is clear in this environment of new challenges that a good labor–management relationship is essential. Learning the skills to develop solid relationships should be a regular part of a fire officer's curriculum and not an ancillary elective.

Fire Service Joint Labor–Management Wellness-Fitness Task Force

The IAFF and the IAFC joined forces to develop a comprehensive and mutually beneficial wellness-fitness program. Released in 1997, the program is designed to serve a fire fighter throughout their career. Ten fire departments and their IAFF local unions participated in the task force to develop a comprehensive and nonpunitive program:

- Austin, Texas, and IAFF Local 975
- Calgary, Alberta, and IAFF Local 255
- Charlotte, North Carolina, and IAFF Local 660
- Fairfax County, Virginia, and IAFF Local 2068
- Indianapolis, Indiana, and IAFF Local 416
- Los Angeles County, California, and IAFF Local 1014
- Metro-Dade County, Florida, and IAFF Local 1403
- New York City and IAFF Locals 94 and 854
- Phoenix, Arizona, and IAFF Local 493
- Seattle, Washington, and IAFF Local 27

The completed program package was released at both the IAFF Redmond Symposium and the IAFC Conference in August 1997. The Wellness-Fitness Initiative (WFI) task force released the third edition of the Wellness-Fitness Initiative in 2008. In addition, the WFI task force developed the Candidate Physical Ability Test (CPAT), a standardized physical ability test that is validated as an assessment of a fire fighter candidate's physi-

Company Tips

The Phoenix Way: Relations by Objectives
In 1984, the Phoenix Fire Department started relations by objectives (RBO). The RBO process had been used in other labor–management disputes in which the two sides had significantly different positions on values, goals, and trust for one another. The Phoenix Fire Department and the United Phoenix Fire Fighters used the RBO process with each major issue. This includes a means of analysis, decision, education, implementation, revision, and review. Each step has specific parameters in which labor and management cooperatively work on issues.

With RBO, the Phoenix Fire Department has been able to collaborate on issues such as incumbent drug testing, apparatus specifications and deployment, safety, and employee wellness. Because both labor and management participated in the development of many programs and issue resolution, both sides of the employee/employer environment shared equal successes and failures.

This led to the culture of the Phoenix Fire Department changing into an organization where each member, including civilian staff, feels like a part owner of the department. Whether this cultural shift was intended or not, the result was the development of a work environment in which each member has the opportunity to make significant contributions.

"The Phoenix Way" was severely tested after the March 12, 2001, line-of-duty death of fire fighter Bret Tarver at a supermarket fire. The recovery process took over a year and resulted in significant operational changes within the Phoenix Fire Department and a national awareness of the problem of applying residential firefighting assumptions when operating at a commercial fire.

cal ability to perform the critical tasks of a fire fighter. Finally, the WFI task force developed the IAFF/IAFC/ACE Peer Fitness Trainer certification program and initiated changes to the traditional critical incident stress management (CISM) behavior health model.

The Fire Officer's Role as a Supervisor

One of the fundamental duties of a fire officer is to supervise the activities of subordinate fire fighters. The basic authority of a supervisor and the duties of subordinates are defined by the personnel rules of the city or governmental organization, as well as the specific rules, regulations, and procedures of the fire department. In addition, when a collective bargaining agreement is in effect, additional details of the relationship are spelled out in the contract or MOU. Supervisors are expected to follow all of the established rules and procedures in assigning duties and all other aspects of the relationship with their subordinates. In many cases, it is a significant challenge for a newly promoted fire officer to learn which rules and regulations apply to which situation and how they are interpreted and applied.

In most organizational structures, there is a clear distinction between labor (the workers) and management (the managers and supervisors). The managers and supervisors represent the organization, and the union represents the workers. If there is any doubt or disagreement about the application or interpretation of the contract, a process should be in place for labor and management representatives to meet and resolve the problem.

This line between labor and management is more complicated in fire departments because the first-level supervisors are often members of the same collective bargaining unit as the fire fighters they supervise. A fire officer's relationship to the organization is often covered by the same contract that the officer has to follow and enforce. The formal line between labor and management is often at a higher level, such as administrative fire officer. In some cases, the supervisory and managing fire officers are members of a bargaining unit that is separate from that of the fire fighters.

As a supervisor, a fire officer is generally the first point of contact between the workers and the fire department organization **Figure 5-4 ▼** . If there is a disagreement relating to an interpretation or application of a work rule that is covered by the contract, the first-level supervisor is the individual who should have the first awareness of the problem and the first opportunity to resolve it. An officer who is a member of the same bargaining unit as an individual who is dissatisfied must clearly understand the established problem-solving processes.

Figure 5-4 A fire officer is generally the first point of contact between the workers and the organization.

Application Example: A Grievance Procedure

A **grievance** is a dispute, claim, or complaint that any employee or a group of employees may have about the interpretation, application, and/or alleged violation of some provision of the labor agreement or personnel regulations. A **grievance procedure** is a formal structured process that is employed within an organization to resolve a grievance. In most cases, the grievance procedure is incorporated in the personnel rules or the labor agreement and specifies a series of steps that must be followed in order. If the problem cannot be resolved in a mutually acceptable manner at one level, it can be taken to the next level, up to some ultimate level, where an individual or body has the final authority to impose a binding decision.

Whether the grievance is related to the labor contract or to the agency's personnel regulations, the grievance procedure should specify a sequential process and a time line to move through the steps. The grievance can be resolved at any point by management's accepting the complaint and the corrective action requested by the grievant or by both sides reaching a negotiated settlement that is acceptable to each. If management rejects the claim, the grievance can be taken to the next level. The time line ensures that a grievance will not be stalled at any level for an excessive time period.

An employee can contact a union representative at any time to discuss a situation, including how the union interprets the rule in question and whether a grievance should be submitted. The employee's union representative usually becomes formally involved at either the first or the second step of the grievance process. The union representative acts as an advocate for the individual or group that submitted the grievance. The union becomes more involved as the process moves through the steps, particularly in cases in which the problem has broad impact within the organization.

The objective should always be to resolve the problem at the lowest possible level and in the shortest possible time. Grievances that have to be processed through multiple steps are disruptive, time consuming, and costly to both sides. The ability to resolve problems at a low level is an indication of a healthy organization with a good labor–management relationship, whereas a steady stream of grievances moving up to the highest levels is a symptom of major problems in the relationship.

The grievance procedure outlined in the following steps is an example that comes from one particular fire department. The detailed procedures employed by different organizations usually include some variations of this model. A similar process can be established to resolve disputes in fire departments where there is no formal labor contract, even in volunteer organizations. The most important responsibility for a fire officer is to know and follow the procedures that apply in his or her organization.

Step One

The grievant presents his or her complaint verbally to a supervisor, shortly after the occurrence of the action that gave rise to the grievance. In some organizations, this nondocumented verbal notification is called an "informal grievance," or step zero. Even this informal step requires the grievant to provide three important pieces of information:

- The article and section of the labor agreement or personnel regulation alleged to have been violated

- A full statement of the grievance, giving facts, dates, and times of events, as well as specific violations
- A statement of the desired remedy or adjustment

Step Two

The second step initiates the formal part of a grievance procedure. If the problem is not resolved at step one, the employee may prepare and submit a written grievance **Figure 5-5 ▾**. This is usually submitted on a specified grievance form document (often a carbonless copy form). The employee, the employee's supervisor, and the personnel office each receive a copy of the grievance.

The supervisor has 10 calendar days to reach a decision and provide a written reply to the grievant. Failure to respond to the grievance within 10 days means that the supervisor has denied the grievance. The grievant can immediately go to step three.

Step Three

A step-three grievance is written out on another specific grievance form and again specifies the article and section of the contract or personnel regulation alleged to have been violated; the dates, times, and specific violations that are alleged to have taken place; and the desired remedy or adjustment. Copies of the step-two grievance form and the supervisor's response are attached.

A step-three grievance is submitted to a second-level supervisor, typically an administrative fire officer, who has 10 calendar days to respond. If the grievance is denied or the administrative fire officer does not respond within the specified time, the grievant can move to step four and present his or her grievance to the fire chief.

Step Four

If the grievance remains unresolved, the grievant can present it to the fire chief or his or her designee as the fourth step. The same written information must be submitted, along with all of the documentation from the previous steps. The fire chief has 10 days to respond to a step-four grievance.

If the fire chief does not respond within the time frame, or if the grievance remains unsettled, the process moves out of the fire department to a mediator, personnel board, or civil service board for resolution. In this example, the grievance goes to the county administrator and, if not resolved at that level, goes to federal arbitration.

SAMPLE GRIEVANCE FORM

STEP TWO: TO BE COMPLETED BY UNION OR EMPLOYEE

Date of filing: _____

From: _____   _____   _____
        Employee   Rank   Assignment/shift

Grievance form must be submitted within 15 calendar days of the incident being grieved.

STATEMENT OF GRIEVANCE Must (1) contain a statement, as complete as possible under the circumstances, of the grievance and the facts upon which it is based, including dates, times, locations, names of witnesses, and other appropriate information; (2) identify the section(s) of the Contract Agreement that affect this grievance; (3) state requested remedy or corrective action.
Additional pages may be attached to the Sample Grievance Form.

(*this is where the employee enters the statement*)

Original copy of the completed Sample Grievance Form shall be delivered to the employee's immediate supervisor, with a copy to the Human Resources office and the Union representative.

_____   _____
        Employee   Date

TO BE COMPLETED BY IMMEDIATE SUPERVISOR within 10 calendar days of receipt of Sample Grievance Form

_____

  Supervisor's Name   Rank   Work Location/Shift   Date Grievance received

(*this is where the immediate supervisor enters a response*)

_____   _____
   Supervisor Signature   Date

If employee is satisfied with Supervisor's answer, sign the original Sample Grievance Form acknowledging agreement and submit it to the Human Resources Director for placement in your employment records. If employee is NOT satisfied, shall sign the original Sample Grievance Form acknowledging disagreement and immediately notify the Union in writing. The original Sample Grievance Form shall then be submitted by the employee to the Deputy Chief within ten (10) calendar days of the decision of the immediate supervisor.

Agree_____   Do not agree: _____

Figure 5-5 Example of a grievance form.

Summary

The fire officer is in a unique position regarding labor issues within the fire station. The fire officer is a working supervisor and is often both the designated representative of fire department administration and a dues-paying member of an IAFF local union. It is important that the fire officer knows the grievance procedures, labor contract provisions, and personnel regulations that create the limits to, and framework of, the scope of supervisory work.

One goal of this chapter was to demonstrate that the federal status of labor and management swings like a pendulum. Labor had an advantage with the Norris-LaGuardia and Wagner-Connery Acts and the 1970 postal service strike. Management had the advantage with the Taft-Hartley and Landrum-Griffin Acts and the 1981 air traffic controller strike.

Another goal was to show the changing methods of organized labor effectiveness. An old labor model would have seen fire fighters striking to keep their pay levels and jobs when the economy slumped. Instead, members are involved in the local and national political process, providing sweat equity and cash contributions to the elected officials who best support the needs of organized labor.

You Are the Fire Officer: Conclusion

You learn that the labor contract overrides the city personnel regulations. The personnel regulations state that a supervisor may change an employee's work hours with 1 day notice. The labor contract specifically requires a minimum of 72 hours notice. The labor contract also requires that employees be provided with the appropriate educational resources (books, practice tests, and study guides) 9 work days before they are scheduled to attend an employer-mandated certification class.

The situation is resolved at the first step of the grievance process after you call the battalion chief and ask for guidance. Following his advice, you apologize to Fire Fighter Trammel and admit that you made a mistake. He is not ordered to attend the pump operator certification course. The union representative calls Trammel later that day to confirm that the problem has been resolved.

The battalion chief clarifies the situation. He was looking to see if any of the fire fighters would volunteer to attend the pump operator course next week because the academy had a last-minute vacancy. The labor contract allows employees to voluntarily attend an employer-mandated certification course without the required advance notice.

The labor contract was not one of the items you had to memorize for the promotional exam; however, you need to know what the contract says in order to succeed as a fire officer. Your ability to quickly correct the problem and apologize for having been wrong will create a positive impression with the members of your company.

Wrap-Up

Chief Concepts

- The Norris-LaGuardia Act of 1932 made yellow dog contracts unenforceable.
- The Wagner-Connery Act of 1935 established the National Labor Relations Board and the procedures commonly called collective bargaining.
- Good faith bargaining and a 60-day "cooling off" period were created by the Taft-Hartley Labor Act of 1947.
- Public sector bargaining rules vary by state.
- The grievance procedure is an incremental, multistep process that requires a timely response.

Hot Terms

Arbitration Resolution of a dispute by a mediator or a group rather than a court of law. Any civil matter may be settled in this way; some labor–management agreements include a binding arbitration clause.

Binding arbitration The resolution of a dispute by a third and neutral party that is not personally involved in the dispute and may be expected to reach a fair and objective decision based on an informal hearing, at which the disputants may argue their cases and present all relevant evidence. It is usually agreed in advance that such a decision will be binding and not subject to appeal.

Collective bargaining Method whereby representatives of employees (unions) and employers determine the conditions of employment through direct negotiation, normally resulting in a written contract setting forth the wages, hours, and other conditions to be observed for a stipulated period (e.g., 3 years). Term also applies to union–management dealings during the terms of the agreement.

Fair Labor Standards Act (FLSA) A 1938 act that provides the minimum standards for both wages and overtime entitlement and spells out administrative procedures by which covered work time must be compensated. Public safety workers were added to FLSA coverage in 1986.

Good faith bargaining A legal requirement of both the union and the employer arising out of Section 8(d) of the National Labor Relations Act. Enforced by the National Labor Relations Board, the parties are required: "To bargain collectively . . . to meet at reasonable times and confer in good faith with respect to wages, hours, and other conditions of employment, or the negotiation of an agreement or any question arising thereunder, and the execution of a written contract incorporating any agreement reached if requested by either party, but such obligation not to compel either party to agree to a proposal or require the making of a concession. . . ."

Grievance A dispute, claim, or complaint that any employee or group of employees may have in relation to the interpretation, application, and/or alleged violation of some provision of the labor agreement or personnel regulations.

Grievance procedure A formal structured process that is employed within an organization to resolve a grievance. In most cases, the grievance procedure is incorporated in the personnel rules or the labor agreement and specifies a series of steps that must be followed in order.

Impasse Occurs when the parties have reached a deadlock in negotiations. Also described as the demarcation line between bargaining and negotiation. A declaration of an impasse brings in a state or federal negotiator that will start a fact-finding process and will lead to a binding arbitration resolution.

Mediation The intervention of a neutral third party in an industrial dispute. The object is to enable the two sides to reach a compromise solution to their differences, which the mediator usually does by seeing representatives of both sides separately and then together.

Negotiation Mutual discussion and arrangement of the terms of an agreement.

Political action committee (PAC) An organization formed by corporations, unions, and other interest groups that solicit campaign contributions from private individuals and distribute these funds to political candidates.

Right to work A worker cannot be compelled, as a condition of employment, to join or not to join or to pay dues to a labor union.

Shop steward A union member appointed or elected to be the first line of labor representation at the workplace. The steward enforces the contract, collective agreement, or memorandum of understanding and represents the union members at that fire station or work location.

Wrap-Up

Strike A concerted act by a group of employees, withholding their labor for the purposes of effecting a change in wages, hours, or working conditions.

Unfair labor practice An employer or union practice forbidden by the National Labor Relations Board or state/local laws, subject to court appeal. It often involves the employer's efforts to avoid bargaining in good faith.

Yellow dog contracts Pledges that employers required workers to sign indicating that they would not join a union as long as the company employed them. Declared unenforceable by the Norris-LaGuardia Act of 1932.

Fire Officer *in Action*

The city is suffering from declining revenues and increasing costs. This morning, local television reports that the city plans to reduce the public safety budget by 35%. That would mean disbanding fire companies and the loss of fire fighter jobs. There are many angry voices at the firehouse kitchen table during morning line-up, some talking about strikes.

You tell your crew that you will try to provide more information. You call the administrative fire officer. She is just as surprised as you are about the news item. You suggest to the senior fire fighter that he call the shop steward or one of the local labor leaders. You start working your contacts in fire department administration.

By noon you and the senior fire fighter share with the rest of the crew what is known so far. It appears to be a trial balloon launched by the city auditor, one of many budget-reducing ideas being shared with the citizens this month. Economic times are grim. You decide to have the crew develop a list of cost-saving suggestions that the fire department can use to reduce the size of the operating budget without laying off fire fighters.

The fire officer is in the middle of the labor–management relationship. While a managing or supervising fire officer represents the management of the department at the fire station, the officer is living with the crew and often affected by the same work conditions. Understanding how labor interacts with management improves the effectiveness of the fire officer.

1. A fire fighter is asked to identify the pro-union members of the fire company to the administrative fire officer. This is an example of:
   A. collective bargaining.
   B. yellow dog contracts.
   C. unfair labor practices.
   D. right to work.

2. What does right to work mean?
   A. Employees are required to be a member of a union.
   B. Employees have the right to bargain their positions.
   C. Workers cannot be compelled to join a union as a condition of employment.
   D. Employees are not permitted to strike.

3. If the contract dispute continues, some states would require binding arbitration. What is this?
   A. Giving a specific time period to agree on contract
   B. Dispute resolution by a third, neutral party
   C. Opening the ability to strike
   D. Mediation

4. What is included in the third step of the grievance procedure?
   A. The grievance is presented verbally to a review board.
   B. A statement is made of the desired remedy or adjustment.
   C. The economic impact of granting the relief is analyzed.
   D. The grievance is submitted to the personnel or human resources office.

Safety and Risk Management

NFPA 1021 Standard

Fire Officer I

4.2.1 Assign tasks or responsibilities to unit members, given an assignment at an emergency incident, so that the instructions are complete, clear, and concise; safety considerations are addressed; and the desired outcomes are conveyed. [p 108, 110–111]

(A) Requisite Knowledge. Verbal communications during emergency incidents, techniques used to make assignments under stressful situations, and methods of confirming understanding. [p 110–111]

(B) Requisite Skills. The ability to condense instructions for frequently assigned unit tasks based on training and standard operating procedures. [p 110]

4.2.3 Direct unit members during a training evolution, given a company training evolution and training policies and procedures, so that the evolution is performed in accordance with safety plans, efficiently, and as directed. [p 102, 105–110]

(A) Requisite Knowledge. Verbal communication techniques to facilitate learning. [p 103, 106–107, 110]

(B) Requisite Skills. The ability to distribute issue-guided directions to unit members during training evolutions. [p 99, 111–112]

4.6.3 Develop and conduct a post-incident analysis, given a single unit incident and post-incident analysis policies, procedures, and forms, so that all required critical elements are identified and communicated, and the approved forms are completed and processed in accordance with policies and procedures. [p 115–116]

(A) Requisite Knowledge. Elements of a post-incident analysis, basic building construction, basic fire protection systems and features, basic water supply, basic fuel loading, fire growth and development, and departmental procedures relating to dispatch response tactics and operations and customer service. [p 115–116]

(B) Requisite Skills. The ability to write reports, to communicate orally, and to evaluate skills. [p 115–116]

4.7* **Health and Safety.** This duty involves integrating health and safety plans, policies, and procedures into daily activities as well as the emergency scene, including the donning of appropriate levels of personal protective equipment to ensure a work environment that is in accordance with health and safety plans for all assigned members, according to the following job performance requirements. [p 99–116]

4.7.1 Apply safety regulations at the unit level, given safety policies and procedures, so that required reports are completed, in-service training is conducted, and member responsibilities are conveyed. [p 99–116]

(A) Requisite Knowledge. The most common causes of personal injury and accident to members, safety policies and procedures, basic workplace safety, and the components of an infectious disease control program. [p 99–116]

(B) Requisite Skills. The ability to identify safety hazards and to communicate orally and in writing. [p 99–116]

4.7.2 Conduct an initial accident investigation, given an incident and investigation forms, so that the incident is documented and reports are processed in accordance with policies and procedures of the AHJ. [p 110, 114–115]

(A) Requisite Knowledge. Procedures for conducting an accident investigation and safety policies and procedures. [p 114–115]

(B) Requisite Skills. The ability to communicate orally and in writing and to conduct interviews. [p 114–115]

4.7.3 Explain the benefits of being physically and medically capable of performing assigned duties and effectively functioning during peak physical demand activities, given current fire service trends and agency policies, so that the need to participate in wellness and fitness programs is explained to members. [p 111–113]

(A) Requisite Knowledge. National death and injury statistics; fire service safety and wellness initiatives; agency policies. [p 99–101]

(B) Requisite Skills. The ability to communicate orally. [p 111]

Fire Officer II

5.6.1 Produce operational plans, given an emergency incident requiring multiunit operations, the current edition of NFPA 1600, and AHJ-approved safety procedures, so that required resources and their assignments are obtained and plans are carried out in compliance with NFPA 1600 and approved safety procedures resulting in the mitigation of the incident. [p 116]

(A) Requisite Knowledge. Standard operating procedures; national, state/provincial, and local information resources available for the mitigation of emergency incidents; an incident management system; and a personnel accountability system. [p 106, 108]

(B) Requisite Skills. The ability to implement an incident management system, to communicate orally, to supervise and account for assigned personnel under emergency conditions, and to serve in command staff and unit supervision positions within the Incident Management System. [p 108, 110–111]

5.7 **Health and Safety.** This duty involves reviewing injury, accident, and health exposure reports; identifying unsafe work environments or behaviors; and taking approved action to prevent reoccurrence, according to the following job requirements. [p 99–116]

5.7.1 Analyze a member's accident, injury, or health exposure history, given a case study, so that a report including action taken and recommendations made is prepared for a supervisor. [p 106, 114–115]

(A) Requisite Knowledge. The causes of unsafe acts, health exposures, or conditions that result in accidents, injuries, occupational illnesses, or deaths. [p 99–104]

(B) Requisite Skills. The ability to communicate in writing and to interpret accidents, injuries, occupational illnesses, or death reports. [p 114–115]

Additional NFPA Standards

NFPA 472 *Standard for Competence of Responders to Hazardous Materials/Weapons of Mass Destruction Incidents*

NFPA 1451 *Standard for a Fire Service Vehicle Operations Training Program*

NFPA 1500 *Standard on Fire Department Occupational Safety and Health Program*

NFPA 1521 *Standard for Fire Department Safety Officer*

NFPA 1561 *Standard on Emergency Services Incident Management System*

NFPA 1581 *Standard on Fire Department Infection Control Program*

NFPA 1583 *Standard on Health-Related Fitness Programs for Fire Department Members*

NFPA 1600 *Standard on Disaster/Emergency Management and Business Continuity Programs*

NFPA 1971 *Standard on Protective Ensembles for Structural Fire Fighting and Proximity Fire Fighting*

NFPA 1975 *Standard on Station/Work Uniforms for Emergency Services*

Introduction to Fire and Emergency Services Administration (FESHE) Course Outcomes

7. Identify roles and responsibilities of leaders in organizations. [p 99, 103]
9. Identify and assess safety needs for both emergency and nonemergency situations. [p 105, 107–108, 110–116]

11. Identify the role of a company officer in Incident Command System (ICS). [p 99–100, 105–108, 110–111]
13. Identify and analyze the major causes involved in line of duty firefighter deaths related to health, wellness, fitness, and vehicle operations. [p 99–105, 107, 115]

Knowledge Objectives

After studying this chapter, you will be able to:

- Describe the most common causes of personal injury and deaths to fire fighters.
- Define incident safety officer.
- Describe safety policies and procedures and basic workplace safety.
- Describe the components of an infectious disease control program.
- Describe procedures for conducting an accident investigation and safety policies and procedures.

Skills Objectives

After studying this chapter, you will be able to:

- Identify safety hazards.
- Implement an incident management system and ensure the safety of personnel under emergency conditions.
- Interpret accident reports.

Tower 5 is responding to assist companies already on the scene of a working fire within a three-story commercial building. The incident commander has reported an explosion on the second floor with partial collapse of the structure. The regional urban search and rescue team, eight fire companies, and dozens of ambulances are on the way. Tower 5 is assigned to the Rescue Group in Division B, working with other greater alarm companies assigned to search the collapsed area for victims.

1. What are the common causes of fire fighter deaths and injuries on the fire ground?
2. Whose responsibility is it to conduct a risk/benefit analysis, and what are the key elements of that process?
3. What are the primary responsibilities of a rapid intervention crew?
4. What are the primary responsibilities of an incident safety officer?
5. What are the typical tasks expected of an incident safety officer?

Introduction to Safety

Fire department operations often include high-risk situations that can occur under any weather conditions at any time of the day or night. The fire officer is responsible for ensuring that every fire fighter completes every incident without serious injury, disability, or death. This is expressed as a fire officer's special obligation to ensure that "everyone goes home" at the end of a workday.

Just because firefighting is a dangerous occupation does not make death or disability acceptable. Fire officers have a personal and professional obligation to prevent deaths and injuries to the fire fighters who are working under their supervision.

The responsibilities of a fire officer include identifying hazards and mitigating dangerous conditions in order to provide a safe work environment for fire fighters. The fire officer must also identify and correct behaviors that could lead to a fire fighter's injury or death. The officer should set a good example because the crew members will follow their officer's lead. Safe practices must be the only acceptable behavior, and good safety habits should be incorporated into all activities.

Fire Fighter Death and Injury Trends

The fire officer develops an **incident action plan** that addresses and minimizes the chances of harm by identifying and controlling the factors that lead to fire fighter injury or death. Fire fighters must be fully prepared to work safely in high-risk situations.

Understanding the causes of fire fighter deaths and injuries is the first step in developing an incident action plan. Prevention depends on the ability to break the cascade of events that leads to a serious injury or death. The National Fire Protection Association (NFPA), the U.S. Fire Administration (USFA), the National Institute for Occupational Safety and Health (NIOSH), and the International Association of Fire Fighters (IAFF) publish reports and statistical analyses that provide information about the causes and circumstances of these events.

In June 2007, the NFPA Fire Analysis and Research Division published a 30-year retrospective analysis of fire fighter fatalities. It notes that the number of fire fighters dying on duty has declined 30 percent, from an average of 151 deaths annually in the 1970s to an average of 99 deaths per year in the 2000s. The two primary drivers of this decline are the reduction of the number of fire fighters dying of sudden cardiac death and the decline in the number of structure fires. However, the rate of fire fighters dying while operating within structure fires is higher than it was in the 1970s. With the significant improvements in protective clothing, breathing apparatus, training programs, and incident management, fire fighter safety within burning structures is an area that needs focused attention from the fire officer.

The U.S. Fire Administration, through the National Fallen Firefighters Foundation (NFFF), has adopted a goal of 50 percent reduction in fatalities within a decade. The U.S. Fire Administration study noted that during the 1990s, the number of structure fires declined significantly, whereas the number of fire fighter fatalities increased. The rate of fire fighter fatalities per 100,000

Getting It Done

Introduction to Developing an Incident Action Plan

An incident action plan (IAP) provides a concise, coherent means of capturing and communicating the overall incident priorities, objectives, and strategies in the contexts of both operational and support activities. Every incident must have an action plan.

Most initial response operations are not captured with a written IAP. However, if an incident involves hazardous materials, or is likely to extend beyond one **operational period**, become more complex, or involve multiple jurisdictions and/or agencies, a written IAP will be required to maintain effective, efficient, and safe operations.

The basic elements of a single unit incident action plan are the following:

- Specifies the incident objectives
- States the activities to be completed
- Covers a specified time frame, known as an operational period

Consider a single fire company response to an activated automatic fire alarm. The company officer verbalizes the incident action plan. The incident objectives are to first investigate the source of the alarm and then restore the alarm system.

The activities are for the crew to proceed to the alarm panel to identify the location of the alarm. With one radio-equipped member remaining at the alarm panel, a crew of two, wearing full personal protective equipment (PPE) and carrying a radio and hand tools, will proceed to the location of the activated alarm. The crew will report back their findings when they get to the activated device. If the alarm is not due to an incipient fire, the member at the alarm panel will attempt to reset the alarm system.

being overwhelmed by a flashover or becoming trapped due to structural collapse **Figure 6-1 ▼**.

When the NFPA released its 2007 annual report on fire fighter fatalities, it noted that of the 89 fire fighters who died in 2006, 38 died on the fire ground, the land or building where a fire occurs. The data trend indicates that fewer fire fighters are dying on the fire ground, both in actual numbers and in relative terms; however, more are being killed while responding to emergency incidents or performing duties other than fighting fires **Figure 6-2 ▶**. Each time a company rolls out of the station responding to a call, the fire fighters are putting their lives at risk.

NFPA estimated that 80,100 fire fighters were injured in the line of duty in 2007. Nearly half occurred during fire-ground operations. The most frequent injury was strain or sprain (45.1 percent), followed by bleeding or bruises (18.2 percent). Burns and smoke inhalation each counted for about 6 percent of the injuries. The balance of the injuries were split between nonfire emergency incidents and other on-duty activities, such as training, physical fitness, cleaning, and maintenance.

■ Sudden Cardiac Arrest

Heart attacks are the leading cause of fire fighter line-of-duty deaths, accounting for 44 percent of the deaths from 1995 to 2004. As an occupational group, fire fighters are more likely to die of a heart attack while on duty than other U.S. workers. Although the term *occupation* is used to refer to fire fighters, the category includes a full range of affiliations, including career, volunteer, contract, and wildland fire fighters.

This heart attack rate for fire fighters is related to the nature of the work, which can suddenly change from low activity to an episode of high stress and intense exertion. This type of sudden stress can trigger a heart attack, particularly in an individual who is predisposed to cardiovascular disease. Security guards also have a high rate of heart attacks, which

fire events rose by 25 percent in the 1990s. The primary cause of death was asphyxiation while still in the burning building. The root cause was running out of self-contained breathing apparatus (SCBA) air, sometimes because the fire fighter was trapped by debris. In this decade the most common causes of death are

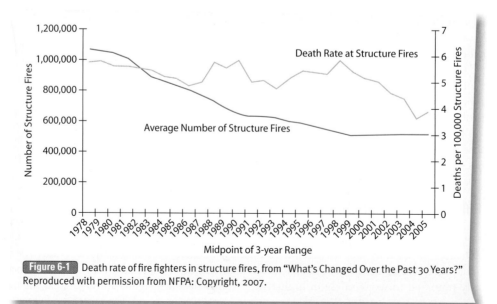

Figure 6-1 Death rate of fire fighters in structure fires, from "What's Changed Over the Past 30 Years?" Reproduced with permission from NFPA: Copyright, 2007.

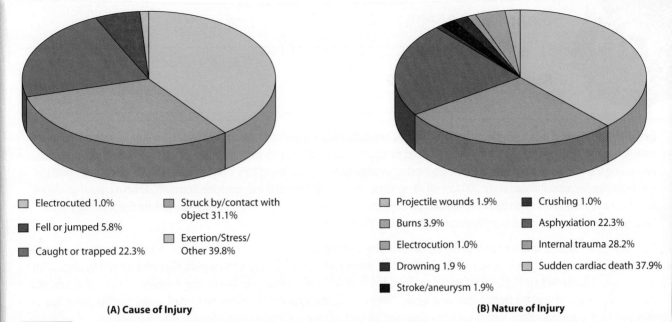

Electrocuted 1.0%

Struck by/contact with object 31.1%

Fell or jumped 5.8%

Caught or trapped 22.3%

Exertion/Stress/ Other 39.8%

Projectile wounds 1.9%

Crushing 1.0%

Burns 3.9%

Asphyxiation 22.3%

Electrocution 1.0%

Internal trauma 28.2%

Drowning 1.9 %

Sudden cardiac death 37.9%

Stroke/aneurysm 1.9%

(A) Cause of Injury

(B) Nature of Injury

Figure 6-2 Fire fighter deaths by cause (A) and nature (B) of injury in 2007. Adapted from "The U.S. Fire Service: Firefighter deaths by cause and nature of injury (2008)," www.nfpa.org.

represent 25 percent of their on-duty deaths. Construction workers, who routinely perform strenuous physical labor, and police officers, who experience similar variations between low-intensity and high-stress periods, both have lower heart attack rates than fire fighters. Among workers in all occupational categories, only 15 percent of on-duty deaths are caused by heart attacks.

Regular medical examinations and physical fitness programs are the most significant factors in preventing heart attacks. The data accumulated by the NFPA included the postmortem or medical history of 713 of the 1117 sudden cardiac arrest deaths that occurred in a 25-year period. The data showed that 84.6 percent had suffered prior heart attacks, had severe arteriosclerotic heart disease, had undergone bypass surgery or angioplasty/stent placement, or were diabetic.

Struck by or Contact with an Object

Traumatic injuries are the second leading cause of fire fighter fatalities. Most of these deaths are caused by vehicle accidents, falls, or structural collapse of a burning building. Trauma deaths resulting from motor vehicle collisions represented 22 percent of annual line-of-duty deaths. The most common fatal motor vehicle collision scenario involves a fire fighter responding to an emergency incident in a personal vehicle. The second most common fatal collision is a tanker, or water tender, rollover. Tankers have a higher and more disproportionate fatality rate than any other type of fire apparatus.

Fire Marks

National Fallen Firefighters Memorial Weekend
The National Fire Academy in Emmitsburg, Maryland, has hosted the National Fallen Firefighters Memorial Weekend every year since 1981. This is a national event to honor and remember the fire fighters who died in the line of duty during the previous year. The memorial lists the names of all of the fire fighters who have died in the line of duty each year.

The memorial service, like the image of a fire fighter's funeral procession, is not easily forgotten. A fallen fire fighter is honored as a hero, buried with great pomp and circumstance, and long remembered on plaques and memorials. Although we will always honor our fallen comrades, we must also remember that most fire fighter fatalities can be prevented. The fire service has a tremendous responsibility to its own members, to their families and friends, and to our communities to protect the lives of all fire fighters.

Whenever a fire fighter is lost, a painful and lasting impact is felt by many individuals, beginning with the members of the fire fighter's immediate family. A spouse must pick up a shattered life and try to make sense of the loss. Children will grow up without their father or mother. Parents have to bury a son or daughter. Personal friends and co-workers of the fallen fire fighter feel a similar type of loss. The loss is also felt throughout the fire department or agency and throughout the fire service whenever a fire fighter is killed or seriously injured.

A fire officer who experiences the loss of a member working under his or her supervision often feels a sense of responsibility for having allowed the loss to occur. This is one of the most difficult situations that a fire officer can experience.

Near Miss REPORT

Report Number: 07-0000759

Event Description: A motor vehicle collision with heavy extrication at a major intersection in town was in progress, with our main patient requiring air transport due to injuries sustained. The scene had been secured, with multiple agencies involved in blocking and redirecting traffic. Scene perimeters were clearly blocked by engines and police squad cars with emergency lights going. The helicopter had begun its descent, and was about 200 feet off the ground, when a red blazer broke through the perimeter. Driving around an engine, the driver was motioned to stop by two firefighters on the scene. The driver hit the two firefighters, drove around the advancing ambulance with the patient inside, and proceeded into the landing zone.

Someone gave the order to abort, but the pilot either did not hear, or was on the wrong channel. The vehicle drove under the helicopter's rear rotor, while the helicopter was approximately 15 feet off the ground. Upon speaking with the pilot after the incident, he stated that he only saw the vehicle's tail lights off his passenger side. The pilot did then communicate that he saw a vehicle in the landing zone and that he could not land until the landing zone was secure. [During incident debriefing] the air transport staff stated that the scene was very well lit, the communication was smooth and clear, and the landing zone was clearly marked and in perfect position.

Lesson Learned: Who could have foreseen this driver barreling through the perimeter and ignoring signals to stop? The only thing we saw as preventable was the fact that the pilot did not hear the order to abort. He stated that he did not receive the message, but it was clearly broadcast. It is unknown why the message did not get through. Fortunately, this particular pilot said he has a habit of stopping in the middle of his descent for just a moment. If not for his habit, the damage from such a crash would be huge.

■ Caught or Trapped

The third most frequent category of fire fighter fatalities includes asphyxiation and burns. In these situations, the fire itself is the direct cause of the fire fighter's death. The fatal injury can involve either burns or asphyxiation or a combination of both.

The risks of respiratory injuries and burns are inherent in a fire fighter's work environment. Firefighting is routinely performed in situations that involve high temperatures and contaminated atmospheres. Protective clothing and SCBA are designed to reduce the risk of injury from these causes; however, their effectiveness is limited. When the capabilities of the protective ensemble are exceeded, a serious or fatal injury can occur quickly.

Fire Marks

FDNY Deaths in Structure Fires: 1994 to 2008
While commanding Division 3 in mid-town Manhattan, Deputy Chief Vincent Dunn wrote his first book, *Collapse of Burning Buildings*, in 1988. The book was dedicated to "... the forty-six FDNY chiefs, company officers and fire fighters who have been killed by burning buildings which collapsed 1956–1986." Dunn provides an analysis on the 28 Fire Department of New York (FDNY) members who died in burning buildings from 1994 to 2008:

- Seventy-five percent of them were members of ladder companies. Three were from rescue companies and four were from engine companies.
- There were 17 fire fighters, 7 lieutenants, and 4 captains. Eighty-six percent of the deaths occurred while searching the structure. Four fire fighters were killed while operating a hose line.
- Multiple-family dwellings counted for 15 of the fatalities, followed by 8 commercial structures, and 3 vacant and 2 single-family homes. Thirteen were killed operating above the fire floor, 10 on the fire floor, and 3 in the basement.
- There were more deaths during the fire growth stage (64 percent) than when the fire was fully developed.

■ Everyone Goes Home

Everyone Goes Home is a program developed by the National Fallen Firefighters Foundation (NFFF) to prevent line-of-duty death and injuries. The NFFF held a Firefighter Life Safety Summit in 2004 that resulted in 16 initiatives:

1. Define and advocate the need for a cultural change within the fire service relating to safety, incorporating leadership, management, supervision, accountability, and personal responsibility.
2. Enhance the personal and organizational accountability for health and safety throughout the fire service.
3. Focus greater attention on the integration of risk management with incident management at all levels, including strategic, tactical, and planning responsibilities.
4. All fire fighters must be empowered to stop unsafe practices.
5. Develop and implement national standards for training, qualifications, and certification (including regular recertification) that are equally applicable to all fire fighters based on the duties they are expected to perform.

6. Develop and implement national medical and physical fitness standards that are equally applicable to all fire fighters based on the duties they are expected to perform.

7. Create a national research agenda and data collection system that relates to the initiatives.

8. Utilize available technology wherever it can produce higher levels of health and safety.

9. Thoroughly investigate all fire fighter fatalities, injuries, and near misses.

10. Grant programs should support the implementation of safe practices and/or mandate safe practices as an eligibility requirement.

11. National standards for emergency response policies and procedures should be developed and championed.

12. National protocols for response to violent incidents should be developed and championed.

13. Fire fighters and their families must have access to counseling and psychological support.

14. Public education must receive more resources and be championed as a critical fire and life safety program.

15. Advocacy must be strengthened for the enforcement of codes and the installation of home fire sprinklers.

16. Safety must be a primary consideration in the design of apparatus and equipment.

A follow-up summit was held in 2007 to develop key recommendations for each of the initiatives. A standard approach to safety incorporates best practices that should be part of every operational situation.

Fire fighters must work in teams at emergency incidents, and fire officers must maintain accountability at all times for the location and function of all members working under their supervision. Every company must operate within the parameters of an incident action plan, under the direction of the incident commander.

Reliable two-way communications must be maintained through the chain of command. Adequate back-up lines must be in place to ensure that a safe exit path is maintained for crews working inside a fire building and any sudden flare-ups can be controlled. Rapid intervention crews must be established to provide immediate assistance if any fire fighter is in danger. Air supplies must be monitored to ensure that fire fighters leave the hazardous area before their low air alarm activates and they run out of air. All fire fighters must watch for indications of impending building collapse.

National Fire Fighter Near-Miss Reporting System

In August 2005, the International Association of Fire Chiefs (IAFC) launched a web-based system to report near-misses. The goal of http://www.firefighternearmiss.com is to track incidents that

Getting It Done

Habits to Improve Safety
A fire officer can demonstrate three simple habits to improve the safety of his or her crew:
- Be physically fit.
- Wear seat belts.
- Maintain fire company integrity at emergency incidents.

avoided serious injury or death, to identify trends, and to share the information with other fire fighters in a confidential and nonpunitive way. The program is based on the Aviation Safety Reporting System, which has been gathering reports of close calls from pilots, air traffic controllers, and flight attendants since 1976.

Examples of near-miss reports are provided throughout this book. Every August, a working group of fire fighters and officers are assembled to analyze the reports using a tool modified from the U.S. Navy's Human Factors Analysis and Classification System (HFACS). The results of the analysis are provided in an annual report.

HFACS Level 1: Unsafe Acts

This level includes two categories: errors and violations. Errors are considered unintentional and can be based on decisions, skills, or perceptions. Decision-based errors include flaws in communicating information to decision makers. Skill-based errors include attention failure (lack of situational awareness), memory failure (forgotten or missed step in a procedure), or technique failure (lack of training). Perception-based errors may include visual illusions.

Violations are considered intentional and are classified as either routine or exceptional. Routine violations include failure to use safety equipment, failure to follow recommended tactile best practice (sounding the floor before entering), or failure to follow recommended cerebral best practices (conducting a risk/benefit analysis). Exceptional violations include not being qualified to perform an action.

HFACS Level 2: Preconditions to Unsafe Acts

This level analyzes substandard conditions and practices of the individuals involved. Substandard conditions include factors contributing to adverse mental states, psychological states, and physical limitations. Substandard practices include failure to use elements of crew resource management and personal readiness.

HFACS Level 3: Unsafe Supervision

Unsafe supervision is broken down into four categories: inadequate supervision, allowing inappropriate operations, failure to correct known problems, and supervisory violations. This level is intended to examine the role (if any) of supervision in a near-miss event Table 6-1 .

Table 6-1 HFACS Level 3: Unsafe Supervision Factors			
Inadequate Supervision	**Allowing Inappropriate Operations**	**Failure to Correct Known Problems**	**Supervisory Violations**
• Guidance not provided	• Personnel not adequately briefed	• Unidentified and unqualified	• Authorization of unnecessary hazards
• Training not provided	• Understaffed	• Failure to provide training to unqualified personnel	• Failure to enforce department rules
• Qualifications not tracked	• Unnecessary hazards permitted	• Failure to correct inappropriate or unsafe behaviors	• Authorization of unqualified personnel to perform tasks

Source: 2007 Near Miss Annual Report. Reprinted with permission of Elsevier Public Safety, Copyright 2008.

HFACS Level 4: Organizational Influences

This is the most difficult level to analyze in a near-miss report. The operating culture of the fire department is often as significant to the near miss as the individual's action. This level examines resource management (staffing, training, budget resources, and equipment/facility resources) and organizational climate (chain of command, delegation of authority, risk management programs, and safety programs).

Reducing Deaths from Sudden Cardiac Arrest

Heart attacks are the leading cause of death for fire fighters. Although most line-of-duty fatalities for fire fighters under the age of 35 years are from traumatic injuries, fire fighters above the age of 35 years are more likely to die from a medical cause. The NFFF Life Safety Summit noted that a disproportionate number of fire fighters over the age of 49 die of cardiac arrest while on duty.

Before authorization to begin training as a fire fighter, every candidate should be examined by a qualified physician. Regular physical examinations should be scheduled for as long as the fire fighter is engaged in performing emergency duties, with an emphasis on identifying risk factors that could lead to a heart attack under stressful conditions. Fire officers should always look for indications that a member is in poor health or is unfit for duty and, if necessary, arrange for a special evaluation by a fire department–approved physician.

Physical fitness activities should be considered an essential component of every fire fighter's training regimen. Changes in lifestyle can often reduce the risk of a fatal heart attack and can include stopping smoking, lowering high blood pressure, reducing high blood cholesterol level, maintaining a healthy weight, and managing diabetes. Fire officers must understand that these changes are as important to the body as wearing full protective equipment.

The NFPA, the IAFC, the NFFF, the National Volunteer Fire Council (NVFC), and the IAFF have developed resources to help the fire officer encourage healthy living. Fire officers should advocate methods of making positive lifestyle changes to increase fire fighter safety. NFPA 1583, *Standard on Health-Related Fitness Programs for Fire Department Members*, provides a structure and resources for a fire officer to develop a health-related fitness program. The International Association of Fire Chiefs partnered with the IAFF to develop the Fire Service Joint Labor Management

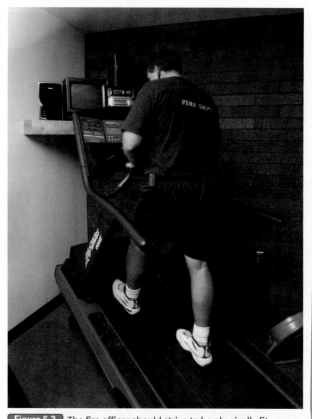

Figure 6-3 The fire officer should strive to be physically fit.

Wellness Fitness Initiative. This initiative produced the Candidate Physical Aptitude Test, as well as a peer fitness training certificate program with the American Council on Exercise.

Regardless of what the fire department mandates, every fire officer should strive to be physically fit **Figure 6-3**. Fitness should be a personal priority and an expression of leadership.

Reducing Deaths from Motor Vehicle Collisions

Collisions account for the largest percentage of traumatic fire fighter deaths. The NFPA data from 1977 to 2007 show that three-quarters of the 406 fatalities from collisions were volunteer fire fighters. Almost 40 percent of the fire fighters died in their personal vehicles. More than three-quarters of the fatalities were not wearing a seat belt. Excessive speed for road conditions and operator error are frequently cited causes of these fatal collisions. Obeying traffic laws, using seat belts, driving sober, and controlling speed would prevent most of the fire fighter fatalities in road collisions.

Placing an untrained driver behind the wheel of an emergency vehicle places the occupants and the general public in immediate danger. The driver of an emergency vehicle is legally authorized to ignore certain restrictions that apply to other vehicles, but only when operating the emergency vehicle in a manner that provides for the safety of everyone using the roadways. Specific procedures for safe emergency response must be practiced. The fire officer is responsible for ensuring that the driver consistently follows the rules of the road for emergency response.

Safety Zone

Physical Fitness Tips

The NFFF summarizes the importance of physical fitness with the following items that support fire fighter maintenance:

- Regular medical check-ups: Yes, they can be a pain, but if you do not do it for you, do it for those who need you.
- Regular exercise: Even walking makes a *big* difference! Walk a mile a day and watch the changes.
- Eat healthy: Think about what you are eating, and then picture operating interior at a working fire 30 minutes later. Now, what do you want to eat?

Driver minimum qualifications are established in NFPA 1002, *Standard for Fire Apparatus Driver/Operator Professional Qualifications*, and NPFA 1451, *Standard for a Fire Service Vehicle Operations Training Program*. The fire officer should set high expectations for driver training and performance. The apparatus operator should be required to have nonemergency driving experience with a specific piece of apparatus before being assigned to emergency response driving duties. There are tremendous differences between the handling characteristics of fire apparatus versus automobiles or light trucks. Stopping distances are directly related to the weight of the vehicle. Heavy fire apparatus can turn over easily in a collision. Driving a ladder truck is vastly different from driving an engine, and a tanker has a different set of handling characteristics from either of these. All of the necessary skills must be learned and practiced under controlled conditions before an operator is qualified to drive under emergency response conditions.

Requiring fire fighters to wear seat belts is a simple requirement that could prevent 10 to 15 fatalities every year. Many of the fire fighters killed in motor vehicle crashes are thrown from the vehicle. A fire officer who allows company members to respond unrestrained has no excuse if a fatal accident results.

The mandatory use of seat belts by fire fighters, like many other safety procedures, may require a change in the culture. A fire officer must be prepared to accept the responsibility and provide the leadership to bring about positive changes.

Reducing Deaths from Fire Suppression Operations

The hazardous nature of fire suppression operations is easily recognized. Asphyxiation or burns are the prime factors in the direct cause of death while operating in burning buildings. In spite of protective clothing and SCBA, remember that fire fighters routinely operate in situations where a problem, a procedural error, or an equipment failure could result in a fatality. The fire officer must follow standard safety practices.

Fire officers must fully understand all local policies and procedures to guide their actions. The established standard operating procedures (SOPs) and safety practices should always be followed in situations that meet their criteria. If a situation does not fall under an established SOP, the fire officer must determine an appropriate course of action and give specific directives to subordinates to clearly indicate how the situation is to be handled. In these situations, the fire officer must provide an even greater level of supervision to ensure that the crew members understand and follow the plan. The fire officer must always be prepared for changing conditions and unanticipated hazards.

Getting It Done

Vehicle Safety Tips
The NFFF summarizes the issue of vehicle safety with the following items that support the driver response plan:
- It's not a race.
- Safe is more important than fast.
- Stop at red lights and stop signs! *No Excuses!*
- If they do not get out of your way, do not run them over! *Think* and *react carefully!*

Maintaining Crew Integrity

A consistent challenge for every fire officer is to maintain the integrity of the fire company under his or her command while operating at an emergency incident. The officer must know the location and function of every crew member at all times. The U.S. Fire Administration retrospective study shows that 82 percent of the fatal fire suppression incidents involved the death of a single fire fighter. Many of these fire fighters became lost or disoriented and died before the fire officer or incident commander was aware that a fire fighter needed help.

NIOSH investigates fire fighter line-of-duty deaths as part of its Fatality Assessment and Control Evaluation program. The lack of an effective incident management system that includes a fire fighter accountability component is a common scenario in many of the single fire fighter deaths investigated by NIOSH.

Operating in an IDLH Environment

The Occupational Safety and Health Administration (OSHA) establishes federal workplace safety regulations in the United States, including 29 CFR 1910.134 (Respiratory Protection), which applies to the use of SCBA by the fire service. This rule applies when members are operating in an environment that is **immediately dangerous to life and health (IDLH)**. NFPA defines IDLH as any condition that would: (1) pose an immediate or delayed threat to life, (2) cause irreversible adverse health effects, or (3) interfere with an individual's ability to escape unaided from a hazardous environment. The interior of a fire building, where fire fighters are using SCBA, is considered to be an IDLH atmosphere.

The OSHA regulation establishes specific requirements for fire fighters operating in an IDLH environment. This is known as the "two-in, two-out rule" and includes:
- A designated officer-in-charge
- Two fire fighters who enter the IDLH area together and remain in visual or voice contact with one another at all times, while wearing SCBA Figure 6-4 ▾
- Two properly equipped and trained fire fighters who must:
  - Be positioned outside the IDLH environment
  - Account for the interior team
  - Remain capable of rescue of the interior team

Figure 6-4 Two fire fighters should enter an area together and remain in visual or voice contact with one another at all times, while wearing self-contained breathing apparatus.

The two-in, two-out rule leads to the requirement of a **rapid intervention crew (RIC)**. NFPA 1710, *Standard for the Organization and Deployment of Fire Suppression Operations, Emergency Medical Operations and Special Operations to the Public by Career Fire Departments* defines RIC as "A dedicated crew of firefighters who are assigned for rapid deployment to rescue lost or trapped members." Most fire departments add an additional fire company to the first alarm assignment for a structure fire, either as part of the initial dispatch or as soon as a working fire is reported to function as the four person RIC.

To comply with OSHA regulations, an initial rapid intervention crew (IRIC) is required to be assembled prior to operations within IDLH environments. Two fire fighters from the initial attack crew will be assigned as the IRIC until the RIC company is deployed at the incident. Both the IRIC and RIC members should meet the requirements of NFPA 1407, *Standard for Training Fire Service Rapid Intervention Crews*.

In order to ensure that the RIC can recognize in a timely manner that a fire fighter is missing, a **personnel accountability system** is needed to track the identity, assignment, and location of all fire fighters operating at the incident scene **Figure 6-5 ▼**. An accountability system typically provides the following:

- A method to identify all personnel that are on the scene
- A method to identify personnel that are in the hazard area
- A process to quickly account for personnel when an evacuation order or unusual event occurs (e.g., a flashover, backdraft, or collapse)
- A process to ensure that personnel do not go unaccounted for an extended period of time
- A standardized method of notification and reaction to a may-day event

An accountability system must function at multiple levels at an incident scene. Although every fire officer must be accountable for all assigned company members, command officers are accountable for the location and function of full companies. An officer who is assigned to manage a sector for a division at an incident must know where every assigned company is operating and what they are doing at all times. The incident commander must know which companies are assigned to each sector or division.

Figure 6-5 A personnel accountability system.

For an effective accountability system, crews must routinely receive training on and follow the procedures. During an emergency situation, fire fighters will revert to habit. The proper use of an accountability system must be fully integrated as one of those habits.

A personnel accountability system is required by NFPA 1500, *Standard on Fire Department Occupational Safety and Health Program*. Some departments require a personnel accountability report (PAR) roll-call every 20 minutes during fire attack or other high-hazard operations. Each sector chief or company officer must account for the personnel operating under his or her command. It is also used after a sudden change in operating conditions, after a flashover or collapse, or when moving from offensive to defensive fire attack.

Air Management

Self-contained breathing apparatus (SCBA) provide a safe and reliable air supply that allows a fire fighter to operate in an IDLH atmosphere for a finite length of time. The length of time depends on the rate at which the individual consumes the available air supply, which varies considerably, depending on the user and the nature of the work being performed. The rated time is based on a standard consumption rate for an individual at rest and does not reflect typical experience in firefighting operations. Running out of air in an IDLH atmosphere can result in an unconscious fire fighter who will die of asphyxiation.

Some fire departments have re-evaluated how they track SCBA use during incidents. Some departments have replaced their 30-minute rated SCBA air tanks with 45- or 60-minute air tanks to increase the time a fire fighter can operate in IDLH atmospheres. Other departments have established or strengthened incident management procedures to more closely follow individual fire fighter air use.

Low-pressure warning devices, which provide an indication when the remaining air supply reaches a set point, are effective only if the fire fighter is able to exit to a safe atmosphere in the time that is provided. The low-pressure alarm might not provide sufficient exit time for a fire fighter who is deep inside a burning building or unable to exit immediately.

Many departments monitor the air levels of fire fighters on entry and exit. If a fire fighter fails to exit within the expected time period, an accountability roll-call is taken.

Teams and Tools

The minimum size of an interior work team is two fire fighters. Every team member must have full personal protective equipment,

Fire Marks

Rapid Intervention Is Not Rapid

Two fire fighters became disoriented when the Phoenix Fire Department moved from offensive to defensive operations at a supermarket fire in 2001. The incident commander sent two companies into the building as the RIC. Despite heroic efforts by the on-scene crews, one of the disoriented fire fighters died in the building. Twelve fire fighters involved in the rescue effort ran into problems and declared their own may-days during the 53 minutes it took to remove Fire Fighter Tarver from the building.

The Phoenix Fire Department embarked on a comprehensive recovery effort to learn what must be done to avoid another fire fighter death in a commercial structure fire. Assistant Chief Stephen Kreis coordinated the evaluation of the rapid intervention effort. To validate the proposed changes in fire fighter rapid intervention and rescue procedure, Phoenix conducted over 200 multiple company drills in three large single-story commercial buildings. Arizona State University was a partner, with Dr. Ron Perry making sure that the drills were valid for academic research purposes.

The training drill scenario was similar to the supermarket fire and designed to reinforce RIC procedures. Two fire fighters would be located within 100' (30.5 meters) of a building exit and not be entrapped in any way. The fire attack hose line was not obstructed by debris or convoluted in any way. There was no smoke, heat, or fire in the drill building. One fire fighter was alert, separated from his partner, and running out of air. The other fire fighter was unconscious, about 40' (12 meters) away from the first fire fighter with a Personal Alert Safety System (PASS) alarm activated.

The vision through participants' SCBA facepieces was obscured by a laminated piece of window tint fitted within the mask. Depending on ambient conditions and handlights, the participants could see 5' to 20' (1.5 to 6 meters) ahead of them. A recording with fire-ground sounds was played during the rescue effort. At that time there were few thermal image cameras in Phoenix and they were not used in these drills.

The drill objectives were to (1) find the alert fire fighter, assess level of air, and transfill SCBA air (buddy breath) while assisting the fire fighter out of the building; and (2) find, transfill SCBA air, and remove the unconscious fire fighter. The results from the 2002 drills showed that it takes 21 minutes to rescue a downed fire fighter using a team of 12 fire fighters.

During the rescue effort, 20 percent of the RIC team will get into some trouble themselves, usually running out of air. A 3000-psi SCBA bottle in the Phoenix drill provided 18.7 minutes of air, plus or minus 30 percent. In this straightforward and noncomplex scenario, the RIC members were running out of air before completing the removal of a downed fire fighter. Add stairs, heat, adrenaline, or entrapment and you can speculate that most of the RIC team will have profound difficulties in rescuing and removing a downed fire fighter.

Figure 6-6 Thermal imaging devices are a critical component of a scene safety program.

to navigate, locate victims, evaluate fire conditions, locate hazards, and find escape routes in smoke-filled buildings or total darkness **Figure 6-6 ▲** .

NIOSH line-of-duty death investigations show that some departments with thermal imaging devices did not use them. The fire fighters did not know how to use the imager, the device was inoperable, or the device was not used. Thirty years ago, the same observation could have been made for breathing apparatus, when SCBAs were left sitting in the apparatus compartments while fire fighters made interior attacks. The fire officer must provide frequent training to integrate the thermal imager into interior firefighting operations.

Situation Awareness

Every fire officer is expected to maintain a continual connection between the functions being performed by the company and the overall situation. This is one of the most important reasons for establishing and maintaining an effective incident command structure at every incident. The fire officer maintains situation awareness by staying oriented, making observations, providing and receiving regular updates within the incident command system, listening to fire-ground radio communications, and continually assessing the risk/benefit model.

It is easy to become distracted or preoccupied with a particular task or function and to lose track of a larger situation that is occurring and changing simultaneously. This can be a problem at a dynamic emergency incident, particularly for an officer who is supervising a crew that is performing a complex task deep inside a smoke-filled building. Conditions can change quickly, and crews might have no way of observing what is going on around them.

Risk/Benefit Analysis

A **risk/benefit analysis** is based on a hazard and situation assessment that weighs the risks involved in a particular course of action against the benefits to be gained from taking those risks. Fire fighters sometimes use the axiom, "Risk a little to save a little. Risk a lot to save a lot." However, this is a simplified description

including SCBA and a Personal Alert Safety System (PASS) device unit. At least one member of every team should have a radio.

Every interior operating team should also have a thermal imaging device, which is a critical component of a comprehensive safety program. The thermal imaging device allows crews of the risk/benefit process. Risk/benefit analysis must always be approached in a structured and measured manner.

Life safety is a paramount goal, including the lives of fire fighters. If the situation requires placing fire fighters in extreme danger with little chance of success, the operation should not be undertaken. There is no justification for risking the lives of fire fighters to save property that is already lost or has no real value. Conducting an interior attack on a fire in an unoccupied abandoned building needlessly places fire fighters at risk. Similarly, entering a burning building to search for missing occupants who could not possibly be alive cannot be justified.

Every interior fire attack operation and many other types of situations and circumstances expose fire fighters to a set of unavoidable inherent risks. These risks are managed within a measured and controlled system that is based on training, coordination, and the use of protective clothing and equipment. The nature of the mission requires fire fighters to be able to work safely in situations that are inherently dangerous. Fire officers are responsible for keeping this system in balance.

The only situation that truly justifies exposing fire fighters to a high level of risk is one where there is a realistic chance that a life can be saved. Even in this circumstance, fire fighters must use all of the resources at their disposal to limit the risks. All of the fire department's training, equipment, and systems are designed to enable fire fighters to be effective in the face of such challenging and dangerous situations.

The fire officer starts the risk/benefit analysis by preparing a **pre-incident plan**, which is a written document that provides information that can be used by responding personnel to determine the appropriate actions in the event of an emergency at a specific facility. Building construction, occupancy, use, contents, and condition are all factors that should be used to develop the risk/benefit analysis.

When operating at an emergency incident, the fire officer reviews the pre-incident plan and makes observations about current conditions. These two inputs are combined to produce an incident action plan, which is developed by the incident commander and incorporates the overall incident strategy, tactics, risk management evaluation, and organization structure for that particular situation. Incident action plans are updated throughout the incident, based on updates from operating crews and observations by the incident management team.

Company Tips

Situational Awareness

The NFFF summarizes situational awareness with the following items that support the interior firefighting plan:

- Work as a team.
- Stay together.
- Stay oriented.
- Manage your air supply.
- Get off the apparatus with tools and a thermal imager for *every* interior operating team.
- Provide a radio for *every* member.
- Provide regular updates.
- Constantly assess the risk/benefit model.

Incident Safety Officer

An **incident safety officer** is a designated individual at the emergency scene who performs a set of duties and responsibilities that are specified in NFPA 1521, *Standard for Fire Department Safety Officer*. The incident safety officer functions as a member of the incident command staff, reporting directly to the incident commander. The incident commander is personally responsible for performing the functions of the incident safety officer if this assignment has not been assigned or delegated to another individual.

Many fire departments assign a designated officer to respond to emergency scenes to fill this position. In the absence of a designated safety officer, this position may be assigned to a qualified individual by the incident commander. NFPA 1521 also specifies that the fire department must have a standard operating procedure to define the criteria for the response or appointment of an incident safety officer. The same principles should be applied to any situation, whether a designated safety officer has been dispatched or an officer has been assigned to this function by the incident commander.

The fact that a safety officer has been assigned does not relieve any officer or fire fighter of the responsibility to operate safely and responsibly. The incident safety officer is an additional resource to ensure that the safety priorities of the situation are being addressed. Every fire officer shares the responsibility to act as a safety officer within his or her scope of operations.

Incident Safety Officer and Incident Management

The incident safety officer is a key component of the incident management system. The **incident command system (ICS)** is the standard organizational structure that is used to manage assigned resources in order to accomplish stated objectives for an incident. Every incident requires someone, known as the incident commander, to be in charge at all times to coordinate resources, strategies, and tactics. This begins with the initial arriving officer, who functions as the incident commander until he or she is relieved by a higher-ranking officer. The incident management structure can become larger and more complex, depending on the nature and the magnitude of the incident.

The incident safety officer reports to the incident commander. The incident safety officer is required to monitor the scene, identify and report **hazards** (the potential for harm to people, property, or the environment) to the incident commander, and if necessary, take immediate steps to stop unsafe actions and ensure that the department's safety policies are followed. In most situations, the exchange of information between the incident commander and the incident safety officer is conducted verbally and quickly at the command post.

Safety officers have specific authority and a special set of responsibilities under the incident command system. One of the primary responsibilities of an incident safety officer is to identify hazardous situations and dangerous conditions at an emergency incident and recommend appropriate safety measures. In most cases, the safety officer acts as an observer, monitoring conditions and actions and evaluating specific situations. When an unsafe condition is observed that does not present an imminent

Voices of Experience

Early on in my career as a fire fighter I was fortunate to participate in many working incidents at both fires and other emergencies. One call in particular that had a profound impact on me and the way I conduct myself today, more than a decade later, occurred at a historic site which through the years has been a hospital, school, orphanage, church, shelter, and convent. At the time of the fire it was in disrepair and unoccupied, awaiting a government-funded restoration.

We received the call at the end of a quiet night shift for bells ringing at the abandoned home. We had been to the building before, but only in the portion that was a modern add-on being utilized as a college.

We arrived on scene to find light smoke in the air and bells ringing in the occupied attached college building. As we made entry we found light smoke in the building and no access to the abandoned older side. Our captain began reconnaissance and determined that the smoke was coming from the abandoned portion of the building. We immediately began forcing entry to the historic building, awaiting our next due apparatus.

My captain informed us even before gaining entry that if there was a working fire, we were not to go offensive, but rather fall back and set up defensive positions to protect the college. We found this order to be rather inappropriate considering that we didn't know what we had, and that this was a prominent building in the community where we worked and lived.

Upon gaining entry we found heavy fire involvement on the interior corridors and were ordered not to enter. This order was questioned because we felt that we had an opportunity to aggressively attack this fire and save the building along with the attached exposure. Our captain explained that with this particular building and its history of renovation and lack of occupancy it would be too dangerous to enter and we would go defensive to protect the surrounding exposures. Also, because of less than optimal hydrant supply, we had only a small window to save the surrounding exposures.

> "My captain informed us even before gaining entry that if there was a working fire, we were not to go offensive."

We followed orders, feeling that we were going to lose a part of our civic history in order to play it safe, when the captain's predicted outcome began to occur. Something within the structure gave way and part of the roof was lost, the fire began to grow rapidly and vent from the roof area distal to where we had gained entry. Over the next 30 minutes, this became one of the largest fires in our city's history.

While feeling a level of guilt for losing a historic site, as a young fire fighter I was amazed that we were able to save the attached college and the surrounding structure with little or no damage, despite a complete loss of the old building.

If the first-arriving officer, a quiet, non-confrontational station captain, had not exercised his rank and made the decision that he did, not only would we have lost the structure and exposures, but certainly would have endangered the lives of the fire fighters on scene.

Command is not just a certification you can get by taking a course; it is a presence and state of mind that comes with experience and training, which is part of the burden of wearing the rank and insignia of officer.

Keith Stahl
Training Officer
Calgary Fire Department
Calgary, Alberta, Canada

danger, the safety officer consults with the incident commander and with other officers to determine a safe course of action.

If a situation creates an imminent hazard to personnel, the incident safety officer has the authority to immediately suspend or alter activities. When this special authority is exercised, the incident safety officer must immediately inform the incident commander of the hazardous situation and his or her actions. It is ultimately up to the incident commander to either approve or alter the action taken by the safety officer.

Qualifications to Operate as an Incident Safety Officer

NFPA 1521 outlines the criteria for an incident safety officer. Every fire officer should be trained to perform the basic duties of an incident safety officer and be prepared to act temporarily in this capacity if he or she is assigned by the incident commander.

According to NFPA 1521, the incident safety officer must be a fire department officer and as a minimum must meet the requirements for Fire Officer I as specified in NFPA 1021, *Standard for Fire Officer Professional Qualifications*. The incident safety officer also must be qualified to function in a sector officer position under the local incident management system.

The general knowledge requirements for an effective incident safety officer are as follows:

- Safety and health hazards involved in emergency operations
- Building construction
- Local fire department personnel accountability system
- Incident scene rehabilitation

Incident safety officers at a special operations incident require additional specialized knowledge and experience. Special operations are emergency incidents to which the fire department responds that require specific and advanced training and specialized tools and equipment, such as water rescue, extrication, confined space entry, hazardous materials situations, high-angle rescue, and aircraft rescue and firefighting. For example, the incident safety officer at a hazardous materials incident needs to have an advanced understanding of this situation, perhaps by training to the Hazardous Materials Technician level of NFPA 472, *Standard for Competence of Responders to Hazardous Materials/Weapons of Mass Destruction Incidents.*

In many fire departments, specialized teams have their own designated safety specialists who are trained to work directly with the incident management team.

Typical Incident Safety Officer Tasks

The specific duties an incident safety officer must perform at an incident depend on the nature of the situation. The following is a partial listing of functions that may need to be addressed at incidents:

- Ensure that safety zones, collapse zones, and other designated hazard areas are established, identified, and communicated to all members present on scene.
- Ensure that hot, warm, decontamination, and other zone designations are clearly marked and communicated to all members.
- Ensure that a rapid intervention crew is available and ready for deployment.
- Ensure that the personnel accountability system is being used.
- Evaluate traffic hazards and apparatus placement at roadway incidents.
- Monitor radio transmissions and stay alert to situations that could result in missed, unclear, or incomplete communication.
- Communicate to the incident commander the need for assistant incident safety officers because of the need, size, complexity, or duration of the incident.
- Immediately communicate any injury, illness, or exposure of personnel to the incident commander and ensure that emergency medical care is provided.
- Initiate accident investigation procedures and request assistance from the health and safety officer in the event of a serious injury, fatality, or other potentially harmful occurrence.
- Survey and evaluate the hazards associated with the designation of a landing zone and interface with helicopters.
- Ensure compliance with the department's infection control plan.
- Ensure that incident scene rehabilitation and critical incident stress management are provided as needed.
- Ensure that food, hygiene facilities, and any other special needs are provided for members at long-term operations.
- Attend strategic and tactical planning sessions and provide input on risk assessment and member safety.
- Ensure that a safety briefing, including an incident action plan and an incident safety plan, is developed and made available to all members on the scene.

Additional duties the incident safety officer must perform when fire has involved a building or buildings are as follows:

- Advise the incident commander of hazards, collapse potential, and any fire extension in such buildings.
- Evaluate visible smoke and fire conditions and advise the incident commander, tactical-level management units officers, and officers on the potential for flashover, backdraft, or any other fire event that could pose a threat to operating teams.
- Monitor the accessibility of entry and egress of structures and the effect it has on the safety of members conducting interior operations.

Assistant Incident Safety Officers at Large or Complex Incidents

Some incidents, based on their size, complexity, or duration, require more than one safety officer. Assistant incident safety officers can be assigned to subdivide responsibilities for different areas and functions at incidents such as high-rise fires, hazardous materials incidents, and special rescue operations. In these cases, the incident safety officer should inform the incident commander of the need to establish a **safety unit** as a component of the incident management organization. Under the overall direction of the incident safety officer, assistant incident safety officers can be assigned to various functions, such as scene monitoring, action planning and risk management, interior operations, or special operations teams. During extended incident operations, a relief rotation can be established to ensure that safety supervision is maintained at all times.

Incident Scene Rehabilitation

Rehabilitation is the process of providing rest, rehydration, nourishment, and medical evaluation to members who are involved in strenuous or extended-duration incident scene operations **Figure 6-7**. Part of the incident safety officer's role is to ensure that an appropriate rehabilitation process is established. **Incident scene rehabilitation** is the tactical-level management unit that provides for medical evaluation, treatment, monitoring, fluid and food replenishment, mental rest, and relief from climatic conditions of the incident.

Fire fighters are aggressive by nature. They want to do a good job and want to be where the action is occurring. This action orientation makes them susceptible to exceeding the physical limitations of their own bodies. Dehydration contributes to sudden cardiac arrest on the fire ground. As a fire officer, you must constantly monitor the health and welfare of your crew. The incident safety officer is a third line of defense, after the individual and the fire officer, for ensuring that fire fighters obtain appropriate rehabilitation.

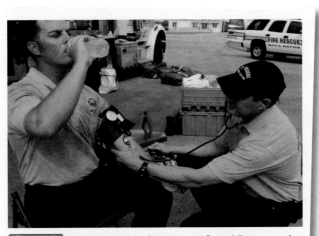

Figure 6-7 Rehabilitation is the process of providing rest, rehydration, nourishment, and medical evaluation to members who are involved in extended or extreme incident scene operations.

Creating and Maintaining a Safe Work Environment

Although the highest priority of a fire officer is to ensure that every subordinate member goes home alive from a call, the reality is that for every fire fighter death in the line of duty, nearly 1000 fire fighter injuries occur. A safety program that is based on preventing fatalities is far from adequate. Every injury, or near miss, should be viewed as a potentially fatal or disabling situation, and injury prevention should be an equally important concern of the fire officer. The fire officer must model good behavior to help develop the subordinates' attitudes about injury prevention.

Safety Policies and Procedures

Most fire departments have policies that regulate safety practices at the company level. These policies are designed to address routine circumstances, and many have been developed in reaction to a previous accident or injury. The fire officer is on the front line in ensuring compliance with all safety policies. The fire officer needs to fully understand each policy, follow all safety policies and procedures, and ensure that all subordinates fully understand and follow them.

There are a number of methods of ensuring that fire fighters understand safety policies and procedures. Many departments require each member to sign a document to acknowledge they have read and understand each policy and any new or amended policy. Some departments leave it to the officer to read and explain each policy to the crew members. This can be done at the morning briefing; however, there may be little retention at this time. A more effective method would be to have the members individually read the policy and then lead a group discussion to ensure that it is understood.

One method to reinforce safety policies is to watch videos of incidents and critique them based solely on safety policies. This often reveals significant differences between the way things are supposed to be done and what actually occurs at an emergency incident. This can be a significant learning experience for everyone involved.

Four other sources to review safety policies are the "Report of the Week" from the National Fire Fighter Near Miss Program, "The Secret List" sent by the Firefighters Close Calls website, information posted by the Emergency Responder Safety Institute (ERSI), and NIOSH case studies. In addition to reporting on incidents and trends, the NFF Near Miss Program, Firefighters Close Calls, and ERSI provide training programs and teaching resources on their websites.

The fire officer must conscientiously ensure that all safety policies are followed in training activities. Training provides the luxury of time to learn and correct errors. On emergency scenes, when there is only one opportunity to do things properly, fire fighters will perform the way they have been trained.

It is impossible to write a policy to cover every conceivable hazard. For this reason, the fire officer should use good judgment to identify hazardous situations and implement mitigating measures. This might include requiring fire fighters to use safety glasses while mowing the lawn or ensuring that food preparation surfaces are properly cleaned to prevent the spread of germs. In the absence of good leadership, activities that are not regulated by an official departmental policy are likely to be ignored. The fire officer should ensure that company members understand the need for safe work practices and develop an attitude that internalizes safety, rather than relying on the fire officer to be a "safety police officer."

Emergency Incident Injury Prevention

Many of the same techniques that are used to prevent fire fighter deaths also help prevent fire fighter injuries; however, they are not necessarily the same. The following principles should be implemented to prevent injuries as well as deaths.

Physical Fitness

Fire fighters who are in good physical condition are less prone to injury, as well as being at reduced risk for heart attack. Through strength and flexibility training, the body is more resistant to sprains, strains, and other injuries and is more capable of responding to critical situations. Sprains and strains are the leading type of fire fighter injury. A comprehensive physical fitness program that includes cardiovascular, strength, and flexibility training is needed to reduce fire fighter deaths and injuries.

Personal Protective Equipment

The capabilities of fire fighter personal protective equipment (PPE) have improved tremendously over the past 30 years. Most fire departments provide their members with protective gear that meets the requirements of NFPA standards governing PPE. The basic protective ensemble for structural firefighting includes a turnout coat and pants, boots, gloves, hood, helmet, SCBA, and a PASS. NFPA standards define the minimum performance standards for each item.

The best protective equipment is of no use if it is not worn. The fire officer should monitor the proper use of protective clothing and take immediate action to correct any deficiencies. In some departments, fire fighters routinely enter buildings to investigate fire alarm activations and other seemingly minor situations without fully donning their PPE and sometimes without tools, hose packs, or extinguishers. This practice has proven deadly when the situation turned out to be more serious than anticipated. The officer must set a good example and require every crew member to treat every call like an actual fire.

The need for protective clothing and equipment does not end when the fire is extinguished. SCBA must be used until carbon monoxide and particulate matter levels have been reduced. Coats, pants, gloves, goggles, and helmets are needed during overhaul to prevent wounds. The incident commander and the safety officer are responsible for determining whether it is safe to reduce the level of PPE that will be used at any time.

In addition to the protective clothing and equipment, fire fighters operating in hazardous areas should carry several safety-related items. A radio is needed to maintain contact, to receive instructions, or to request assistance in an emergency; at least one member of each team should have a radio, and, preferably, every fire fighter should have one. A flashlight is needed to help prevent becoming lost or disoriented and as a signal for help. A forcible entry tool should be carried to provide a method of escape in an emergency. Personal wire cutters can also prove valuable if a fire fighter becomes entangled in wires. A personal escape rope should also be carried by every fire fighter. Each item provides an additional measure of safety, and the fire officer should always set the example.

There are nonfire incidents that also require the use of appropriate protective equipment. The *2008 Near Miss Annual Report* included a focus on five topics after reviewing the 590 reports submitted in 2008. Operating hydraulic rescue tools and power saws were two of the topics. In both cases, fire fighters must pay attention to the task at hand and use proper eye protection during forcible entry or vehicle extrication.

Although station uniforms are not included in most definitions of personal protective equipment, the clothing that is worn under turnout gear can provide an additional level of protection. NFPA 1975, *Standard on Station/Work Uniforms for Emergency Services*, provides a set of performance requirements for station clothing. The use of clothing that meets this standard under turnout gear will further reduce the risk of burn injuries in a severe situation.

With EMS responses accounting for more than 70 percent of responses in many fire departments, the fire officer must also understand the requirements for medical PPE. In many cases, this requirement is as simple as wearing appropriate gloves. Today, most health care providers routinely wear gloves for all patient contact situations. When blood or other body fluids are present, eye and face protection as well as full body protection may be needed. An effective education program is essential for fire fighters to understand the importance of wearing the appropriate protection for every situation.

Fire Station Safety

In addition to emergency situations, safety considerations also apply to the fire station environment. The fire station and other fire department facilities are a workplace, and the fire department is fully responsible for maintaining a safe work environment. Every fire department should have a comprehensive workplace safety program that applies to all fire department facilities. The station setting allows the fire officer to be even more proactive in enforcing safety policies than is possible at an incident scene.

Safety hazards at the fire station can have the same consequences as hazards encountered at emergency incidents. The most important difference is that the fire department has control over the fire station environment at all times and has the ability, as well as the responsibility, to identify and correct any safety problems that are present. The fire department does not have the same type of control over the conditions that are typically encountered at the scene of an emergency incident.

Clothing

Protective clothing can become contaminated at incident scenes and should never be worn in the living quarters of the fire station. Turnout gear that is known to be contaminated with fire residue, chemicals, or bodily fluids should be cleaned as soon as possible. When washing turnout gear, be sure to follow the manufacturer's recommendations. The cleaning requires the use of a commercial extractor-type washer rather than a residential washer. Many fire departments provide special washing machines for protective clothing at the fire station or send gear out for professional cleaning and decontamination. Duty uniforms should also be washed at the fire station or sent out to a commercial laundry to prevent cross-contamination of the family laundry.

Protective clothing should be inspected regularly by the fire officer, and any items that are worn or damaged should be repaired or replaced. Properly fitting gear is also important, both for comfort and for protection. Gear inspection should include an exterior evaluation while the clothing is being worn and a close examination of the condition of all interior layers.

Housekeeping

General housekeeping around the fire station is also important in injury and accident prevention. Apparatus bay floors that become slippery when wet have caused many fire fighters to slip and fall. A squeegee should be used to remove standing water, and "wet floor" warnings should be used. The same is true for interior station floors that require mopping.

Leaving equipment lying around can present a tripping hazard to fire fighters when they are in a hurry to respond to an alarm. The walking traffic flow areas should always be kept clear.

The same fire hazards that exist in other types of occupancies can be found in many fire stations. Leaving food unattended on the stove and improperly storing flammable liquids can lead to a publicly embarrassing situation. Every fire station should have the appropriate number and types of fire extinguishers, which must be properly maintained. The fire officer should ensure that working smoke alarms are present and checked regularly. Automatic fire suppression sprinklers are highly recommended.

Food preparation activities are also responsible for many relatively minor injuries to fire fighters. Inappropriate use of kitchen knives has caused many injuries that have resulted in lost time and worker's compensation claims. In addition, proper decontamination of all food preparation surfaces is often overlooked. Regular hand washing while cooking is important to prevent the transmission of illnesses.

Many fire departments have installed diesel exhaust systems in apparatus bays to reduce the exposure of fire fighters to this contaminant. If the fire station does not have an exhaust system, the fire officer should ensure that vehicles are not left running inside the building. Daily checks should be performed outside the apparatus bay. In inclement weather conditions, some departments allow apparatus checks to be conducted inside the bay, but only with the apparatus bay doors and windows fully open to quickly remove the exhaust from the interior of the fire station.

Lifting Techniques

Back injuries are frequently caused by improper lifting techniques. Back injuries can occur while lifting and moving patients, dragging fire hoses, setting ladders, or working at the station. No matter the cause, back injuries are serious and potentially career ending.

Proper lifting techniques should always be used to reduce the risk of back injuries. A fire fighter should never bend at the waist to pick up items or patients. Instead, the fire fighter should bend at the knees and lift by standing straight up. The back should be kept in a natural position rather than locked in a hyperextended manner. When an object is too bulky, in an awkward position, or too heavy for one fire fighter, additional help should be found. Fire officers should always reinforce the use of proper techniques and procedures to avoid injuries.

Many injuries have resulted from horseplay around the station. This is an area where the fire officer must exercise supervisory control to reduce the risk of unjustifiable injuries.

Infection Control

Every fire department should have an infection control program that meets NFPA 1581, *Standard on Fire Department Infection Control Program*. This standard identifies six components of a program:

1. A written policy with the goal of identifying and limiting exposures
2. A written risk management plan to identify risks and control measures
3. Annual training and education in infection control
4. A designated infection control officer
5. Access to appropriate immunizations for employees
6. How exposure incidents are to be handled

Proper decontamination procedures are followed after emergency medical incidents or any situation where equipment could have become contaminated. Disposable items, gloves, and equipment that has been contaminated should be disposed of in a specially marked bag that is designed for that use.

Patient compartments in ambulances should be properly decontaminated after every transport. Patient cots should also be decontaminated after every call. Equipment that is designed to be decontaminated and reused should be cleaned only in an approved decontamination sink, typically in a designated area of the fire station. Contaminated equipment should *never* be taken into the living area of the station or cleaned in a sink where food is prepared.

Always follow written procedures and manufacturer's guidelines for decontamination of medical equipment. A 1 percent bleach and water solution is typically used for this purpose; however, this solution should *not* be used on turnout gear. Many of the materials commonly used in protective clothing are seriously damaged or destroyed by this type of cleaning agent.

NFPA 1581 provides specific information on how an infection control program should be established, including equipment cleaning and storage, facility requirements, and methods of protection. The fire officer should also consult departmental guides and policies for further details.

Infectious Disease Exposure

NFPA 1581 also provides a model program for situations where a fire fighter has been exposed to an infectious or contagious disease. Most fire departments have established policies to guide postexposure procedures. The fire officer must be prepared to fulfill the duties of the initial supervisor when a member has had an infectious disease exposure.

The most important first step for any exposure is to immediately and thoroughly wash the exposed area with soap and running water. If soap and running water are not available, waterless soap, antiseptic wipes, alcohol, or other skin cleaning agents can be used until soap and running water are obtained.

The fire department infection control officer should be notified, typically within 2 hours of the exposure incident. The designated infection control officer arranges for the member who has experienced an exposure to receive medical guidance and treatment as soon as is practical, but at least within 24 hours. An exposure can be a stressful event for a fire fighter. It is important to provide confidential postexposure counseling and testing. The fire fighter may not seek out this information on his or her own, so the department should proactively inform the individual of this service.

It is important to document all exposures as soon as possible using a standardized reporting form. At a minimum, the record should include the following:

- Description of how the exposure occurred
- Mode of transmission
- Entry point
- Use of personal protective equipment
- Medical follow-up and treatment

The record of an exposure incident will become part of the member's confidential health database. A complete record of the member's exposure incidents must be available to the member on request. Data on incident exposures should also be maintained by the department but without personal identifiers to allow for analysis of exposure trends and to develop strategies to prevent them.

Due to the hazardous nature of some communicable diseases, individuals who have been exposed are often required to report to the infection control officer. This information must be maintained with the strictest of confidence. It is up to the fire department physician to determine fitness-for-duty status after reviewing documentation of a member's exposure.

Accident Investigation

The fire department **health and safety officer** is charged with ensuring that all injuries, illnesses, exposures, fatalities, or other potentially hazardous conditions and all accidents involving fire department vehicles, fire apparatus, equipment, or fire department facilities are thoroughly investigated. The initial investigation of many situations is often delegated to a local fire officer.

An **accident** is any unexpected event that interrupts or interferes with the orderly progress of fire department operations. This definition includes personal injuries as well as property damage. An accident investigation should determine the cause and circumstances of the event in order to identify any corrective actions that are needed to prevent another injury, accident, or incident from occurring.

Ensure that all required federal, state, and local documentation is complete and accurate. The result of an accident investigation should always include recommended corrective actions that are presented to the fire chief or the chief's designated representative.

Accident Investigation and Documentation

Most fire departments have established procedures for investigating accidents and injuries. The fire officer is usually responsible for conducting an initial investigation and for fully investigating many minor accidents. Accidents that involve serious injuries, fatalities, or major property damage are usually investigated by the health and safety officer or by other qualified individuals.

An officer could be expected to conduct the full investigation for a simple ankle sprain, a broken pike pole, or a dented rear step on the apparatus. In each case, it is the fire officer's responsibility to protect the physical and human resources of the department. To accomplish this, the officer must understand the basic principles of investigation.

An investigation normally consists of three phases:

1. The identification and collection of physical evidence
2. The interviewing of witnesses
3. The written documentation phase (at the end of an investigation)

Each step is essential for the development of a comprehensive report that can be used to initiate safer work practices and conditions. The investigation report can also provide evidence that could be used to support or refute future claims. If a department has a standardized procedure for investigating accidents, the fire officer should always use the recognized procedure.

Examination of the physical evidence includes documenting the time, day, date, and conditions that existed at the time of the incident. This is factual information that is verifiable and should not include any interpretation. The weather conditions should be noted. The scene of the accident should be fully documented, with drawings noting the locations of relevant artifacts that could provide information about how or why the accident occurred. These details are needed to help determine prevention methods.

Witnesses to an accident should also be interviewed, including all of the individuals who were directly or indirectly involved, as well as anyone who simply observed what happened. When interviewing a witness, explain the rationale for the interview and that this is a fact-finding process. Advise the person that it is okay to not know the answer to every question. Begin with having the individual give a verbal chronology of whatever occurred, being sure to not interrupt. Watch for nonverbal communication by the witness. Once the opening report has been given, ask questions to clarify the detail, but do not ask leading or misleading questions. If the level of PPE being worn at the time of the incident has not been mentioned, it is important to determine what was being worn, how it was being worn, and whether it performed as designed. Last, restate previous questions in a different order and from a different perspective to verify previous answers.

The last step in an accident investigation is the documentation of the findings as well as the conclusions and recommendations. Most departments have a standardized form to ensure that all information is included. The facts should be presented in a logical sequence with the results of the interview or interviews attached. A determination as to the most likely cause or causes of the accident should be included, along with a recommendation on how this type of accident could be prevented. Some departments require a determination about the likely frequency as well as the likely severity of this type of accident. This optimizes prevention efforts.

You have a duty to perform an initial accident investigation that is fair and unbiased. Both the department and the employee have a vested interest in the report. An employee who applies for a disability pension based on an accident that occurred several years earlier could be denied the pension if adequate documentation is not maintained. For this reason, most departments require a report to be filed whenever there is a potential exposure to an infectious disease, when injury has occurred to an employee or a citizen, and when property damage has occurred to either private property or departmental property.

Data Analysis

The fire department is required to maintain records of all accidents, occupational deaths, injuries, illnesses, and exposures in accordance with federal, state, and local regulations. In addition to meeting regulations, these reports can assist the fire officer in identifying trends and safety-related issues.

<u>Risk management</u> is the identification and analysis of exposure to hazards, selection of appropriate risk management techniques to handle exposures, implementation of chosen techniques, and monitoring of results, with respect to the health and safety of members. Accident and injury reduction should be a major concern of every fire officer. The fire officer should review all injury, accident, and health exposure reports in order to identify unsafe acts and work conditions.

When reviewing an accident report, consider what was the root cause of the accident. Were unsafe acts being committed? If so, what must be done to change the behavior or attitude? If the cause was an unsafe condition, how can this condition be corrected? An evaluation of all unsafe acts and conditions and how to address each is required to prevent future occurrences. Some corrective actions can be made at the company level, whereas others require action at a higher level.

Once the cause has been identified, a report should be filed with your supervisor that outlines the problem, actions taken to correct the problem, and any additional actions that should be taken. Some departments have these reports sent directly to the department's health and safety officer.

The fire officer may be required to complete a longitudinal study of accidents, injuries, and exposures for the company. The officer must review the data to determine what areas are causing the greatest number of incidents as well as the ones that have the greatest potential for loss. The review should look for trends, such as an increase in the number of slips, trips, and falls on the apparatus bay floor. This might point to the condition of the floor surface being a contributing factor, which could be corrected. A series of exposures to blood by a few individuals may indicate a need for additional training for the entire company.

FIRE OFFICER II

Company Tips

Enhancing Fire Fighter Safety

Gordon Graham is a risk management expert. He has shared his success as a supervisor in a large police agency, where his systematic approach to risk management reduced organizational liability and injuries.

He advocates that emergency service organizations concentrate on the high-risk/low-frequency events, when the fire officer or fire fighter does not have any discretionary time to evaluate the situation or consult with others. Those events need to have clear policy and frequent training, so that the response is instinctive when a high-risk/low-frequency event is encountered with no time to think about the course of action.

Graham applied a systems approach in considering safety in public safety agencies. In 2002, he provided *Graham's Rules for Enhancing Firefighter Safety* (GREFS). Here is a condensed version of the rules:

1. People are your most critical element of the safety process. Select the best. Screen out those with a history of serious medical conditions and inappropriate vehicle operations.
2. Management has leadership responsibility for preventing injuries. Identify those high-risk/low-frequency events that have a probability of causing the greatest problems. Have a workable fire fighter accountability system in place.
3. Employees must be continuously trained to work safely.
4. Safety is a condition of employment. Every fire fighter has to recognize that they have an ongoing obligation to work safely. Supervisors and managers must both monitor safety performance in the workplace and follow the same rules themselves.
5. When safety rules are not followed, address the issue with prompt, fair, and impartial discipline.
6. Answer this question: "Does the benefit of fighting this fire justify the risks involved in fighting this fire?"
7. Women and men who work safely are more likely to be productive.
8. Safety extends beyond the job to be part of every person's life.
9. Safety is a business responsibility. Ethically, all management teams have an affirmative obligation to make sure that fire fighter safety is being taken seriously.
10. Most things that go wrong are highly predictable. Predictable is Preventable.

Ideally, fire companies should train on every aspect of the job. The reality is that there is not enough time to do it all. Risk management principles suggest that fire officers should focus attention on the high-risk activities that are performed infrequently. An example would be training on locating and removing lost or trapped fire fighters.

Postincident Analysis

One method of identifying unsafe situations is to conduct a postincident analysis. The incident safety officer provides a written report for the department that includes pertinent information relating to safety and health issues involved with the incident. This would

FIRE OFFICER II

include information about the use of protective clothing and equipment, personnel accountability system, rapid intervention crews, rehabilitation operations, and other issues that directly affect the safety and welfare of members at the incident scene.

Mitigating Hazards

The real value in a postincident analysis is the learning process that results from the information obtained during the process. Here are a few tips to help a fire officer to maximize the value of a postincident analysis:

1. Determine what the preferred and safer response or activity should be.
2. Develop a procedure, practice, or equipment list needed to deliver the preferred and safer response or activity.
3. Consult with supervisors for approval and formal adoption.
4. Train your peers and subordinates in the preferred and safer response or activity. Training should address recommendations arising from the investigation of accidents, injuries, occupational deaths, illnesses, and exposures and the observation of incident scene activities.
5. Make it part of an updated standard operating procedure or directive.
6. Integrate the new response or activity into the existing continuing education and basic program.

The fire officer has four roles to provide a safe environment for the fire company and ensure that everyone goes home:

1. Identify unsafe and hazardous conditions.
2. Mitigate or reduce as many problems as possible.
3. Train and prepare for the remaining hazards.
4. Model safe behavior.

Summary

Promoting fire fighter safety and health is one of the most important roles of a supervising or managing fire officer. Since the start of the NFFF Life Safety Initiatives in 2004, many have provided outstanding efforts to raise the awareness and change some procedures to improve fire fighter safety. Unfortunately, the goal of reducing the number of fire fighters dying in the line of duty by 50 percent in a decade has not been reflected in the annual reports.

The Second National Summit was held in 2007 by the NFFF to develop an action plan for each of the 16 Life Safety Initiatives. In the discussion of the efforts and experiences since the first summit, it is clear that the company officer has the most impact on reducing fire fighter death and disability. The best training program, national safety stand down, or awareness campaign lacks the lasting effort that a company officer has when requiring every member to use a seat belt before a rig responds to an emergency. Every year we still are losing a dozen fire fighters who are ejected from fire apparatus or personal vehicles.

The courage to be safe may require the company officer to do the unpopular but correct action. This includes requiring a fire fighter who has suffered a heart attack, undergone cardiac surgery, or is under medical care for severe arteriosclerotic heart disease to be evaluated by an occupational medicine physician before returning to duty.

Wearing seat belts and removing heart-damaged fire fighters from the front line would reduce the number of line-of-duty deaths by half.

You Are the Fire Officer: Conclusion

Division B of the fire ground is the area where an explosion on the second floor has blown out an exterior wall and created a lean-to collapse of the third floor. The administrative fire officer assigned to Division B has assigned you to be the Rescue Group safety officer, working under a supervising fire officer from the urban search and rescue (USAR) team. The rescue group is composed of four companies that are searching the second floor for fire fighters who were conducting a primary search when the explosion occurred. You make sure that adequate lighting and ventilation are provided in the search area and establish medical and rehab areas in the cold zone. The USAR supervising fire officer identifies the unstable areas on the second floor. Two fire fighters and three civilians are located in the debris and moved off the second floor. In addition, civilian fatalities are located in the rubble. They are left until the investigation is completed and the unsafe areas are secured.

The primary function of a rapid intervention crew (RIC) is to rescue any fire fighter who is lost, disoriented, trapped, or injured. The RIC should be strategically positioned, standing by ready for immediate action and monitoring fire-ground radio traffic. The incident or sector safety officer is responsible for verifying that operations are conducted safely. This includes identifying imminent and potential hazards, providing a risk assessment to the incident commander, and ensuring that all safety policies and procedures are followed.

Wrap-Up

Chief Concepts

- About 100 fire fighters die in the line of duty each year. More than 40 percent die of heart attacks, and more than 20 percent die in motor vehicle crashes.
- Ten to 15 deaths could be avoided each year if every fire fighter wore a seat belt while responding to and returning from incidents.
- Almost one-third of the fire fighters who die in a motor vehicle crash are ejected from the vehicle.
- Three fire officer responsibilities for safety:
  - Be physically fit.
  - Wear seat belts.
  - Maintain fire company integrity at emergency incidents.
- Every fire fighter operating in the interior should have a radio.
- A fire officer may be required to function as an incident safety officer.
- General knowledge requirements for the incident safety officer:
  - Safety and health hazards involved in emergency operations
  - Building construction
  - Local fire department personnel accountability system
  - Incident scene rehabilitation
- The incident safety officer monitors the scene and reports the status of conditions, hazards, and risks to the incident commander.
- A member with an infectious disease exposure must immediately and thoroughly wash the affected area.
- The infection control officer must be notified within 2 hours of the exposure.
- Infectious disease reporting information:
  - Description of how the exposure occurred
  - Mode of transmission
  - Entry point
  - Use of personal protective equipment
  - Medical follow-up and treatment
- There is an accident investigation on all occupational injuries, illnesses, exposures, fatalities, or other potentially hazardous conditions involving fire department members and all accidents involving fire department vehicles, fire apparatus, equipment, or fire department facilities.
- The conclusion of an accident investigation includes identifying corrective actions needed to prevent another accident, injury, or incident from occurring.

Hot Terms

Accident An unplanned event that interrupts an activity and sometimes causes injury or damage. A chance occurrence arising from unknown causes; an unexpected happening due to carelessness, ignorance, and the like.

Hazards Any arrangement of materials and heat sources that presents the potential for harm, such as personal injury or ignition of combustibles.

Health and safety officer The member of the fire department assigned and authorized by the fire chief as the manager of the safety and health program.

Immediately dangerous to life and health (IDLH) Any condition that would do one or more of the following: (1) pose an immediate or delayed threat to life, (2) cause irreversible adverse health effects, or (3) interfere with an individual's ability to escape unaided from a hazardous environment.

Incident action plan The objectives reflecting the overall incident strategy, tactics, risk management, and member safety that are developed by the incident commander. Incident action plans are updated throughout the incident.

Incident command system (ICS) A system that defines the roles and responsibilities to be assumed by personnel and the operating procedures to be used in the management and direction of emergency operations; also referred to as an incident management system (IMS).

Incident safety officer An individual appointed to respond to or assigned at an incident scene by the incident commander to perform the duties and responsibilities specified in NFPA 1521, *Standard for Fire Department Safety Officer*.

Incident safety plan The strategies and tactics developed by the incident safety officer based on the incident commander's incident action plan and the type of incident encountered.

Incident scene rehabilitation The tactical-level management unit that provides for medical evaluation, treatment, monitoring, fluid and food replenishment, mental rest, and relief from climatic conditions of an incident.

Wrap-Up

Operational period A term used with a written incident action plan identifying a period of time during a long-term incident that a specific incident action plan covers. For federally funded incidents, the operational period is 12 hours; local incidents may use operational periods of 8 hours.

Personnel accountability system A method of tracking the identity, assignment, and location of fire fighters operating at an incident scene.

Pre-incident plan A written document resulting from the gathering of general and detailed data to be used by responding personnel for determining the resources and actions necessary to mitigate anticipated emergencies at a specific facility.

Rapid intervention crew (RIC) A dedicated crew of four fire fighters who are assigned for rapid deployment to rescue lost or trapped members.

Rehabilitation The process of providing rest, rehydration, nourishment, and medical evaluation to members who are involved in extended or extreme incident scene operations.

Risk/benefit analysis A decision made by a responder based on a hazard and situation assessment that weighs the risks likely to be taken against the benefits to be gained for taking those risks.

Risk management Identification and analysis of exposure to hazards, selection of appropriate risk management techniques to handle exposures, implementation of chosen techniques, and monitoring of results, with respect to the health and safety of members.

Safety unit A member or members assigned to assist the incident safety officer. The tactical-level management unit that can be composed of the incident safety officer alone or with additional assistant safety officers assigned to assist in providing the level of safety supervision appropriate for the magnitude of the incident and the associated hazards.

Fire Officer *in Action*

Once the rescues are accomplished on the second floor of the commercial building, Tower 5 is sent to rehab. When ready for reassignment, Tower 5 is assigned to overhaul operations on the third floor of Division B. The incident safety officer points out that the third floor processes chemical samples and requires special procedures. This includes wearing disposable medical exam gloves under the fire gloves and remaining on SCBA air while working in the area. After completing overhaul, everyone will go through the decontamination station before leaving the work area.

The fire officer has three roles when it comes to safety and risk management. The first role is to model appropriate behavior and work habits. The second role is to positively ensure that the members under the officer's supervision are not subjected to avoidable risks and unsafe work conditions. The third role is to fully participate in the local, regional, and national effort to reduce fire fighter deaths, disabilities, and injury.

1. What is the most common characteristic of fire fighters who died from a sudden heart attack while at work?
- **A.** Pre-existing cardiovascular disease or cardiac surgery history
- **B.** Cigarette smoker for more than 5 years
- **C.** Body fat represents over 20 percent of weight
- **D.** Family history of high cholesterol and diabetes

2. "Failure to correct known problems" describes what level of HFACS classification?
- **A.** Level 1: Unsafe Acts
- **B.** Level 2: Preconditions to Unsafe Acts
- **C.** Level 3: Unsafe Supervision
- **D.** Level 4: Organizational Influences

3. General knowledge requirements to operate as an incident safety officer include:
- **A.** history of the incident management process.
- **B.** safety and health hazards involved in emergency operations.
- **C.** certification as a hazardous materials response team technician.
- **D.** human anatomy and physiology.

4. Who is authorized to submit a report to the National Fire Fighter Near-Miss Reporting System?
- **A.** Chief of department or union president
- **B.** Incident commander
- **C.** Supervising fire officers or higher
- **D.** Anyone who experienced a near miss

Training and Coaching

NFPA 1021 Standard

Fire Officer I

4.2.3 Direct unit members during a training evolution, given a company training evolution and training policies and procedures, so that the evolution is performed in accordance with safety plans, efficiently, and as directed. [p. 125–129, 132–135]

(A) Requisite Knowledge. Verbal communication techniques to facilitate learning. [p 124–128]

(B) Requisite Skills. The ability to distribute issue-guided directions to unit members during training evolutions. [p 124–128]

Fire Officer II

5.2.3 Create a professional development plan for a member of the organization, given the requirements for promotion, so that the individual acquires the necessary knowledge, skills, and abilities to be eligible for the examination for the position. [p 135–136]

(A) Requisite Knowledge. Development of a professional development guide and job shadowing. [p 132, 135–136]

(B) Requisite Skills. The ability to communicate orally and in writing. [p 135–136]

Additional NFPA Standards

NFPA 1000 *Standard for Fire Service Professional Qualifications Accreditation and Certification Systems*

NFPA 1001 *Standard for Fire Fighter Professional Qualifications*

NFPA 1041 *Standard for Fire Service Instructor Professional Qualifications*

NFPA 1142 *Standard on Water Supplies for Suburban and Rural Fire Fighting*

NFPA 1403 *Standard on Live Fire Training Evolutions*

NFPA 1410 *Standard on Training for Initial Emergency Scene Operations*

Introduction to Fire and Emergency Services Administration (FESHE) Course Objectives

1. Identify career development opportunities and strategies for success. [p 128, 135–136]
2. Explain the need for effective communication skills both written and verbal. [p 125–128]
7. Identify roles and responsibilities of leaders in organizations. [p 122, 124–125, 128–129, 132]
9. Identify and assess safety needs for both emergency and nonemergency situations. [p 129, 133–134]

Knowledge Objectives

After studying this chapter, you will be able to:

- Explain what accreditation means for fire fighter certification programs.
- Discuss the four-step method of job instruction training.
- Explain the difference between competence and confidence in individual skill sets.
- Identify the federal regulations that affect the training of every fire fighter.
- Describe the importance of NFPA 1403, *Standard on Live Fire Training Evolutions*.
- List the steps in developing a training program.

Skills Objectives

After studying this chapter, you will be able to:

- Direct fire company members in proper completion of a prepared training evolution.

You Are the Fire Officer

N ew fire fighters in your department spend their probationary period at the fire station by completing a field manual of procedures, specialized knowledge, and competencies. The company officer reviews the progress on a regular basis. The administrative fire officer or battalion commander reviews the progress quarterly. The recruit fire fighter has 1 year to complete the field manual and pass an end-of-probation knowledge and skill test.

You are assigned a recruit fire fighter who has completed 9 months of the probationary period at another work location. You discover that the recruit has completed less than half of the field manual. Failing to complete the field manual within a year or failing the end-of-probation test is grounds for immediate dismissal.

The recruit wants the job and assumed that the former supervising fire officer was providing adequate training. That officer and battalion commander retired last month. Two challenges in this recruit's manual are demonstrating proficiency in operating a pumper and knowledge of the first due district.

1. What should you do when a fire fighter has a performance problem?
2. How much time should you spend with individual fire fighter performance issues?
3. How can you develop a work improvement plan?

Introduction to Training and Coaching

Training and coaching have been core fire officer tasks since the establishment of the first organized fire departments. **Training** is defined as the process of achieving proficiency through instruction and hands-on practice in the operation of equipment and systems that are expected to be used in the performance of assigned duties. Fire service training has evolved in complexity and sophistication at a rapid pace as new areas of expertise have been added to the list of services performed by fire departments. **Coaching** is a method of directing, instructing, and training a person or group of people with the aim to achieve some goal or develop specific skills.

NFPA 1041, *Standard for Fire Service Instructor Professional Qualifications*, defines the standard and describes the requirements for three levels of instructor. Certification as an Instructor I is a prerequisite for Fire Officer I candidates.

Training versus Education

The relationship between training and education is confusing. **Education** is the process of imparting knowledge or skill through systematic instruction. Education programs are conducted through academic institutions and are primarily directed toward an individual's comprehension of the subject matter. Training is directed toward the practical application of education to produce an action, which can be an individual or a group activity. There is an important distinction between these two types of learning.

Training is an essential fire service activity. The emphasis on fire fighter academic education is a recent development. The First Wingspread Conference on Fire Service Administration, Education and Research was sponsored by the Johnson Foundation and held in Racine, Wisconsin, in 1966. This conference brought together a group of leaders from the fire service to identify needs and priorities. They agreed that a broad knowledge base was needed and that an educational program was necessary to deliver that knowledge base. This became the blueprint for the development of community college fire science and fire administration programs as well as the bachelor's-level Degrees at a Distance program.

The U.S. Fire Administration hosted the first Fire and Emergency Services Higher Education (FESHE) conference in 1998. That conference produced *The Fire Service and Education: A Blueprint for the 21st Century*, which started a national effort to update the academic needs of the fire service. FESHE participants developed a model fire science curriculum that would go from community college through graduate school. By the 2008 conference, the FESHE model undergraduate fire science curriculum was adopted by most academic institutions, with each course supported with textbooks available through two or more publishers.

Academic Accreditation

Universities and schools participate in a voluntary regional accrediting process to issue degrees and academic transcripts that are acceptable to other educational institutions. Working under the Council for Higher Education Accreditation (CHEA), there are nine regional organizations that provide academic **accreditation**:

- Middle States Association of Colleges and Schools
- Middle States Commission on Higher Education
- New England Association of Schools and Colleges' Commission on Institutions of Higher Education
- New England Association of Schools and Colleges' Commission on Technical and Career Institutions
- North Central Association of Colleges and Schools' Higher Learning Commission
- Northwest Commission on Colleges and Universities
- Southern Association of Colleges and Schools' Commission on Colleges
- Western Association of Schools and Colleges' Accrediting Commission for Community and Junior Colleges
- Western Association of Schools and Colleges' Accrediting Commission for Senior Colleges and Universities

For an academic institution to issue credit that will be accepted by other academic institutions, it needs to have accreditation from one of those regional organizations. Coursework, transcripts, and degrees from academic institutions without accreditation from one of these organizations will not be recognized by other colleges, universities, or employers.

There are universities that offer degrees entirely through distance education, delivering academic coursework either from an interactive web-based service or through weekly assignments delivered by mail. Some have accomplished regional academic accreditation, including those that have no brick-and-mortar campus.

Fire Fighter Certification Programs

Chief Engineer Ralph J. Scott of the Los Angeles Fire Department (LAFD) is one of the fathers of fire fighter certification training, creating a fire college in 1925. The LAFD training staff researched and documented every task a fire fighter might be required to perform. The list of almost 2000 entries evolved into a document that became known as *The Trade Analysis of Fire Fighting*. While functioning as president of the International Association of Fire Chiefs in 1928, Scott convinced the U.S. Department of Vocational Education to accept this list as an official definition of fire fighter tasks.

The National Fire Protection Association (NFPA) helped standardize fire fighter training by publishing the inaugural edition of NFPA 1001, *Standard for Fire Fighter Professional Qualifications*, in 1974. For the first time, there was a national consensus on the knowledge and skills a fire fighter should possess. This started a trend of developing national consensus standards on a wide range of fire service occupations **Table 7-1 ▶**.

Accreditation of Certification Programs

Every state or commonwealth in the United States has some type of fire service professional certification system in place. The particular programs range from local to national, depending on local history, government structure, legislation or regulation, and funding. Most of these systems are based on the NFPA professional qualifications standards.

Development of the NFPA professional qualifications standards during the 1970s created a need to validate the many different certification systems that were already in use and to establish criteria for new systems. The NFPA standards define the minimum qualifications that an individual must demonstrate to be certified at a given level, such as Fire Officer I and Fire Officer II. Accreditation establishes the qualifications of the system to award certificates that are based on the standards. A certification that is awarded by an accredited agency or institution is generally recognized by other accredited agencies and organizations.

Table 7-1   **Standards for Fire Service Operations**

- NFPA 472, *Standard for Competence of Responders to Hazardous Materials/Weapons of Mass Destruction Incidents*
- NFPA 1001, *Standard for Fire Fighter Professional Qualifications*
- NFPA 1002, *Standard for Fire Apparatus Driver/Operator Professional Qualifications*
- NFPA 1003, *Standard for Airport Fire Fighter Professional Qualifications*
- NFPA 1005, *Standard for Professional Qualifications for Marine Fire Fighting for Land-Based Fire Fighters*
- NFPA 1006, *Standard for Technical Rescuer Professional Qualifications*
- NFPA 1021, *Standard for Fire Officer Professional Qualifications*
- NFPA 1031, *Standard for Professional Qualifications for Fire Inspector and Plan Examiner*
- NFPA 1035, *Standard for Professional Qualifications for Public Fire and Life Safety Educator*
- NFPA 1037, *Standard for Professional Qualifications for Fire Marshal*
- NFPA 1041, *Standard for Fire Service Instructor Professional Qualifications*
- NFPA 1051, *Standard for Wildland Fire Fighter Professional Qualifications*
- NFPA 1061, *Standard for Professional Qualifications for Public Safety Telecommunicator*
- NFPA 1071, *Standard for Emergency Vehicle Technician Professional Qualifications*
- NFPA 1081, *Standard for Industrial Fire Brigade Member Professional Qualifications*
- NFPA 1404, *Standard for Fire Service Respiratory Protection Training*
- NFPA 1451, *Standard for a Fire Service Vehicle Operations Training Program*
- NFPA 1521, *Standard for Fire Department Safety Officer*
- NFPA 1584, *Standard on the Rehabilitation Process for Members During Emergency Operations and Training Exercises*
- NFPA 1670, *Standard on Operations and Training for Technical Search and Rescue Incidents*

Accreditation is a system whereby a certification organization determines that a school or program meets with the requirements of the fire service. A group of impartial experts is assigned to thoroughly review a given program and determine whether it is worthy of accreditation. Two organizations provide accreditation to fire service professional certification systems: the National Board on Fire Service Professional Qualifications and the International Fire Service Accreditation Congress.

National Board on Fire Service Professional Qualifications

The Joint Council of National Fire Service Organizations created the National Professional Qualifications System (also known as the "Pro Board") in 1972. When the Joint Council dissolved in 1990, the Pro Board evolved into the independent National Board on Fire Service Professional Qualifications (NBFSPQ). The board of directors consists of representatives from national fire service organizations that have an interest in training and certification. The NBFSPQ has accredited the certification programs that are operated by 49 states, provinces, and other agencies.

International Fire Service Accreditation Congress

Established by the National Association of State Directors of Fire Training, the International Fire Service Accreditation Congress (IFSAC) provides accreditation to certificate-issuing entities. IFSAC also accredits fire-related degree programs at the college and university levels. The IFSAC process follows the CHEA format and operates through Oklahoma State University. In 2007, there were 59 IFSAC-accredited fire service training programs and 35 fire science degree programs.

Overview of Training

Training ensures that every fire fighter can perform competently as an individual and every fire company is prepared to operate as a high-performance team. Fire service training must anticipate high-risk situations, urgent time frames, and difficult circumstances. A wide variety of methods and practices is included in the overall category of training.

Initial training leads to basic skill certifications, such as NFPA Fire Fighter I and EMS first responder. These certifications are often required before a fire fighter is authorized to participate in emergency operations. Certification usually involves a formal training program, conducted by a training academy or equivalent organization, that includes both classroom and skills practice Figure 7-1 ▶ . The trainee must pass both skill and knowledge evaluations to be certified.

After initial certification, most fire fighters work toward achieving progressively higher levels of certification and additional specialty qualifications. Often, a fire fighter is required to achieve additional qualifications, such as Fire Fighter II, driver/operator, or emergency medical technician (EMT), within a specified time period. Some departments require higher-level certifications, such as Fire Officer I or II, for promotion or advancement. Many certifications require periodic refresher or update training, and some expire after a set period unless the fire fighter completes a refresher training requirement.

Several important components of training occur at the fire station or company level under the supervision of fire officers. This type of training is directed toward practicing basic skills and improving both individual and team performance. Many departments have a standard set of evolutions that each company is expected to be able to perform without flaw, such as advancing a fire attack line over a ground ladder and into a third-floor window. Additional training at the company level often includes learning how to use new tools and equipment as well as refreshing, reinforcing, or updating knowledge and skills that are related to different aspects of the firefighting craft.

Additional company-level training often includes pre-incident planning and familiarization visits to different locations in a company's response area. These activities sometimes include a group of companies that would normally respond together to that location. Conducting these activities as a group is an excellent method of improving coordination between companies. Periodic multicompany drills should also be conducted for the same reason.

Figure 7-1 A formal training program includes both classroom and skills practice.

Fire Officer Training Responsibilities

A basic responsibility of every fire officer is to provide training for subordinate fire department members. The specific training responsibilities assigned to fire officers vary, depending on the organization and the available resources. At a minimum, a fire officer must be prepared to conduct company-level training exercises and evolutions to ensure that the company is prepared to perform its basic responsibilities effectively and efficiently.

One of the enduring products of the 20th century is the four-step method of skill training. This prepare-present-apply-evaluate method originated during World War I, when the armed services were teaching farmers how to fly biplanes, drive tanks, and operate ships. The process was updated and renamed as <u>job</u>

Assessment Center Tips

Four-Step Fire Fighter Development
The four-step method remains the method of choice if a candidate must describe how to develop a work improvement plan or explain a new procedure or device. When developing a written or oral response to such an issue, the candidate should identify each of the four steps in the training process.

<u>instruction training</u> when more than a million men and women received technical skill training during World War II.

The four-step method is the foundation of Fire Instructor I. Most fire officers use standardized curricula and training packages that are commercially prepared or developed by the local fire academy. Occasionally a company officer needs to start from scratch to develop a training program.

■ Review of the Four-Step Method

Step 1: Preparation

The fire officer conducts training to maintain proficiency of core competencies. Crews should be able to catch a hydrant, raise a ground ladder, and deploy attack lines so well that the task is automatic. They are <u>unconsciously competent</u> with these tasks.

Sometimes the fire officer or the department determines the need for focused instruction. Three indicators that training is needed would be a near miss, a fire-ground problem, or an observed performance deficiency. For example, during a structure fire, the fire officer might observe that the fire fighters are having difficulty placing a 35' (10.5-meter) extension ladder at an upper-floor window. Or, the deployment of an attack line from a standpipe connection might have created a tangle of hose in the stairwell. These problems need to be addressed through additional training and practice.

The fire officer begins by obtaining the necessary material and teaching aids. If needed, the fire officer writes a lesson plan **Figure 7-2 ▶** . Components of a lesson plan include the following:

1. Break the topic down into simple units.
2. Show what to teach, in what order to teach it, and exactly what procedures to follow.
3. Use a guide to help accomplish the teaching objective.

If this is a new lecture or topic, the fire officer should practice delivering the lesson to make sure that the important items are covered in a timely fashion. In addition, the fire officer should preview any audiovisual items and check all of the equipment that will be used during the presentation.

The final preparation activity is to check the physical environment. Make sure you have an environment that is conducive to adult learning. This includes taking every reasonable effort to reduce distractions and student discomfort **Figure 7-3 ▶** .

Step 2: Presentation

This is the lecture or instructional portion of the training. The objective here is to introduce the students to the subject matter, explain the importance of the topic, and create an interest in the presentation. The fire officer could be demonstrating or showing a skill or explaining a concept.

When the fire officer is teaching a skill, he or she should present it one step at a time, delivering a perfect demonstration of how it should be performed **Figure 7-4 ▶** .

Figure 7-2 Fire officers often write lesson plans in preparation for training activities.

When presenting a concept or idea, recommended lecture practices should be followed. The objective of a lecture is to provide knowledge and develop understanding so that the fire fighter will be able to perform the skills properly. The overall goal is to increase fire company efficiency.

Fire Marks

Conscious Competence Learning Matrix

Dr. Thomas Gordon developed the Conscious Competence Learning Matrix in the 1970s. There are four levels:

Level 1: Unconsciously Incompetent: Fire fighter is not aware of the existence or relevance of a skill area. They are not aware that they have a deficiency in this area and might deny the relevance or usefulness of the skill.

Level 2: Conscious Incompetence: Fire fighter becomes aware of the existence and relevance of a skill. In attempting to perform the skill they are aware of their deficiency. Training and practicing on the skill will improve their performance and get them into Conscious Competence.

Level 3: Conscious Competence: Fire fighter can perform the skill reliably at will and without assistance. Fire fighter will still need to concentrate and think in order to perform the skill. Repeated practice in the skill will get the fire fighter to Unconscious Competence.

Level 4: Unconscious Competence: The skill becomes so practiced that it enters the unconscious parts of the brain—it becomes second nature. Fire fighter will be able to perform other tasks while performing this skill, like conducting a search and rescue while operating a fire attack line inside a burning structure. The tasks involved in wearing self-contained breathing apparatus (SCBA) and operating a fire attack line are second nature. The fire fighter can focus attention on performing the search and locating victims.

Figure 7-3 Check the physical environment to make sure it is conducive to adult learning.

The instructor should use simple but appropriate language. Begin with simple concepts and move on progressively to more complex information, relating the new material to old ideas. Lecture only on what is important at this time to achieve the teaching objective, and do not teach alternative methods. Teach in positive terms and avoid telling the fire fighter what *not* to do.

A lesson plan allows the fire officer to stay on topic and emphasizes the important points **Figure 7-5 ▶**. Increasing the

Figure 7-4 Fire officers should deliver a perfect demonstration of how skills are performed.

Tools for the Instructor's Toolbelt
By David Hall

Rationale: At some point in their career in the fire service, most everyone is going to be required to teach others about a topic. To effectively do this, the instructor must understand the basic principles of learning and have an understanding of the material that they are going to present. This course will give an overview of those principles and then focus on two methods that may be particularly effective in teaching many firefighting topics.

Purpose: The purpose of this course is to provide participants with the knowledge required to analyze the content to be presented, the target audience, and the desired outcomes, and be able to select the appropriate teaching method. The course will also teach the instructor how to effectively utilize games and case studies to present material appropriate for these methods.

Objectives:
At the conclusion of this course, the participant will be able to:
1. Describe how to choose the right instructional tool
2. Identify rote information, hard skills, and soft skills
3. Describe how to evaluate the audience
4. Identify the outcomes of a course
5. List the advantages of using games
6. Demonstrate the use of the "Bingo" game
7. Demonstrate the use of the "Generic Game Board"
8. Demonstrate the use of a case study
9. Develop a case study without external resources
10. Develop a case study using external resources

Qualifications for Attendance: This course is designed for all members of the fire service that instruct other members. The course is primarily aimed at individuals that are the Fire Service Instructor I or greater. This course will also be particularly beneficial to fire officers that instruct at the company level.

Summary of Subject areas:	Hours:
Introduction and Welcome	.50
Choosing the Right Tool	.50
Games	1.50
Case Studies	1.50
Summary and Evaluation	.50
TOTAL	4.5

Figure 7-5 A sample lesson plan. Courtesy of Assistant Fire Chief David Hall.

number of senses engaged in the training session helps the fire fighter to retain more of the material. Audiovisual aids and training props should be used to enrich the presentation. Fire fighters retain information more effectively if they actually perform the skill in the process of learning it.

Step 3: Application

The fire fighter should now demonstrate the task or skill under the fire officer's supervision. The objective is a correct demonstration of the task, safely performed. A good reinforcing technique is to have the fire fighter explain the task while demonstrating the skill. The fire officer should provide immediate feedback, identifying omissions and correcting errors **Figure 7-6 ▼**. Success is achieved when the student can perform the task safely without input from the supervisor.

Step 4: Evaluation

There should be an evaluation of the student's progress at the end of the lesson or program. Training that is related to a certification program always includes an end-of-class evaluation. Depending on the skill and knowledge sets involved, the evaluation may be a written or practical examination.

The fire officer can be certain that training has occurred only when there is an observable change in the fire fighter's performance when responding to a real situation where that task or skill is applied. For example, Fire Fighter Jones has not been coming to a complete stop at red traffic lights or stop signs when driving fire apparatus to emergency incidents. The fire officer provides a training session to address this behavior. If, after the training session, Fire Fighter Jones always comes to a complete stop at red traffic lights and stop signs, there has been an observable change of behavior. If the behavior does not change, the training objective has not been accomplished.

Figure 7-6 The fire officer should provide immediate feedback to the fire fighter.

Ensure Proficiency of Existing Skill Sets

The fire officer needs to ensure that every fire fighter is proficient in skill performance. This is both an individual and a company-level requirement. Some departments have a standard set of evolutions that have to be performed proficiently in which both task and time requirements must be met. The fire officer must invest some of the available training time practicing and reviewing these standard evolutions.

It is important that some of these practice sessions be performed while the fire fighters are wearing full personal protective clothing, operating within a realistic fire-ground situation. This may require construction of training props, such as an assembly that allows the fire fighters to practice opening a roof using a power saw. The fire officer should always be on the alert for opportunities to acquire abandoned structures where realistic fire-ground skills can be practiced.

Mentoring

Mentoring is a developmental relationship in which a more experienced person, or mentor, helps a less experienced person, referred to as a protégé. The unique organization of fire departments, where company officers function as working supervisors within a high hazard environment, makes mentoring easy.

Mentoring is a one-on-one process in which the more experienced person provides a deliberate learning environment through instructing, coaching, providing experiences, modeling, and advising. Mentoring may be provided through feedback while picking up after an emergency incident. Failures as well as successes make up the mentoring experience.

The mentoring process extends beyond a particular fire company assignment and rank, usually lasting for a long period of time. Boston Fire Commissioner Leo Stapleton, writing in *Thirty Years on the Line*, described how he would get feedback from a senior fire fighter decades after Stapleton was his rookie.

Qualities that make an effective mentor include:
- A desire to help
- Current knowledge
- Effective coaching, counseling, facilitating, and networking skills

Provide New or Revised Skill Sets

On occasion, the fire officer is required to provide the initial training for a new or revised skill set. This is often related to a new device that has been acquired, such as a new type of breathing apparatus or a thermal imaging camera. The fire fighters need to become familiar with the new equipment and proficient in its use. In the case of a thermal imaging camera, the fire fighters have to learn how it works and how to maintain it, as well as how to incorporate its use into fire-ground operations. This type of training could also be required to introduce a change in standard operating procedures. Sometimes procedures are changed or additional training is mandated in response to a near-miss incident.

Teaching new skills takes more time than maintaining proficiency of existing skills. The fire officer should obtain as much information as possible about the device or procedure, especially identifying any fire fighter safety issues. The emphasis of fire station–based training should be on the safe and effective use of the device or procedure. The fire officer should plan to spend a couple of training periods developing competency and showing how it relates to existing procedures. Avoid spending more than 15–20 minutes on any lecture or video presentation; the adult learner will lose attention in longer sessions.

Practicing new skills is facilitated by encouraging adventure, challenge, and competition. Using a thermal imaging camera to complete a scavenger hunt for heat sources around the fire station is one way to reinforce the capability of the device.

Ensure Competence and Confidence

The fire officer works as a coach when providing training to an individual or a team. After team members have learned the basic skills and can appropriately demonstrate them, the coach has to work with them to build competence and confidence. The coach has to provide the guidance that advances them from being capable of performing the basic required skills to being able to perform them effectively, efficiently, and consistently.

Many fire fighter tasks involve psychomotor skill sets. Psychomotor skill levels fall into four categories. The levels can be described using the following example of a new driver/operator who is required to know, from memory, every address in the fire district.

- Initial: The driver/operator knows the main streets and has a basic understanding of how the street grid and numbering system work. The fire officer has to help by reading the map when responding in subdivisions and office parks.
- Plateau: The driver/operator can drive to more than 85 percent of the streets in the company's response area without assistance from the officer. At this level, the driver/operator is competent.
- Latency: The driver/operator can remember the route to an area of the district where the engine company has not had a response for several months.
- Mastery: The driver/operator can easily drive to any address in the district and knows at least two or three alternative routes to each area. The driver knows all of the subdivisions, the layout of every office park, and the locations of all hydrants and fire protection connections. At this level, the driver/operator is confident in his or her knowledge of the district.

To bring fire fighters up to the mastery level, the fire officer must work every day to reinforce their skills. Many fire fighter skill sets are infrequently used, yet when that task is needed, the fire fighter has to deliver a near-perfect performance under urgent or critical conditions. It is dangerous to ask a fire fighter to deliver a rusty skill set in a critical emergency situation.

The fire officer needs to provide enough repetition and simulations to maintain fire fighter confidence in seldom-used skill sets **Figure 7-7 ▶**. The continuing expansion of fire fighters as all-hazard mitigation specialists increases the fire officer challenge. Fire fighters must become proficient as first responders to terrorist incidents involving chemical, biological, or nuclear elements.

When New Member Training Is On-the-Job

Many departments require that a new fire fighter attend a recruit school and obtain NFPA Fire Fighter I certification before responding to emergencies. Fire officers have special responsi-

Figure 7-7 Skills practice may require training props.

bilities when operating with inexperienced fire fighters and fire fighters in training. At the very first meeting with the new fire fighter trainee, the fire officer should explain the procedures in the fire station when the company receives an alarm, assign a senior fire fighter to function as a mentor, and describe any restrictions that are placed on fire fighters in training.

Skills That Must Be Immediately Learned

Four federal regulations affect every fire fighter. These topics need to be covered as part of any emergency service training program:

- Bloodborne pathogens: Even "suppression only" fire fighters are at risk of being exposed to blood and other bodily fluids. The Occupational Safety and Health Administration (OSHA) has issued regulation 29 CFR 1910.1030, "Occupational Exposure to Bloodborne Pathogens." This requires all fire fighters to be trained about the department's exposure control plan, the personal protective equipment used by the fire fighter, and the reporting requirements if there is an exposure. Usually, this training takes about 4 hours.
- Hazardous materials awareness and operations: Every public safety member is required to have training at the hazardous materials awareness and operations levels. OSHA regulation 29 CFR 1910.20, "Hazardous Waste Operations and Emergency Response (HAZWOPER) Training," describes these requirements. The awareness

Voices of Experience

In December of 1988, I met Brian Crandell, who was then serving as Deputy Director of Montana State Fire Training (MSFT), at a conference in Florida. Brian and his boss, MSFT Director "Butch" Weedon, were vocal and passionate advocates of an approach to training they called "training in context," or "practice."

Practice is a simple concept that makes a ton of sense: When we prepare people for important jobs, we ought to bring as much of the job context (or environment) into the training environment as we can. Seems like common sense to me.

Not only was MSFT putting raw recruits into 3-member fire companies led by experienced company officers from Day 1 of the training program, they were putting them on the drill-ground in full PPE, in NIMS incident and accountability systems, under orders, etc. These recruits were "on the job" starting Hour 1 of their training program.

And the training methodology MSFT used was different, too—almost backwards from the traditional approach. Instead of spending many hours in the classroom building foundation knowledge before introducing recruits to hands-on skills, Montana was mapping out "standard plays" for each fire-ground evolution in advance and then putting recruits into the approximate "real world" environment (with appropriate safety considerations) and addressing the cognitive components of the training content as the recruits needed to know. Crandell liked to say of the recruits, "When they need to know, they'll ask." Retention of cognitive information in these recruits was nearly 100 percent.

> **"[They were] mapping out 'standard plays' for each fireground evolution in advance, putting recruits into the approximate 'real world' environment, and addressing the cognitive components as needed."**

To build solid fire-ground skills in recruit fire fighters, Montana used a series of practice repetitions:

1. **Practice for sequence:** Recruits learned an evolution by walking through the steps repeatedly. Early repetitions were simply role playing, then with limited PPE and mock tools. Walk-throughs kept the recruits fresh and focused. Coaches (as instructors were termed) reinforced key points and big picture "macros" as repetitions progressed.
2. **Practice for technical skills:** When they were ready, recruits moved into a more realistic environment and perfect practice—starting from an apparatus or resource area in companies, donning appropriate PPE for the assignment, and accomplishing the assignment start (from resource) to finish (out of the hazard zone) over and over to refine the technical skills.
3. **Practice for standard:** Additional repetitions from start to finish were implemented to build confidence and shorten the time. Evolutions that were time-sensitive in the real world were timed during training. Again, common sense.

Maine Fire Training & Education, which I administered at the time, adopted much of the Montana approach and saw impressive results. While it takes more resources—more realistic facilities, more tools, more apparatus on the drill-ground—and a corps of instructors who are willing to try something new and put the candidates' needs first (in other words, become coaches), we found the benefits to be dramatic. Recruit school graduates were much better prepared for assignment and were more confident, more "worldly," and safer.

Voices of Experience

The practice methodology is being applied in many occupational and human development areas. Athletic teams and the military use it extensively.

One caution: Trainees prepared this way will be conditioned to operate this way. Under stressful conditions, they will act the way they've been trained. That puts a tremendous onus on the training agency and coaches to get it right.

Steve Willis
Fire Science Instructor and Student Live-In Coordinator
Southern Maine Community College
South Portland, Maine

level enables first responders to recognize a potential hazardous materials emergency, isolate the area, and call for assistance. The operations level enables first responders to recognize a potential hazardous materials incident, isolate and deny entry to other responders and the public, evacuate persons in danger, and take defensive actions such as shutting off valves and protecting drains without having contact with the product.

- SCBA fit testing: OSHA regulation 29 CFR 1910.134, "Respiratory Protection Training," requires that anyone who uses respiratory protection during job tasks must be provided with appropriate training, be fit-tested for a mask, and be subject to a health monitoring program. This training is usually part of the Fire Fighter I training program.
- National Incident Management System: Homeland Security Presidential Directive (HSPD) 5, "Managing Domestic Incidents," requires incident management training. The National Incident Management System (NIMS) provides a consistent framework that operates at all jurisdictional levels regardless of the cause, size, or complexity of the incident. The NIMS creates an all-hazard template for fire fighters to operate within a multiple jurisdiction domestic incident within the federal National Response Plan.

In addition to the federally required training, the fire department needs to provide emergency scene awareness training to reduce the risk that a probationary fire fighter will be injured at the emergency scene. This training would include such items as how to avoid being struck by a car when operating on an interstate highway. The fire officer should spell out the expected behavior of the trainee when operating at emergencies. This may include assigning the trainee to shadow, or work with, a senior fire fighter on the team.

The expected behavior falls into three areas: responding to alarms, on-scene activity, and emergency procedures. Responding to alarms includes the trainee behavior that should occur when an alarm is received at the fire station. Be specific about the appropriate actions necessary when a fire fighter is preparing to respond to an alarm. Emphasize the importance of not running in the station, donning protective clothing, and always wearing a seat belt when riding in a fire vehicle.

On-scene activity includes clearly defining the expected location, activities, and behavior of the trainee at an emergency incident. Make sure that the trainee knows the safe locations when working at an incident on a roadway. Identify off-limits activity. The fire officer should know all of the restrictions that apply to each individual and ensure that the trainee is equally aware of those limitations. A trainee who has not completed SCBA training and fit testing must not wear breathing apparatus or enter an immediately dangerous to life and health (IDLH) environment. If the trainee is a teenager, additional restrictions might apply; allowing a 15-year-old to operate inside a burning building is prohibited in many jurisdictions.

Finally, the officer should clearly identify the trainee's expected behavior when operating under an emergency situation, such as a may-day (fire fighter down or lost) or a potentially violent situation. The fire officer should ensure that the trainee will be in a safe location and stay out of the way until the emergency situation is controlled.

Skills Necessary for Staying Alive

Once the skills that must be immediately learned are covered, the fire officer should concentrate on the skills the fire fighter in training needs to know in order to stay alive. Most states and commonwealths require the trainee to pass a knowledge exam and a skill test for Fire Fighter I before engaging in interior structural firefighting. Some communities start with the trainee limited to "outside only" activities at fire scenes. Examples of skills and topics that should be taught first in a training program include:

- Fire-ground tasks that emphasize teamwork and require mastering the location of all of the equipment on the apparatus:
  • Supply line evolutions
  • Ropes and knots
  • Laddering the fire building
  • Lights, fan, and power deployment
- Crashes and medical emergencies
  • CPR and semiautomatic defibrillator training
  • Outside circle activities on a crash extrication
  • Helicopter landing zone procedure
  • Assisting the paramedics on a medical emergency

The trainee has received enough training and demonstrated enough stay-alive skills to begin responding to emergencies. This allows the student to gain experience functioning in an "outside only" role while completing the Fire Fighter I certification training. Some fire departments do not allow trainees to respond to any emergency incidents until they have achieved the initial Fire Fighter I certification level.

Live Fire Training

Fire fighters must receive appropriate training before participating in any live fire evolutions. NFPA 1403, *Standard on Live Fire Training Evolutions*, provides essential information for any type of live fire training session.

Fire Marks

Fire Officer Convicted of Criminally Negligent Homicide After Live Fire Training

Alan G. Baird III was the instructor-in-charge of a live fire training session in an acquired structure that led to the death of a 19-year-old rookie and severely burned two other fire fighters. The National Institute for Occupational Safety and Health (NIOSH) report identified eight major mistakes made by Baird in setting up and running the evolution.

The Oneida County district attorney filed charges of second degree manslaughter. Part of Baird's defense was that he did not know about NFPA Standard 1403. He was convicted of criminally negligent manslaughter. In the sentencing of Baird to 75 days in jail and 5 years probation, Judge Michael Dwyer said "This was not an accident. This was a series of bad decisions, decisions that never should have been made."

Student Prerequisites

Before participating in any live fire training evolutions, the student must have received training to meet the Fire Fighter I performance objectives from the following sections of NFPA 1001:

- Safety
- Fire behavior
- Portable extinguishers
- Personal protective equipment
- Ladders
- Fire hose, appliances, and streams
- Overhaul
- Water supply
- Ventilation
- Forcible entry

In addition to the prerequisite training, the trainee must be equipped with full protective clothing, a personal alarm device, and self-contained breathing apparatus that is compliant with the relevant NFPA standard.

Fire Officer Preparation Responsibilities

NFPA 1403 provides detailed instructions on how to conduct a safe live fire training evolution under five different scenarios: acquired structures, gas-fired training center buildings, non–gas-fired training buildings, exterior props, and exterior class B fires. The fire officer should read the standard and, if available, consult with a training officer who has received additional instruction in the delivery of live fire training evolutions.

There are general points that apply to all live fire evolutions. The fire officer who is the instructor-in-charge develops a training action plan that includes:

- Preburn plan, including communications, personnel accountability, rehabilitation, and building evacuation procedures. The instructor-in-charge must remain mindful of severe weather conditions that increase the risk of injury or illness.

Safety Zone

Training Injury Patterns
From 1997 to 2007, deaths during training activities represented 7.4 percent of the annual line-of-duty death (LODD) total. This is an unacceptable situation. Deaths and serious injuries that occur during live fire training are caused by:

- Missing, incomplete, or inadequate preparation of the drill site.
- Inadequate training of instructors and assistant instructors.
- Inadequate orientation or training of recruits. Live fire evolutions are not the place to conduct or complete initial SCBA training. The students should be competent in SCBA use and emergency procedures *before* participating in a live fire training exercise.
- Inadequate planning of the training evolutions by the lead instructor.
- Nonexistent incident management system, or no designated safety officer.
- No provision of rehabilitation or EMS services at the drill site.

- Calculate the needed water supply. NFPA 1142 provides a reference for this subject. Each attack and back-up hose line should be capable of flowing a minimum of 95 gallons (360 liters) per minute. The water supply resources should be capable of delivering 150 percent of the minimum needed water supply at the training site from two separate sources.
- Arrange to have a dedicated ambulance or emergency medical services unit assigned on site during the training evolution.
- Establish a designated and appropriately equipped rest and rehabilitation area.
- Inspect the structure to ensure it is safe for use in live fire evolutions. Determine the required class A materials to be used to provide the fire load. Watch out for excessive fire load and for materials that could lead to early and unwanted flashover.
- Assign dedicated positions of safety officer and ignition officer.
- Assign instructors to each functional crew. The maximum size of a functional crew is five trainees. An additional instructor is assigned to each back-up hose line and functional assignment. All instructors have full protective clothing and SCBA. Weather, duration of the evolution, and size of the trainee group determine whether additional instructors are required.
- Conduct a preburn briefing session and walk-through for instructors, safety crew, and trainees. Identify evacuation routes during the walk-through. All facets of each evolution should be discussed and assignments made before the fire is lit.

Prohibited Live Fire Training Activities

NFPA 1403 identifies activities that are absolutely prohibited during live fire training. These prohibitions are the direct result of investigations of prior live fire evolutions that resulted in fire fighter deaths or serious injuries.

- No live "victims" may be used in live fire training evolutions.
- Flammable or combustible liquids cannot be used as fuel.
- Acquired structures have been the source of many line-of-duty deaths. Run just one fire evolution in an acquired structure at a time.

Getting It Done

Performance in Context
Whole-skill training or training in context is an effective tool for helping experienced fire fighters improve their group performance at standard evolutions. Athletes use visualization and sequential task mastery to develop consistent and high-level performances. Members who prepare for the Fire Fighter Combat Challenge, Transportation Emergency Rescue Committee (TERC) Extrication Challenge, or other timed skill events also use training in context to improve performance.

Near Miss REPORT

Report Number: 06-0000270

Event Description: I was participating in a live burn scenario at an acquired structure in which the goal was to observe the benefits of vertical ventilation. We observed a free burning fire in a room from an adjacent hallway. 1 3/4" hoselines were placed with all interior crews and clear escape routes were established prior to ignition. The fire began to grow and roll across the ceiling. The heat increased as expected but there were no signs of off-gassing of anything other than the pallets used for the fire. I was in a crouched-down position on my knees and at no time did I feel any hot spots on my shoulders or arms that I have felt many times before in fires while wearing the same bunker gear.

As the fire continued to grow, I suddenly felt a burning sensation on my forehead and my breathing air got very hot. At that point, I retreated to one of the safe zones in a back bedroom to cool off. After cooling off I returned to the scenario and assisted with removal of hoselines after the fire was extinguished. I then proceeded outside and doffed my facemask. The clear plastic piece of my facemask had almost completely separated from the rubber that seals to my face. There were two firefighters within 3' of me with the same protective equipment and in the same position. Both of them experienced similar levels of exposure to heat and had no equipment failure.

Lessons Learned: 1. Training fires utilizing wood pallets and straw can create considerable heat even without visible off-gassing. Training officers should monitor temperature closely and control the intensity of fire. Remember that training fires injure and kill too! 2. Prolonged exposure to what may be considered "moderate heat" can have a cumulative effect and lead to equipment failure. 3. You can't count on traditional signs, hot spots under gear, to predict equipment failure and must be constantly looking for areas of refuge or escape routes in case of an emergency. 4. Thorough inspection of SCBA equipment must be conducted on a regular basis and always after each use.

Developing a Specific Training Program

On occasion, the fire officer may need to develop a specific training program that is not covered by an existing certification training program or prepared lesson plan. This may be a work improvement plan or training related to a new device or procedure. NFPA Instructor II covers the development of a training program in more detail. Here is an overview of the five steps for developing a training program.

Assess Needs

The fire officer must first confirm that there is a need for a training program. Some performance problems may be better solved through an engineering solution. For example, the effective use of Personal Alert Safety System (PASS) devices was a problem. Many fire fighters were operating in an immediately dangerous to life and health environment while their PASS device was not armed. Despite training programs and, in some departments, progressive discipline, the devices remained unarmed. An engineering solution has resolved this performance issue. The PASS device is now integrated into the SCBA and is armed every time the high-pressure hose is charged.

Establish Objectives

Training has occurred when there is an observable change in behavior. Identify the specific behavior you want the fire fighters to exhibit after the training. The desired behavior could be that

fire fighters will demonstrate the proper procedures for deploying a ground ladder. The description must include the conditions under which the behavior will be demonstrated. For example, if an expectation is that a ground ladder will be deployed while fire fighters are wearing full protective clothing and SCBA, then that condition must be part of the expected fire fighter behavior.

The final part of the objective is the measure of performance. Fire fighter evolutions are often timed events, so the measure of performance is usually described in terms of the time required to properly complete a task.

The completed behavioral objective could look like this:

> Given a fire department pumper and a building two or more stories high, a crew of two fire fighters will exit the pumper in full protective personal clothing and SCBA, select and remove the appropriate ground ladder from the pumper, and properly deploy the ground ladder to the assigned window within 2 minutes.

Develop the Training Program

Various methods exist for developing the training program. If training is needed for a new device, the manufacturer or vendor may have a training package available. If training is needed for a new procedure or company evolution, the department may have a template the fire officer can use. Other departments may have developed training programs, props, or resources that they will share. Organizations like Fire Fighter Near Miss, Responder Safety, and special interest websites may post resources available for downloading. A few minutes networking may provide a rich response.

The fire officer also needs to consider how the training will be delivered. Will it require a skill drill using multiple companies or can the skill be practiced by an individual fire fighter?

Deliver the Training

New programs should have a pilot class or trial run before finalization. This provides the opportunity to tweak the program, identify any problems, and correct any unforeseen issues.

Although the use of a lesson plan is part of the four-step method, it is important that the fire officer develop lesson plans for every training program. A good lesson plan:

- Organizes the lesson
- Identifies key points
- Can be reused
- Allows others to teach the program

Evaluate the Impact

Have you accomplished the change of behavior? Was the training program worth the fire fighters' or instructor's time? Was the instructional method appropriate for the learning objective? What can the instructor-developer do to improve the training program?

Many successful national fire service training programs, like the "Saving Our Own" fire fighter rescue class or the "Car Busters" extrication seminars, have been developed by fire officers and fire fighters. They have the perspective, the understanding, and the need to develop and share vital emergency activity training.

Building a Professional Development Plan

The International Association of Fire Chiefs (IAFC) defines professional development as ". . . the planned, progressive, life-long process of education, training, self-development and experience." The IAFC's *Officer Development Handbook* provides a guide for progressing from Supervising Fire Officer to Executive Fire Officer, building on the work done by the U.S. Fire Administration/National Fire Academy's Fire and Emergency Services Higher Education (FESHE) network. The National Professional Development Model shows how education and training fit into a path that starts at Fire Fighter I and can end at Chief Fire Officer **Figure 7-8 ▾** .

Supervising Fire Officer Preparation

Training to become a supervising fire officer includes achieving NFPA Fire Officer I, Incident Safety Officer (NFPA 1521), Hazardous Material at Operations Level (NFPA 472), and the equivalent of NFPA Instructor I (NFPA 1041) and Inspector I (NFPA 1031).

Required lower level college courses to become a supervising fire officer include English composition, public speaking, business communications, biology, chemistry, psychology, sociology, finite math (or algebra), business computer systems, health/wellness, American government, and human resource management. Specialized lower level college courses include fire behavior, building construction, and fire administration.

The supervising fire officer candidate needs 3 to 5 years of experience in agency operations and 200 hours of experience as an acting unit officer handling emergency responses and nonemergency activities. He or she must develop and deliver training classes, participate in the planning process, participate in mass casualty training exercises, and be capable of operating as a single unit supervisor within an incident management system.

At this level, the supervising fire officer candidate has performed a personal and professional inventory. The goal of the inventory is to establish a career map that identifies personal traits, strengths, and areas for improvement.

Figure 7-8 The National Professional Development Model.

Managing Fire Officer

Training to become a managing fire officer includes becoming an NFPA Fire Officer II, completing the National Fire Academy Leadership Development and Managing Company Tactical Operation series, and training for Unified Command for Multi-Agency and Catastrophic Incidents, Public Information Officer, the equivalent of NFPA Investigator I (NFPA 1033) and Fire and Life Safety Educator I (NFPA 1035).

Lower level college courses needed include statistics, interpersonal communication, philosophy, critical reasoning, professional ethics, professional report writing, accounting analysis, introduction to law, and introduction to planning. Specialized lower level college courses are needed in fire service management, prevention, and education; fire protection systems; and fire protection hydraulics.

The managing fire officer candidate needs 2 to 4 years of experience as a qualified supervising fire officer. He or she must function as an acting officer for multicompany operations that include emergency and nonemergency activities. He or she must also be the supervisor or an aide to the incident commander of a multicompany operation, deliver performance appraisals and discipline, and participate in the development or updating of local emergency plans.

The managing fire officer candidate is expanding the career map to explore areas of special interest and to seek a mentor.

Chief Fire Officer and Chief Medical Officer

The Center for Public Safety Excellence (CPSE) provides the terminal focal point in fire officer professional development. It builds upon the NFPA *Fire Officer Professional Qualification* certification training and the IAFC *Officer Development Handbook* and provides ". . . personal guidance for career planning and development and recognizes lifelong career excellence and achievement."

Under CPSE, the Commission on Professional Credentialing awards the designation of Chief Fire Officer or Chief Medical Officer to candidates who have completed an application and validation process. Successful candidates have demonstrated a strategy for continued career improvement and development. They also have met the education, certification, training, professional contributions, professional affiliation, and community involvement requirements. There were 630 Chief Fire Officers and 34 Chief Medical Officers as this book went to press.

Summary

Providing and assessing skill and knowledge training are some of the most frequent fire officer activities. Making fire fighters competent and confident in their emergency service duties has a powerful impact on fire fighter safety and operational effectiveness.

Digital video technology is reshaping the way emergency service collects information and will influence how a fire officer provides training. You can download pictures and videos of incidents as they are occurring. There are websites available to download training materials, including Firefighter Close Calls, Everyone Goes Home, commercial, and individual websites.

The four-step method works when training rookie fire fighters, providing proficiency training, preparing for a competitive challenge, or developing a work improvement plan. The company officer should plan two training activities for every fire station workday: a 5-minute drill on a high-risk/low-frequency item and a longer proficiency session.

The company officer should consult the IAFC model when preparing a career development guide, vetting the training, education, experience, and self-development tasks with the requirements of the local organization.

You Are the Fire Officer: Conclusion

The personnel regulations prohibit extending this recruit's probationary period. After conferring with the administrative fire officer and your senior apparatus operator, you meet with the recruit.

You describe how the recruit's workday will progress until the probationary field manual is completed. The recruit will be working with the senior apparatus operator to complete the pump operator proficiency knowledge and skills. In order to master the knowledge of Station 5's district, the recruit will use easel pad maps, driving drills, and Google Earth graphics. Other members of the shift will develop flash cards to test the specialized knowledge and procedures that are part of the probationary exam. You lay out the recruit's on-duty work schedule for the next 10 weeks:

- 0700: Line up, flash-card pop quiz
- 0730–0830: Apparatus and equipment check
- 0830–1030: Physical training, station maintenance, and pumping drill in rear of fire station
- 1030–1100: Street drill with easel pad maps (recruit and senior apparatus operator)
- 1100–1200: Driving drill (find five locations in the district) and lunch run
- 1300–1700: Scheduled company training, productivity, and flash-card pop quiz
- 1700–1900: Dinner and station clean-up
- 1900–2000: Recruit works on drawing personal map of district, target hazard overview, or making new knowledge flash cards

You encourage the recruit to do additional work during the scheduled days off. You will meet with the recruit weekly to assess her progress and start providing practice tests to prepare for the end-of-probation exam.

Wrap-Up

Chief Concepts

- Education is the process of imparting knowledge or skill through systematic instruction. Training is directed toward the practical application of education to produce an action, which can be an individual or a group activity.
- The NBFSPQ has accredited the certification programs operated by 35 states, provinces, and other agencies. Accreditation is issued for certification based on 16 different standards and 67 recognized levels of fire service–related competencies.
- The four-step method is a core part of most Fire Instructor I certification programs. The components of the four-step method are:
  - Preparation
  - Presentation
  - Application
  - Evaluation
- When presenting tasks, break the topic down into simple units.
- Ensuring proficiency of existing skill sets includes having fire fighters practice their craft in full personal protective clothing and perform fire-ground tasks, such as cutting a ventilation hole.
- In-station lectures should be no more than 15 or 20 minutes long.
- The fire officer has to provide the guidance that advances fire fighters from being capable of performing the basic required skills to being able to perform them effectively, efficiently, and consistently.
- In psychomotor skill development,
  - At the initial level, the fire fighter can recall the sequence but the individual skill activities are uneven or inconsistent.
  - At the plateau level, the fire fighter is competent.
  - At the latency level, the fire fighter can recall skill steps without significant effort.
  - At the mastery level, the fire fighter is confident in his or her ability to perform the skill.
- Four federal regulations that require fire fighter training are:
  - OSHA regulation 29 CFR 1910.1030, "Occupational Exposure to Bloodborne Pathogens"
  - OSHA regulation 29 CFR 1910.20, "Hazardous Waste Operations and Emergency Response (HAZWOPER) Training"
  - OSHA regulation 29 CFR 1910.134, "Respiratory Protection Training"
  - Homeland Security Presidential Directive (HSPD) 5, Managing Domestic Incidents
- Live fire training must always comply with NFPA 1403.
- A good lesson plan:
  - Organizes the lesson
  - Identifies key points
  - Can be reused
  - Allows others to teach the program

Hot Terms

Accreditation A system whereby a certification organization determines that a school or program meets with the requirements of the fire service.

Coaching A method of directing, instructing, and training a person or group of people with the aim to achieve some goal or develop specific skills.

Education The process of imparting knowledge or skill through systematic instruction.

Job instruction training A systematic four-step approach to training fire fighters in a basic job skill: (1) prepare the fire fighters to learn, (2) demonstrate how the job is done, (3) try them out by letting them do the job, and (4) gradually put them on their own.

Mentoring A developmental relationship between a more experienced person, a mentor, and a less experienced person, referred to as a protégé.

Training The process of achieving proficiency through instruction and hands-on practice in the operation of equipment and systems that are expected to be used in the performance of assigned duties.

Unconsciously competent The highest level of the Conscious Competence Learning Matrix developed by Dr. Thomas Gordon in the 1970s. At the unconsciously competent level the skill becomes so practiced that it enters the unconscious parts of the brain—it becomes second nature.

Fire Officer *in Action*

Your department provides a wide range of services: fire suppression, hazardous materials, technical rescue, public education, and emergency medical services. There are dozens of highly critical activities that are infrequently used. You consider the recommendation made by public safety expert Gordon Graham: spend 5 minutes each day covering one high-risk/low-frequency knowledge or skill item.

You have the crew start a list of items that need to be covered in a 5-minute drill that will be delivered as part of the morning line-up. Some items are part of the monthly training program. Some topics come from Fire Fighter Near Miss, Everyone Goes Home, and Firefighter Close Calls. Other topics come from your department's standard operating procedures and regulations.

Fire fighter training and development remains a core fire officer task since the Officer's School was established in New York City in 1869. While the apparatus, tools, and educational techniques have changed, the special responsibility of training and coaching fire fighters remains the same.

1. When can a person be utilized as a victim in a live fire training session?
   A. Never
   B. Only in gas-fired training center buildings
   C. Only when there is a safety officer assigned to the victim sector
   D. Only in non–gas-fired training center buildings

2. There are four federal regulations that affect every fire fighter. Which of the following is one of them?
   A. Fire Fighter I
   B. Rescue Systems I
   C. SCBA fit testing
   D. WMD awareness

3. Using a video to show an incident or situation is an example of the _____ step in the four-step process.
   A. preparation
   B. presentation
   C. application
   D. evaluation

4. "The apparatus operator can remember the route to an area of the district where the engine company has not had a response for several months" is a description of the _____ level of psychomotor skill.
   A. initial
   B. plateau
   C. latency
   D. mastery

Evaluation and Discipline

NFPA 1021 Standard

Fire Officer I

4.2 Human Resource Management. This duty involves utilizing human resources to accomplish assignments in accordance with safety plans and in an efficient manner. This duty also involves evaluating member performance and supervising personnel during emergency and nonemergency work periods, according to the following job performance requirements. [p 142–144, 147]

4.2.4 Recommend action for member-related problems given a member with a situation requiring assistance and the member assistance policies and procedures, so that the situation is identified and the actions taken are within the established policies and procedures. [p 142, 148–152]

(A)*Requisite Knowledge. The signs and symptoms of member-related problems, causes of stress in emergency services personnel, adverse effects of stress on the performance of emergency service personnel, and awareness of AHJ member assistance policies and procedures. [p 142, 154–155]

(B) Requisite Skills. The ability to recommend a course of action for a member in need of assistance. [p 154–155]

4.2.5 Apply human resource policies and procedures, given an administrative situation requiring action, so that policies and procedures are followed. [p 143–144, 151–152, 154]

(A) Requisite Knowledge. Human resource policies and procedures. [p 143–144, 151–152, 154]

(B) Requisite Skills. The ability to communicate orally and in writing and to relate interpersonally. [p 142–144, 147–149, 151–152, 154]

Fire Officer II

5.2 Human Resource Management. This duty involves evaluating member performance, according to the following job performance requirements. [p 144–147]

5.2.1 Initiate actions to maximize member performance and/or to correct unacceptable performance, given human resource policies and procedures, so that member and/or unit performance improves or the issue is referred to the next level of supervision. [p 151–155]

(A) Requisite Knowledge. Human resource policies and procedures, problem identification, organizational behavior, group dynamics, leadership styles, types of power, and interpersonal dynamics. [p 144–147, 151–155]

(B) Requisite Skills. The ability to communicate orally and in writing, to solve problems, to increase teamwork, and to counsel members. [p 144–147, 151–155]

5.2.2 Evaluate the job performance of assigned members, given personnel records and evaluation forms, so each member's performance is evaluated accurately and reported according to human resource policies and procedures. [p 144–146]

(A) Requisite Knowledge. Human resource policies and procedures, job descriptions, objectives of a member evaluation program, and common errors in evaluating. [p 144–147]

(B) Requisite Skills. The ability to communicate orally and in writing and to plan and conduct evaluations. [p 144–146]

Introduction to Fire and Emergency Services Administration (FESHE) Course Objectives

4. Recognize appropriate appraising and disciplinary actions and the impact on employee behavior. [p 142–154]

7. Identify roles and responsibilities of leaders in organizations. [p 142–143, 147]

12. Describe the benefits of documentation. [p 154]

13. Identify and analyze the major causes involved in line of duty firefighter deaths related to health, wellness, fitness and vehicle operations. [p 154–155]

Knowledge Objectives

After studying this chapter, you will be able to:

- Describe the special requirements for supervising a probationary fire fighter who has a structured in-station training program.
- Describe how to use a performance log or T-account to document fire fighter work performance.
- List the activities associated with a mid-year review.
- Describe the requirements of an advanced notice of a substandard employee evaluation.
- Describe the concept of progressive discipline.
- List and describe the components of a written reprimand.
- Describe the services available through an employee assistance program (EAP).

Skills Objectives

After studying this chapter, you will be able to:

- Use a performance log or T-account.
- Issue an oral reprimand, warning, or admonishment when a fire fighter demonstrates a substandard or unacceptable behavior.

The fire station phone rings at 6:50 A.M. Sam, one of the members of your crew, is asking for Frank, one of the members of the crew that is finishing their 24-hour shift. After a couple of minutes, Frank asks you to pick up the phone. Sam says that Frank will stay until Sam gets to the fire station, at about 10 A.M. You agree to this last-minute coverage.

The department regulations require fire fighters to notify their supervisor at least 1 hour before the reporting time if they cannot get to work due to illness or unplanned vacancies. This is the seventh or eighth time in 3 months that Sam is both late in reporting for work and failing to notify the supervisor by 6:00 A.M. The first few times went undocumented, but the last time generated a written reprimand. You thank Frank for covering the position and notify the administrative fire officer.

1. How should you handle a fire fighter who has an existing disciplinary issue?
2. What should you do about this morning's late reporting to work?
3. Should you be involved in resolving off-duty problems that affect on-duty fire fighter performance?

Introduction to Evaluation and Discipline

Evaluation and discipline are essential components of a fire fighter's development. The fire officer plays a key role in developing fire fighter success. For probationary fire fighters, their first fire officer sets the stage for a 20- to 40-year career.

Supervision of fire fighters requires that the fire officer conduct regular evaluations to provide feedback on job performance, on-duty behavior, and problem resolution. Regular evaluations of employees are required and should be approached in a standard, professional manner.

The officer is also responsible, to a certain extent, for supervising off-the-job behavior where it reflects on the fire department. Performance and behavior are two different issues, but they are linked in the sense that the officer has to monitor, evaluate, and deal with both types of problems.

Discipline can be either positive or negative. Positive discipline is intended to help the employee recognize problems and make corrections to improve performance or behavior. Negative discipline is intended as punishment for unsatisfactory performance, unacceptable behavior, or failure to respond appropriately to positive discipline.

Discipline is progressive, moving from positive to negative and from minor to major. The nature and the seriousness of the problem and the employee's efforts to correct reoccurring problems influence the starting point of the process. Progressive discipline could begin with verbal counseling, then move on to a verbal reprimand with a written note to be put in the employee's file, then to a formal written reprimand, and finally to suspensions, reductions in pay, or demotion. Dismissal is considered as the ultimate level of negative discipline for an employee.

Some problems are so serious that the earlier steps in progressive discipline might be bypassed. Most fire departments or municipalities have a description of offenses that are considered to be so serious that they can lead to immediate suspension. Although these steps and actions are spelled out in written policies, a fire officer still uses judgment and discretion when performing evaluation and disciplinary functions.

Evaluation

A fire officer is expected to conduct regular evaluations of fire fighter performance. Evaluations are performed to ensure that each fire fighter knows the performance that is expected while on the job and where he or she stands in relation to those expectations. This process helps the fire fighter set personal goals for professional development and performance improvement and provides positive motivation to perform at the highest possible level Figure 8-1 ▸.

Figure 8-1 The evaluation process helps the fire fighter set personal goals for professional development and performance.

Dodd J. Miller Training Academy
Graduates from the Dodd J. Miller Training Academy at the Dallas, Texas, Fire-Rescue Department have completed a 15-month training program that includes fire suppression and paramedic certifications. A unique characteristic of the Dallas program is that the graduates are ready to assume all fire fighter job responsibilities, including paramedic in charge of the ambulance and apparatus driver/operator.

Most career fire departments require a supervisor to conduct an annual performance evaluation for each assigned employee. The annual performance evaluation is a formal written documentation of the fire fighter's performance during the rating period. This permanent personnel record is important to both management and the fire fighter, but it cannot replace the ongoing supervisory actions that maintain and improve employee performance throughout the year.

Volunteer fire departments provide an equivalent form of periodic evaluation for each member by a higher-ranking individual. The documentation procedures are often less structured; however, every member deserves evaluation. The responsibilities of a supervisor are just as important in a volunteer fire department as in a full-time career organization.

Starting the Evaluation Process with a New Fire Fighter

Fire officers have a special responsibility when starting an evaluation process with a probationary fire fighter. The fire officer is helping to shape that individual's fire department career. The officer will be shepherding the probationary fire fighter through his or her first real-life emergency experiences, providing feedback and guidance. Most fire fighters have vivid memories of their first supervisor and continue to respond to the expectations established by that officer.

New fire fighters start with wide variations in the range and depth of their skills. In some departments, probationary members can start to ride the apparatus after receiving a few hours of orientation. Many fire departments require new members to achieve NFPA 1001 Fire Fighter and EMT certification before they are authorized to respond to any alarms. It is the fire officer's responsibility to determine each individual's skills, knowledge, aptitudes, strengths, and weaknesses and then set specific expectations for each new fire fighter.

Recruit Probationary Period

Most fire departments have structured in-station training that is part of the recruit fire fighter's probationary period. Regardless of the new fire fighter's level of certification or pre-employment experience, the fire officer is responsible for evaluating each individual during the probationary period. Often there is a classified job description that specifies all of the required knowledge, skills, and abilities that a fire fighter is expected to master within a specified time period in order to complete the probationary requirements. The fire fighter should be provided with a copy of the specific requirements to use as a checklist Figure 8-2 ▶.

During the probationary period the recruit fire fighter is expected to obtain experiences and demonstrate competencies related to the classified job description. This period often includes assignments on different types of companies and specialty units operated by the department, such as an engine company, a ladder company, an ambulance, and even the 9-1-1 center. Competencies may include demonstrating all of the skill sets required for NFPA 1001 Fire Fighter II certification.

Structured probationary programs require the fire officer to complete a monthly evaluation of each probationary fire fighter. This evaluation typically assesses the probationary fire fighter's progress in four areas:

- Fire fighter skill competency, including proficiency as an apparatus operator
- Progress in learning job-specific information not covered in basic training, such as the local fire prevention code and the department's rules, regulations, and standard operating procedures
- Progress in learning the fire company district, including streets and target hazards
- Performance of other job tasks, such as recording deliveries, performing housework, conducting in-station tours, and completing reports

In volunteer fire departments, it is common for different individuals to certify that a probationary fire fighter has met the requirements for different components of the program. One officer should be specifically assigned to oversee the progress of each individual probationary member to ensure that the overall program requirements are being accomplished successfully.

The fire fighter probationary period creates a foundation for a long career. The fire officer has a special opportunity to prepare future departmental leaders by providing a comprehensive and effective probationary period.

City of Charlottesville Fire Department

Fire Fighter Job Description

General Definition of Work
The fire fighter performs responsible service work in fire suppression and prevention; does related work as required. Work is performed under the regular supervision of a company and/or shift commander.

Typical Tasks
Responds to alarms, drives and operates equipment and related apparatus, and assists in the suppression of fires, including rescue, advancing lines, entry, ventilation and salvage work, extrication, and emergency medical care of victims.
Performs cleanup and overhaul work, establishes temporary utility services.
Assists in maintaining and repairing fire apparatus and equipment, and cleaning fire stations and grounds.
Checks fire hydrant flows.
Makes fire code inspections of business establishments and prepares pre-fire plans.
Responds to emergency and nonemergency calls, pumps out basements, inspects for gas leaks, secures vehicle accidents, inspects chimneys, etc.
Participates in continuing training and instruction programs by individual study of technical material and attendance at scheduled drills and classes.
Conducts station tours for the public, school, and community demonstrations and programs.
Backs up for dispatching personnel, monitors alarm boards, receives and transmits radio and telephone messages.
Performs related tasks as necessary.

Knowledge, Skills, and Abilities
General knowledge of elementary physics, chemistry, and mechanics; general knowledge of technical firefighting principles and techniques, and principles of hydraulics applied to fire suppression; general knowledge of the street system and physical layout of the city; general knowledge of emergency care methods, techniques, and equipment; ability to understand and follow written and oral instructions; ability to establish and maintain cooperative relationships with fellow employees and the public; ability to keep simple records and prepare reports; possess a strong mechanical aptitude; ability to perform heavy manual labor; skill in operation of heavy emergency equipment.

Education and Experience
Any combination of experience and training equivalent to graduation from high school.

Special Requirements
Possession of a valid driver's permit issued by the Commonwealth of Virginia.

Future Requirements at Three Years of Service
NFPA 1001   Fire Fighter Level II
NFPA 1002   Driver/Operator Certification
Commonwealth of Virginia Emergency Medical Technician or greater

Figure 8-2   Sample fire fighter job description.

Annual Evaluations

The personnel regulations in most career fire departments require that every employee who has completed the initial probationary period will receive an annual written evaluation from his or her immediate supervisor. These annual evaluations become a formal part of the employee's work history.

The fire officer may receive an annual evaluation form from the human resource or personnel division a month or two before the employee's annual evaluation is due to be submitted. Some organizations expect the officer to keep track of the dates and submit a completed form when the evaluation for each employee is due.

The annual evaluation process requires four steps. First, the supervisor fills out a standardized evaluation form. The form asks the supervisor to evaluate the subordinate on a number of knowledge areas, skills, and abilities appropriate for the subordinate's rank and classified job description. There is usually an opportunity for the supervisor to add comments in a narrative section. The supervisor's manager is required to review the evaluation before it is issued to the subordinate.

In the second step the subordinate is allowed to review and comment on the officer's evaluation. The subordinate has an opportunity to provide feedback or additional information that would affect the performance rating.

The third step is a face-to-face feedback interview between the supervisor and the subordinate to discuss the evaluation. This is the opportunity to clarify points and review the performance

> March 11, Incident 3467. Trouble positioning apparatus to hook up soft suction and hydrant. Charged wrong attack line. First structural fire as engine driver.

Figure 8-3 Sample notation in a performance log.

over the last rating period. A goal of the feedback interview is that both the supervisor and the subordinate understand the results of the evaluation.

The final step, usually completed at the end of the feedback interview, is to establish goals for the subordinate to accomplish during the next evaluation period. These are usually specific, measurable goals that allow the subordinate to improve on the knowledge, skills, and abilities of the current job or prepare for the next higher job level.

Conducting the Annual Evaluation

The completion of the annual evaluation forms should be viewed as a formality. The fire officer should be providing continual evaluation and feedback to the fire fighter throughout the year, so the information that goes onto the form should not come as a surprise. Unless there is a problem with the employee's performance, the scheduled evaluation should be used primarily as an opportunity to discuss future goals and objectives.

The methods outlined in the following section are designed to track an employee's performance and progress throughout the year. Using the information gathered from a year of entries in a performance log or T-account, the fire officer can provide an objective and well-documented annual performance evaluation. This documentation is essential, regardless of whether the fire fighter receives an "outstanding" rating or an "unsatisfactory" rating.

Keeping Track of Every Fire Fighter's Activity
In a **performance log**, the fire officer maintains a list of the fire fighter's activities by date, along with a brief description of performance observations **Figure 8-3**.

The **T-account** is a slightly more sophisticated documentation system, similar to an accounting balance sheet listing credits and debits. A single-sheet form is used to list the assets on the left side and the liabilities on the right side, appearing like a letter "T" **Figure 8-4**.

Using either method, the fire officer compiles an extemporaneous "when, what, and how" record of each fire fighter's work history throughout the evaluation period. This record can be a powerful evaluation and motivational tool.

Establishing Annual Fire Fighter Goals
The fire officer should require all fire fighters who have completed probation to identify three work-related goals they want to achieve during the next evaluation period. This provides focus beyond the minimum day-to-day activities and helps fire fighters prepare for future promotional examinations or assignments. Examples of individual fire fighter goals might include:

- Enrolling in and completing a building construction course at the local community college
- Becoming qualified as an aerial operator
- Learning how to use the computer-aided design software in order to develop tactical pre-incident plans

The fire officer should track the progress toward these goals during the evaluation period, recording progress in a T-account or a performance log.

Informal Work Performance Reviews
Most civil service and personnel regulations require an annual evaluation. Many departments encourage fire officers to conduct informal performance reviews with each fire fighter throughout the year. During these informal sessions, the fire officer can review the T-account or performance log with the fire fighter and see what the officer can do to assist the fire fighter in meeting the established goals. Together they can identify situations or work conditions that impede progress. For example, a fire fighter who is spending 70 percent of the time detailed to cover positions at other stations will have difficulty qualifying as a back-up aerial operator. The fire fighter and officer could discuss changing the goal or adjusting assignments so that the fire fighter would have more time at the station to work on meeting the goal.

Mid-Year Review
The mid-year review is another informal performance review that requires a higher level of documentation. The officer should have each fire fighter write a self-evaluation in preparation for the session. This allows the fire fighter to focus on the job description and the personal goals that were set at the beginning of the year. The officer and the fire fighter should review this document together and discuss how well expectations are being met.

If necessary, the officer can identify resources or assistance that would help the fire fighter meet the mutually established goals. For example, the officer might suggest meeting with another fire fighter who is experienced at working with the pre-incident planning software or point out that only one semester remains to enroll in the building construction course at the college during the evaluation period.

+	−
March 11, Incident 3467. Rescue of elderly female from hallway outside burning bedroom in a one-story, single-family dwelling.	March 11, Incident 3467. Had to return to Engine 4 to retrieve forcible entry tools and firefighting gloves during initial actions.

Figure 8-4 Sample notation in a T-account.

This is also a time when the personal goals can be adjusted because of changes in the work environment. It would be unrealistic to maintain the goal to attend a building construction class at the community college if the department has selected the fire fighter to attend paramedic school for the next 9 months.

Advance Notice of a Substandard Employee Evaluation

A fire fighter who is not meeting expectations should know there is a problem long before the annual evaluation. The subordinate should be given adequate time to change the behavior or improve the skill, particularly if it is jeopardizing a scheduled pay raise or continuing employment.

Municipalities require a supervisor to provide a formal notification to an employee who is likely to be evaluated below the level necessary to obtain a pay increase or a "satisfactory" rating. The notice must identify the aspect of the job performance that is substandard and what the employee must do to avoid receiving a substandard annual evaluation. A common practice is to require such a notification 10 weeks before the annual evaluation is due.

If the employee receives a substandard annual evaluation, then the municipality might require a **work improvement plan**. The plan should cover a specific period of time, such as 120 calendar days, that would be designated as a special evaluation period. The employee is expected to fully participate in a work improvement plan during this special evaluation period. He or she is given an opportunity to demonstrate the desired workplace behavior or performance no later than the end of the special evaluation period.

The following is an example of the human resource division requirements that could be applied to a work improvement plan:

- The work improvement plan shall be in writing, stating the performance deficiencies and listing the improvements in performance or changes in behavior required to obtain a "satisfactory" evaluation.
- During the special evaluation period, the employee shall receive monthly progress reports.
- If, at the end of the special evaluation period, the employee's performance rating is "satisfactory" or better, the time-in-rank pay increase will start at the first pay period after the work improvement period.
- If, at the end of the special evaluation period, the employee rating remains unsatisfactory, then no time-in-grade pay increase will be issued. In addition, the supervisor will determine whether additional corrective action is appropriate.

Regardless of the outcome, a special evaluation period performance evaluation is filled out and submitted as a permanent record in the employee's official file.

The employee's performance is officially evaluated during the annual review. Conducting informal reviews enables a fire officer to identify an underperforming fire fighter early in the annual evaluation period. The advanced-notice procedure should provide the fire fighter with enough time to change the behavior or improve the necessary skills before the formal evaluation is scheduled to occur. If the employee has still not improved, a substandard evaluation would not be a surprise to either the fire fighter or the fire officer's supervisor.

Six Weeks Before the End of the Annual Evaluation Period

The fire fighter should conduct another self-evaluation approximately 6 weeks before the official annual evaluation is due. The goal of this self-evaluation is to identify how well the fire fighter has met the organizational expectations and personal goals for the year. The fire officer should review this document and provide feedback. Together, the fire officer and fire fighter develop the final, formal evaluation report for this rating period, which includes developing three new work-related personal goals for the next evaluation period.

Completing annual evaluation reports is an important fire officer responsibility. It is an opportunity to identify and evaluate the work performed by subordinates. The fire officer should recognize outstanding accomplishments, encourage improvement, and, in some cases, identify those who should be considering another career option.

The fire officer who maintains a performance log or T-account, conducts bimonthly informal reviews, conducts a mid-year review, and has the fire fighter complete a self-evaluation 6 weeks before the annual evaluation will have a wealth of information to present an accurate, well-documented, and comprehensive annual evaluation.

■ Evaluation Errors

Using a T-account or performance log to assemble a detailed list of work behavior observations over the evaluation period provides excellent background when the fire officer is preparing a fire fighter's evaluation. Evaluation is a largely subjective process that is vulnerable to unintentional biases and errors.

Leniency or Severity

Some fire officers tend to rate all of their fire fighters either higher or lower than their actual work performance. This is called leniency or severity, respectively. Leniency reduces conflict. A positive evaluation is likely to make the evaluation a more pleasant experience and avoids confrontation. Leniency is common when the fire officer is required to conduct a face-to-face meeting with the fire fighter to review the evaluation.

Some fire officers lean in the opposite direction and rate all fire fighters with "needs improvement" or "unsatisfactory." Some officers think that low ratings will cause fire fighters to be motivated to work harder. Newer fire officers sometimes make this mistake when they have not received adequate training in preparing performance evaluations.

Personal Bias

Personal bias is an evaluation error that occurs when the evaluator's perspective skews the evaluation such that the classified job knowledge, skills, and abilities are not appropriately evaluated. Fire officers must not allow an evaluation to be slanted by race, religion, gender, disability, or age.

Recency

Recency is an evaluation error in which the fire fighter is evaluated only on incidents that occurred in the last few weeks rather than on all of the events that occurred throughout the evaluation period. Fire fighters who are aware of this tendency are on their best behavior in the weeks leading up to the fire officer's evaluation.

Central Tendency

Most evaluation systems involve some type of rated scale, ranging from unsatisfactory to outstanding. A fire officer demonstrates a **central tendency** when a fire fighter is rated in the middle of the range for all dimensions of work performance.

A central tendency evaluation provides little value to the fire fighter or the evaluation process. Being rated as "OK in all areas" is not very informative or helpful.

Frame of Reference

In a **frame of reference** evaluation error, the fire fighter is evaluated on the basis of the fire officer's personal ideals instead of the classified job standards. For example, a fire officer who spent hours every day fixing, improving, and polishing the engine when he was an apparatus driver/operator might issue "unsatisfactory" or "needs improvement" ratings to apparatus operators who are meeting all of the departmental standards but are not meeting his personal ideals.

Halo and Horn Effect

Like life, fire fighters' performances are not just black and white. The fire officer may concentrate on only one aspect of the fire fighter's performance, which is either exceptionally good or bad, and apply that perception across the board to all aspects of the individual's work performance. This is called the **halo and horn effect**.

Contrast Effect

Contrast effect is an evaluation error that can occur when the fire officer compares the performance of one subordinate with the performance of another subordinate instead of against the classified job standards.

Providing Feedback After an Incident or Activity

Performance evaluation should be a continual supervisory process, not a special event that is performed only when a scheduled rating has to be submitted. Regular feedback from the fire officer should keep fire fighters aware of how they are doing, particularly after incidents or activities that present a special challenge.

Performance feedback is most effective when delivered as soon as possible after an action or incident. That means providing essential feedback before the ashes get cold after a structure fire. The fire fighter is intensely aware of the event and wants to know how well he or she performed. The fire officer needs to be ready to provide specific information in order to recognize or improve fire fighter performance **Figure 8-5 ▶**.

Near Miss REPORT

Report Number: 07-0001182

Event Description: Dispatched to a report of a heavily involved structure fire on an early Sunday morning. It was unknown if residents were in the house; however, two vehicles were in the driveway. Prior to arrival, it was stated that … the family was believed to be inside. On arrival, we had heavy fire conditions. The front door was breached and we had immediate fire. To the right was a stairway to the upstairs and the stair case was fully involved. We made our way toward the back of the home and knocked down the majority of fire. We then called for a second crew to help extinguish while we went upstairs to extinguish and search. Knocked down stairway area where there was extremely high heat and very uncomfortable.

At top of steps my partner yelled that the fire had started again behind us. I turned to extinguish the fire and the fire at the top of the steps burned me. I exited the fire when the second crew arrived not knowing what happened. I just knew that I was exhausted. I returned to the house and did virtually nothing. Fire was knocked down prior to my return. Victims were found and then we exited.

It was later learned through debriefings that a Lieutenant asked a FF to knock down the fire in the gable end of the home with a 2 1/2 from the exterior. When questioned by the junior the Lieutenant exclaimed, "Just knock it down real quick." It had to have been at that moment that I was burned because the fire shot down the steps like a blow torch. I was treated and released with 1st- and 2nd-degree burns on my arm, shoulder, and face. I had some third-degree burns on my ear and I was wearing a Nomex hood. It was also learned that it was arson and up to 5 gallons of gas was used in the home.

Lesson Learned: No notification was made to the interior crew as to the 2 1/2 being used outside. Remember the basics. No exterior attack while FFs are inside. Correction: Even officers must go through command to make a decision to spray water into an occupied house (although in my opinion it should NEVER be done).

Figure 8-5 Performance feedback should be delivered as soon as possible after an action or incident.

Discipline

Discipline is a moral, mental, and physical state in which all ranks respond to the will of the leader. The fire officer builds discipline by training to meet performance standards, using rewards and punishments judiciously, instilling confidence in and building trust among team leaders, and creating a knowledgeable collective will.

Within the fire department, discipline divides into positive and negative sides. Positive discipline is based on encouraging and reinforcing appropriate behavior and desirable performance. Negative discipline is based on punishing inappropriate behavior or unacceptable performance. Both positive and negative discipline can be applied to a full range of activities, including emergency incidents and administrative functions.

Positive discipline should be used before negative discipline. Progressive discipline refers to starting out to correct a problem with positive discipline and then increasing the intensity of the discipline if the individual fails to respond to the positive form, perhaps by using mild negative discipline. Increasingly, negative discipline might have to be used in a situation in which an individual fails to respond in an appropriate manner to correct the problem. There are exceptions to this rule; some actions or behaviors are so unacceptable that they must result in immediate negative discipline.

Positive Discipline: Reinforcing Positive Performance

Positive discipline is directed toward motivating individuals and groups to meet or exceed expectations. The key to positive discipline is to convince them that they want to do better and are capable and willing to make the effort. A fire officer provides positive discipline by identifying weaknesses, setting goals and objectives to improve performance, and providing the capability to meet those targets.

The starting point for positive discipline is to establish a set of expectations for behavior and performance. Once the expectations are known, there must be a consistent and conscientious effort to meet them. The expectations have to apply to the entire team as well as to each individual team member. Positive discipline is reinforced by recognizing improved performance and rewarding excellent performance.

If an individual fire fighter's performance needs improvement in a particular area, the fire officer should coach that person, providing guidance and extra opportunities to correct the problem. Sometimes, simply pointing the individual in the right direction and offering encouragement can achieve the objective. In many cases, the officer can arrange for another fire fighter to work with the individual to correct the problem.

Teamwork is a key factor for fire companies. In addition to each individual having the required knowledge and skills, the company must be able to work efficiently and effectively as a team. All of the company members have to work, learn, and practice together to become capable and confident. The fire officer has to provide the leadership to make it happen.

An officer sets the stage for positive discipline by setting clear expectations and by "walking the talk." Fire officers are working supervisors; they supervise while directly participating in firefighting activities and performing nonemergency duties. The officer should demonstrate a personal commitment to the department's goals, objectives, programs, rules, and regulations by participating in all of the activities that are expected of fire fighters, such as physical fitness training and regular company drills. Fire fighters can gauge an officer's level of commitment by observing the characteristics of self-discipline **Figure 8-6 ▶**.

Competitiveness can also be used as a stimulant in positive discipline, particularly at the company level. Fire fighters are naturally competitive, and most fire companies work hard to prove that they can be better, faster, more skillful, or more impressive than a rival company. The officer has to point that competitive energy in a positive direction, making sure that the ultimate objective is high performance.

Figure 8-6 Fire officers set the stage for positive discipline by "walking the talk."

Empowerment

Empowerment is one of the most effective strategies of positive discipline. Fire fighters often complain that they have little control over their work environment; they are told where to go, what to do, and when to do it. Fire officers can help make fire fighters feel stronger by learning how to control their own destiny. An officer who identifies an area where improvement is needed can often empower fire fighters to correct the problem on their own. It is important for the officer to identify the target and provide the resources, but doing the work on their own and demonstrating their capabilities can be a very positive motivator for the fire fighters.

Providing information to help fire fighters learn more about the job can support the empowerment process. The first component can be described as "Local Government 101." This information helps fire fighters learn more about how the fire department and local government work. It could include reviewing the approved budget for the next fiscal year or discussing which functions are performed by each part of the organization.

Company Tips

Hands-On Skill Drills
A commitment to in-station training produces competent and confident fire fighters and fire companies. A fire officer should provide frequent hands-on opportunities for fire fighters to practice their emergency service delivery skills, such as laying out hose, throwing ladders, forcing doors, and cutting roofs. Activity should be scheduled every day.

Many fire departments have developed training props for practicing with hand and power tools. As a confidence builder, fire fighters should practice working with hose lines and tools in simulated fire-ground scenarios while wearing full personal protective clothing **Figure 8-7 ▾**. Fire officers should also look out for opportunities to conduct realistic exercises in buildings scheduled for demolition.

The more fire fighters understand about how the organization and local government work, the more they can feel a sense of participation.

The second component could be described as "Success 102." This information would identify the tools that others have used to achieve success within the fire department. If some companies are consistently recognized for positive performance, what are they doing right? When a few individuals achieve high scores on every promotional examination, what is their secret method? Do fire fighter self-study groups improve promotional examination results? Answering these questions may take some research by the fire officer, such as identifying the individuals or groups that have performed well in the promotion process and asking them how they prepared. An officer can often help fire fighters succeed by identifying successful practices.

Figure 8-7 Fire fighters should practice working with hose lines and tools in simulated fire-ground scenarios while wearing full personal protective clothing.

Voices of Experience

Let me start by saying that employee counseling and discipline is an area that I really had major issues with when I first got promoted to Lieutenant. I had learned that it was important to be firm, fair, and able, but really wanted to be liked and struggled with being firm.

When faced with employee conflict, I often took the easy way out by ignoring or avoiding the issues and people causing the problems. I had good intentions and I would tell myself I needed to take action, but the thought of confronting crewmembers often made me feel physically ill. I can still remember the headaches and upset stomach at times.

Like it or not, what you permit…you promote! I quickly discovered that problems that are not dealt with usually linger and have a tendency to become bigger issues.

I can honestly say that I turned a serious weakness into one of my greatest strengths. One of the best things I did to become better at counseling and discipline was to develop steps on how to go through the process. Since I had never received training on how to conduct a counseling session, I sought out a friend who did counseling on a regular basis. I started by developing a list of the most common problems and then we did a series of role playing exercises. The first one was pretty difficult for me, but it didn't take long to become comfortable with the steps that were developed.

> **"Like it or not, what you permit… you promote!"**

Employee Counseling and Discipline steps:
1. Identify problem(s) and concerns/effects—One at a time; focus on job performance.
2. Ask for reason and allow response—Crew members will have issues, but you can't solve them; offer available resources such as Employee Assistance Program (EAP).
3. State expectations—Desired performance; be positive and avoid negative "don't…"
4. Develop a plan—Ask for suggestions; include time parameters and follow-up.
5. State consequences—"If the issues continue, then…(time off, demotion, or even loss of employment)."
6. Have a positive ending—State something positive and reassure them that you know they are capable of working through this.
7. Document—Take notes as needed during the interview, but full documentation is done after the session.

Although the disciplinary process is viewed by most as being very negative, most of my experiences with employee counseling and discipline are now very positive. I would say that over 90 percent of the employees I have counseled or disciplined have walked out of the office with a smile on their face. I imagine that some might find that hard to believe, but it is true. If you approach the issue calmly and work with the person to improve the situation, most individuals will respond in a similar manner.

The last important lesson I would like to share is that while it is most important to have empathy toward those you are counseling it is also important not to personalize it. It is *their* actions that created the situation, and it is your responsibility *in the role of fire officer* to deal with the issue.

Randy Keirn
District Chief
Lealman Fire District
St. Petersburg, Florida

FIRE OFFICER II

Negative Discipline: Correcting Unacceptable Behavior

Whereas positive discipline is directed toward encouraging desirable behavior and high performance, negative discipline is aimed at discouraging unacceptable behavior and poor performance. Negative discipline is a stronger force.

Sometimes, an effort that begins with positive discipline (motivation, training, and coaching) evolves into negative discipline because of the continuing inability or unwillingness of a fire fighter to meet the required performance or behavioral expectations. If an individual does not respond to positive efforts to correct a problem, the next logical step is to punish continuing unsatisfactory performance. **Progressive negative discipline** moves from mild to more severe punishments if the problem is not corrected. The ultimate goal remains fire fighter performance improvement. In an extreme case, in which progressive discipline fails to solve the problem, it may become necessary for the organization to terminate an employee who is ineffective and unwilling to improve.

The disciplinary process is designed to be consistent and well-documented. Typical steps in a progressive negative discipline system include:

- Counsel the fire fighter about poor performance and ensure that he or she understands the requirements. Ascertain whether there are any issues contributing to the poor performance that are not immediately obvious to the supervisor. Resolve these issues, if possible.
- Verbally reprimand the fire fighter for poor performance.
- Issue a written reprimand and place a copy in the fire fighter's file.
- Suspend the fire fighter from work for an escalating number of days.
- Terminate the employment of a fire fighter who refuses to improve.

Some employee behaviors require the fire officer to immediately implement negative discipline. This is a common policy when there is willful misconduct, as opposed to inadequate performance. Personnel regulations usually provide a list of behaviors that will lead to immediate negative discipline, such as:

- Knowingly providing false information affecting an employee's pay or benefits or in the course of an administrative investigation
- Willfully violating an established policy or procedure
- Being convicted of a criminal offense that affects the ability of the employee to perform his or her job
- Displaying insubordination
- Behaving in a careless or negligent way that leads to personnel injury, property damage, or liability to the municipality
- Reporting to work when under the influence of alcohol or a controlled substance
- Misappropriating fire department property or funds

As the penalty increases, negative discipline requires the participation of higher levels of supervision. A fire officer is often required to consult with his or her supervisor before issuing formal negative discipline. Here is an example of how the negative discipline process might proceed:

- Informal oral or written reprimand: Reprimand is issued by a supervising or managing officer. Reprimand remains at the fire station level and expires after a time period no longer than 1 year.
- Formal written reprimand: Document initiated by a fire officer (usually after consulting with his or her supervisor). Some organizations require a battalion chief or other command officer to issue a written reprimand. A copy of the letter goes into the individual's official personnel file. It expires after a set time period, usually 1 year.
- Suspension: May be initiated and/or recommended by a fire officer. The suspension notice is usually issued by a battalion chief or a higher-level command officer after consultation with the fire chief or designate. The record of a suspension remains permanently in the employee's official file. Occasionally, the record is removed after a grievance, arbitration, or civil service hearing, or a lawsuit.
- Termination: Although a termination is usually recommended by a lower-level command officer, the fire chief issues the formal termination notice after consulting with the personnel office.

Oral Reprimand, Warning, or Admonishment

An **oral reprimand, warning, or admonishment** is the first level of negative discipline, considered "informal" by many organizations. By informal, it means that this discipline stays with the fire officer and does not become part of the employee's official record. For example, a new fire officer observes that the apparatus driver does not comply with the department regulation requiring that units responding to an emergency come to a complete stop when encountering a red traffic light. When this occurs, the fire officer should have a private face-to-face meeting with the apparatus driver. In this meeting, the fire officer would determine why the driver did not stop at a red traffic light and would clearly state the policy and expectation for all subsequent responses. If the fire officer determines the reason was willful noncompliance with the department regulation, he or she would use the following steps:

- Tell the driver that the officer expects compliance with the regulation requiring that rigs stop at red lights before proceeding through an intersection. Provide the driver with a printed copy of the regulation.
- Inform the driver that this is a verbal reprimand and that the fire officer will maintain a written record of the reprimand for whatever time period is required by the authority having jurisdiction. (A common practice is to maintain the written record for 366 days. At the end of that period, if there have been no further problems related to the same individual or issue, the paper record is removed and destroyed.)
- Inform the driver that continued failure to comply with the regulation will result in more severe discipline.

In most situations, these actions correct the fire fighter's behavior. If the fire fighter continues to have difficulty complying with the regulation, this is the first step in progressive negative discipline.

Informal Written Reprimand

Some fire departments require the fire officer to use a standard form when issuing an oral reprimand, warning, or admonishment.

Metro County Fire Department
Memorandum

March 4, 2013

To: Fire Fighter Arthur Johnson
             Engine 6

From: Captain William Schwartz
             Engine 6

Subject: Written Reprimand: Driving through red light intersection

On March 2, 2013, you were driving the pumper in response to a second alarm commercial fire at 16171 Enterprise Avenue, incident #0164. While driving north on Waterway Boulevard with emergency lights on and siren sounding the traffic light at Edgerly Street turned red for Waterway. While you slowed down, you did not come to a complete stop for the red light signal as required in Standard Operating Procedure 3.4.1.b: **Emergency Vehicle Response.**

This is the fourth time you have failed to come to a complete stop when encountering a red light intersection when responding to a working structural fire.

Jan 17:   Verbal reprimand by Captain Schwartz for incident #1009
Dec 22:   Counseling by Lieutenant Scopes for incident #0789
Sept 03:   Counseling by Captain Schwartz for incident #2084

Continued failure to come to a complete stop at red light intersections may result in more severe proposed discipline, including suspension or termination. This reprimand will remain in your permanent personnel folder for 366 days.

Chapter 8 of the Metro County Personnel Regulations explains your rights and options if you wish to appeal this discipline.

I have read and received a copy of this written reprimand:

_____
Fire Fighter Arthur Johnson   Date

cc: Personnel file
             Battalion Chief Andrea Walters, 3rd Battalion

Figure 8-8 Sample written reprimand.

The form covers the three items described earlier and provides a space for the employee to respond to the reprimand. The form ensures that the fire officer covers all of the requirements of an informal reprimand. In addition, the form allows the fire fighter to clearly understand that this is a disciplinary issue and to have an opportunity to respond. This type of contemporaneous record is valued by the personnel office if the issue becomes a grievance or results in a civil service hearing.

Formal Written Reprimand

A **formal written reprimand** represents an official negative supervisory action at the lowest level of the progressive discipline process. Even when the department requires a command-level officer to issue a formal written reprimand, the document is often prepared by a fire officer. The fire officer is the closest person to the issue and will be involved in the remediation effort. The fire officer should consult the personnel regulations and departmental guidelines when preparing a written reprimand.

In departments in which a fire officer can issue a written reprimand, the fire officer's supervisor should review and approve the document before it is issued to the fire fighter.

The written reprimand should contain the following:
- Statement of charges in sufficient detail to enable the fire fighter to understand the violation, infraction, conduct, or offense that generated the reprimand
- Statement that this is an official letter of reprimand and will be placed in the employee's official personnel folder
- List of previous offenses in cases in which the letter is considered a continuation of progressive discipline
- Statement that similar occurrences could result in more severe disciplinary action, up to and including termination

A written reprimand starts the formal paper trail of a progressive disciplinary process **Figure 8-8 ▲**. If the behavior is not repeated, most written reprimands are removed from the files after a designated time period, usually 1 year after the reprimand is issued.

This written reprimand has a wide potential audience. If the reprimand is appealed or becomes part of a larger disciplinary action, the fire chief, labor representatives, civil service commissioners, and attorneys read the reprimand. Fire officers must focus on the work-related behaviors and clearly explain the behavior or action that generated the reprimand.

Suspension

Suspensions are the next step of a progressive negative discipline path but have many different forms. A **suspension** is a negative disciplinary action that removes a fire fighter from the work location and prohibits the fire fighter from performing any fire department duties. A disciplinary suspension usually results from a willful violation of a policy or procedure or another specific act of misconduct. For career fire fighters, suspension is a personnel action that places an employee on a leave-without-pay (LWOP) status for a specified period. For volunteer fire fighters, a suspension means they are not allowed to respond to emergencies and, in some cases, are prohibited from entering the fire station or participating in other fire department activities. Suspensions usually run from 1 to 30 days.

A fire officer usually recommends a suspension to a higher-level officer, providing the required documentation. In some organizations, a fire officer has the authority to immediately suspend a fire fighter for the balance of a work period, pending a formal disciplinary action. Depending on the nature and the severity of the offense, career fire fighters can also be suspended with pay or placed on restrictive duty while an administrative investigation is being conducted. **Restrictive duty** is usually a work assignment that isolates the fire fighter from the public, often an administrative assignment away from the fire station environment.

In a few situations, the municipality can suspend a fire fighter before an investigation is completed. This can occur when the employee is being investigated by the fire department or a law enforcement agency for an offense that is reasonably related to fire department employment, or when an employee is waiting to be tried for an offense that is job related or a felony.

Termination

Termination means that the organization has determined that the employee is unsuitable for continued employment. In general, only the top municipal official, such as the mayor, county executive, city manager, or the civil service commission, can terminate an employee. Many senior municipal officials delegate this task to the agency or department heads, such as the fire chief. Terminations are high-stress events, with labor representatives, the personnel office, and the fire chief's office all involved in the process.

Predetermined Disciplinary Policies

Some common disciplinary issues may have a predetermined policy for specific offenses. For example, here is how one department deals with fire fighters who are late reporting to work, without mitigating circumstances:

- First offense: Oral admonition and LWOP covering the period of time employee was late. Admonition remains active for 366 days with immediate supervisor.
- Second offense (within 366 days of first offense): Written reprimand and LWOP for the period of time employee was late.

- Third offense (within 366 days of second offense): Suspension for one workday (8 hours for day-work staff or 12 hours for shift workers).
- Fourth offense (within 366 days of third offense): Suspension for four workdays (32 hours for day-work staff, 48 hours for shift workers).
- Fifth offense (within 366 days of fourth offense): Suspension for 10 workdays (80 hours for day-work staff, 120 hours for shift workers).
- Sixth offense (within 366 days of fifth offense): Proposed termination.

Alternative Disciplinary Actions

Depending on the nature and the severity of the offense, a variety of penalties can be imposed, such as:

- Extension of a probationary period: If the fire fighter is in a probationary period, as occurs with a recruit or after a promotion, the probationary period can be extended until the work performance issue is resolved.
- Establish a special evaluation period: An incumbent fire fighter might be given a **special evaluation period** to resolve a work performance/behavioral issue. For example, a fire fighter who fails a required recertification examination could be placed in a special evaluation period until he or she is recertified.
- Involuntary transfer or detail: An **involuntary transfer or detail** is when a fire fighter is transferred or detailed to a different or a less desirable work location or assignment.
- Make financial restitution: For example, the fire fighter might be required to pay the insurance deductible after a property damage incident. One department has a $2500 deductible insurance policy on fire and EMS vehicles and equipment. If the fire fighter is judged to be responsible for a department vehicle crash, he or she is required to repay the deductible amount. The restitution payment period can take up to a year through payroll deductions.
- Loss of leave: A fire fighter might lose annual or compensatory leave. This is equivalent to the practice of paying a cash fine.
- Demotion: A **demotion** occurs when an individual is reduced in rank, with a corresponding reduction in pay. Demotions are more common in the supervisory ranks.

Predisciplinary Conference

In most cases, a predisciplinary conference or hearing must be conducted before a suspension, demotion, or involuntary termination can be invoked. The degree of investigative effort and the opportunities for fire fighter response before these punishments are issued are set higher than for less severe levels of discipline. Most fire departments require a formal disciplinary hearing before a suspension is issued in order to provide an opportunity for the fire fighter to formally respond to the charges.

Known as a **Loudermill hearing**, this process resulted from a 1985 U.S. Supreme Court decision in *Cleveland Board of Education v. Loudermill*. The Board of Education hired James Loudermill as a security guard in 1979. Loudermill received a letter from the Board on November 3, 1980, dismissing him because of dishonesty in filling out his employment application. He had not listed a 1968 felony conviction of grand larceny on his application. The U.S. Supreme Court combined the *Loudermill* case with a similar appeal filed by Richard Donnelly, a Parma Board of Education bus mechanic fired in August 1997 after failing an eye examination.

The court ruling indicated that a **pretermination hearing**, including a written or oral notice, in which the employee has an opportunity to present his or her side of the case, and an explanation of adverse evidence were essential to protecting the worker's due process rights. The hearing is a check against a possible mistaken decision, to determine whether there are reasonable grounds to believe that the charges against the employee are true and whether they support the proposed action.

Not all disciplinary actions require a Loudermill hearing before suspension. The rules vary from state to state. In *Gilbert v. Homar*, the U.S. Supreme Court ruled in 1997 that East Stroudsburg University did not violate the constitutional due process rights of a university police officer when it immediately suspended him without pay after learning he had been charged with a drug felony.

A predisciplinary hearing can be conducted by a disciplinary board, by the fire chief or another ranking officer, or by a hearing officer. The designated individual or board reviews the case and makes a recommendation to the fire chief. Before the hearing, the fire fighter receives a letter that outlines the offense and the results of the investigation. The proposed duration of the suspension without pay could also be stated in the letter.

Here is a typical example of the process:

- Fire officer investigates alleged employee offenses promptly and obtains all pertinent facts in the case (time, place, events, and circumstances) including, but not limited to, making contact with persons involved or having knowledge of the incident.
- Fire officer prepares a detailed report outlining the offense, the circumstances, the individual's related prior disciplinary history, and recommended disciplinary action. The report is submitted to a higher-ranking officer in the chain of command.
- Fire department representative consults with the Human Resources Director or his or her designee if necessary when suspensions are contemplated.
- Disciplinary board hearing is scheduled, and an advance notice letter is prepared.
- Disciplinary board considers the charges and hears the employee's response.
- Disciplinary board makes a recommendation to the fire chief, who issues the final decision.

At the hearing, the accused individual has an opportunity to refute the charge or present additional information about the mitigating circumstances. In most career fire departments, a union representative is present at the hearing to advise the individual. An alternative or lower level of discipline might be proposed, based on past department practices. The disciplinary board makes a final recommendation to the fire chief.

The fire fighter might or might not have the ability to appeal the disciplinary action through the grievance procedure, depending on the personnel rules of the organization. The final resolution of a disciplinary action usually resides with the civil service commission or the city personnel director. Some municipalities use arbitration to resolve these issues.

Documentation and Record Keeping

Municipal personnel rules usually require all of the official records of an employee's work history to be in a secured central repository. This repository includes all of the official documents accumulated during employment, including:

- Hiring packet (application, Candidate Physical Aptitude Test score, and medical examination results)
- Tax withholding, I-9 status, and insurance forms
- Personnel actions (changes in rank and pay)
- Evaluation reports (probationary, annual, and special)
- Grievances
- Formal discipline

Some fire departments maintain a second personnel file at fire headquarters. That file may include letters of commendation, transfer requests, protective clothing record, work history, copies of certifications, and other fire department–specific information that does not need to be kept in the secured personnel file.

Employee Assistance Program

An **employee assistance program (EAP)** is designed to deal with issues such as substance abuse, emotional or mental health issues, marital and family difficulties, or other difficulties that affect job performance. EAPs help the employees cope with underlying issues that might be affecting workplace performance. Fire department EAPs are comprehensive programs that deal with a wide range of issues that can affect fire fighters. When an EAP is available, it is a resource that fire fighters can turn to when in crisis.

One important characteristic of an EAP is its ability to maintain the value to the organization of highly trained emergency service professionals. Earlier in this chapter, a six-step progressive discipline process, followed by how one department handled chronic tardiness, was discussed. If the fire fighter is unable to correct this behavioral problem, and the process is followed, then termination is inevitable. The underlying problem could be an off-the-job issue that the individual cannot solve without assistance. Termination of an otherwise good employee would be a tremendous waste in resources because it could cost the department up to $100,000 to find and train a replacement. If EAP involvement can help solve the problem, it is worth the effort.

Consider some of the reasons a fire fighter would continue to report late for work:

- Child or elder care issues; employee is late due to unanticipated coverage problems
- Family crisis, such as a divorce proceeding or a dying family member
- Alcoholism or substance abuse
- Coming from a second job that is needed to handle a financial crisis, such as a child with a significant health problem not covered by insurance, a crushing debt, or gambling losses
- Employee is suffering from a psychological condition or chemical imbalance

Often, these situations cause stress for the employee. A referral to the EAP may assist in addressing the issues. For an EAP to be successful, the fire officer must be able to recognize stress in an employee. Signs that may be noticed at work could include absenteeism, unexplained fatigue, memory problems, irritability, insomnia, increased use of products with caffeine/nicotine, withdrawal from the crew, resentment toward management/co-workers, stress-related illnesses, moodiness, or weight gain or loss. Although any of these individually may not indicate the employee is stressed, multiple signs may indicate that the fire officer should ensure that the employee is aware of the EAP.

If the stressful situation goes unresolved, it can seriously affect performance on the job. Obviously, an employee who is absent reduces the efficiency of a fire company. However, more subtle are the symptoms of stress such as physical exhaustion from stress-related fatigue or lack of sleep. The fatigue or the indecision caused by it can have serious safety consequences to every person on scene. It may also affect the group dynamics of the crew. A crew member who is stressed may not intend on taking it out on the crew at work, but the tension created is often very evident to the rest of the company.

The goal of the EAP is to provide counseling and rehabilitation services to get the employee back to full productive duty as soon as possible. Fire department EAPs have been successful in lowering employee turnover and reducing absenteeism, tardiness, accidents, and injuries. In addition, there are fewer employee grievances and severe disciplinary actions when an EAP is in place.

Successful EAPs place a high value on confidentiality and require that fire fighters enter the program voluntarily. Although a fire officer can recommend or suggest that a fire fighter consider seeking assistance from an EAP, the fire officer cannot know the details of any fire fighter/EAP interaction. The fire officer's focus is on the fire fighter's job performance.

Summary

A "well-disciplined" fire company means that the members are competent and confident in their tasks and duties through frequent training and practice. The fire officer plays a key role in creating and maintaining that discipline.

Evaluating work performance in order to assist members in improving their individual and group performance during emergency events and routine activities is a core supervisory activity. Much of the fire officer's time and energy will be spent in this area.

The formal fire organization also tasks the fire officer with ensuring that all members comply with the rules and regulations of the department. The fire officer represents the formal organization when providing positive and negative discipline to subordinates.

The fire officer is a working supervisor of a small team within a 24-hour organization. The team works together for long duty shifts, with minutes of explosive action in a high hazard environment. It is difficult to issue discipline in this environment. What is more difficult, and unfair to the fire fighters, is to bump the issuing of the discipline up the chain of command or to personalize the issue.

As the designated agent for the formal department, the fire officer is expected to understand and properly follow the procedures to issue positive and negative discipline as defined by administrative law, agency regulations, and departmental policy.

You Are the Fire Officer: Conclusion

When Sam arrives for work you arrange a private meeting. You review the Work Hours regulation and point out that this incident will generate a proposed 12-hour suspension without pay. You review the progressive discipline process and point out that continued tardiness and late notifications could result in termination. You ask what is causing his continuing difficulty in reporting for work.

Sam explains that he recently separated and is a single parent. This morning the babysitter did not show and he lost track of time while finding a replacement. He is trying not to use sick leave unless it is re-ally needed, because most of his sick leave was used last year when one of his kids was hospitalized.

You acknowledge that the situation is difficult and explain the department's reliance on fire fighters showing up for work on time. You provide information about the employee assistance program and ask if a temporary detail to a day work position would help. You conclude the meeting by pointing out the excellent work he does as a fire fighter, including a unit citation he received a few months ago, and reinforcing the need to comply with the Work Hours regulation.

Wrap-Up

Chief Concepts

- Fire fighter evaluations are an ongoing process throughout the year. The annual performance evaluation is a formal written documentation of the fire fighter's performance during the rating period.
- New fire fighters should receive a copy of the classified fire fighter job description.
- The fire officer is responsible for overseeing any structured probationary fire fighter in-field training and experience program.
- Performance feedback is most effective when delivered as soon as possible after an action or incident. It should be done in private and be specific enough for the fire fighter to understand what needs to improve.
- Fire officers should maintain an extemporaneous performance log or T-account on every fire fighter under the officer's command.
- Informal work performance reviews should be regularly scheduled with each fire fighter under the fire officer's command.
- A mid-year review includes a fire fighter–prepared self-evaluation that is reviewed by the fire officer.
- Some municipalities require that an employee receive a formal notification if the annual evaluation is below "satisfactory." It is usually issued 10 weeks before the annual evaluation is due.
- Six weeks before the end of the evaluation period, the fire fighter completes another self-evaluation that is reviewed by the fire officer. Together, they develop the final, formal evaluation report for the rating period.
- Progressive discipline is a process for dealing with job-related behavior that does not meet expected and communicated performance standards.
- Positive discipline makes fire fighters strong and confident.
- Negative discipline is aimed at correcting individual behavior.
- An oral reprimand, warning, or admonishment is an "informal" negative discipline, which stays at the fire officer level and expires after a year.
- A written reprimand requires review by the fire officer's supervisor and is placed in the fire fighter's personnel folder for a specified period of time. This starts the paper trail of progressive discipline.
- Suspension is a personnel action that places a career fire fighter on a leave-without-pay (LWOP) status for a specified period of time.

- Negative disciplinary actions may include:
  - Financial restitution
  - Extension of a probationary period
  - Involuntary transfer or detail
  - Loss of annual or compensatory leave
  - Establishment of a special evaluation period
  - Demotion
- An EAP provides counseling and rehabilitation services to get the fire fighter back to full duty as soon as possible.

Hot Terms

Central tendency An evaluation error that occurs when a fire fighter is rated in the middle of the range for all dimensions of work performance.

Contrast effect An evaluation error in which a fire fighter is rated on the basis of the performance of another fire fighter and not on the classified job standards.

Demotion A reduction in rank, with a corresponding reduction in pay.

Discipline A moral, mental, and physical state in which all ranks respond to the will of the leader.

Employee assistance program (EAP) An employee benefit that covers all or part of the cost for employees to receive counseling, referrals, and advice in dealing with stressful issues in their lives. These may include substance abuse, bereavement, marital problems, weight issues, or general wellness issues.

Formal written reprimand An official negative supervisory action at the lowest level of the progressive disciplinary process.

Frame of reference An evaluation error in which the fire fighter is evaluated on the basis of the fire officer's personal standards instead of the classified job description standards.

Halo and horn effect An evaluation error in which the fire officer takes one aspect of a fire fighter's job task and applies it to all aspects of work performance.

Involuntary transfer or detail A disciplinary action in which a fire fighter is transferred or assigned to a less desirable or different work location or assignment.

Loudermill hearing A predisciplinary conference that occurs before a suspension, demotion, or involuntary termination is issued. The term refers to a Supreme Court decision.

Oral reprimand, warning, or admonishment The first level of negative discipline. Considered informal, it remains with the fire officer and is not part of the fire fighter's official record.

Performance log Informal record maintained by the fire officer listing fire fighter activities by date and with a brief description. Used to provide documentation for annual evaluations and special recognitions.

Personal bias An evaluation error that occurs when the evaluator's perspective skews the evaluation such that the classified job knowledge, skills, and abilities are not appropriately evaluated.

Pretermination hearing An initial check to determine if there are reasonable grounds to believe that the charges against the employee are true and support the proposed termination.

Progressive negative discipline A process for dealing with job-related behavior that does not meet expected and communicated performance standards. Discipline increases from mild to more severe punishments if the problem is not corrected.

Recency An evaluation error in which the fire fighter is evaluated only on incidents that occurred over the past few weeks rather than on the entire evaluation period.

Restrictive duty A temporary work assignment during an administrative investigation that isolates the fire fighter from the public and usually is an administrative assignment away from the fire station.

Special evaluation period A designated period of time when an employee is provided additional training to resolve a work performance/behavioral issue. The supervisor issues an evaluation at the end of the special evaluation period.

Suspension A negative disciplinary action that removes a fire fighter from the work location; he or she is generally not allowed to perform any fire department duties.

T-account A documentation system, similar to an accounting balance sheet, listing credits and debits, in which a single-sheet form is used to list the employee's assets on the left side and liabilities on the right side, appearing like a letter "T."

Termination The organization ends an individual's employment against his or her will.

Work improvement plan A written document that is part of a special evaluation period. The plan identifies performance deficiencies and lists the improvements in performance or changes in behavior required to obtain a "satisfactory" evaluation.

Fire Officer *in Action*

As a new supervising fire officer you are starting the process of informal performance reviews for your crew, in anticipation of completing the formal annual evaluations later this year. In reviewing earlier annual evaluations, it appears that the last supervising fire officer created a one-size-fits-all template that met the department's requirement but said nothing specifically positive or negative about each fire fighter. You want your reviews to paint an accurate picture of each individual.

The fire officer is the starting point of fire fighter evaluation or discipline actions. Striving to provide an accurate and complete picture of each person under supervision provides an excellent foundation for building confident and competent fire fighters. The required tools include a detailed understanding of the AHJ's personnel regulations, labor agreements, and departmental regulations.

1. The 12 hours of suspension that Sam will receive for a third incident of violating the Work Hours regulation is an example of:
   A. informal discipline.
   B. compliance reinforcement.
   C. formal discipline.
   D. an event not covered by progressive discipline.

2. You have had a private counseling session with a rookie fire fighter about standing in the traffic lane while operating at a vehicle crash incident. Your instructions are to remain within the lane blocked by fire apparatus or police to avoid getting struck by a vehicle. While assisting on a patient extrication on the interstate, you observe the rookie backing into an active traffic lane. You should:
   A. issue a written reprimand after returning to the fire station.
   B. make an entry into your T-account.
   C. use your cell phone to get a digital clip documenting the behavior.
   D. immediately order the fire fighter to get out of the traffic lane.

3. Sam is the first person for whom you need to complete a formal annual evaluation report. The reporting for work issue started 3 months ago. To focus on that issue as the main theme of an annual report narrative is an example of:
   A. central tendency.
   B. halo and horn effect.
   C. contrast effect.
   D. personal bias.

4. When should positive or negative feedback be delivered?
   A. At the next beginning-of-shift line-up
   B. At the next informal work performance review session
   C. As soon as possible after the incident or event
   D. As part of the annual performance review

Leading the Fire Company

NFPA 1021 Standard

Fire Officer I

4.2.1 Assign tasks or responsibilities to unit members, given an assignment at an emergency incident, so that the instructions are complete, clear, and concise; safety considerations are addressed; and the desired outcomes are conveyed. [p 162, 165, 167–169]

(A) Requisite Knowledge. Verbal communications during emergency incidents, techniques used to make assignments under stressful situations, and methods of confirming understanding. [p 165, 167–169]

(B) Requisite Skills. The ability to condense instructions for frequently assigned unit tasks based on training and standard operating procedures. [p 162–163]

4.2.2 Assign tasks or responsibilities to unit members, given an assignment under nonemergency conditions at a station or other work location, so that the instructions are complete, clear, and concise; safety considerations are addressed; and the desired outcomes are conveyed. [p 165]

(A) Requisite Knowledge. Verbal communications under nonemergency situations, techniques used to make assignments under routine situations, and methods of confirming understanding. [p 165, 169–170]

(B) Requisite Skills. The ability to issue instructions for frequently assigned unit tasks based on department policy. [p 165]

Fire Officer II

5.2.1 Initiate actions to maximize member performance and/or to correct unacceptable performance, given human resource policies and procedures, so that member and/or unit performance improves or the issue is referred to the next level of supervision. [p 164–165]

(A) Requisite Knowledge. Human resource policies and procedures, problem identification, organizational behavior, group dynamics, leadership styles, types of power, and interpersonal dynamics. [p 164–165]

(B) Requisite Skills. The ability to communicate orally and in writing, to solve problems, to increase team work, and to counsel members. [p 164–165]

Introduction to Fire and Emergency Services Administration (FESHE) Course Objectives

2. Explain the need for effective communication skills both written and verbal. [p 169]

5. Examine the history and development of management and supervision. [p 164–165]

7. Identify roles and responsibilities of leaders in organizations. [p 162, 165, 167–170]

8. Compare and contrast the traits of effective versus ineffective supervision and management styles. [p 163–165]

9. Identify and assess safety needs for both emergency and nonemergency situations. [p 167]

11. Identify the role of a company officer in Incident Command System (ICS). [p 163, 165, 167, 169]

Knowledge Objectives

After studying this chapter, you will be able to:

- Describe leadership styles.
- Describe how to motivate.
- Describe leadership in routine situations.
- Describe emergency scene leadership.
- Describe the fire officer challenges in the 21st century.
- Compare and contrast the fire station as a municipal work location versus a fire fighter home.

Skills Objectives

There are no skills objectives for this chapter.

I t has been a year since you started as a brand new supervising fire officer at Tower Ladder 5. You have gotten into the rhythm of how The Nickel works, but your helmet is still the cleanest in the fire station. You meet Captain Torres and the administrative fire officer to complete your end-of-probation report.

At the end of the review, the administrative fire officer drops a bombshell. Two senior fire fighters are retiring and Captain Torres will be on an extended detail to Special Operations. For the next 6 to 11 months, you will be the boss of the firehouse. You will also be getting two rookie fire fighters.

1. How can you develop trust as the new boss?
2. How can you develop the new fire fighters into an effective firefighting workforce?

Introduction to Leadership

Leadership is a complex process by which a person influences others to accomplish a mission, task, or objective and directs the organization in a way that makes it more cohesive and coherent. An alternative way to describe leadership is the art of getting someone else to do what you want them to do, but they do it because they want to do it. A person carries out this process by applying his or her leadership attributes, such as beliefs, values, ethics, character, knowledge, and skills. Although your position as a fire officer gives you the authority to accomplish certain tasks and objectives in the organization, this power does not make you a leader—it simply makes you the supervisor. Bosses tell people to accomplish a task or objective, whereas leaders make them want to achieve high goals and objectives. Your goal should be to lead, not just to be a boss.

The Fire Officer as a Follower

Leaders can be effective only to the extent that others are willing to accept their leadership. This is sometimes described as **followership**. It is self-evident that the leader cannot lead if others will not follow. An effective leader uses persuasiveness and motivation to overcome resistance. In some cases, for a variety of reasons, others simply refuse to follow.

The fire officer has to be both a leader and a follower. This means that the officer leads the fire company to achieve the goals and objectives that have been established by the department or the jurisdiction. The officer has to follow leadership that comes from a higher level, even if it is not always pointing in a direction where the officer would prefer to go. In many cases, the officer is the messenger who has to deliver unpopular news to the company members and then provide the leadership to ensure compliance with instructions that came from a higher level. Chapter 3, Fire Fighters and the Fire Officer, discussed how the fire officer handles an unpopular order.

Followership is particularly important for a fire officer because subordinates are always aware of what the fire officer does. If the fire officer demonstrates selective following of orders from the fire chief, deciding which ones to follow and which ones to ignore, the officer sends a clear message to the company members: It is acceptable for the fire fighters to be selective in following orders from their fire officer. Followership is an important character trait that will serve the officer well in the future when dealing with others who have not followed the rules because they don't seem fair.

Fire Marks

The Role of the Officer Is to Keep Calm

David Halberstam wrote *Firehouse*, a book about his Upper West Side neighborhood fire station, Engine 40 and Ladder 35. Only one of the 13 fire fighters returned from the World Trade Center.

During a June 3, 2002, online interview conducted by *The Washington Post*, Halberstam discussed the role of the fire officer. It is an observation not found in the book.

One of the key things in a firehouse is the role of the officer and the need on their part in whatever danger, in the worst kind of crisis, to stay calm. Because if they're not, and if they show fear, it will ripple right through the men. And so the men will say of an officer, "He's a very good officer. He always stayed calm." . . . the captain is the first in, by tradition, and the last out.

Leadership Styles

There are many ways to look at leadership, including how to describe a leader, what leaders do, and how leaders act. One approach to better understanding a leader is to look at the different styles of leadership. Most agree that an effective leader changes the style of leadership based on the specific situation. Three leadership styles are traditionally identified as autocratic, democratic, and laissez-faire.

Autocratic

This iron-hand approach is used when the fire officer needs to maintain high personal control of the group. The fire officer is telling subordinates what to do and is expecting immediate and complete adherence to the issued instructions. This style is required in two situations. The first situation occurs when the fire company is involved in a high-risk, emergency scene activity, such as conducting a primary search during a structure fire Figure 9-1 ▼ . There is no time for discussion, and this is not the situation to experiment with alternative approaches.

The second situation occurs when the fire officer needs to take immediate corrective supervisory activity, such as during a "control, neutralize, command" response to a confrontation. In this situation, the officer has to be firmly in control of the situation.

Democratic

This consultative approach uses all of the ingenuity and resourcefulness of the group in determining how to meet an objective or complete a task Figure 9-2 ▶ . The officer should use the democratic style when planning a project or developing the daily work plan of the company. This approach can also be used in some low-risk emergency scene operations.

Specialized and highly technical fire companies often use the democratic approach when faced with a complex or unusual emergency situation. They depend on the skills and experience of the individual team members to analyze the situation, consider the alternatives, and develop the incident action plan. Execution of the plan often involves an autocratic command style, however.

Figure 9-2 The democratic approach uses the resourcefulness of the group in determining how to meet an objective.

Company Tips

Delegating Routine Activities

A goal of an effective fire officer is to push decision making to the lowest possible level. Because a fire officer gains experience as a leader and builds confidence and trust with a team, many routine activities can be delegated. For example, a fire fighter could be assigned to manage in-station training activities, another could oversee routine station maintenance, and a third could keep pre-incident plan files updated. The fire officer continues with the morning line-up and evening check-up, but the individual fire fighters perform their assignments autonomously. This allows the officer to focus on activities that require his or her personal attention.

Laissez-Faire

This free-rein style moves the decision making from the fire officer to the individual fire fighters. The fire officer depends on the fire fighters' good judgment and sense of responsibility to get things done within basic guidelines. This is an effective leadership style when working with experienced fire fighters and when handling routine duties that pose little personal hazard.

Getting It Done

Transitioning from Democratic to Autocratic
The increase in specialized and highly technical emergency services creates a unique challenge for the fire officer. The best way to size up a complicated situation and develop an incident action plan is to use the democratic style, using the knowledge and experience of all responders to develop the best approach. Once the operation begins, the fire officer needs to assume the autocratic role to ensure the safe execution of the plan.

Figure 9-1 An autocratic approach to leadership is required in a high-risk, emergency scene activity.

Power

Power is the capacity of one party to influence another party. One of the first works on the management of power was *The Prince*. Niccolo Machiavelli's masterpiece was posthumously published in 1532.

French and Raven describe types of power in the 1959 *Studies of Social Power* as the result of the "target person's" response from the "agent" making a request.

- Legitimate power: The target person believes that the agent has the right to make the request and the target person has the obligation to comply. For example, under the incident management system, the incident commander has the legitimate power to reassign the ventilation sector.
- Reward power: The target person complies in order to obtain rewards believed to be controlled by the agent.
- Expert power: The target person complies due to a belief that the agent has special knowledge.
- Referent power: The target complies due to admiration of or identification with the agent and seeks approval.
- Coercive power: The target person complies in order to avoid punishment believed to be controlled by the agent.

Professor Gary A. Yukl at the University of Albany researches leadership, power, and influence. He updated the French and Raven taxonomy into two types, personal and positional. Personal power, which includes expert and referent power, reflects the effectiveness of the individual. Positional power is defined by the role an individual has within the organization. Legitimate, reward, and coercive are the three examples of positional power. Yukl provides two additional position-based power descriptions:

- Information power: Control over information. Unlike expert power, information power is based on the target person's assessment of the agent's ability to rapidly and efficiently discover or obtain relevant information, usually through a cultivated network of sources.
- Ecological power: Control over the physical environment, technology, or organization of work. The target person's behavior is based on perceptions of opportunities and constraints.

Motivation

One of the key components of leading is the ability to motivate. Chapter 4, Understanding People: Management Concepts, introduced some of the background for motivation. The fire officer must be able to apply methods to inspire subordinates to achieve their maximum potential. Although each method has a slightly different focus, the overriding principles are:

- Recognize individual differences.
- Use goals.
- Ensure goals are perceived as attainable.
- Individualize rewards.
- Check for system equality.

Reinforcement Theory

Probably the best-known motivational theory is the reinforcement theory. This theory suggests that behavior is a function of its consequences. To motivate employees to perform, the officer must provide reinforcers to encourage the employee to act in the desired manner. Reinforcement must immediately follow an action in order to increase the probability that the desired behavior will recur. There are four types of reinforcers:

- Positive reinforcement: Giving a reward for good behavior
- Negative reinforcement: Removing an undesirable consequence of good behavior
- Extinction: Ignoring bad behavior
- Punishment: Punishing bad behavior

Whereas positive and negative reinforcement increase the likelihood of good behavior, extinction and punishment decrease the likelihood of bad behavior. Using punishment and extinction does not guarantee good behavior, but rather only reduces the specific bad behavior. It may be replaced with a different undesirable behavior.

Most fire officers quickly learn to use positive reinforcement. It is easy to pat someone on the back for a good job. Positive reinforcement must be sincere and deserved. Most people are also familiar with and may use punishment in order to correct an employee who does not perform as desired. Most parents can identify with using extinction. Frequently, this strategy is used when a child is acting up. The parent simply ignores the behavior in order to decrease the chance of the child doing it again.

Negative reinforcement is often underutilized. This involves removing the negative consequences of good performance. For example, if a company completes their yearly inspection list early, many departments increase the number of inspections they are required to complete the following year. In effect, the crews are punished for good behavior. Negative reinforcement would remove the undesirable consequence.

Motivation-Hygiene Theory

One method of motivation was described by Frederick Herzberg, a psychologist who developed the concepts of job enrichment and the motivation-hygiene theory. His 1968 *Harvard Business Review* article "One More Time, How Do You Motivate Employees?" had sold 1.2 million copies by 1987 and remains the magazine's most requested reprint.

Motivation-hygiene theory breaks the motivational process into two parts: hygiene factors and motivation factors. Hygiene factors are conditions that are external to the individual, such as pay and work conditions. Motivation factors are the individual's internally determined motivators, such as the desire for recognition, achievement, responsibility, and advancement.

Hygiene factors do not motivate individuals, but if the person is not satisfied with any of the external conditions, he or she will not be motivated. In order to achieve maximal performance, the officer must ensure that the fire fighters are satisfied with their work environment, pay, supervision, and company policy and then motivate them by giving recognition for performance, providing promotional opportunities, and allowing for added responsibilities.

Probably the most significant lesson to be learned from this theory is that employees who are dissatisfied with external conditions will not be motivated to achieve maximal performance for the company. Alternatively, once fire fighters are satisfied with the external conditions, improving these will not increase performance. No two fire fighters will consider the same conditions to be at the same satisfaction level. The officer must consider the individual's perception. The fire officer must have open and honest communication with the individual fire fighters.

Goal-Setting Theory

Another method of motivating fire fighters is by the use of goal setting, which relies on the natural competitiveness of people. In this theory, the key to motivation is for the officer to set specific goals that will increase performance. The goals must be difficult, but attainable. If they are easy to achieve, there is little motivation to work hard. If they are perceived to be impossible to achieve, the fire fighter will not try hard because failure is inevitable. The fire fighter must honestly believe that performing well can result in success.

Clear, specific, and measurable goals are essential for motivation. Vague goals, such as "get to the apparatus as quickly as you can," do not give the fire fighter clear enough expectations to determine whether he or she succeeded. Instead, a goal might be to "reduce your turnout time to less than 1 minute." The fire fighter in the latter example would be much more motivated than the fire fighter in the former example.

The most significant lesson in the goal-setting theory that a fire officer can use to motivate fire fighters is to carefully consider what actions are needed to improve the organization and then to develop clear, specific, and challenging but attainable goals for the individual fire fighter.

Equity Theory

This motivational process suggests that employees evaluate the outcomes they receive for their inputs and compare them with the outcomes others receive for their inputs. Outcomes range from pay and benefits to recognition, achievement, and promotion. Inputs include educational level, performance level, risks taken, and special skills.

This theory explains why the fire chief is paid more than a fire fighter or why a fire fighter is paid more than a restaurant dishwasher. Conversely, according to this theory, most fire fighters would not understand why they are paid in the lowest one third in a survey of comparable cities whereas the fire chief's pay is in the top one third in the survey.

If a fire officer wants to motivate the fire fighter, he or she must determine where the fire fighter believes an inequity exists. Although the fire officer would likely not have the ability to change the organization's pay structure, he or she may be able to provide other outcomes that are desired by the fire fighters. This might include a more flexible time trade policy or recognition.

Expectancy Theory

This motivational theory is based on the premise that people act in a manner that they believe will lead to an outcome they value. According to expectancy theory, the fire officer must address three considerations in order to motivate the individual:

- The employee's belief that his or her effort will achieve the goal
- The employee's belief that meeting the goal will lead to the reward

- The employee's desire for the reward or the reward's value to the employee

Using this method, the fire officer might indicate to the fire fighter that if he or she has a turnout time of less than 1 minute for the next month, the officer will make sure that this performance is reflected on the fire fighter's evaluation, which is used for the upcoming promotional process in which the fire fighter is interested.

Leadership in Routine Situations

Most fire officer leadership activity is directed toward accomplishing routine organizational goals and objectives in nonemergency conditions. This includes being well prepared to perform in emergency situations. The fire officer accomplishes most of these goals and objectives through the efforts of fire fighters. The role of the officer involves influencing, operating, and improving to ensure that those efforts achieve the desired results.

Fifty years ago, fire officers used an autocratic style of leadership in both emergency and nonemergency duties. The officer decided what was to be done and who was to do it, and the fire fighters responded without question. Today, the workforce is very different. All employees demand to be included in the decision-making process. Effective fire officers today provide a more participative form of leadership in routine activities.

Although the fire officer is commonly given specific assignments that the company must complete, there can be much discretion about when, how, and by whom activities are carried out. Some officers sit down with their crews at the start of the shift and cover the assignments, if any, that the administrative fire officer has indicated need to be completed for the day. He or she may also outline the status of other duties that are long term that will need to be completed but are not due as of yet. An open question is then addressed to the crew as to what they think should be done, including any areas that they believe should be addressed. Depending on the outcome of the discussion, the officer then decides or has the group decide the plan for the day on how to accomplish the activities.

Some tasks are to be completed every day. In most departments, station cleaning is performed daily, making a trip to the grocery store is required, and equipment must be checked. The fire officer may decide that these duties are easiest if they are assigned in advance. Typically, in conjunction with the crew, a determination is made about what duties will be completed every day based on the work position of the individual. The driver is usually assigned to check all equipment and clean the apparatus bay. The rookie fire fighter may be assigned to the bathroom and kitchen detail, whereas the senior fire fighter is assigned to sweeping and vacuuming floors.

Emergency Scene Leadership

A core responsibility of a fire officer is to handle emergencies effectively. Chapter 15, Managing Incidents, covers structural components of incident management in more detail. This chapter examines some of the leadership issues that apply specifically to a fire officer.

A fire officer always has direct leadership responsibility for the company that he or she is commanding. The first-arriving fire

Voices of Experience

Knowing when to lead the company and when to help get the work done is often one of the most difficult decisions for a new fire officer. On one hand, you are responsible for supervising the company and ensuring their safety, so you need to oversee their actions. However, a new officer does not want the reputation of standing around while everyone else is working.

I have taught officer training for many years. In my experience, new officers err on the side of doing too much of the work. There have been many times when I have asked an officer trainee, who was down on his knees connecting hoses, what his fire fighter was doing. The response is almost always, "I don't know." As the officer leading the company, you must always know where your personnel are located and what they are doing. This type of situation is easily resolved just by the officer asking a fire fighter to connect the hoses instead of doing it himself. This way, the work still gets done and the officer is better able to fulfill his responsibilities as a supervisor.

A good rule of thumb is that an officer should not perform task-level work until all the other members of the company have been assigned a task. This allows the officer to first supervise and lead. However, there are times when an officer must help get the work done. For example, when an engine arrives first on the scene at a working structure fire, everyone must help get the hose deployed, the door forced, ventilation performed, etc. There are too many critical tasks that must be done for the officer to stand back and just supervise. Officers in this type of situation must be able to assist with tasks but still step back often enough to perform their supervisory functions. One technique that I often use is taking time out to size up the scene periodically. For example, I can help deploy the hose for a structure fire but then I will take a few seconds to size up the scene again and account for my personnel before donning my face mask to move in for fire attack. This is an example of switching back and forth between getting the work done and supervising.

A company officer must balance leading the company with helping to get the work done. With practice it is possible to accomplish both. If you start to get the feeling that you are supervising too much and not helping enough with the work, find times to pitch in and help when supervision is less critical, such as loading hose or checking equipment. Your company members will respect you both for leading during crucial times and for working with them when they can use an extra helping hand.

Chris Watson
Battalion Chief
Austin Fire Department
Austin, Texas

officer has additional important responsibilities to establish command of the incident and to provide direction to the rest of the responding units. The first-arriving fire officer also has leadership responsibilities that relate to the communications center.

Methods of Assigning Tasks

The fire officer's primary responsibility is to the team of fire fighters under his or her direct supervision. The fire officer is responsible for their safety, their actions, and their performance at the incident scene. The fire officer should develop a consistent approach to emergency activities, in which all of the department's operating procedures are followed.

Most departments have some form of standard operating procedures (SOPs) that specify what is expected of each company and what to do in a given situation. Some are very specific and detailed, whereas others are more general. For departments with very detailed procedures, the task that each fire fighter is to perform may already be determined. The SOPs may indicate what tools the fire fighter is to carry based on whether the company has three or four persons. In these cases, the fire officer is responsible for ensuring that the fire fighters know and follow the policy.

For departments that have broad SOPs or none at all, the fire officer has two choices in assigning tasks: pre-assigning them or assigning them as needed on the scene. The advantage of pre-assignment is that it relieves the officer of having to make as many decisions during the emergency. The advantage of assigning tasks on the scene is that it reduces unnecessary efforts.

When giving assignments on the scene, the fire officer must be clear and concise and ensure that the fire fighter understands. In this situation, the fire officer uses an autocratic style of leadership because the emergency scene does not allow for participative decision making. For this method of operation to be effective, the fire fighter must clearly understand the reason for the different leadership styles before the incident. One method that may help a fire officer make assignments under pressure is to develop a standard method of handling situations. Making standard decisions in a consistent manner will assist in the decision-making process. If you can't use a written checklist, then a mental checklist that is rapidly gone through every time will create a predictable outcome. There are so many factors that come into play when determining tactics, it is crucial that officers be trained in the critical decision-making process. For example, if you need to assign someone to rescue, always consider first whether you have a rescue company available; if not, then a truck company and lastly an engine company can be used. This allows a sequential thought process to develop, making assignments easier.

A method of assigning tasks that allows for more participation in the process yet reduces the number of decisions the officer must make on scene is the broad standard operating procedures, in which the fire officer discusses with the crew the various "routine" emergencies that they respond to. The officer and crew can discuss the needs of the incident and what must be done to mitigate it. Finally, the officer and crew decide who will be responsible for doing what.

For example, the crew may decide that on any life-threatening medical incident, the fire fighter is responsible for checking for respirations and securing an airway and then providing ventilation.

The driver is responsible for checking for a pulse and uncontrolled bleeding and then providing compressions. The officer is responsible for setting up and using the semiautomatic external defibrillator. This plan ensures that the crew covers all the basics without fumbling and waiting to be directed.

Critical Situations

Dangerous situations develop suddenly during incident operations. A fire officer must use the autocratic leadership style when immediate action is required. Orders to evacuate a building and a fire fighter may-day are two critical events that require immediate autocratic action.

When command issues an evacuation order, the fire company should immediately leave the fire building. Evacuation may mean that the company leaves a fire hose in the building and the fire fighters quickly leave the building. The fire officer is responsible for accounting for the fire fighters under his or her supervision. The fire officer provides a head count to the sector command officer or the incident commander.

A may-day requires a complex response from the company operating within a hot zone or burning structure. The first obligation is to maintain radio discipline so that command can determine the may-day location and situation. The second obligation is to maintain company or sector integrity. Incident control evaporates if every fire fighter rushes to assist the fire fighter in trouble. The fire officer needs to maintain company integrity to facilitate the may-day rescue and continue to fight the fire or control the hazard. Changes in assignment come from the sector incident commander.

The fire officer may directly encounter a dangerous situation, such as a collapsing building, a person with a gun, or a careening vehicle. The officer must make an immediate, clear, and autocratic command to protect the fire fighters under his or her supervision by removing them from the danger area. Then, the fire officer needs to inform command.

After every incident, the fire officer should briefly review the event while still at the scene or as soon as possible after returning to quarters. This is an opportunity to clarify issues and answer questions. The fire officer can reinforce good practices and immediately identify any unacceptable performance.

Near Miss REPORT

Report Number: 07-0001110

Event Description: We were dispatched to a reported structure fire. The first arriving officer had firsthand knowledge of the dwelling from previous runs there. Companies engaged in a primary search due to the possibility of squatters. After forcing the front door, two members with a 1 3/4" handline and thermal imaging camera (TIC) made their way into the front living area. They encountered a "Collyer Mansion Syndrome."

There was wall-to-wall debris approximately 3' high. The fire was noted to our right and to Side "C" of the dwelling. Advancing the line was slowed due to the debris. About 10' into the room the floor felt like it was sagging. Command was notified of the conditions. After advancing approximately 3' or 4' more the floor dropped even more significantly. Command was notified of the more significant sagging by the handline crew. Recommendations were made to Command to evacuate and go defensive.

Crews immediately backed out and a tower ladder went into operation with numerous handlines. The building was a 1 1/2 story Cape Cod packed with everything from books, monitors, furniture and even a cache of weapons. . . . a total of 24 personnel to the scene. All members are equipped with portable radios. All companies have TICs. Attack line experience: Nozzle-13 years; Back-up less than 6 months. Full PPE was in place.

All companies were advised of the building's condition by the first arriving unit. Having live fire training and experience in reading a building's conditions is key. The floor which we were advancing our attack line on had its joists burned completely away. We were prevented from falling into the basement by the carpet being held in place by all of the wall-to-wall debris. This was not discovered until overhaul from the exterior was accomplished. ICS was used and is used on all incidents by this F.D. Weather was cold and dry.

Lesson Learned: Know building construction past and present. Example: a spongy roof 15–20 years ago means something different than a spongy roof today. All firefighters should focus on building construction during training, as it assists us with every aspect of a job (i.e., investigative work, inspections, fire attack, overhaul and salvage, forcible entry, and ventilation). Fires are burning hotter and faster than ever before. This creates fire conditions in which you cannot advance a line fast enough or the 1 3/4" line will be ineffective. The fire may overrun you causing confusion and separation of your crew leading to other problems.

Experience or inexperience can certainly make a difference. Keep experienced fire fighters with inexperienced firefighters when possible. Preferably keep inexperienced fire fighters with an experienced officer. This will ensure that fire fighters learn from experience, not from trial and error. The error part is unforgiving. The command decision to go defensive after hearing the condition report was a key. Having a company back out prematurely is a philosophy of some commanders, fortunately not on this night. Command had complete confidence in his crews and was able to trust in their decisions. Salty aggressive commanders need to learn from this. Technology is not the end all. Falling back on basic fire fighter training should always be considered. Crawling into this fire using a TIC for guidance might have been a big mistake. A thermal imager does not provide you depth perception nor does it feel for you. Using all of your senses, touch, feel and hearing, are senses which are available during a fire attack. Don't forget them.

The Dispatch Center

One of the responsibilities of the first-arriving fire company–level officer at an incident scene is to concisely describe conditions to the dispatch center. The content of the initial radio report needs to meet the departmental operating procedures. In addition, this verbal picture establishes the tone of the incident. A fire officer who provides a calm and complete description of the situation on arrival demonstrates leadership and ability to control the action that will occur. An officer who gives the impression of confusion, indecision, or uncertainty fails the first test of leadership.

Without a specific requirement, a radio report would include:

- Identification of the company arriving at the scene.
- A brief description of the incident situation. This may include providing information about building size, occupancy, hazardous chemical release, and multiple-vehicle accident.
- Obvious conditions: working fire, multiple patients, hazardous materials spill, or dangerous situation.
- Brief description of action to be taken: "Engine 2 is advancing an attack line into the first floor."
- Declaration of strategy to be used: offensive or defensive.
- Any obvious safety concerns.
- Assumption, identification, and location of command.
- Request or release resources as required.

A calm, concise, and complete radio report helps the dispatch center, the chief officer, and other responding companies to understand the situation at hand. It is important to use radio terminology that meets the department's operating procedures and is clear to everyone who is listening. Close compliance with the communications standard operating procedures is an important component of effective fire-ground command.

A visit to the dispatch center is a valuable experience for every fire officer. It is important to understand the dispatch center's operating conditions and how dispatchers depend on fire officers at the incident scene to provide good information.

Other Responding Units

The first-arriving fire officer provides leadership and direction to responding units by implementing the incident management plan. The fire officer must demonstrate the ability to take control of the situation and provide specific direction to all of the units that are operating and arriving. This requires mastery of the autocratic style of leadership.

Fire Officer Challenges in the 21st Century

The fire officer works in a dynamic environment with changing conditions and evolving organizational needs. The environment is continually changing around basic leadership concepts. The fire officer faces two challenges: the fire station as a work location and the special challenges of leading a volunteer fire company.

Fire Station as Municipal Work Location versus Fire Fighter Home

Most fire fighters do not think of the fire station simply as the place they go to work. Fire fighters work rotating shifts together

---

Look Official When Performing Formal Fire Officer Duties
Many departments provide two different uniforms that can be worn by the fire officer. The station work uniform could be a golf shirt, or "job" shirt. The more formal "Class A" uniform generally includes a white shirt with insignia, badge, and tie.

One way the fire officer can ensure that the fire fighters understand when he or she is performing an official supervisory task is to wear the formal uniform shirt. If the officer always puts on a dress shirt and tie to conduct employee evaluations or counseling sessions and to issue commendations or disciplinary actions, the serious and official nature of these activities is clearly visible. This increases the fire officer's effectiveness by alerting the fire fighters that a formal supervisory task is at hand Figure 9-3 ▾ .

and prepare and share meals, often spending days of dull routine, punctuated by episodes of intense excitement. These factors create a powerful and special community among fire fighters that is difficult to compare with any other workplace environment. The fire station becomes a "home away from home," and the fire fighters become an extended family.

Although this workplace environment produces a special type of bonding among fire fighters, it can easily produce a variety of productivity problems, as well as behavioral and personality traits that are often associated with a dysfunctional family. The resulting situations can be quite different from the problems that occur in a "normal" work environment, and the behavioral issues are often more severe. Case law and administrative actions reinforce the expectation that the fire station is essentially a place of business, subject to the same rules and expectations as any other workplace. They expect the fire fighters who are in the station at 3 A.M. to behave the same way administrative services staff behaves at 3 P.M.

A fire officer must balance the expectations of the employer with the realities of a fire station work environment and the

Figure 9-3 Wearing a uniform alerts the fire fighters that a formal supervisory task is at hand.

desire to create an effective team. A certain amount of spirited behavior can be healthy, as long as some basic rules are observed. The fire officer has to maintain order and ensure that whatever occurs can be explained and would be acceptable to a rational observer.

The fire officer should establish house rules. Two general rules for nonemergency activities are as follows:

1. Do not compromise the ability of the fire company to respond to emergencies in its district.
2. Do not jeopardize the public's trust in the fire department.

Responding to emergencies and preserving the public trust are two values that every fire fighter needs to demonstrate. There remains wide latitude for team-building extracurricular activities.

Leadership in the Volunteer Fire Service

Fire officers leading volunteer fire companies have to rely on their leadership skills even more than their municipal counterparts do. Pride, group identity, and personal commitment are the factors that keep volunteers active and loyal to the organization—there is no biweekly paycheck that compels a volunteer to endure an unpleasant situation. The volunteer fire officer must pay attention to the satisfaction level of every member and be alert for issues that create conflict or frustration.

The relationship between the fire officer and fire fighters is more like that of a mom-and-pop family business than that of a municipal agency. Effective leadership is often the strongest force that influences their performance and commitment to the organization. If the negative aspects outweigh the positive aspects, an individual can step aside or drop out.

Some of the unique issues that a volunteer fire officer must consider are as follows:

- Changes in employment, family situations, or child/elder care profoundly affect the time a volunteer is able to devote to the fire department.
- Extensive training requirements can have an impact on volunteer availability. It is easy to assign a municipal employee to attend a 40-hour training class that occurs during the regular workweek. It is difficult to accomplish the same training when the opportunities are restricted to evenings and weekends.
- Interpersonal conflicts can develop between members. Conflicts can drive away members and erode fire company preparedness. The volunteer officer must act quickly when such problems are identified. The rights and reasonable expectations of individual members must be considered, without compromising the mission or the good of the organization.

A special concern in volunteer organizations is the political balance that results from electing officers. A conscientious volunteer officer has to use strong leadership and the courage of conviction to implement an unpopular policy.

Summary

Leadership requires the fire officer to provide purpose, direction, and motivation to fire fighters. These ingredients are produced through a complex process that has to maintain balance between factors. The fire officer applies personal attributes to the leadership process. This is a lifelong endeavor as the fire officer develops and refines the beliefs, values, ethics, character, knowledge, and skills that make up the sum of a person.

You Are the Fire Officer: Conclusion

On your next day off, you meet Captain Torres for lunch. Torres introduces you to his mentor, a retired command officer who ran The Nickel when Torres was a fire fighter. The chief suggests having frequent hose-and-ladder drills to hone the crew's ability to advance hose lines and throw ladders. He said that The Nickel's secret has been superior training. He had the engine and tower crews train every day in the basic fire suppression skills. He said that they took pride in getting out of the station quickly on structure fire calls. He finished lunch with the recommendation that every little win should be recognized.

Wrap-Up

■ Chief Concepts

- Effective leaders are also good followers, supporting the fire department leadership.
- Three situational leadership styles are:
  - Autocratic
  - Democratic
  - Laissez-faire
- Dangerous emergency scene situations, such as an evacuation order or a fire fighter may-day, require an immediate and autocratic response from the fire officer.
- Leadership in routine situations is accomplished through influencing, operating, and improving.
- Fire officer house rules are as follows:
  - Do not affect the ability of the fire company to respond to emergencies in their district.
  - Do not jeopardize the public trust of the fire department.

■ Hot Terms

<u>Followership</u> Leaders can be effective only to the extent that followers are willing to accept their leadership.

<u>Leadership</u> A complex process by which a person influences others to accomplish a mission, task, or objective and directs the organization in a way that makes it more cohesive and coherent.

<u>Power</u> The capacity of one party to influence another party.

Fire Officer *in Action*

Engine and Tower Ladder 5 are first due to a late night townhouse fire. Initial operations do not go very smoothly as you encounter severely injured occupants outside of the front and back of the building. The two rookies are energetic but ineffective as they try to help the injured occupants and perform their assigned fire-ground tasks. The inexperienced Engine 5 apparatus operator has trouble getting water to the tower. You note that five of the six members from The Nickel are in new assignments since recent retirements and details.

Standing at the command vehicle, the administrative fire officer reviews the incident with the supervising and managing fire officers. The chief notes the quick removal of remaining occupants in the townhouse of origin and the adjacent exposures. She asks the unit officers to write a narrative of what their crews observed and did. Reports are due at the battalion office within 24 hours. She says that it appears two adults and three children were severely burned and may not survive.

1. The rookies were overwhelmed on their first working fire. What is the preferred way to deal with the emotions that may arise after a stressful incident?
   A. Not bringing them up is the best way; talking about things that have obviously bothered the crew is just another way to add tension.
   B. Everyone has different ways of dealing with stress, and the crew should respect that some individuals will want to talk about the incident while others will not.
   C. By being forced to talk about the issues in a round-table-like discussion.
   D. Ignoring the issues if they are brought up.

2. On the next workday you have a meeting with the crew to see what can be done to improve company performance on the fire ground. This is an example of _____ leadership style.
   A. democratic
   B. reinforcement
   C. expectancy
   D. laissez-faire

3. During the meeting, a senior fire fighter complains about the loss of tradition and the poor job the academy does teaching new recruits. When he is done you decide to make no response. This is an example of _____ in reinforcement theory.
   A. positive
   B. extinction
   C. punishment
   D. negative

4. The crew agrees that they will practice a standardized fire-ground evolution every day, with the goal to complete the tasks quicker than the target time identified in the drill manual. This is an example of _____ motivational theory.
   A. motivation-hygiene
   B. goal-setting
   C. equity
   D. expectancy

Working in the Community

NFPA 1021 Standard

Fire Officer I

4.3 **Community and Government Relations.** This duty involves dealing with inquiries of the community and communicating the role, image, and mission of the department to the public and delivering safety, injury, and fire prevention education programs, according to the following job performance requirements. [p 176–185]

4.3.1 Initiate action on a community need, given policies and procedures, so that the need is addressed. [p 177–185]

(A) Requisite Knowledge. Community demographics and service organizations, as well as verbal and nonverbal communication, and an understanding of the role and mission of the department. [p 177]

(B) Requisite Skills. Familiarity with public relations and the ability to communicate verbally. [p 178, 184]

4.3.3 Respond to a public inquiry, given policies and procedures, so that the inquiry is answered accurately, courteously, and in accordance with applicable policies and procedures. [p 180–181]

(A) Requisite Knowledge. Written and oral communication techniques. [p 180–181]

(B) Requisite Skills. The ability to relate interpersonally and to respond to public inquiries. [p 180–181]

Fire Officer II

5.3 **Community and Government Relations.** This duty involves dealing with inquiries of allied organizations in the community and projecting the role, mission, and image of the department to other organizations with similar goals and missions for the purpose of establishing strategic partnerships and delivering safety, injury, and fire prevention education programs, according to the following job performance requirements. [p 185–189]

5.3.1 Explain the benefits to the organization of cooperating with allied organizations, given a specific problem or issue in the community, so that the purpose for establishing external agency relationships is clearly explained. [p 185–189]

(A) Requisite Knowledge. Agency mission and goals and the types and functions of external agencies in the community. [p 185–187]

(B) Requisite Skills. The ability to develop interpersonal relationships and to communicate orally and in writing. [p 185–189]

5.4 **Administration.** This duty involves preparing a project or divisional budget, news releases, and policy changes, according to the following job performance requirements. [p 187–188]

5.4.4 Prepare a news release, given an event or topic, so that the information is accurate and formatted correctly. [p 187–188]

(A) Requisite Knowledge. Policies and procedures and the format used for news releases. [p 187–188]

(B) Requisite Skills. The ability to communicate orally and in writing. [p 185–189]

Introduction to Fire and Emergency Services Administration (FESHE) Course Outcomes

2. Explain the need for effective communication skills both written and verbal. [p 178, 180–181, 185–189]
9. Identify and assess safety needs for both emergency and nonemergency situations. [p 178, 180–185]

Knowledge Objectives

After studying this chapter, you will be able to:

- Discuss community demographics.
- Discuss risk reduction.
- Discuss public inquiries.
- List four objectives of a public safety education program.
- Describe Fire Prevention Week.
- Describe the National Fire Protection Association (NFPA) Risk Watch program.
- Describe the goal of a Community Emergency Response Team.
- Describe developing a local public education program.
- Describe the three steps NFPA recommends in developing a relationship with the media.
- Define proactive media communications.
- List the four NFPA guidelines to use when conducting an interview with the media.

Skills Objectives

After studying this chapter, you will be able to:

- Develop a public education program using a five-step program.
- Prepare a media release that conforms to local format.
- Conduct a media interview as the fire department representative.

I t is almost noon and your company is returning from a greater alarm structure fire that was started by discarded fireplace ashes placed in a plastic trash can that was inside a garage. Three hours after the homeowner left for work, a neighbor called 9-1-1 to report smoke coming from a garage. As the fire companies started to arrive, the fire ignited improperly stored gasoline, creating an impressive explosion and pushing the fire into the home.

As you arrive at the fire station, you notice a private vehicle parked at the front door. A citizen approaches as soon as you climb down from the cab and asks if you can take care of her problem right away. She has been waiting for 2 hours for someone to help her install a child car seat. She had an appointment and is unhappy because no one was at the station when she arrived.

Before you can ask your driver/operator, who is also the designated car seat technician, to help the citizen, you see the local day care center minivan arriving for a scheduled station tour. You look at the two adults and six preschoolers in the van and see how excited they are to visit a fire station.

1. What are your options for dealing with the station tour at this point?
2. How do you respond to the citizen who is unhappy about having to wait 2 hours for your return?
3. What will you do with your visitors if you get another alarm within the next 30 minutes?
4. The departmental regulations require you to be clean and dressed in uniform when conducting public education activities. At this moment you are wet and dirty; what should you do?

Introduction to Working in the Community

Volunteer fire departments were often established by community members after a local disaster. During the 1800s, fire stations were frequently used for public meetings and assemblies, and membership in the local volunteer fire company was often a stepping stone for politicians and power brokers.

This still exists in rural and small town fire departments. Even in cities where the 20th century saw the transition to career firefighting, the bonds between local communities and their fire fighters have remained strong. The apparatus doors of many fire stations are left open, and the proverbial outside bench is still a popular location for fire fighters to maintain close contact with the neighborhood.

The situation is different in most urban areas, where the benches have been moved inside and the apparatus doors are always closed. Decay and crime have changed the nature of the relationship in hundreds of communities.

Even in large cities, the fire station continues to be widely viewed as an important community member. Citizens tend to think of the fire department in relation to the local fire station. The financial crises of the 1980s showed the value of strong community ties, as residents in major urban centers, such as New York City, Baltimore, and Toronto, rallied to keep their local fire stations open in the face of budget cuts, measuring their personal perceptions of safety by the proximity of the closest fire station.

The trend of local government in the early 21st century is to move toward more community-based local government. Next to schools, the fire department is the most decentralized and community-based function of local government. Some cities use their neighborhood fire stations as a primary point of contact for local government services.

Company Tips

Child Safety Seats

Demographics can assist the fire department in planning for new safety programs, such as child safety seat installations. In the late 1990s, public safety officials in many communities identified a need to provide public education and assistance regarding proper child car seat installations. The sudden popularity of this service overwhelmed some fire departments that were not adequately prepared.

The proper installation of a child car seat is complex. A 40-hour class is required for a public safety official to achieve certification as a child car seat technician. It makes sense to provide this training to fire fighters in the stations, where a high demand for the service can be anticipated.

The demand for this type of service is predictably concentrated in areas where there are families with automobiles and young children. The fire department can use demographic information to identify the fire stations that serve communities where there are concentrations of young families who would take advantage of the service.

Fire officers can then survey those communities to identify the days and times when the demand for the service would the highest. To meet the public's need, specific days and times might be designated to install child safety seats. Dedicated staffing could be assigned on those days to ensure that a certified child car seat technician would be available to perform the installation, regardless of the emergency workload. This information can be disseminated in press releases and posted on the department's website and on signs at every fire station **Figure 10-1 ▶** .

The fire officer in each fire station is the official fire department representative and also ensures that the community's needs are being addressed by the department. The fire company is also a neighbor. This chapter provides the fire officer with information about how to learn more about the community, how to provide fire and life-safety education to meet local needs, and how to work with the news media to disseminate information clearly and appropriately.

Understanding the Community

Each community has special needs and different characteristics that should be considered in relation to every service and program provided by the fire department. The most significant information comes from understanding what type of person lives or works in the community. The fire officer should develop a good understanding of the population and demographics of the particular areas where the company responds.

The federal government undertakes a nationwide census once every decade. This information is readily available and provides an excellent starting point to begin an analysis of the local community. The census collects and identifies a massive amount of information about the **demographics** of the nation. Demographic data describe the characteristics of human populations and population segments. The 2000 and 2010 census data can be analyzed with the use of sophisticated database software tools and digitized mapping to profile many characteristics of local populations down to the neighborhood level.

There is a variety of demographic analysis techniques. Demographic data are often used to identify and analyze consumer markets, to help retailers predict what consumers will buy, and to give advertisers insight into the messages that will be most effective in particular markets. Politicians make extensive use of demographic data to predict the voter response to policies and messages in each area. Messages and programs are fine tuned to reach certain segments of the population based on detailed analysis of what different types of people like, what they value, what they believe, and where they live. The same approach can be used to ensure that the fire department is delivering the appropriate services and information to the local community.

Understanding the cultural factors that influence particular behaviors will increase the effectiveness of fire department messages. You can identify groups with special needs to improve the delivery of emergency services. An effective program should be designed to meet the needs of the particular community, and the message has to be formulated to reach the target audience.

The United States is a land of immigrants and becomes more multiethnic with each census. There is a continual flow of new immigrants, who bring with them a variety of languages, cultures, religions, traditions, and beliefs. The President's Council of Economic Advisors reported that from 1996 to 2003, nearly 60 percent of net economic growth and 50 percent of the growth of working age population was due to recent immigrants. There are school districts where students speak more than 100 different languages.

One dimension of demographics is the identification of communities where people share the same cultural background and language. Within a small neighborhood, there may be thousands of people who speak one particular language and share a culture that is totally different from that of the adjoining neighborhoods. Highly diverse populations can be found in many rural and suburban communities. This type of diversity presents tremendous challenges to the fire service in trying to meet the needs of different groups.

Safety information should be applicable to the particular community and delivered in a format that the community can understand and act on. The format and the method of delivery should be different when fire fighters are meeting with a community of new immigrant families who understand very little English versus a community of university students living in fraternity houses.

Both emergency service and public education efforts have to be fine-tuned to identify the needs and meet those needs in particular communities. The fast and loud aspects of a typical Western world response to a fire emergency could be overwhelming, terrifying, or embarrassing to a recent immigrant who needs assistance. The customs and traditions of another culture may clash with the training and recognized response to a cardiac arrest. A shop owner may be angered by the arrival of a fire company to perform a fire safety inspection. Cultural sensitivity is required to help the customers appreciate the services the fire department is providing.

City of
Hazelwood

415 Elm Grove Lane
Hazelwood, MO 63042-1917
Phone: 314-839-3700
Fax: 314-839-0249

Hazelwood Fire Activities   **02/02/2003**

The Hazelwood Fire Department would like to remind the residents of our community that we continue to serve you with our new child car safety seat installation program. This is a free service residents can use to get information and assistance with learning how to properly install your child's car seat. Please understand that it is important to call ahead to make an appointment with the car seat installer for the day you would like to come in. Unfortunately, our work schedule does not always allow for "walk-ins" to be immediately accommodated. We almost always have at least one certified car seat installer on duty at any given time, so it is likely that you may be able to come in the same day as you call. It just may not be at the same time as your call or random visit. Mornings are our busiest time, but we are willing to be very flexible in scheduling installations including evenings, weekends, etc. We want you to be sure that your child is as safe as he or she can possibly be while riding in your car. And we are willing to do whatever we can to achieve that goal. Thank you.

If you would like to schedule a car seat installation appointment, call the Hazelwood Fire Department at 731-3424.

Courtesy of the City of Hazelwood, Missouri.

Figure 10-1 Description of a car seat installation program.

Getting It Done

Se Habla Espanol?

In some communities, fire fighters have to be able to communicate in a second language. There are communities where English is not the primary language, so it is important to be able to communicate in the language of the neighborhood. Many departments encourage their members to learn additional languages, and some even provide special classes or offer pay incentives for bilingual fire fighters. A few departments have experimented with a language immersion program, making the primary language of the neighborhood (e.g., Spanish) the primary language at a designated fire station. The fire fighters assigned to that station answer the telephone, speak to each other, and handle emergencies while speaking another language.

▌Risk Reduction

The fire service has a rich history of striving to reduce the risk of fire in the community. This has been achieved through emergency response, fire safety education, adoption of fire codes, and enforcement of those codes. Fire departments have recognized that their primary goal is to save lives and property from more than just fires. This broader vision has been evident as the fire service has been responding to extrications, drownings, heart attacks, strokes, hazardous materials releases, trench collapses, building collapses, high-angle rescues, swift water rescues, underwater and ice rescues, and weapons of mass destruction incidents.

With this broadened scope, the fire service has taken on the role of prevention and mitigation of these types of incidents. The best method of preventing fire injuries and deaths is by preventing the fire or reducing the severity of the fire. The same

Voices of Experience

There are lots of reasons why fire fighters and EMS personnel may not be able to live in the community they serve, but it is always helpful to live there if possible. If not, then you certainly must be part of local organizations and community groups whenever you can, and certainly shop locally, eat in the restaurants, and be known as the "fireman" who comes in all the time.

The people in the community will get to see the human side of you rather than just the professional business side. When people can associate with you as a regular person, the public support for your organization becomes cemented. This becomes critical when budgets are being discussed and new programs are being implemented.

From the days of old, the local volunteer fire department was a vital core of the community, and in many parts of the country this great tradition continues.

A department where I was fire chief was trying to incorporate EMS into the fire service, but there did not seem to be much public support for it. In fact, we had failed for a couple of years. I was going to work early one morning and was in the local restaurant having breakfast about 6:30 A.M., when someone I did not know approached me and asked if he could sit down. He recounted that he had followed my story in the local press. He said he supported our efforts but that we might have been going about it all wrong.

I was very puzzled and asked, "How so?"

He went on to explain that the form of government in this community was very driven by the needs and wants of the community. Although the people seated in government were in power, they would take their lead from prominent community leaders. This man suggested I begin to speak to certain groups in the community and just factually explain our needs, attend some dinner meetings, and let folks see how genuine I was in my concern for getting the EMS back into the fire department.

He assisted me by making some introductions to folks I certainly would not have ever met had I been sitting in my office doing nontraditional thinking.

After several meetings, many more breakfasts, and about 6 months of time, it suddenly became suggested by others that the EMS belonged back in the fire department.

Fire departments will exist in most communities. They will not thrive, flourish, or grow unless the department's members—of all ranks—become part of the community.

> **"He assisted me by making some introductions to folks I certainly would not have ever met had I been sitting in my office doing nontraditional thinking."**

Peter J. Lamb
Fire Chief
North Attleboro Fire Department
North Attleboro, Massachusetts

is true for the other types of incidents to which the fire department is called.

This change in thinking often requires a shift in the culture of the fire department. Members joined to fight fires—not to check car seats. The company officer challenge is to promote the concept that the fire fighter role is to save lives and property from every type of incident to which we might be called. A link must be made between the incidents to which the fire department responds and preventing those types of incidents.

The fire officer should use incidents that are encountered to reinforce the need for risk-reduction programs. For example, the fire company responds to a child who is seriously injured from head injuries in a bicycle accident that could have been prevented by wearing a helmet.

There are two levels of needs: systemic and individual. Systemic needs can be addressed through the development of programs to eliminate or reduce risk, such as a bicycle helmet safety program directed toward children. Often, these programs are a community-wide response and involve many organizations and agencies. Many of these needs are identified at the departmental or community level rather than the fire company level. Fire officers may identify systemic needs from the calls to which the fire company responds. If a fire officer identifies preventable incidents, he or she should report this to fire administration for consideration.

The fire officer is also in the unique position to identify the needs of the individual. The fire department is one of the few community resources that regularly respond into private homes. These needs can vary from reporting signs of child abuse to heating assistance programs. Typically, programs are already in place to address the need. The company officer's role is to connect the citizen in need with the community resource.

This requires the fire officer to proactively learn what community programs are available. Many programs are run by the health department and the department of social services. Often, these departments know what assistance programs are available for low-income families and the elderly. The area safety council and hospitals are valuable resources for information on the prevention of accidents and injuries. The police department and the highway department have a vast amount of information on vehicle accident and injury prevention. Faith-based and civic groups also have many programs to assist in community needs.

A fire officer needs to seek out this information before he or she needs it. Often, brochures are a good method of having the information available when it is needed; this allows the fire officer to address the need on the scene. For example, during a call, the fire officer notices a pool with an unlocked gate. Discussing the issue with the resident and leaving a pool safety information packet may prevent the fire department from responding to a child drowning at the location. Other issues may not be able to be addressed on the scene. An elderly couple who lives alone and is unable to care for themselves will likely need to be referred to the appropriate community agency for a follow-up visit.

It is important to follow departmental policies for these situations. Sometimes, the fire officer is allowed to call directly to the agency that can provide the assistance. Fire officers are likelier to take action in fire departments that empower the fire officer to initiate action directly.

Getting It Done

Reducing Alpha Calls

There has been an explosion of 9-1-1 ambulance calls, with many of them not requiring emergency medical care or an emergency room visit. Grady Hospital provides emergency ambulance service for Atlanta, Georgia, and responded to 24,000 Alpha calls in 2006. Alpha, the least urgent medical calls in the Medical Priority Dispatch System, accounted for 27 percent of the 9-1-1 workload, with 9000 of those runs coded as Alpha 26-1, no apparent problem or symptom.

The Memphis Fire Department was able to eliminate 3000 9-1-1 dispatches in 2007 by assigning a fire officer/paramedic to "reduce call volume from 16 of the top 25 addresses." Initial success was due to meeting the needs of the five most frequent users of the ambulance.

As the Memphis 9-1-1 Alternatives coordinator, the company officer identified that mega-users were in an emergency department every day. The fire officer connected the mega-users with a primary care physician, social services, or whatever was needed. Transportation was the most common issue, with the mega-users using 9-1-1 ambulances as their method of transportation. During the first year of the program, four of the top five mega-users stopped using the ambulance as their taxi. The 3000 fewer 9-1-1 responses represented 3 percent of the Memphis Alpha calls.

Tulsa Fire Department established Alpha trucks. Two fire fighter/EMTs in a defibrillator-equipped pick-up truck respond to nonemergency Alpha medical calls. They stay until the issue is resolved, about 45 minutes per incident. The first unit handled 3000 calls in a pilot program that covered the busiest fire district.

Both Memphis and Tulsa were able to reduce ambulance and fire company workload by analyzing the problem and considering innovative solutions.

■ Responding to Public Inquiries

Sometimes, the public makes an inquiry to the fire officer, for example, for general information, such as a request for a description of the services the fire department or city provides or a request to remove a cat from a tree. Both of these situations have the potential to leave the citizen satisfied or for the citizen to go away with a negative view of the fire department.

The fire officer must treat all requests professionally and with respect. Even though you may find the request less than critical, the citizen believes that it is valid and important. Failure to approach the request with sincerity can have a lasting negative impact.

Every effort should be made to answer each question fully and accurately. The fire officer may not know all the information that the individual wants to know. For example, the fire officer may not know where the closest flu shot clinic is located. When this occurs, the fire officer should seek out the information immediately rather than just expressing that he or she does not know. A phone call or two can usually lead to the answer and leaves the citizen with a positive image of the fire department.

Some requests that citizens make may not be within the fire officer's authority. In these situations, the fire officer should provide a method of moving the request to the level where it can be resolved. If time allows, the best method is to get the citizen's

Getting It Done

"My Cat Is Stuck in the Tree. Can You Get It Down?"

Many citizens have the image of a fire fighter rescuing a cat from a tree. Most fire departments do not provide this service. There is a trend to reconsider this position in light of risk reduction and public relations. First, if the cat does not come down, the owner experiences a loss. Second, if the resident attempts to retrieve the cat, he or she can be placed in grave danger.

One method is to send a fire company to the scene to evaluate the situation. Most cats are perfectly able and will come down in a matter of time. An empathetic explanation to the owner, with a referral to animal control or a veterinarian, may allay concerns. If the cat is actually stuck in the tree, has remained there for several days, and is in jeopardy of dying, the fire officer may consider whether the fire crew can retrieve the cat in a reasonably safe manner. The fire officer must balance the safety of the fire crew against the seriousness of the situation, the risk to the citizen from the fire department's inaction, and the potential benefits of a positive result.

Figure 10-2 Stop, Drop, and Roll teaches young children what to do if their clothing catches fire. © National Fire Protection Association.

contact information, write up a summary of the discussion, and forward it to the appropriate administrative fire officer at headquarters. The fire officer should also follow up to ensure that the citizen is contacted in a reasonable amount of time. Some citizens may prefer to contact fire administration themselves. In these situations, the fire officer should give the citizen specific information on whom to contact. The fire officer should also contact the administrative or executive fire officer before the citizen does to provide information on the situation.

Many departments have specific policies that address how citizen inquiries are to be handled. The fire officer should understand and follow these policies. Failure to do so may lead the citizen to believe that he or she is being treated unfairly. It can also leave the fire officer open to disciplinary action.

Public Education

The goals, objectives, content, and delivery mechanisms for public education programs vary greatly among fire departments, depending on their resources and circumstances. Some large fire departments have staff bureaus that specialize in the development and delivery of public education programs. Other fire departments adopt national programs or adapt a program that was developed by some other department. In many cases, a public education program is developed at the local level to meet local needs. Public fire safety education programs include:

- Learn Not to Burn
- Risk Watch
- Stop, Drop, and Roll **Figure 10-2 ▶**
- Getting to Know Fire
- Change Your Clock—Change Your Battery
- Fire safety for babysitters
- Fire safety for seniors
- Wildland fire prevention programs

The delivery of public education programs often involves fire suppression companies and depends on fire officers to transmit the message to the intended audience. Fire officers can also be assigned to positions that involve specific responsibilities for public education program planning and development. Every fire officer should understand some of the basic principles of public education programs.

The fundamental goal of a public safety education program is to prevent injury, death, or loss due to fire or other types of incidents. A public safety education program can have four objectives:

- Educate target audiences in specific subjects in order to change their behavior.
- Instruct target audiences on how to perform specific tasks, such as Stop, Drop, and Roll or operate fire extinguishers.
- Inform large groups of people about fire safety issues.
- Distribute information on timely subjects to target audiences.

An educational presentation is successful when it causes an observable change of behavior. The fire officer should have a specific goal in mind when educating the public, just as there should be a goal when providing training for fire fighters **Figure 10-3 ▼**.

Figure 10-3 The goal of public safety education programs is to prevent injury, death, or loss due to fire.

National and Regional Public Education Programs

The NFPA, the United States Fire Administration (USFA), and other specialized associations and industry groups have developed national and regional public education programs that can meet local community needs. Some of these programs are designed for general outreach, such as Change Your Clock—Change Your Battery and Fire Prevention Week themes. Programs targeted toward particular problems and population groups are also available.

Fire Prevention Week

The history of Fire Prevention Week has its roots in the Great Chicago Fire, which began on October 8, 1871, but continued into and did the most damage on October 9, 1871. In just 27 hours, this tragic conflagration killed 300 people, left 90,000 homeless, and destroyed more than 17,400 structures.

On the 40th anniversary of the Great Chicago Fire, the Fire Marshals Association of North America (now known as the International Fire Marshals Association) decided that the date of this occurrence should be observed in a way that would keep the public informed about the importance of fire prevention.

President Woodrow Wilson issued the first National Fire Prevention Day proclamation in 1920. National Fire Prevention Week has been observed on the Sunday-through-Saturday period in which October 8 falls, every year since 1922. In addition, the President of the United States has signed a proclamation announcing the national observance of Fire Prevention Week every year since 1925. NFPA has officially sponsored Fire Prevention Week since the observance was first established.

Risk Watch

Risk Watch is a comprehensive program directed at injury prevention. It was developed by the NFPA, with co-funding from the Lowe's Home Safety Council and in collaboration with a panel of respected safety and injury prevention experts. A school-based program that links teachers, safety experts, and parents, Risk Watch gives children and their families the skills and knowledge they need to create safer homes and communities.

The curriculum is divided into age-appropriate lessons, each of which addresses the following topics:

- Motor vehicle safety
- Fire and burn prevention
- Choking, suffocation, and strangulation prevention
- Poisoning prevention
- Firearms injury prevention
- Bike and pedestrian safety
- Water safety
- Natural disasters

Community Emergency Response Team

The Community Emergency Response Team (CERT) concept was developed and first implemented by the Los Angeles City Fire Department (LAFD) in 1985. The CERT program helps citizens understand their responsibilities in preparing for disaster and increases their ability to safely help themselves, their families, and their neighbors in many types of situations.

The concept of the program was to provide basic training to residents and employees in local communities, as well as to government workers, that would allow them to function effectively during the first 72 hours after a catastrophic event. Experience had shown that in the event of an earthquake or similar event,

Near Miss REPORT

Report Number: 09-0000313

Event Description: An engine and a rescue from our department went to an elementary school to do a program for fire prevention month. My partner and I were assigned to the rescue and we set up a high-angle rescue demonstration for the kids. We placed a 14' roof ladder on the roof of the gym and built a raise/lower system with a Stokes basket. We put a 175-pound "Rescue Randy" in the basket as our victim. One of us would go up on the roof to sit on and "heel" the roof ladder while the other demonstrated the operation of the raise/lower rescue system.

At lunch time, we went into the cafeteria to eat. When I was walking back to our display area, I saw one of the rookie firefighters about to hoist our rope to raise the basket. I yelled at him to stop but I was too far away for him to hear. When he pulled the rope, the 14' roof ladder came off the top of the gym, fell approximately 25', and struck the ground next to him. He was not injured and the ladder was not damaged. He was not wearing a helmet and if the ladder had struck him on the head, it would have killed or seriously injured him.

Lesson Learned: I learned that as the person assigned to the rescue equipment I am responsible for everyone's safety in regards to the equipment. I learned that everyone should be cross trained as to the rescue equipment uses and hazards. I learned not to leave my equipment in an unsafe manner. I learned not to be complacent about safety issues when performing non emergency evolutions.

Getting It Done

Public Safety Education Messages

In the "You Are the Fire Officer" scenario, the fire station tour for the day care center provides an opportunity to deliver a public safety education message that is suitable for a preschool audience. The NFPA has developed age-appropriate fire prevention or safety messages that can be delivered as part of a fire station tour. For preschoolers, knowing how to Stop, Drop, and Roll if clothes are on fire is a very appropriate safety message. The tour should include the opportunity for the children to demonstrate how to Stop, Drop, and Roll.

Fire officers should develop scripts to guide fire station visits for different age groups. These scripts can include the safety message or messages that are appropriate for the particular audience. The USFA Kids website provides general lesson plan examples. Audiovisual aids, handouts, or souvenirs should be available for planned visits Figure 10-4 ▾ . Do not forget the adults who accompany the children. In addition to the fire safety messages, the adults might be interested in a handout describing how the department serves the community, including a breakdown to show what the department has purchased from tax revenues and community donations.

Figure 10-5 CERT groups can provide immediate assistance to victims and assist professional responders with prioritization and allocation of resources.

emergency services would be overwhelmed with serious incidents and unable to respond promptly to every problem. The CERT program was developed to train citizens to help themselves.

CERT groups can provide immediate assistance to victims in their area and collect disaster intelligence that assists professional responders with prioritization and allocation of resources after a disaster Figure 10-5 ▸ . The training also teaches the CERT members how to organize spontaneous volunteers who have not undergone the training.

The Whittier Narrows earthquake in 1987 underscored the area-wide threat of a major disaster in California as well as the effectiveness of the CERT program. As predicted, during that event, there were so many individual incidents over such a wide area that fire department responders could not quickly get to every location where assistance was needed. Where the CERT program had been implemented, the value of having trained groups of citizens was clearly demonstrated. As a result, the LAFD Disaster Preparedness Division was established, and the CERT program was expanded.

The Emergency Management Institute and the National Fire Academy adopted and expanded the CERT materials. This training was made available nationally by the Federal Emergency Management Agency (FEMA) in 1993. Since the September 11, 2001, terrorist attacks, many more fire departments have started similar CERT training programs in their communities. CERT was moved to the Citizen Corps section of the federal government in 2004. In 2008, Citizen Corps listed 3091 CERT programs.

Figure 10-4 Audiovisual aids, handouts, or souvenirs should be available for planned visits. © National Fire Protection Association.

Safety Zone

A Risk Watch Save

A sixth grader from Wishek, North Dakota was walking home from a friend's house one January when he heard an elderly man calling for help. The man had slipped on ice. "He told me he couldn't move. He thought he had a broken bone or something. He was just sitting there shivering." The child ran to a nearby home, told the people there what had happened, and asked them to call 9-1-1. He returned to the man and waited with him until an ambulance and other authorities arrived.

"I'm very proud . . . of his heroic effort to help someone in need," says his teacher. "His good deed may have saved a man's life." The child says the Risk Watch component in his school health class helped him know what to do. "I saw he was hurt and realized right away that someone had to call 9-1-1," he says. "I would have done the same thing for anybody."

CERT Course Schedule

The CERT course is delivered in the community by a team of first responders. The instructors should complete a CERT Train-the-Trainer Program, conducted by their State Training Office for Emergency Management.

Training is usually delivered in nine 2.5-hour sessions and delivered one evening per week, covering the following:

- Session 1, Disaster Preparedness: Addresses the different hazards to which people are vulnerable in their community. Materials cover actions that participants and their families can take before, during, and after a disaster. As the session progresses, the instructor begins to explore an expanded response role for civilians to become disaster workers. The CERT concept and organization are discussed, as are applicable laws governing volunteers in that jurisdiction.
- Session 2, Disaster Fire Suppression: Briefly covers fire chemistry, hazardous materials, fire hazards, and fire suppression strategies. The thrust of this session is the safe use of fire extinguishers, sizing up of the situation, control of utilities, and extinguishing of a small fire.
- Session 3, Disaster Medical Operations, Part I: Participants practice diagnosing and treating airway obstruction, bleeding, and shock by using simple triage and rapid treatment techniques.
- Session 4, Disaster Medical Operations, Part II: Covers evaluating patients by performing a head-to-toe assessment, establishing a medical treatment area, performing basic first aid, and demonstrating medical operations in a safe and sanitary manner.
- Session 5, Light Search-and-Rescue Operations: Participants learn about search-and-rescue planning, size-up, search techniques, rescue techniques, and most important, rescuer safety.
- Session 6, CERT Organization: Addresses organization and management principles and the need for documentation.
- Session 7, Disaster Psychology: Covers signs and symptoms that the disaster victim and worker might experience.
- Session 8, Terrorism and CERT: Provides an overview on terrorism and weapons of mass destruction.
- Session 9, Course Review and Disaster Simulation: Participants review their answers from a take-home examination. Finally, they practice the skills they have learned during the previous six sessions in disaster activity.

Since the move to the Citizen Corps there has been an increased emphasis on assessing community needs and developing CERT response goals that address local needs. The appendix in the CERT training manual includes lesson plans for 13 hazardous situations.

Maintaining CERT Involvement

CERT members should receive recognition for completing their training. Some communities issue identification cards, vests, and helmets to graduates.

It is important to keep the graduates involved and practiced in their skills after they have completed the basic training course. Trainers should offer periodic refresher sessions to reinforce the basic training. CERT teams can sponsor events such as drills, picnics, neighborhood clean-ups, and disaster education fairs, which keep them involved and trained.

Getting It Done

Catastrophe Planning

Some fire departments have developed media communication/public education plans to assists their communities in responding to natural and man-made catastrophes. For example, the San Francisco Fire Department accepted the services of a retired local television anchorperson to develop preassembled public safety messages and citizen instructions in the event of a wide variety of major emergencies.

Fire departments have to be able to respond to dynamically changing conditions. For example, national concerns, such as the methicillin-resistant *Staphylococcus aureus* (MRSA) outbreak, required fire departments to quickly develop specific public information messages, using information updates from the Centers for Disease Control and Prevention. The sniper attacks in the Washington, D.C., area in 2003 required more than a dozen federal, state, and local public safety agencies to coordinate their efforts to provide a consistent message to the general public.

Locally Developed Programs

Like the CERT program, many regional and national public education programs started within local fire departments. Brush abatement programs for wildland fire areas and fireplace ash disposal programs in the northern states are two examples of regional programs that meet specific needs.

In large fire departments, the development of new public education programs is usually assigned to an individual or a work group specializing in this area. In a smaller department, a fire officer could be given this assignment as a special project.

The five-step planning process can help local fire departments create and develop programs. This process was created by the U.S. Fire Administration in the 1970s and updated in 2003 by FEMA.

Identify the Problem

The first task is to identify a fire or life-safety problem. In many cases, a public education project is initiated to address a problem that demands attention, such as an increasing number of fire deaths in a community. The process sometimes works in the opposite direction, with the decision to undertake a public education effort being made first, before the objective is identified. In either case, the fire officer should attempt to clearly identify the problem before developing the program.

The fire officer should look at local emergency incident data to identify the types of events that generate deaths or serious injuries as well as the most frequent causes of fires. The analysis should consider both the frequency and the severity of different risks. For example, the fire data for a particular city might indicate an average of 100 food-on-the-stove fires and 25 bedroom fires a year. The food-on-the-stove fires result in three serious burn injuries each year and one fatality every 3 years. The bedroom fires cause an average of three fatalities and six serious burn injuries annually. This analysis indicates that bedroom fires are the more critical life-safety problem in this community. There are more kitchen fires than bedroom fires, but the bedroom fires cause more deaths and serious injuries. (Chapter 6, Safety and Risk Management, discusses the concept of hazard and risk in more detail.)

Data from another city might point out that the total number of fire deaths is lower than the number of young children who drown in backyard swimming pools in an average year. A public education project that is directed toward the drowning problem would be a higher priority than a program designed to prevent fire deaths.

Public education programs are, in many cases, developed in response to a single, high-profile incident. Sometimes the high-profile incident directs public attention toward a problem that has been ignored in the past and creates an opportunity to provide valuable public education. Unfortunately, in other cases, this type of program is more reactive than effective and causes resources to be wasted on low-priority problems.

Getting It Done

Engineering, Education, or Enforcement?

- **Fireplace ashes:** Fire departments in cold climates often experience large-loss fires due to the improper disposal of fireplace ashes. In many cases, the ashes are placed in combustible containers and left in a garage or on a deck. The smoldering ashes eventually ignite the container and cause a fire. An engineering solution for this problem might be for the fire department to provide free metal ash cans to the community as part of a seasonal fire safety education program.
- **Dead smoke detectors:** Many fire departments encounter single-station smoke alarms with dead or missing batteries in residences. Some departments make it a practice to check the smoke alarm during every fire company response to a residential occupancy in the community. After stabilizing the initial incident, the fire officer may request permission to check the smoke alarms in the home. A simple engineering solution can quickly solve this problem. Every fire company carries a supply of 9-volt batteries that are used to replace any dead or missing batteries. Some fire departments even issue complete smoke alarms to their fire companies with instructions to install one in any home that does not have a functioning smoke alarm. The fact that the fire department takes this problem so seriously conveys an added public education lesson.
- **Brush abatement:** Failure to maintain a clear area around a building is a common cause of structure fire losses in wildland fires. Los Angeles County uses all three factors—education, engineering, and enforcement—in efforts to reduce the impact of brush fires on the built environment. Education starts with seasonal public safety messages through the media in multiple languages. Engineering includes a description of the size of the required clear area and a program to certify brush clearance contractors. Enforcement includes local government inspection of properties, as well as issuing of fines and work orders if the brush has not been cleared.

Select the Method

After identifying the problem, the fire officer should select the most effective method to address it. Public life-safety education is not necessarily the best or the only response for every type of problem. The fire officer should try to determine whether an engineering solution or an enforcement solution would be more efficient.

Design the Program

When designing a public education program, the fire officer should review the ABCDs of course preparation, just as in fire fighter training:

- Identify the *audience*.
- Explain the desired *behavior* the student should demonstrate after training.
- Under what *conditions* will the student perform the task?
- What *degree* of proficiency is expected?

The fire officer has to identify the specific objectives of the educational program. The program format should be matched to the message, the audience, and the available resources. Demographic information assists in describing the audience. The development process should include delivery of a pilot class to a target audience to see how well it works. Most new programs benefit from revision after the pilot is evaluated.

Implement the Program

Implementation occurs when the program is actually delivered to the target audience. Timing is often an important consideration when a public education program is implemented. For example, a fire safety program for university students would be most effective during the first couple of weeks in the fall semester in order to reach all of the new residents. The objective could be to reach every on- and off-campus student housing location at the start of the academic year.

Evaluate the Program

There are two types of evaluation—immediate feedback and long-term evaluation. Immediate feedback is obtained from the students at the conclusion of the training, usually from a short survey form. Immediate feedback can help the fire officer determine whether the specific objectives were met by the presentation. This type of immediate feedback is most effective in evaluating the mechanics of the class presentation, such as interaction between the speaker and the audiovisual media.

A longer-term evaluation should focus on the effectiveness of the program in relation to the desired effect. Success is accomplished when the students actually demonstrate the desired behavior in the anticipated situation. For example, the goal of a program at a university could be to reduce the number of malicious false alarms from pull stations. To evaluate the effectiveness of the program, the number of malicious false alarms in a time period before the training would be compared with the number during an equivalent period after the training.

Media Relations

Most fire departments have frequent interactions with the local news media, including during emergency incidents, in which reporters want to obtain information about a situation that has just occurred or is still occurring, as well as in newsworthy situations that are related to other fire department programs or activities. In addition, in many situations, the fire department has information or an important message to communicate to the public.

Emergency incidents are high-profile news events that capture public attention. The local print, radio, and television media can be expected to appear at major emergency incidents or to call the fire department seeking information. Many other fire department activities, ranging from inspection and public education programs to promotions, retirements, awards ceremonies, or delivery of a new fire truck, also generate media interest.

The news media have an important and legitimate mission to obtain and report information to the general public. The fire department is an organization that performs a public function, operates in the public view, and is usually supported by public funds. It is always in the best interest of the fire department to maintain a good relationship with the local media. Providing information to the news media should be a normal occurrence.

Although the news media have a legitimate purpose in seeking information, there are limitations on the information that can be released in many circumstances. The fire officer should know the applicable departmental procedures and guidelines and how to respond to a request for information that the fire officer is not authorized to release. The fire officer should understand that the interaction that occurs between the fire department and the news media is likely to have a direct impact on the department's public image, reputation, and credibility. The fire department depends on public trust and confidence to be able to perform its mission.

The Fire Department Public Information Officer

Some fire departments have a full-time or part-time public information officer (PIO) who functions as the media contact person and the source of official fire department information. Where this position exists, most interactions with the news media should go through this individual. Other officers have limited responsibilities and opportunities to interact with the news media. There may be occasions when the PIO refers a reporter to another fire officer to obtain information about a particular subject or situation.

Many smaller departments rely on the local fire officer or a staff officer to function as the PIO and interact with the local newspaper, radio, and television media as the need arises. In these situations, the individual should be guided by the same basic principles as a regularly assigned PIO. NFPA 1035, *Standard for Professional Qualifications for Public Fire and Life Safety Educator,* covers the job requirements of a public information officer.

This section presents basic recommended practices for a PIO who has to establish and maintain an ongoing relationship with the local media. This aspect is particularly important when the fire department has information it needs to release to the general public. Reaching out to a community is much easier for an organization that has already established solid contacts and relationships with the local media Figure 10-6 ▶ .

The NFPA provides a "Get Your Message Out" media primer to help the fire service publicize community outreach campaigns and other events associated with public safety education. The basic concepts are valuable for any fire officer who has to be prepared

Figure 10-6 Reaching out to a community is easier with solid contacts and relationships with members of the media.

to interact with media representatives. The NFPA publication recommends three steps when working with the media:

- Build a strong foundation.
- Use a proactive outreach.
- Use measured responsiveness.

Step 1: Build a Strong Foundation

Look at your relationship with the media as a business arrangement: the media have something you need (an audience and the means to communicate with them), and you have something they need (news and information). If you work collaboratively, you will both achieve your objectives. If you expect too much or if you do not anticipate the media outlet's needs, someone is going to end up disappointed, and that could be counterproductive.

The most important media asset you will ever have is a good relationship with the media. Members of the local media need to be familiar with you and need to be comfortable working with you; they must have confidence in the information you provide and, most importantl they need to trust you. Whatever else you do, always tell the truth to the media, even if the truth is "I do not know" or "I cannot release that information at this time" or "It is premature to answer that question." Do not guess, speculate, or lie.

Do not wait until you need a reporter to establish media contacts; lay the groundwork in advance for a collaborative working relationship. Build a list of local media contacts and keep it up-to-date. Obtain the names, titles, and contact information for key individuals. Learn who has responsibility at the newspapers, radio stations, and television stations.

Find out about the deadlines for different news outlets and how to reach reporters with late-breaking information. Television reporters need to have the information before news time, and print reporters need it before the paper is printed.

Be a consumer of your local media. Read the local newspapers, watch television, surf the online news sites, visit appropriate blogs, and listen to the radio [Figure 10-7 ▾]. Pay attention to the kinds of stories they tend to cover, the angles they use, and the personal style of individual reporters. If a reporter does a good job reporting a fire department story, make a thank-you call. If a story contains inaccurate information, make a call to provide the correct information, keeping the tone positive.

Make sure that the news representatives know how to contact you whenever they need information. If you are not going to be available, make sure that someone else is equally accessible. Your objective is to make sure that the news media representatives know that your department is interested in working with them and in responding to their inquiries, that you are the person to contact, and that they can have confidence in any information you give them.

Step 2: Proactive Outreach

If you have the resources, a proactive media communications plan can be much more beneficial than a reactive approach. Simply responding to media inquiries as they come to you is a reactive strategy. You should be as helpful as possible, providing information within the resource capabilities of your department.

Being proactive means that in addition to being responsive, you actively look for opportunities to use the media to communicate your department's objectives and mission. The media need factual, timely stories to report, and the fire department generates broad public interest. You can help them by providing announcements of newsworthy issues, topics, and events.

Once the working relationship has been established, reporters begin to look to you for story ideas, resources, and quotes. You can call a reporter or editor and suggest a story that could be interesting to him or her. Learn the types of stories that are most interesting to different reporters and call that person when you have something that fits. They are not going to automatically accept every story idea, but they will listen to what you have to suggest.

When you have information that you want to distribute to several news outlets, use a press release or hold a press conference. A press conference is a staged event in which you are inviting the news media to come and hear something important that you have to announce. Press conferences should be reserved for topics and situations that definitely have broad interest; on these occasions, you know that the media will want to attend and ask questions.

Step 3: Measured Responsiveness

When dealing with the news media, remember that one of your responsibilities is to present a positive image of the fire department. In certain situations, this is difficult, particularly if something negative has occurred. It is no use denying that something has happened when the facts are clearly evident. Sometimes, the best that you can do is to be honest and factual to maintain the credibility of the organization.

Be wary of situations in which someone could be trying to misrepresent a situation in a manner that reflects badly on the department. The reporter who comes to you may have already been given a highly inaccurate or slanted viewpoint on a story by some other source. Do not allow yourself to be placed in a situation in which you make a statement that is inaccurate, misleading, or damaging to the fire department. Make sure that your response is accurate and that the information is appropriate for release before you react.

■ Press Releases

A press release is used to make an official announcement to the news media from the fire department PIO. It could be an announcement of a special event, a promotion or appointment, an award ceremony, a fire station opening, a retirement, or any similar occurrence that the department wants to make known to the public. A press release could also provide information about a program, such as the launching of a new juvenile fire setter counseling program or the fact that the annual report has been released and statistics show a 34 percent decrease in residential fires since a major public education program was implemented.

A press release should be dated and typed on department stationery, with the PIO contact information at the top. The release should be as brief as possible, ideally one to two pages, using an accepted journalistic writing style. Make the lead paragraph powerful to entice the reporter to read on and answer the basic "who, what, when, where, and why" questions [Figure 10-8 ▸].

Depending on the preference of your local media contacts, you can mail, fax, or email press releases. Reporters and editors are likelier to read your releases if they receive only those that

Figure 10-7 Be a consumer of your local media.

Civilians Rescued From Burning Home

Twenty-eight fire fighters responded to an early morning house fire in the Grandview district. The first 9-1-1 call was at 10:47 pm on Monday, December 10, 2012; the caller reported smoke coming from a three-story townhome.

Metro County fire fighters arrived at 11:07 pm; they encountered smoke and flames coming from the first floor windows at 1928 Braniff Boulevard. Two elderly females were found on the third floor. They were taken out of the building through a window via an aerial tower, treated by paramedic/fire fighters, and transported to University Hospital.

It took 17 minutes of aggressive fire suppression before Battalion Chief Devon Jones declared the fire "under control." The first floor of the townhouse was extensively damaged, with heat and smoke damage to the adjacent townhouses.

The cause of the fire remains under investigation. Preliminary results show that the fire started in the kitchen. Smoke detectors were in the home, but batteries were removed.

The monetary loss has not been calculated.

Submitted by T. L. Gaines, Metro County Fire Department

Figure 10-8 Example of a press release.

Company Tips

Grammar Counts
Press releases are public documents that are read by reporters, the public, and local elected officials. The presentation and the content both reflect on the image of the fire department. Invest in a dictionary, thesaurus, and grammar book or style guide to improve your written presentation skills. Check the document's spelling, particularly names, and be sure that ranks, titles, and telephone numbers are correct.

pertain to their particular beat or to topics in which they are interested. Personalize the release to an individual at each destination by name and title. For example, a release announcing the official opening of a new fire station could go to the assignment editor at a television station and to the newspaper reporter who usually covers the fire department. A release on cooking fire safety could be directed to the food editor.

The Fire Officer as Spokesperson

Every fire officer should be prepared to act as a spokesperson for the fire department. Even if the department has a PIO who is the designated official spokesperson, there are likely to be occasions when other fire officers have to use these skills.

The most common situation in which a fire officer would appear as an official spokesperson is an interview. The NFPA provides these guidelines for the fire officer conducting an interview with the media:

- Be prepared.
- Stay in control.
- Look and act the part.
- It is not over until it is over.

Be Prepared

Before agreeing to be interviewed, make sure you are authorized to speak on the subject as a departmental spokesperson and

be sure you have the appropriate information. Determine the reporter's story angle and what he or she already knows about the topic. Ask with whom he or she has already spoken or plans to contact and what he or she hopes to learn from you.

If you are uncomfortable with the situation angle or do not have the information, you do not have to agree to be interviewed. You should be polite but firm if you decline a request for an interview. If possible, refer the reporter to someone else who can provide the information.

Be on time for an agreed-on interview, but do not begin the interview until you are ready. The reporter probably has a deadline that you should try to respect, but not at the expense of your readiness.

When preparing for an interview for which you want to deliver a specific message, determine no more than three key message points in advance and practice saying them in varied ways. Learn to use wording that emphasizes that what you are saying is important, such as, "The number-one thing to remember is _____ " or "More than anything else, people should realize that _____ ." Practice your talking points with a colleague or a tape recorder. Do not memorize your message points but be very familiar with them. Check any statistics, trends, or other information you would not know on a casual basis.

Stay in Control

Be cooperative, but stay focused on your key message points. Listen carefully to the reporter's questions and ask for clarification if you do not fully understand a question. If you need a few seconds to think through your answer, take the time to formulate your answer. If you do not know the answer, say so and offer to find the information and get back to the reporter. Avoid jargon and highly technical language that confuses or distracts your audience. When you have answered the question, stop talking.

Remember that the reporter came to you looking for information and expertise. Be confident and authoritative, steering the interview in the direction that you want it to go. Showcase your message points and your department's mission and work.

Figure 10-9 Make sure your appearance is clean, tidy, and professional, and wear your uniform properly.

Television Interview Behavior Tips

The following recommendations from the NFPA will help if part of the assessment center process includes conducting a television interview.

Look into the reporter's eyes, not into the camera. Keep your gaze steady and avoid rolling your eyes, blinking excessively, or closing your eyes when you are thinking about your answer. If you are standing, do not sway. Plant your feet about 18 inches apart, and keep your hands down to your sides or clasped together loosely in front of you. Do not put your hands in your pockets. If you are sitting, ask for a stationary chair. If you must sit in a swivel chair, plant your feet to help you keep the chair still. Sit with your knees together and your feet flat on the floor or your ankles loosely crossed. Keep your hands folded comfortably in your lap. Do not clench your fists, crack your knuckles, pick your nails, or play with your earrings. These are not only distracting behaviors, but also signal to viewers that you are nervous, which can be interpreted as a lack of confidence about the subject. (Imagine a close-up camera angle of your clenched fists or fidgety fingers.)

Look and Act the Part

When doing a television interview, make sure your appearance is clean, tidy, and professional. Wear your uniform properly, and show pride in the department that you are representing **Figure 10-9 ▲**. Adopt the appropriate demeanor for the subject. If you are on the scene of a fire fatality, a somber yet authoritative tone is appropriate. If you are conducting a television interview from a fire station open house, a more enthusiastic tone is acceptable.

It Is Not Over Until It Is Over

When speaking to a reporter, do not assume that anything you say is "off the record." Assume that everything you say could be quoted, including conversations before and after the actual interview. When you are wearing a microphone, anything you say can be overheard and recorded.

After the interview, double-check to ensure that the reporter has the correct name and spelling for you and your department and accurate information about any program or event that you are promoting. Leave your business card to ensure that the proper name, title, and organization attribution are recorded.

Summary

The fire officer is an ambassador of the fire department. Formal and informal educational sessions can have a tremendous impact on community safety. A fire officer's challenge is to be able to meet the community's needs and respond to emergency incidents in a manner that creates community goodwill and reduces fire-related deaths and injuries.

You Are the Fire Officer: Conclusion

Apologize to the mother who waited 2 hours for you to return to the fire station. Explain to her that you were out fighting a fire and that the driver/operator will install the child safety seat as soon as the apparatus is ready for the next emergency. Have the senior fire fighter assist in getting the engine ready to reduce the delay. Offer to reschedule the installation if this arrangement is inconvenient for her.

Have the fire fighters clean up quickly and prepare for the tour while you greet the children and teachers from the day care center. You can start the tour while the fire fighters put on clean uniforms. Apologize for your appearance and explain that you look the way you do because you have been fighting a fire. When the fire fighters return, you can excuse yourself briefly to get cleaned up. After the fire fighters finish the tour, rejoin the group and help the fire fighters demonstrate the Stop, Drop, and Roll technique to the children.

Wrap-Up

Chief Concepts

- Demographics describe a community through characteristics of human population (e.g., age, race, income, education).
- Community characteristics might call for an innovative public safety response, such as a Spanish-language immersion course at the fire station.
- Effective public fire and life-safety education programs must result in a desired change in behavior.
- Public safety education programs should educate, instruct, inform, and distribute information.
- NFPA's Risk Watch is an age-appropriate, school-based curriculum linking teachers with community safety experts and parents.

- The Los Angeles Fire Department developed Community Emergency Response Teams in 1985 to train citizens to assist each other in the first 72 hours after any calamity that would delay the response of public safety agencies, such as an earthquake.
- Fire officers should use the identify-select-design-implement-evaluate method to develop local public education programs.
- Media relations require the fire officer to build a strong foundation, reach out proactively, and provide measured responsiveness.
- The fire officer should be prepared, remain in control, and have an appropriate demeanor/appearance when participating in a media interview.

Wrap-Up

Hot Terms

Community Emergency Response Team (CERT) A fire department training program to help citizens understand their responsibilities in preparing for disaster and increase their ability to safely help themselves, their families, and their neighbors in the first 72 hours of a catastrophe.

Demographics The characteristics of human populations and population segments, especially when used to identify consumer markets. Generally includes age, race, sex, income, education, and family status.

Risk Watch A comprehensive NFPA school-based program focused on injury prevention.

Fire Officer *in Action*

While reviewing the second alarm garage fire, one of your crew members points out that many of the large-loss fires in the wintertime are due to improperly disposed fireplace ashes. Other members chime in, sharing stories of recent fires and close calls. Many of these incidents are occurring in large multiple-family dwellings in your first due. Many of the apartments have a fireplace and there have been a half-dozen near misses and small fires.

This property has 10 three- and four-story garden-style apartment buildings with combustible vinyl siding and lightweight wood truss frame. The buildings have a residential sprinkler system and underground parking. There is a moderate turnover of tenants. The fire company members remember fireplace ash starting fires on balconies, in the parking garage, and in the apartments.

Last year, a fire on a balcony spread up the vinyl siding and got into the attic, requiring a third alarm assignment to control the noontime fire. You wonder what the outcome would have been if the fire started at 1 A.M. With 10 buildings, this is a significant target hazard.

1. Your first goal is to:
   A. determine whether it is an engineering, education, or enforcement issue.
   B. prepare a press release on the hazards of discarded fireplace ashes.
   C. define the problem.
   D. design the program.

2. About 30 percent of the tenants are recent immigrants who may not be fluent in English. Your message should:
   A. be provided in English to the apartment resident manager for translation.
   B. use pictures and graphics.
   C. provide an opportunity for tenants to learn English.
   D. be in a format and use a method of delivery that is effective and culturally appropriate for this group.

3. How can you measure the effectiveness of a fireplace ash disposal educational effort?
   A. Determine whether there are fewer fireplace ash–initiated fires after delivery of the program.
   B. Count the number of times the program is mentioned in the local media.
   C. Estimate the market penetration of the fire department–provided ash cans.
   D. Base it on the number of requests for presentation of the program to civic groups and organizations.

4. When preparing for a media interview, provide no more than _____ key message point(s).
   A. one
   B. two
   C. three
   D. four

Handling Problems, Conflicts, and Mistakes

NFPA 1021 Standard

Fire Officer I

4.3.2 Initiate action to a citizen's concern, given policies and procedures, so that the concern is answered or referred to the correct individual for action and all policies and procedures are complied with. [p 205–206]

(A) Requisite Knowledge. Interpersonal relationships and verbal and nonverbal communication. [p 203–207]

(B) Requisite Skills. Familiarity with public relations and the ability to communicate verbally. [p 205–206]

4.4 This duty involves general administrative functions and the implementation of departmental policies and procedures at the unit level, according to the following job performance requirements. [p 201–205]

4.4.1 Recommend changes to existing departmental policies and/or implement a new departmental policy at the unit level, given a new department policy, so that the policy is communicated to and understood by unit members. [p 206–207]

(A) Requisite Knowledge. Written and oral communication. [p 206–207]

(B) Requisite Skills. The ability to relate interpersonally and to communicate change in a positive manner. [p 206–207]

Fire Officer II

5.4 **Administration.** This duty involves preparing a project or divisional budget, news releases, and policy changes, according to the following job performance requirements. [p 206–207]

Introduction to Fire and Emergency Services Administration (FESHE) Course Outcomes

2. Explain the need for effective communication skills both written and verbal. [p 198–199, 203, 205–206]

4. Recognize appropriate appraising and disciplinary actions and the impact on employee behavior. [p 206–207]

7. Identify roles and responsibilities of leaders in organizations. [p 196–199, 200–207]

10. Identify the importance of ethics as they apply to supervisors. [p 196–199]

Knowledge Objectives

After studying this chapter, you will be able to:

- Describe how to manage conflict.
- Describe how to deal with citizen complaints.
- Describe how to recommend policies and policy changes.
- Describe how to implement policies.
- Describe the difference between customer service and customer satisfaction.

Skills Objectives

After studying this chapter, you will be able to:

- Develop a policy or procedure.

your fire company responds to a late night "smoke in the building" call during a severe thunderstorm. A police officer on patrol observed smoke at a specialty food store and notified the dispatch center. On arrival you note that the store is on the first floor of a three-story building with apartments above the store. There is an electrical power failure in the neighborhood.

You confirm that there is a strong haze in the building. Before you complete your size-up, the rookie on your crew tries to force the front door. In his enthusiasm he breaks a large glass display window, right next to the fumigation notice and adjacent to the rapid entry key safe.

1. How should you handle this situation?
2. What steps should be taken immediately?
3. What steps should be considered after initial issues are addressed?

Introduction to Problems

A **problem** is the difference between a current situation and the desired situation. If the way we park at a highway incident places us at greater risk than is necessary, it is a problem. Forcing a commercial door and destroying a display window when there is an available rapid-entry key safe is a problem. There is a discrepancy between the current and the desired situation.

Fires and emergency incidents present a unique category of problems that call for specialized problem-solving skills. Non-emergency situations require the application of conventional problem-solving skills and techniques. These situations include supervisory, management, and administrative activities in which the fire officer is directly responsible for solving the problem, as well as initially processing situations that require resolution from an administrative or executive fire officer. These include situations that involve individuals or organizations outside the fire department.

Decision-making skills are used whenever the fire officer is faced with a problem or situation that requires a response. Promotional examinations evaluate the ability of fire officer candidates to exercise good judgment and make sound decisions. Decisions are guided by organizational values, guidelines, policies, and procedures. This is true even if that decision does not coincide with the fire officer's opinion or personal preferences.

Requiring the fire officer to act in the best interest of the department in solving problems and making decisions does not mean that other values and considerations are ignored. A problem can present several possible solutions, some better than others

and some more desirable to one set of interests than another. In most cases, there is a reasonable solution that serves multiple interests and concerns, whereas in other cases, one concern has to prevail over all others. Problem-solving techniques are designed to identify and evaluate the realistic potential solutions to a problem and determine the best decision.

Complaints, Conflicts, and Mistakes

Complaints, conflicts, and mistakes are special categories of problems. One of the key factors in decision making is how to deal with situations that involve conflicts or complaints.

These three terms are defined as follows:

- A **complaint** is an expression of grief, regret, pain, censure, or resentment; a lamentation; an accusation; or fault finding.
- A **conflict** is a state of opposition between two parties. A complaint is often a manifestation of a conflict.
- A **mistake** is an error or fault resulting from bad judgment, deficient knowledge, or carelessness. A mistake can also be a misconception or misunderstanding. Mistakes happen; the issue is how to deal with a mistake, or the perception of a mistake, when someone complains to the fire officer about it.

Sometimes a fire officer has to make a decision or enforce a policy that is not popular with the crew members. A citizen could be frustrated with a fire department action or may be unhappy with a situation. People misbehave and make mistakes. Disagreements and differences of opinion occur. It is not possible to make

Figure 11-1 Dealing appropriately with problems and conflicts requires maturity, patience, determination, and courage.

everyone happy all of the time. A fire officer must deal with all of these situations in a professional manner. Dealing appropriately with problems and conflicts requires maturity, patience, determination, and courage **Figure 11-1 ▲**.

The types of problems that a fire officer could be expected to encounter can be divided into four broad categories:

1. In-house issues: These are situations or decisions occurring at the work location that are within the direct scope of supervisory responsibilities. An example might be a complaint about the assignment of duties to different individuals within a fire station. Most of these conflicts begin and end at the company officer level.

2. Internal departmental issues: Operational policies, decisions, or activities that go beyond the scope of the local fire station are internal departmental issues. An example is a conflict over where a reserve ladder truck is housed and which company is responsible for maintaining it. Another example is a dispute between companies over the tasks of a rapid intervention team in a high-rise fire. The resolution usually requires action by command officers at a higher level on the organizational chart.

3. External issues: These are fire department activities that involve private citizens or another organization; for example, a citizen making a complaint about an inappropriate remark uttered during an EMS incident. External issues require the fire officer to perform one additional task early in the conflict resolution process: making sure that the fire officer's supervisor is not surprised. It is poor form for the battalion chief to learn about a fire department incident from the local media.

4. High-profile incidents: These include any issue that is likely to become a major event. An example might include a fire fighter arrested while on duty. The department must take immediate actions to respond to these events. Senior

fire administrators often become directly involved in these situations or keep a close watch on how they are handled.

A problem should be solved at the lowest possible level within an organization. A fire officer is expected to manage problems within the level of authority for a supervising or managing fire officer. At the same time, the fire officer should recognize problems that need to be handled at a higher level and make the appropriate notifications without delay. If there is any doubt, it is a wise policy to discuss the situation with the next higher level officer in the chain of command.

General Decision-Making Procedures

A fire officer is called upon to make many different types of decisions about a wide variety of subjects. At the fire officer level, most of the problems are fairly uncomplicated, although they are not necessarily easy to solve. Moving up the ranks means an exponential increase in decision-making situations. Often the problems are more complex, requiring the participation of multiple organizations.

The following provides a systematic approach to high-quality decision making:

1. Define the problem.
2. Generate alternative solutions.
3. Select a solution.
4. Implement the solution.
5. Evaluate the result.

Although the five-step technique appears to be designed for situations where there is plenty of time, the same basic approach is used for emergency incidents. A fire officer should be able to move quickly through the steps. Training and experience will prepare one to quickly identify the pertinent problem, generate realistic solutions, and select the best option.

■ Define the Problem

The first step in solving any problem is to closely examine and carefully define the problem. A well-defined problem is one that is half solved. Poorly defined problems waste tremendous time and effort.

Pay Attention

Tom Peters is a management consultant who popularized the concept of seeking out and emulating best practices. Peters used the phrase "management by walking around" to describe a practice he observed in the successful businesses he profiled in the 1982 book *In Search of Excellence*. An effective manager should know what is going on within the organization and address most issues before they become major problems. The best way to prevent major problems is to deal with issues before they reach the crisis stage **Figure 11-2 ▶**.

Ask Basic Questions

Peter Drucker was a famed management consultant. He served as a Professor of Management at the Graduate Business School of New York University for more than 20 years. He is best known

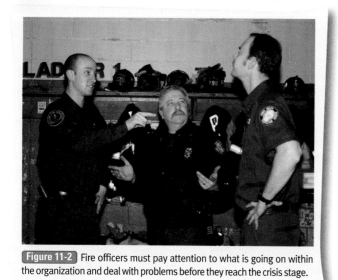

Figure 11-2 Fire officers must pay attention to what is going on within the organization and deal with problems before they reach the crisis stage.

for writing dozens of management books. Drucker encouraged managers to question the value of each organizational activity once a year. What may have been vital last year may have minimal importance this year.

Fire departments have a strong inclination to keep on doing the same things in the same ways. The fire officer should identify activities that can be changed, improved, or updated. If there is a better way to do something, give it appropriate consideration. If we are doing something that is no longer worth doing, why not use the time to do something more productive?

How Quickly Do You Get Bad News?

Few things damage a new fire officer's reputation quicker than not finding out about a situation that is going poorly until it is too late to fix it. Effective fire officers create a work environment that encourages subordinates to report bad news immediately. The sooner the officer receives the bad news, the quicker he or she can implement corrective action. Fire fighters should be encouraged to tell their supervisor whenever an event or performance is out of balance with expectations **Figure 11-3 ▾**.

Figure 11-3 Fire fighters should be encouraged to report problems immediately.

This includes immediately reporting injuries or broken equipment, regardless of the time of day.

This principle also applies to bad news during an emergency. Fire officers who create a barrier to receiving administrative bad news can suffer catastrophic results on the emergency scene. From the fire fighter's perspective, the fire officer who does not appreciate hearing bad news in the fire station will probably not want to hear it at the emergency scene either. There are cases where a fire fighter was afraid to point out a problem that could have saved a life or prevented a serious injury.

Fear versus Trust

Every fire officer should foster a trusting relationship with their employees. Employees who do not trust their boss, or each other, are unlikely to make good decisions when faced with a problem. Employees who do not feel that their input is valuable will stop passing vital information to the fire officer. Effective problem solving requires good information.

Many fire officers believe that because they have been promoted, they can identify the problem. This is not true. The only way the fire officer can best define the problem is with the best information. Eliminating sources of valuable information because of fear and mistrust will cause the fire officer to make an inaccurate assessment.

■ Generate Alternative Solutions

Involve Anyone with Direct Knowledge of the Problem

The best people to solve a problem are usually those who are directly involved in the problem. The fire officer who is struggling with a new incident management clipboard in the rain has a valuable perspective on what might improve the clipboard's performance. The fire fighters who make roof ventilation openings have ideas about how they could be accomplished more efficiently. Company-level problems are most likely to be solved by involving the members of the company.

Brainstorming

Brainstorming is a method of shared problem solving in which all members of a group spontaneously contribute ideas. A typical fire company is a good size and a natural group for brainstorming. Fire fighters are usually very adept at solving problems.

The following eight steps will assist the fire officer when brainstorming alternative solutions:

1. Using a flip chart, whiteboard, or chalkboard, write out the problem statement. Everyone should agree with the words used to describe the problem.

2. Give the group a time limit to generate ideas. Although the scope of the problem is a factor, 15 to 25 minutes seems to work for groups of 4 to 16.

3. The fire officer should function as the scribe. The scribe writes down the ideas and keeps the group on task.

4. Tell everyone to bring up alternative solutions. At this point, there is no commenting on ideas. All suggestions are welcome.

5. Once the brainstorming time is up, have the group select the five ideas they like the best.

6. Write out five criteria for judging which solution best solves the problem. Criteria statements start with "should."

For example, "It should be legal," or "It should be possible to complete in 6 months."

7. Have every participant rate the five alternative solutions, using a 0 to 5 scale. The value 5 means the solution meets all of the criteria, and 0 means the solution does not meet any of the criteria.

8. Add up the scores for each idea. The idea with the highest score is the one that provides the best solution for your problem.

There are some constraints to brainstorming. The process assumes that the problem statement is accurate and the criteria are valid. On occasion, the best solution that comes out of a fire station may crash at headquarters because of a criterion or restriction that is unknown to the fire officer, such as a new federal or state regulation.

Should the Fire Chief Participate?

Fire company group dynamics are different when the administrative or executive fire officer is in the station. The presence of a senior command officer significantly influences a group that is brainstorming alternative solutions. The verbal and nonverbal signals sent by the chief officer influence the group, and the participants self-censure their suggestions and limit the range of possible and plausible solutions because of how they think the chief will react to them. The best ideas occur when they come from the fire fighters directly affected by the problem or issue. Each layer of hierarchy added to the group reduces the range of options.

Do Fire Fighters Feel Comfortable Sharing Ideas?

Fire fighters are naturally competitive. Even when the decision group is restricted to the fire fighters who are directly affected, such as in determining how fill-in apparatus drivers will be trained, they may be reluctant to share ideas. The fire officer can encourage participation by creating a positive and nonhostile work environment.

Is the Process Legitimate?

A legitimate problem-solving process has to be reasonable and based on logic and organizational values. Fire fighters should be able to anticipate that their decision will be implemented if it meets the criteria. Going through a process that results in no change or provides no feedback to the fire company members is the quickest way to destroy fire fighter participation in the decision-making process. A legitimate process reinforces the trust between the fire company and the fire administration.

■ Select a Solution

You have defined the problem, generated solutions, and ranked them based on criteria. One factor in deciding on the best solution is the core value system of your department. For example, if participation in local neighborhoods is one of the core values, a solution that increases the fire department's involvement in local neighborhoods would be preferred.

■ Implement the Solution

Once the decision has been made, the solution still has to be implemented. The implementation phase is often the most challenging aspect of problem solving, particularly if it requires the coordinated involvement of many different people. One of the reasons for involving as many players as possible in making the decision is to capture their commitment to the plan when it is time for implementation.

Consider a plan for reducing the number of fire fatalities in residential occupancies. After careful analysis, the decision has been made that the best strategy would be a community outreach program to check every smoke detector in every multiple-family dwelling in a community over a 4-month period. The plan supports the fire department's mission statement and reflects the organization's values. To accomplish this objective, each fire company will have to invest 2 hours every evening for the next 4 months.

Before this plan can be implemented, buy-in is required. The plan has to be "sold" to the people who will perform the task, particularly to those who were not involved in the decision-making process. Rearranging schedules to do something different for 2 hours every evening involves a significant behavioral change, even if it is for a limited time and for a good cause. It would be possible to simply order everyone to "just do it"; however, this is not the best implementation strategy for a program that involves extensive public contact. Willing participation works better than involuntary compliance.

Fire Marks

Dispatch and Alarm Assignments
New York City hired the RAND Corporation in the late 1960s to analyze city services. Using operations research methods, RAND provided recommendations to Mayor Lindsay to increase the efficiency and effectiveness of city services. RAND published the research and New York City publicized the successes.

One project was a computer simulation model that projected Bronx fire workload. RAND recommended a reduced dispatch of a single engine company to all pulled fire alarm street boxes. The simulation showed there was a less than 1 percent probability that a pulled fire alarm box, with no other information, would turn out to be a structure fire. The Fire Department of New York (FDNY) continued to send a full first alarm assignment if there was a second call reporting a fire at the same location.

Other fire departments mimicked FDNY's changes in dispatch and alarm assignments. One small city fire department, which was similar to New York City in density and demographics, adopted the reduced response to pulled fire alarm street boxes with disastrous results. A single company responding to a fire alarm street box would often discover a working structure fire, sometimes requiring civilian rescue.

The small city revisited the problem-solving process. They discovered that about 30 percent of their serious fires were reported only through a pulled fire alarm street box, with no telephone call or other notification. They also obtained information from the local telephone company showing that many of the small apartments in the city lacked residential phone service. There were few cellular phones in the mid-1970s. The small city restored the response of two pumpers, an aerial, and a command officer to all activated street box alarms.

Fiscal Year 2014 Capital Improvement Training Room   *11/30/2013*

Goal	Due Date
Update and rehabilitate fire station training room	2/1/2014

Task	Assigned to	Due Date	Status
Complete media center design	Station commander	8/16/2013	Done
Issue Request for Proposal for audiovisual	Resource Management	9/4/2013	Issued
Review RFP responses	Station commander/resource management	10/18/2013	Done
Award RFP	Resource Management	11/4/2013	Done
Remove old furniture, tables, and carpet	B shift	11/13/2013	Done
Strip wall coverings and prepare to paint	A shift	11/22/2013	Done
Install fiber optic and intranet services	Technical services	12/2/2013	Done Dec 13
Install media center	Contractor	12/20/2013	Started
Paint and wallpaper party	C shift and volunteers	1/4/2014	
Install carpet	Facilities maintenance	1/17/2014	scheduled Jan 14
Wire projector, speakers, and podium	Technical services	1/23/2014	need to schedule
Move new furniture into training room	A shift	1/25/2014	

Figure 11-4 Example of a project control document for a fire station work project.

Who Does What When?

The fire officer must clearly assign tasks to individuals or teams. The work group benefits by using a project plan, which lists tasks, responsibilities, and due dates **Figure 11-4**. Most of the projects supervised by a company-level officer do not require a sophisticated planning system. A simple project control document can be used to divide a project into segments, with milestones to identify progress. Complex and long-range tasks may require a formal project management plan and a designated coordinator, particularly if they require the coordinated activity of multiple agencies.

No Deadline Means No Implementation

An implementation plan must include a schedule to ensure that the goals are met. Deadlines focus effort and help prioritize activities. If the solution requires activity by other organizations, such as changing a local ordinance or submitting a budget request, then the fire officer must determine the time it will take to accomplish that task and incorporate that time into the schedule. The schedule is valuable only if it is followed and someone ensures that the deadlines are met.

Plan B

Many problems remain unsolved, long after the problem has been clearly defined and a good solution has been selected. The problem is not truly solved unless the solution is implemented. There are many reasons why good solutions are not implemented, including cases where the required approvals cannot be obtained or the necessary resources are not available. Quite simply, the organization might not have the capability to solve the problem or to implement the solution that was selected.

Fire officers should consider a "plan B" if the original solution cannot be implemented. Plan B could be an extended implementation schedule, a modified plan, or a completely different solution to the problem.

■ Evaluate the Results

After implementing the solution, the fire officer must assess whether or not it produced the desired results. Evaluation should be a standard part of the process of any problem-solving activity. The nature of the evaluation depends on the complexity of the problem and the solution; in most cases, an initial evaluation should be performed immediately after implementation, and then follow-up evaluations should be performed at regular intervals.

Determining whether the solution actually solved the problem requires some type of measurement that compares the original condition with the condition after implementation. For example, do the new hose loads really result in quicker deployment of attack lines? The answer to this question requires data on how long the old way took versus how long the new way takes. The evaluation should also look out for situations where the original problem is solved, but another unintended and equally bad situation is created. If the hose is deployed more quickly but it comes out twisted and kinked, the negative could outweigh the positive.

Change the Plan If Necessary

If necessary, the fire officer needs to be prepared to adjust the plan or re-evaluate the original decision. Many problems are solved in stages, with gradual progress being made toward a solution. In spite of the analysis, plan B may turn out to be a better choice. Changing a plan should not be viewed as a failure.

Feedback

Part of the evaluation process is to go back and listen to the people who identified the original problem and ask for feedback. In the example illustrated in the Fire Marks box, the fire chief listened to the fire officers and dispatchers and learned that sending a single fire company to pulled fire alarm street boxes was providing a lower level of service and producing undesirable results. The policy was quickly changed.

Near Miss REPORT

Report Number: 07-0000739

Event Description: Upon arrival at a structure fire, I was assigned to perform search and ventilation on the first floor of the residence with a company captain. Equipped with a TIC, I noticed what appeared to be hot embers in the ceiling. I reported to the captain that I believed there was fire on the second floor. We attempted to locate the second floor. When we located the stairs, my low air alarm on my SCBA activated. I passed the TIC to the crew attempting the stairs and I exited the dwelling. As I exited the structure, the battalion chief assigned as the operations officer asked me about interior nature and conditions. I reported to him that the bulk of the fire was extinguished on the first floor but that I believed fire had extended to the second floor.

As I changed my SCBA cylinder, I heard the IC place the fire under control. I figured that the crews made the second floor and determined that the fire had not extended, or if it had, the extension was minimal. After changing cylinders, another fire fighter and I were assigned to take a 1 3/4" handline to the second floor. Since I knew the location of the stairs, I led the other fire fighter back into the dwelling. As I attempted to go up the stairs and got to the fourth step, I felt the stair begin to give out under my feet and I returned to the main floor. My partner exited the dwelling to report to the operations chief and I reported to the interior battalion chief the fact that we could not make the second floor.

A few minutes later, the incident commander ordered all personnel out of the dwelling. After making sure all interior personnel were notified about the evacuation (the IC did not activate the evacuation signal), the interior battalion chief and I exited the residence. As I exited the residence, I saw heavy fire conditions from second floor windows and from the roof. I later learned that the crew attempting to make the second floor discovered the unstable steps; however, this information was never fully communicated throughout the chain of command.

Lesson Learned: Interior conditions and structural conditions need to be relayed to all personnel operating in and around the structure and especially the command personnel. The operations sector, interior sector, and incident commander need to communicate better so that all pertinent information is relayed up and down the IMS system.

Managing Conflict

A factor that distinguishes a fire officer from a fire fighter is the responsibility to act as an agent of the formal organization. A fire officer is the official first-level representative of the fire department administration when dealing with subordinates and enforcing policies and procedures. This responsibility places the fire officer in the first position to deal with a wide variety of problems, including situations that potentially involve conflict, emotions, or serious differences of opinion.

Situations that involve conflicts and grievances require an additional set of skills that go beyond the general problem-solving model. The general model is designed to focus on solving the problem itself. In conflict situations, the issues are often much more complicated and sensitive. There could be a relatively simple problem that becomes complicated by the ways that different individuals react to it or to each other. In some cases, the problem is centered on the relationship between individuals or groups and is played out in relation to other issues.

Personnel Conflicts and Grievances

The close living relationships within a fire company can produce a variety of tensions, anxieties, and interpersonal conflicts. This is in addition to the common types of conflicts in most workplaces. One of the most difficult situations for a fire officer is an interpersonal conflict or grievance within the company or directly involving a company member.

A fire officer might be faced with four different types of internal situations. A fire fighter might come to an officer with a complaint about:

- A co-worker (or co-workers)
- The work environment, including the fire station, apparatus, or equipment
- A fire department policy or procedure
- The fire officer's own behavior, decisions, or actions

The fire officer is the individual's first point of contact with the formal organization. The official response to the problem begins when the officer becomes aware that a problem exists. The relationship of the fire officer to the conflict and the complainant makes a significant difference in the role the officer can play in resolving the conflict.

Fire officers with staff assignments must also be prepared to deal with problems that involve conflict. Their relationship to the individuals involved is likely to be different; however, their responsibility to officially represent the formal organization is the same.

Voices of Experience

One of the circumstances of my promotion to Captain was a shift reassignment. Being scheduled for an overtime shift, I would return to my old shift not as just another one of the guys on the crew, but as an appointed leader.

After roll-call we began our day as usual by conducting apparatus and equipment checks. About 20 minutes had gone by and I noticed that one of the fire fighters on my crew had not checked his personal equipment or the apparatus he was assigned to check. As I began to leave the truck to go look for the fire fighter, he came through the truck bay carrying brushes and soap, and stated that he was going to wash the command vehicle. I told the fire fighter that he needed to check his equipment immediately. The fire fighter mumbled an acknowledgment and kept going. At this point I wasn't sure if this reply was in discontent for the actual order or from the direction being given by me as an appointed person of authority.

I gave the fire fighter a few minutes to do what I had asked. Seeing that the fire fighter had not yet checked out his personal equipment I was angered. I found the fire fighter outside beginning to check out his assigned truck. In a rather firm tone I told the fire fighter, "You need to go check out your gear and air-pak…now!" The fire fighter replied, "I don't feel like dealing with this today; my Dad died one year ago today." At that time I realized that this was a pivotal point in handling this problem and though still angry myself, I needed to take a couple minutes to address the problem in a calmer state of mind. I told the fire fighter once again to do what I had asked and I returned to the office.

> **I advised the fire fighter that in my position, I am held accountable for all of my crew members as well as myself.**

After a few minutes, I went back outside and asked the fire fighter to come to the office. I asked the acting station chief to come in as well. Understanding that the situation needed to be brought to a resolve with a positive outcome, I began by expressing my sympathy for the fire fighter's situation. At this point, the fire fighter was quite emotional and not wanting to resolve the conflict. The fire fighter stated that I was not his direct supervisor and that we were not going to resolve the issue. He then told the acting station chief that he wanted to go home. Though a bit upset by the fire fighter's venting, I understood the importance of my next reaction. I asked the fire fighter to calm down and hear me out. I told the fire fighter that this was not a personal attack or lack of compassion for his situation. I advised the fire fighter that in my position, I am held accountable for all of my crew members as well as myself and that he was quite simply being asked to fulfill his responsibilities. I re-expressed my sympathy for his personal situation and informed him that though we sometime have difficult times in life, we still have a job to do here as fire fighters. The fire fighter agreed and immediately went to complete the original task.

Brandon Cunningham
Captain
Fort Gordon Fire and Emergency Services
Fort Gordon, Georgia

Conflict Resolution Model

The conflict resolution model is a basic approach that can be used in situations where interpersonal conflict is the primary problem or a complicating factor.

Listen and Take Detailed Notes

The first phase of the conflict management template is to obtain as much information as possible about the problem. The fire officer should encourage the complainant to explain the situation completely. If the details are even slightly complicated, the fire officer should take notes. The person who is making a complaint has a certain perspective on the situation. Whether you agree or disagree with that person's perspective, an important starting point is to find out what the complainant thinks about the situation.

Active Listening

When dealing with an individual who is expressing a concern or a problem, the fire officer should focus on active listening. Engaged or active listening is the conscious process of securing all kinds of information through a combination of listening and observing. The listener gives the speaker full attention, being alert to any clues of unspoken meaning while also listening intently to every word that is spoken. The fire officer should be aware of nonverbal clues that may indicate agreement, dissatisfaction, anger, or other emotions. Often, these nonverbal clues provide great insight as to the disposition of the speaker. The fire officer actively seeks to keep the conversation open and satisfying to the speaker, showing an interest in feelings and emotions as well as raw information Figure 11-5 ▾ .

Paraphrase and Receive Feedback

The first objective should be to understand the issue and why the individual is complaining. After listening, the fire officer

Figure 11-5 Give the speaker your full attention.

should be able to paraphrase the complaint and recite it back to the complainant. Paraphrasing the issue and receiving feedback from the complainant accomplishes two goals: The fire officer finishes this phase with a good understanding of the issue from the complainant's perspective, and the complainant feels that the fire officer really listened.

Do Not Explain or Excuse

In situations where the complaint is directly related to actions taken or policies enforced by the fire officer, it is understandable that the officer would want to immediately respond to the complaint. In this situation, it is important to listen and process the information before deciding on an appropriate response. A reflexive explanation or excuse gives the individual an additional reason to complain. If the complainant feels strongly enough to complain about something the officer has done, that officer's explanation is probably not going to solve the problem.

Investigate

An **investigation** is a detailed inquiry or systematic examination. All complaints should be investigated, even if the foundation for the complaint appears to be weak or nonexistent. Fire department procedures should determine who will conduct the investigation, depending on the nature of the complaint and the relationship of the individuals who are involved. Sometimes the fire officer who received the complaint is assigned to conduct the investigation; however, a fire officer who is directly or personally involved in the problem should never be involved in conducting the investigation. Sensitive matters require an appropriate level of investigator.

The purpose of the investigation is to obtain additional information beyond the original complainant. The investigator

Metro City Fire Department
Internal Memorandum

DATE: February 07, 2012
TO: Deputy Chief Bruce Appleton, Field Forces
THRU: Battalion Chief Jane Stapleton, 2nd Battalion
FROM: Captain Stan Holtz, Engine 7
SUBJECT: Delayed response to a fire emergency
REF: Inquiry 2012-14 Incident #1294, 3177 Usher St, January 26, 2012

Biff Davis, 3177 Usher St, (555) 555-0045, is the owner of the property and made the 9-1-1 calls. Mr. Davis complained to Captain Holtz that it took over 20 minutes for the fire department to arrive. Captain Holtz said that he would investigate and notified Battalion Chief Stapleton before leaving the incident scene.

Investigation:
Data from computer aided dispatch records:
23:36:44: 9-1-1 receives telephone call reporting an odor in a single family home.
23:37:20: Engine 7 dispatched to 3177 Usher Street.
23:38:40: Engine 7 marks EN ROUTE to incident.
23:39:07: Second 9-1-1 call "Where is the fire department?" Reports fire in the basement.
23:39:29: Engines 11, 3, and 9; Truck 11; Paramedic 7 and Battalion 2 dispatched for a structure fire.
23:39:40: Dispatch notifies Engine 7 of upgraded alarm and report of basement fire.
23:42:24: Engine 7 marks ON SCENE, reports "Two-story, single family frame with smoke showing from basement windows."

Engine 7 was on the scene five minutes and four seconds from dispatch.
There was no delay in dispatch (36 seconds), turnout (80 seconds) or response.

Action Taken
Chief Stapleton and Captain Holtz met with Mr. Davis at the Shady Oaks Inn, (555) 555-9703. Discussed his concerns and reviewed the results of the investigation.

Follow-Up
No further action is recommended

Figure 11-6 Example of a fire officer investigation report.

must be impartial in gathering and documenting information. The information could come from other individuals, reference documents, or incident-specific data. When investigating a human resource conflict, such as a payroll or work assignment issue, the fire officer might have to refer to specific departmental directives and regulations.

The product of an investigation is a report, which is provided in an appropriate format for the fire officer's immediate supervisor **Figure 11-6 ▲**. A complete investigative report has three objectives:

1. The report must first identify and clearly explain the issues.
2. The report should then provide a complete, impartial, and factual presentation of the background information and relevant facts.
3. The conclusion should be a recommended action plan, which is based on and supported by the information.

■ Take Action

Once the investigation is completed, the fire officer presents the findings and recommended action to a supervisor at a higher level. There are four possible responses:

1. Take no further action. The investigation concludes that the complaint was unfounded or requires no further action. If the complaint was related to an earlier decision or action, that original decision is affirmed. The response should include the reasons why no further action is recommended.
2. Recommend the action requested by the complainant. In this case, the investigation concludes that the complaint was justified and the requested action is the best solution to the problem.
3. Suggest an alternative solution. The investigation concludes that some alternative action or policy is the best solution to address the complainant's concerns. For example, a citizen might complain that the fire truck blocks several spaces in the parking lot at a local gym and does not want the fire fighters to go there. The fire officer meets with the citizen and proposes a more appropriate parking space for the fire truck. The compromise is acceptable to the citizen and to the department.
4. Refer the issue to the office or person who can provide a remedy. Other members of the fire department or some other municipal agency can provide the relief the complainant seeks. Grievance procedures require that the employee start with the

immediate supervisor for all complaints. If the employee is not satisfied with the response at that level, the grievance can then be taken to a higher level. If the problem involves a paycheck deduction issue, it will probably have to be resolved by the payroll clerk or human resources office. The fire officer's duty is to refer the complaint to the appropriate person.

Follow Up

For many of the conflicts, the fire officer needs to follow up with the complainant to see whether the problem is resolved.

Emotions and Sensitivity

Fire fighters are passionate about their profession and are deeply concerned about issues that affect the job. They live, breathe, eat, and dream about fire operations. They often are emotional when making a complaint. Michael Taigman, a consultant on emergency service performance issues, provides a conflict resolution model that is especially effective when emotions are high. It was effective for Taigman when working with employees during the stressful creation of a paramedic ambulance service under a tight schedule. He continued to refine the model while working as a consultant. The model follows four steps:

1. Drain the emotional bubble.
2. Understand the complainant's viewpoint.
3. Help the complainant feel understood.
4. Identify the complainant's expectation for resolution.

Step 1: Drain the Emotional Bubble

The body reacts to emotional conflict or stress the same way it does when you are the first-arriving company at a working structure fire. Both situations result in dumping of adrenaline to prepare the body to fight or run away. The same adrenaline-induced red haze that reduces fire fighter effectiveness at emergencies also happens to ordinary citizens. It tends to bring complaints to the surface and impedes resolution of conflicts. Adrenaline fills up the prefrontal lobes of the neocortex of the brain. That creates an emotional bubble that interferes with the ability of the complainant to hear the fire officer or consider any response to the issue.

Taigman recommends listening deeply, actively, and empathetically in order to drain this emotional bubble. This is not an easy task for the fire officer, but writing detailed notes and not explaining or excusing helps the process. The fire officer asks questions and encourages responses, draining the emotional bubble by allowing the complainant to completely express grief, regret, pain, censure, or resentment.

This discussion should be in private and should have a couple of ground rules. There should be no physical contact. If this discussion is between a fire fighter and a fire officer, avoid personal attacks and concentrate on the work issues.

Step 2: Understand the Complainant's Viewpoint

The initial complaint or behavior may be a sign or symptom of a larger problem. By draining the emotional bubble through active listening, the fire officer may identify the root cause or issue of the complaint.

Internal conflicts, grievances, or issues occasionally suffer from long memories. It may require some investigating to understand the current issue, which may be related

to something that happened months ago or be wrapped up in history and tradition.

Step 3: Help the Complainant Feel Understood

The "listen and take detailed notes" part of the basic conflict management template includes the recommendation to paraphrase and feed back what you heard from the complainant. Some issues are resolved if the complainant feels that the fire officer understands the issue, conflict, or problem.

Step 4: Identify the Complainant's Expectations for Resolution

By this final step, the complainant has drained the emotional bubble, has been able to describe what is going on, and feels that the fire officer understands the issue, problem, conflict, or grievance. The fire officer should now ask what the complainant expects the department to do to resolve this issue. If the problem is an internal grievance, this is where the employee should be asked to describe the desired action.

Citizen Complaints

The fire officer represents the department in dealing with citizens, public or private organizations, and other governmental agencies. A fire officer could be faced with three different types of citizen complaints. A citizen might complain about:

1. The conduct or behavior of a fire fighter (or a group of fire fighters)
2. The fire company's performance or service delivery
3. Fire department policy

Sometimes, a citizen may want to express an alternative viewpoint on an issue and try to see whether there is a resolution that is mutually agreeable to the department and the citizen. At other times, the citizen is making a formal complaint. When this occurs, the fire officer is functioning as the official recipient of the complaint. In some cases, the fire officer is the subject of the complaint. The first role of the fire officer is to respond to a complaint in a professional manner that effectively obtains the needed information and does not make the situation worse. The methods outlined in resolving conflict within the company also apply to a citizen complaint.

The fire officer must take notes and function as an active listener. By listening attentively and taking detailed notes, the fire officer is demonstrating that the complaint is officially

Safety Zone

Aggression, Anger, and Acting Out
Some people have trouble handling their anger. Some have very abrasive personalities, others launch vicious verbal attacks, and a few may physically act out their feelings. The fire officer needs to monitor the situation and the complainant. If, in attempting to drain the emotional bubble, there is a rise in rage, then take a time-out.

If this situation involves a fire fighter, follow your organization's procedure on workplace violence or hostility. If the conflict is with a civilian at an emergency scene, ask for police assistance. The police have training in handling angry and potentially violent citizens.

considered to be important and is receiving the fire officer's full attention. If the immediate response to a conflict is an explanation or excuse, the complainant will feel that the fire officer is not paying attention, does not care about the issue, or has something to hide. The person who had a complaint will likely stop providing information and feel even more strongly that the complaint was justified.

On some citizen complaints, the fire officer may be able to resolve the problem. On other issues, such as a complaint about using a siren at night, the fire officer should explain the rationale for the use of the siren as well as the laws that regulate its use. Be empathetic and listen to the problem without interrupting. Frequently, these kinds of complaints are a method of venting frustration by the citizen rather than a real expectation that it will change. Allowing the citizen to express the frustration is important to resolving the issue.

If the fire officer does not have the authority to make a decision on the issue or the citizen is dissatisfied with the officer's decision, the officer should ask whether the citizen would like the issue forwarded to the next level within the organization. If the citizen would like further action, the fire officer should determine the appropriate organizational level where the decision can be made. A preferred method is to consider the scope of the complaint. If the complaint is strictly an operational issue, it should be sent to the chief of operations. If it is an issue about codes, it would be sent to the chief fire marshal. If the issue affects all areas of the department, it would be forwarded to the chief of the department. Complaints about personnel should be forwarded to the supervisor of the individual that is involved for action.

All relevant facts should be identified and forwarded, along with the details of the complaint. In some departments, the proper procedure is to follow the chain of command. In other departments, fire officers are encouraged to forward the information directly to the decision maker in order to ensure prompt attention to the matter. Even in these circumstances, the fire officer should inform his or her supervisor about the situation. If there is any doubt, discuss the issue with your supervisor before taking any action.

Policy Recommendations and Implementation

Because the fire officer is in direct contact with fire fighters and citizens, the officer is in the best position to recommend new departmental policies. This change could be the result of a citizen or employee complaint or an identified problem. The problem could be anything that creates a disparity between the actual state and the desired state. For example, not having a standardized location for all equipment on the apparatus reduces efficiency on incident scenes; therefore, it is a problem.

Recommending Policies and Policy Changes

The fire officer must understand the procedure for adopting new policies within the department. Although many fire fighters may believe that the procedure simply consists of getting the chief to put his or her name on it, usually it entails much more. Some departments have a policy on how to implement a new policy.

The most common method to get new policies adopted or to cause changes to existing policies is to outline the problem they have identified to their supervisor with the recommendation that "someone ought to do something about this." This places all responsibility on the supervisor to come up with a solution and get the department to adopt it. The benefit that this method is easy for the officer is far offset by its ineffectiveness. This is why most of the time, nothing changes, and if it does, the change may create more problems than it solved.

To successfully recommend a new policy or a change to an old policy, the fire officer should carefully identify the problem and develop documentation to support the need for a change. This could be with statistical measures, anecdotal evidence, or both. The fire officer's opinion seldom carries enough weight to cause a change because other officers may have different opinions. One effective technique is to discuss the situation with other fire officers initially to determine how widespread the perceived problem is.

Once the supporting evidence has been developed that there is a widespread problem, the problem-solving techniques outlined earlier in the chapter should be used to develop and choose the best alternative. At this point, the fire officer is ready to write out a proposal to administration. The problem should be carefully outlined, along with the proposed policy change that will resolve it. Resources that will be needed for the solution must also be identified, including financial and time commitments. The benefits of the solution should be identified, as should any potential negative effects and how they will be addressed.

Once a written proposal is developed, the fire officer should approach his or her superior if that person has not already been part of the development. Along with the recommendation, the policy should be presented to the appropriate officer whose scope is to oversee the area affected by the policy. It is critical to get the agreement of this fire officer, because without this recommendation, the policy will likely not be implemented.

With this officer's recommendation, the proposal is usually forwarded to the chief of the department for review. Usually, a review committee composed of senior staff officers evaluates the proposal and makes changes and accepts or rejects the proposal. Once it is accepted, a draft policy is usually developed in the proper format. The draft policy is distributed to all personnel for review and comment. After a comment period, all comments are addressed, and a final policy is signed by the chief and distributed to the department.

Implementing Policies

Like the recommendation process, the process for implementing new policies varies widely. The fire officer should follow departmental procedures. In their absence, the officer may use the following methods to improve his or her communication and understanding.

Because policies are the backbone of order for the fire department, every individual should understand the policies that affect him or her. For this to occur, the fire officer must take responsibility to ensure that the fire fighters are informed about the policy and take the time to understand it. The fire officer must also ensure that he or she follows all policies. Failure of the officer to follow all policies undermines their importance, and fire fighters will develop the attitude that they can choose which policies they will follow.

When a new or amended policy is distributed, the fire officer should ensure that it is communicated to the subordinates. At the beginning of each work shift, the new policies should be discussed. The fire officer should identify the points that are most relevant, particularly those areas that change the current practice. For example, if a revised driving policy requires every apparatus responding to an emergency to come to a complete stop at all red signals and stop signs rather than slow to a speed so as to avoid an accident, the fire officer should specifically point this change out.

The fire officer should require that all fire fighters read the policy and sign off that they understand the policy. This helps ensure accountability. Employees tend to follow policies more closely if they believe they will be held accountable for them. To make sure that employees understand the policy, the fire officer should provide fire fighters with situations covered by the policy and ask them to apply the policy. Many departments also require that a copy of the policy be posted on the bulletin board for a period of time. Some also require a notation in the station log book that the policy was distributed.

The fire officer should evaluate the employee's actions against the policy. If a policy is violated, the fire officer should review the policy with the employee involved. For repeated violations or a violation of a safety policy, disciplinary action should be considered. Employees should never get the feeling that it is acceptable to ignore policies.

Last, a regular review of policies should occur. Selecting policies that are not routinely encountered and testing fire fighters' knowledge about them will help keep everyone up-to-date.

Customer Service versus Customer Satisfaction

Customer service is a term that public safety has adopted from the retail business. A focus on customer service fixes problems, straightens out procedural glitches, corrects errors of omission (or commission), and provides information.

Customer satisfaction focuses on meeting the customer's expectations. Fire departments meet customers during one of the worst days of their life. They did not start the day planning to crash the car, melt the stove, or have trouble breathing. Good customer service requires sensitivity on the part of every person involved. Although there is little competition for others to provide public safety services, having satisfied customers is important to the municipality **Figure 11-7 ▶**.

Complainant Expectations

When dealing with a civilian, the fire officer should ask what could resolve the situation. The fire officer needs to take the response from the complainant in order to resolve the problem.

The resolution may be as simple as acknowledging that the complainant has had a bad experience with the department. In some cases, it may be as extreme as recommending that the department terminate the employee or employees involved. Unless it is within the scope of the fire officer's authority, the fire officer should make no promises or imply that certain actions will be taken in the discussion with the complainant. In many cases, the complainant's expectations must be relayed to the fire officer's supervisor as a part of resolving the issue. If the proposed resolution

Figure 11-7 Creating satisfied customers is one of the fire officer's most important activities.

involves discipline, the fire officer is also obligated to protect employee privacy and civil service due process **Figure 11-8 ▶**.

Keep Complainant Informed

If the fire officer needs to do research, consult with others, or obtain direction from supervisors, he or she should keep the complainant informed during the process. This communication demonstrates that the fire officer is not ignoring the issue, and it educates the complainant on the process.

Follow Up

Citizens frequently complain about local government unresponsiveness. By following up with the complainant, the fire officer reinforces the impression that the complainant's issue is important. This is especially important if the fire officer has referred the issue to another individual, agency, or organization.

Follow-ups may be inappropriate if the fire officer was the subject of the complaint or the conflict was handled at a higher supervisory level. The fire officer may consider consulting with a supervisor before conducting a follow-up.

Summary

Chapters 8, 9, and 11 describe the core supervision and managerial competencies required by a supervising or managing fire officer. This chapter describes many tools and practices used to handle complaints, manage conflicts, make decisions, and change departmental policy. Special effort has been made to provide the tools that work well in the unique fire department environment.

You need to use these tools to gain proficiency. There is a natural tendency to push complaints and conflicts further up the

**Metro City Fire Department
Headquarters
July 18, 2010**

Mrs. Caroline Steen
Exotic Food Emporium
437 East Center Street
Metro City

Mrs. Steen

We have completed our investigation into the response to your store on July 7th. During an investigation for smoke in the building, the fire department forced the front door and destroyed the storefront window.

The initial actions taken by our members were not consistent with departmental policies and procedures. The city will pay for repairs to the storefront. Sam Hall from the Mayor's office will continue to work with you during the repair process.

I apologize for the actions taken by our members. The fire department is taking appropriate training and disciplinary actions to ensure that all members know how to properly respond to a fumigation incident in a building with a rapid entry key lock-box.

Sincerely,

Carl Cosgrove

Fire Chief
Metro City

Figure 11-8 Example of a letter to a civilian in response to a complaint.

chain of command. In some departments it can lead to the administrative fire officer handing an issue that should have remained at the supervising fire officer level. The farther up the chain of command you go with an issue or problem, the more likely the resolution will not work to your advantage.

If you wish to promote up through the department, developing mastery in handling problems, conflicts, and mistakes increases your value to the formal organization. Even if you just want to improve the situation where you work, the skills described in this chapter will help you meet that goal.

You Are the Fire Officer: Conclusion

There are four problems to solve after this hasty action.

1. Mitigate the emergency. Limit the exposure of your crew, the police officer, and anyone else from the fumigation chemical cloud. You request the hazardous materials response team and the on-duty command officer. Read the fumigation notice posted on the building and call the contractor for decontamination recommendations. Contact the business owner and/or the building representative and ask for a representative to come to the scene.

2. Document the damage. Once the hazardous condition is resolved, start documenting the damage done to the storefront by taking pictures and notes. This includes names and contact information on everyone exposed at the scene. Make sure to meet with the owner or owner representative and walk through the damaged area.

3. Make the appropriate notifications and documentation. Although the administrative fire officer will complete the internal investigation, as the officer-in-charge you will be doing a lot of writing.

4. Start the fire company problem-solving process. In addition to the eager rookie, the administrative fire officer expressed concern about the company not knowing about the fumigation or that the business had a rapid entry key safe. Issues may include both individual performance and system issues. The chief wants your report within a week.

Wrap-Up

Chief Concepts

- The fire officer has to deal with conflict, emotions, or serious differences of opinion.
- Situations that involve conflicts and grievances require an additional set of skills.
- The fire officer takes or recommends four actions after completing an investigation:
  - Take no further action.
  - Recommend the action requested by the complainant.
  - Suggest an alternative solution.
  - Refer the issue to the office or person that can provide a remedy.
- A citizen might complain about:
  - The conduct or behavior of a fire fighter
  - The fire company's performance or service delivery
  - Fire department policy
- Because the fire officer is in regular direct contact with both fire fighters and citizens, he or she is often in the best position to recommend new departmental policies.
- A fire officer must understand the procedure for adopting new polices within the department.
- The process for implementing new policies varies.
- Customer service is an important part of customer satisfaction.

Hot Terms

Brainstorming A method of shared problem solving in which all members of a group spontaneously contribute ideas.

Complaint Expression of grief, regret, pain, censure, or resentment; lamentation; accusation; or fault finding.

Conflict A state of opposition between two parties. A complaint is a manifestation of a conflict.

Investigation A systematic inquiry or examination.

Mistake An error or fault resulting from defective judgment, deficient knowledge, or carelessness. A misconception or misunderstanding.

Problem A condition in which the desired situation is different from the current situation.

Fire Officer *in Action*

One outcome of the fumigation investigation was discovering that there was an inappropriate procedure to ensure that the local fire companies knew when and where fumigations were occurring. The administrative fire officer who conducted the investigation has tasked your company to recommend a policy change to update the notification procedure. In addition, your company will prepare a field training drill on responding to fumigation incidents.

Within the company you have assigned the rookie to make a list of the address, location, and brand of every rapid entry safe location in the fire station's response area. This information will be added to the property information contained in the computer-aided dispatch program.

The fire officer is the first supervisor or manager to deal with external and internal problems and complaints. Documenting and properly processing each issue is a required activity that frequently is a component in a promotional exam. Beyond complying with the internal process, a fire officer encounters unique opportunities to learn from each problem and complaint as part of a personal professional development journey.

1. Allowing a person with a complaint to completely express grief, anger, pain, or resentment is an example of:
   A. active listening.
   B. paraphrase and feedback.
   C. emphatic listening.
   D. draining the emotional bubble.

2. It has been 6 weeks since implementing a new procedure to reduce the time from dispatch to wheels rolling. The new policy has increased the time it takes to get wheels rolling by 30 percent. You should:
   A. continue the new procedure for another 6 weeks.
   B. re-interview the fire fighters to make them feel understood.
   C. change the plan.
   D. return to the original procedure.

3. While taking notes during a citizen's complaint you realize that the resolution requires action by an executive fire officer. You should:
   A. complete active listening and explain that you will be forwarding the complaint to the appropriate administrative fire officer for investigation.

   B. stop active listening and provide the complainant with the contact information of the appropriate administrative fire officer.
   C. stop the process and ask to reschedule this session when the appropriate administrative fire officer is available.
   D. recommend that the complaint be filled out on a fire department form and sent to the fire chief's office.

4. Which of the following is NOT a response after completing an investigation?
   A. Take no further action.
   B. Recommend additional study.
   C. Suggest an alternative solution.
   D. Refer the issue to the administrative fire officer that can provide a remedy.

Pre-Incident Planning and Code Enforcement

NFPA 1021 Standard

Fire Officer I

4.5* **Inspection and Investigation.** This duty involves conducting inspections to identify hazards and address violations, performing a fire investigation to determine preliminary cause, securing the incident scene, and preserving evidence, according to the following job performance requirements. [p 216–220, 230–236]

4.5.1 Describe the procedures of the AHJ for conducting fire inspections, given any of the following occupancies, so that all hazards, including hazardous materials, are identified, approved forms are completed, and approved action is initiated:

  (1) Assembly
  (2) Educational
  (3) Health care
  (4) Detention and correctional
  (5) Residential
  (6) Mercantile
  (7) Business
  (8) Industrial
  (9) Storage
  (10) Unusual structures
  (11) Mixed occupancies [p 234–236]

(A) Requisite Knowledge. Inspection procedures; fire detection, alarm, and protection systems; identification of fire and life safety hazards; and marking and identification systems for hazardous materials. [p 216–220, 230, 233–234]

(B) Requisite Skills. The ability to communicate in writing and to apply the appropriate codes and standards. [p 220, 233]

4.5.2 Identify construction, alarm, detection, and suppression features that contribute to or prevent the spread of fire, heat, and smoke throughout the building or from one building to another, given an occupancy, and the policies and forms of the AHJ so that a pre-incident plan for any of the following occupancies is developed:

  (1) Public assembly
  (2) Educational
  (3) Institutional
  (4) Residential
  (5) Business
  (6) Industrial
  (7) Manufacturing
  (8) Storage
  (9) Mercantile
  (10) Special Properties [p 222–230]

(A) Requisite Knowledge. Fire behavior; building construction; inspection and incident reports; detection, alarm, and suppression systems; and applicable codes, ordinances, and standards. [p 220–228, 233]

(B) Requisite Skills. The ability to use evaluative methods and to communicate orally and in writing. [p 216–220]

4.6* **Emergency Service Delivery.** This duty involves supervising emergency operations, conducting pre-incident planning, and deploying assigned resources in accordance with the local emergency plan and according to the following job performance requirements. [p 215–237]

Fire Officer II

5.6 **Emergency Service Delivery.** This duty involves supervising multi-unit emergency operations, conducting pre-incident planning, and deploying assigned resources, according to the following job requirements. [p 215–237]

Additional NFPA Standards

NFPA Fire and Life Safety Inspection Manual

NFPA 1 *Fire Code*
NFPA 10 *Standard for Portable Fire Extinguishers*
NFPA 12 *Standard on Carbon Dioxide Extinguishing Systems*
NFPA 101 *Life Safety Code*
NFPA 220 *Standard on Types of Building Construction*
NFPA 291 *Recommended Practice for Fire Flow Testing and Marking of Hydrants*
NFPA 704 *Standard System for the Identification of the Hazards of Materials for Emergency Response*
NFPA 1561 *Standard on Emergency Services Incident Management System*
NFPA 1600 *Standard on Disaster/Emergency Management and Business Continuity Programs*
NFPA 1620 *Recommended Practice for Pre-Incident Planning*
NFPA 2001 *Standard on Clean Agent Fire Extinguishing Systems*

Introduction to Fire and Emergency Services Administration (FESHE) Course Outcomes

2. Explain the need for effective communication skills both written and verbal. [p 233. 236–237]
7. Identify roles and responsibilities of leaders in organizations. [p 215, 225, 230–232]
9. Identify and assess safety needs for both emergency and nonemergency situations. [p 216–218, 225–228, 230, 234–236]
11. Identify the role of a company officer in Incident Command System (ICS). [p 216–220, 237]

Knowledge Objectives

After studying this chapter, you will be able to:

- Discuss how to develop a pre-incident plan.
- Understand built-in fire protection systems.
- Understand fire code compliance inspections.
- Identify the five types of building construction, as used in the fire prevention code.

- Prepare for an inspection.
- Describe general inspection requirements.
- Describe the difference between a pre-incident plan and an emergency management/business continuity plan.

Skills Objectives

After studying this chapter, you will be able to:

- Demonstrate how to conduct an inspection.
- Demonstrate how to utilize an emergency management and business continuity plan.

The new community is as unique as its technology. A "green" manufacturing plant is under construction in your district. It is 2000' (610 meters) long and 600' (183 meters) wide, full of new and innovative technology. Surrounding the plant will be a new company town, with two-story mixed use buildings containing stores, restaurants, and theaters. Also included in the plans will be two 7- story office buildings to be built on the edge of the town center. There also are four planned apartment buildings.

The architect is proudly describing the new environmentally sound technology and techniques. Most of the descriptions involve technical processes and procedures that you have never heard about. As the local fire company commander, the fire chief has appointed you as liaison to the factory management.

1. What is the best way to perform a pre-incident plan for a complex facility?
2. How can you prepare fire fighters for fire inspection duties?
3. How does the fire pre-incident plan interact with the emergency management and business continuity plan?

Introduction to Pre-Incident Planning and Code Enforcement

The fire officer considers the built environment from several viewpoints. If it is burning, damaged, or expelling hazardous materials, the fire officer is expected to command the incident, rescue those in harm's way, mitigate the situation, and render the scene safe.

In order to accomplish those tasks, the fire officer looks at the building from two different perspectives. One way is to prepare to handle an emergency in the building by developing a pre-incident plan. The other way is to perform a life safety inspection to ensure that the building is meeting the appropriate fire prevention code requirements. Although separate activities, both pre-incident planning and code enforcement require similar skill sets, including an understanding of building construction and built-in fire protection systems.

The Fire Officer's Role in Community Fire Safety

Fire departments perform fire prevention, risk reduction, pre-incident planning, and public education. Fire officers play multiple roles in relation to properties within their communities, including:

1. Identifying and correcting fire safety hazards through safety checks or code enforcement

2. Developing and maintaining pre-incident plans
3. Promoting fire safety through public education

In most areas, fire inspectors and fire officers working in staff assignments perform fire inspections and code enforcement duties **Figure 12-1 ▶**. Fire suppression companies are usually involved in pre-incident planning. In some areas it is the local fire suppression company that conducts code enforcement inspections. Public education activities are often performed by a combination of staff personnel and fire companies.

Even where the role of fire companies does not include code enforcement, fire officers and fire fighters should conduct regular visits to properties to develop pre-incident plans and look for correctable fire safety hazards. A fire officer should always try to reduce the impact of any potential fire emergency that could occur, including identifying and correcting conditions that could start a fire and ultimately increase the risk to citizens and fire fighters in the event of an emergency.

Pre-Incident Planning

A **pre-incident plan** is described by NFPA 1620, *Recommended Practice for Pre-Incident Planning*, as a document developed by gathering general and detailed data used by responding personnel to determine the resources and actions necessary to mitigate anticipated emergencies at a specific facility **Figure 12-2 ▶**.

Figure 12-1 Fire officers often conduct pre-incident planning and fire inspection activities during the course of their workday.

The original purpose of a pre-incident plan was to provide information that would be useful in the event of a fire at a high-value or high-risk location. **High-value properties** contain equipment, materials, or items that have a high replacement value. Examples include properties containing agricultural equipment, electronic data processing equipment, or scientific equipment; fine arts centers; and storage or manufacturing sites.

High-risk properties contain the potential for a catastrophic property or life loss in the event of a fire. Examples include nuclear power plants, bulk fuel storage facilities, hospitals, and jails. The pre-incident plans include information that could apply to a variety of situations that could occur at the location.

Facilities that store or handle hazardous materials are required to submit information to the fire department and the Local Emergency Planning Council (LEPC). Some plans include response to natural or man-made catastrophic incidents.

The second purpose of a pre-incident plan is to identify in advance the strategies, tactics, and actions that should be taken if a predictable situation occurs. Pre-incident plans can be particularly useful at the company level for practicing initial operations for buildings in the company's district.

When the fire officer is assessing a facility, the following factors need to be evaluated:

- Construction
- Occupant characteristics
- Fire protection systems
- Capabilities of public or industrial responding personnel
- Availability of mutual aid
- Water supply
- Exposure factors
- Access
- Utility cutoff locations

During this assessment, the fire officer also evaluates the department's ability to respond to and control an incident at the facility. For example, most fire departments need specialized equipment and additional foam supplies in the event of a fuel storage depot fire. The mutual aid can be planned in advance of the incident.

A Systematic Approach

To ensure a systematic approach that collects all of the required data, the fire officer should use a standardized method for each pre-incident plan. NFPA 1620 provides a six-step method:

1. Identify physical elements and site considerations.
2. Identify occupant considerations.
3. Identify fire protection systems and water supply.
4. Identify special hazards.
5. Identify emergency operation considerations.
6. Identify special or unusual characteristics of common occupancy.

Step 1: Identify Physical Elements and Site Considerations

The first step is to evaluate the physical elements and site considerations. **Plot plans** provide a representation of the exterior of a structure, identifying doors, utilities access, and any special considerations or hazards. **Floor plans** are interior views of a building. Rooms, hallways, cabinets, and the like are drawn in the correct relationship to each other.

The building's size and dimensions, including overall height, number of stories, length, width, and square footage, should be determined and included in the plan. Connections between buildings and distances to exposures should be noted. Access routes and points of entry should be clearly indicated. The pre-incident plan should identify concealed spaces and windows that can be used for ventilation and/or rescue.

The pre-incident plan should include detailed information about the construction of the roof, floor, and walls. Include an assessment of their structural integrity. The focus is on factors that could lead to collapse of building components or spread of fire or toxic gases. Any factor that would affect the ability of responding fire fighters to enter and safely perform interior operations should be highlighted. Conditions that affect the safety of fire operations on the roof also should be noted.

This part of the pre-incident plan also includes documenting the location of electrical and domestic water shutoffs, and heating, ventilating, air-conditioning controls. Note the presence of flammable liquids, compressed or liquefied gases, and steam lines. Elevator information should include their location in the building, floors served, type of elevator, and type of recall or override service.

Note conditions that delay access, including weight-restricted bridges, low overhead clearances, and roads subject to natural or man-made blockages. Security information, such as the presence of fences, 24-hour security forces, and guard dogs, should all be part of the plan. The fire officer must also determine the environmental impact of any contaminants that could be released.

During a test of the pre-incident plan, the fire officer should document any interference or poor coverage of the two-way

Tactical Priorities

Address: 1500, 1510, 1520		
Occupancy Name:		

Preplan #: 02-N-01	Number Drawings: 1	Revised Date: 12/2002
District: E275A	Subzone: 60208	By: ACEVEDO

Rescue Considerations: Yes (   ) No (X)

Occupancy Load Day:	Occupancy Load Night:
Building Size:	Best Access:

Knox Box: NONE	Knox Switch: NONE	Opticom: NONE
Roof Type: X	Attic Space: Yes ( ) No ( )	Attic Height: X
Ventilation Horizontal:	Ventilation Vertical:	

Sprinklers: Yes (X)   No (   )   Full (X)   Partial (   )

Standpipes: Yes (X)   No (   )   Wet (   )   Dry (   )

Gas: Yes (   )   No (   )   Lpg (   )

Hazardous Materials: Yes (X) No (   )
DIESEL GENERATORS

1,000 GALLON TANKS

BATTERY ROOM

Firefighter Safety Considerations:
ELEVATOR PIT

Property Conservation And Special Considerations:
VENTILATION: AUTOMATIC SMOKE REMOVAL SYSTEM

3 OFFICE BLDGS; 2 PARKING STRUCTURES

6 FLRS - 1230 W. WASHINGTON ST.
4 FLRS - 1500 N. PRIEST DR.
4 FLRS - PARKING GARAGE

Figure 12-2   An example of a pre-incident plan.

radio system. For example, when conducting a pre-incident plan of a high rise, be sure to test your portable radios to ensure that you can contact dispatch when in the inner core as well as in the basement. These are "dead" spots in many of the radio systems. The problem could be mitigated by having a fixed antenna system or repeaters installed in the building.

Step 2: Identify Occupant Considerations

The fire department is expected to provide a rapid and safe evacuation of a facility or plan for protection in place. Protecting in place would require the officer to determine what areas within the structure are resistant to the magnitude of fire that would be expected to occur.

If the pre-incident plans determine that the occupants should be removed, the plan should identify how that will be accomplished. It should describe the planned process of fire department and building resources that will be dedicated to occupant protection, including the occupant escape routes.

Document the number of occupants and their ages, their physical or mental conditions, and if they need assistance leaving the building. The information should include hours of operation, occupant load, and location of occupants. Each of these points of information has a bearing on how many companies would be needed and what they can expect for removing occupants.

The pre-incident plan should note the location of exits and any special locking devices, such as a delayed release or stairwell unlocking system. The fire officer should ensure that the pre-incident plan reflects coordination between the facility staff and the fire fighters. This might include a meeting between the fire officer and the staff of a nursing facility in which they predetermine where occupants will be moved by the staff. The pre-incident plan should identify the locations of any occupants who need assistance to safely evacuate the building, noting the locations of stair chairs, stretchers, and lifts.

When large numbers of people are relocated, they need basic services, such as food, water, and sanitation facilities. The public school systems can assist the fire department in these types of situations.

A tracking system for the occupants should be established, particularly for locations where the number of occupants varies with the facility's operations and time of day. The American Red Cross has had considerable experience with tracking evacuees.

Some facilities, such as high-rise, health care, or detention facilities, cannot be completely evacuated in a short time. The plan could be to defend in place, providing areas within the structure that are designed to provide occupant protection during a fire. These facilities are normally fire-resistive construction, which allows occupants to remain protected from the fire for significant periods of time.

The products of combustion must be segregated from the occupants. Move those in most peril to areas away from the fire. Close fire doors and seal the individual rooms with wet towels to prevent smoke from filtering into the room. The heating, ventilating, air-conditioning (HVAC) system is often designed to shut down upon activation of a fire alarm. Sophisticated HVAC systems, often found in hospitals and high rises, are divided into zones. The exhaust fans can be redirected to eject the smoke from the zone that is in alarm.

Step 3: Identify Fire Protection Systems and Water Supply

The required water flow is determined by evaluating the size of the building or buildings, contents, construction type, occupancy, exposures, fire protection systems, and any other features that could affect the amount of water needed to control the fire. The adequacy of available water for sprinkler systems, inside and outside hose streams, and any other special requirements or needs should be considered.

Identify the available water supply. Document the location of fire hydrants as well as flow rates and the distribution system. The ideal hydrant would be fed from a large main that is part of a grid that allows water to flow from several directions. A water supply test should be conducted in accordance with NFPA 291, *Recommended Practice for Fire Flow Testing and Marking of Hydrants*. When the demand exceeds the available supply, the pre-incident plan should address an appropriate response to mitigate the deficiency. This might include a water shuttle operation or the initiating of relay operations. Some sites may have a private water system, including their own water tower.

The pre-incident plan should identify the location and details of every fire department connection, fire pump, standpipe system, and automatic sprinkler system Figure 12-3 ▶ . Smoke management and special hazard protection systems should also be detailed on the pre-incident plan. Finally, the pre-incident plan should include data on the protective signaling system.

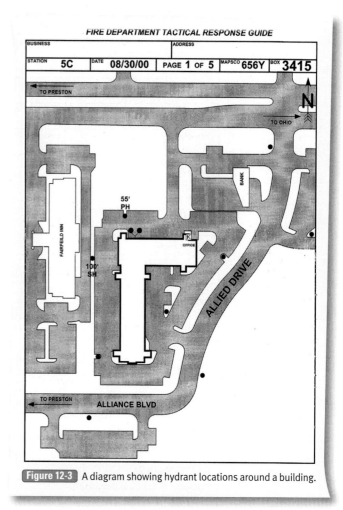

Figure 12-3 A diagram showing hydrant locations around a building.

Step 4: Identify Special Hazards

Document special hazards in the facility and develop a plan to send the proper resources during an emergency. Special hazards include flammable or combustible liquids, explosives, toxic or biological agents, radioactive materials, and reactive chemicals or materials. Request the maximum inventory of hazardous materials and highly combustible products found in the building. Contact information for the facility hazardous materials coordinator and location of material safety data sheets should be on the pre-incident plan.

Some buildings contain specialized operations, processes, and hazards that can pose unique challenges in an emergency. The pre-incident plan should identify any area of the occupancy that contains gases or vapors that could present a hazard to emergency responders. This includes confined spaces, inert atmospheres, ripening facilities, and special equipment–treating atmospheres. Document emergency operating procedures and identify personnel that can provide technical assistance during emergency incident mitigation.

The pre-incident plan should include instructions for de-energizing electrical systems and isolating and securing mechanical systems. These systems should be shut down or monitored to protect emergency responders from electrocution, mechanical entrapment, or other perils. Digital cameras are useful to document systems for training.

Near Miss REPORT

Report Number: 07-0000691

Event Description: We were dispatched to a working structure fire assist at a chemical storage facility. The police department on location reported a working fire in the warehouse storage area. Initial crews were attempting an interior attack from the unburned side of the fire wall. On my way to the interior attack crews, I noticed two crew members exiting the structure. One of them was not paying attention and walked into a dump valve on an acid tank. The valve opened and proceeded to dump acid onto the floor. One of the guys on my crew saw this and quickly shut the valve off. There was less than 1/2 gallon of acid dumped onto the floor. At this point, there was already 3" of water on the floor to dilute the 1/2 gallon of acid that was dumped. After moving past this point, we proceeded to meet up with the attack crew that was staged behind the fire door. The lieutenant from my crew decided to pull everyone back from the burning side of the door. This action was done before my arrival. After finally making it to the nozzle, the fire door came down. It was a good call by my lieutenant to make sure everyone was back behind the door.

Lesson Learned: Always check your surroundings and pay attention to what you are doing. You never know what kind of reaction your actions can cause. Even the little things like walking out of the building can create a dangerous situation if you are not paying attention.

Step 5: Identify Emergency Operation Considerations

The incident action plan should be based on the priorities of life safety, incident stabilization, and property conservation. The pre-incident plan should address the appropriate and adequate departmental response to a working fire or emergency incident. The planned location of the incident command post and emergency operations center, if provided, should be noted.

The number of required fire companies is affected by fuel loading. **Fuel load** refers to the total quantity of all the combustible products that are within a room or space. The fuel load determines how much heat and smoke will be produced by a fire, assuming that all of the combustible fuel in that space is consumed. The size and the shape of interior objects and the types of materials used to create them have a tremendous impact on the objects' ability to burn and on their rate of combustion. The same factors influence the rate of fire spread to other objects.

Fires develop in phases: ignition, growth, fully developed, and decay. The ignition phase of a fire is the starting point of a fire. A fire in this phase usually involves only the object of origin. A pre-incident plan may identify what would be required to handle a fire at this stage. For example, a loaded stream-type portable fire extinguisher may be required for the combustibles present in the building. A fire in the growth phase involves other objects in the fire. The pre-incident plan might indicate that a growth-stage fire would require a minimum of a 1 3/4" (44-mm) handline.

A fire in the fully developed phase has already flashed over. Depending on occupancy and construction factors, the pre-incident plan might recommend that a fire in this stage be fought defensively on arrival, or it might indicate that an attack with large handlines be initiated. When the fire has consumed all of the oxygen but has retained the heat and has fuel, it has entered the decay phase. Plans would indicate how the building would be ventilated when this situation is encountered.

The plan should identify anticipated areas of fire spread. Open pipe chases, elevator shafts, and balloon construction all contribute to vertical fire spread. Large open areas may conceal the magnitude of a fire that has only light smoke showing from the outside yet allows for significant horizontal fire spread. False ceilings and cocklofts may conceal horizontal fire spread.

The fire department's pre-incident plan should be fully coordinated with the internal evacuation or emergency operations plan. The facility should provide the fire department with an on-site liaison as soon as command is established.

Step 6: Identify Special or Unusual Characteristics of Common Occupancy

Particular hazards for each occupancy group should be identified. After gathering data on the specific facility, the fire officer should research the applicable codes, standards, and other information sources to assist in identifying the common and unusual hazards that are associated with that occupancy group. For example, multiple fatality fires are the major risk factor that should be considered in places of assembly.

Effective pre-incident plans for simple sites or facilities can be developed with minimal amounts of data. Additional data may be required for more complex sites or facilities, such as locations with numerous hazards with the potential for high-risk incidents.

Time, owner resistance, and proprietary information are factors that hinder the collection of data. The fire officer might have to prioritize the data collection effort to obtain the most useful information. An incremental process can be used, in which very basic plans are developed and then are updated at a later time as more information is available.

Putting the Data to Use

The goal is to develop a written plan that would be valuable to both the owner and the fire department if an incident occurs at that location. The pre-incident plan should provide critical information that could be advantageous for responding personnel, in a format appropriate for emergency conditions. Many fire departments maintain pre-incident plans in electronic form instead of printing out hard copies.

Data storage systems allow the information to be automatically retrieved when the dispatch system processes an alarm for the location. This information can be printed at the fire station, or on-board mobile data equipment can provide access to the information from a command post or while en route to the incident. Additional detailed information, such as building plans and fire alarm drawings, can be kept in a lock box or other secured area at the site.

The pre-incident plan includes a plot plan. The plot plan should show the relationship of the building to other buildings, streets, hydrants, utility controls, and other features. The plot plan visually represents these objects, allowing the officer to quickly identify relevant facts.

Some departments may also have a drawing of the interior of the buildings. Like the plot plan, the floor plan allows the officer to quickly identify considerations for a fire attack. The floor plan is a drawing of the interior of the structure and is similar to an architect's blueprints. The floor plan notes stairwell locations, elevators, standpipe connections, hazardous material storage areas, alarm panel locations, and points of entry.

A written report also accompanies most pre-incident plot plans. The written report identifies the address, type of occupancy, construction type, size of the occupancy, and required fire flow all in one section. In another section, the report identifies the resources available, including responding units and water sources. Another section includes any special hazards or considerations, such as fire department connection, utility controls, hazardous material storage, and rapid entry key box locations. It is important that these are in a standardized format with standardized information to allow for quick access by the officer, whether the report is in an electronic form or a hard copy.

Understanding Fire Codes

Fire officers often perform inspections to enforce a fire code (or fire prevention code). A fire prevention code establishes legally enforceable regulations that relate specifically to fire safety, although some fire codes also include related subjects, such as regulation of hazardous materials.

Codes can be adopted by different jurisdictions. A state fire code applies everywhere in the state, whereas a locally adopted code can be enforced only within that particular jurisdiction. In many cases, there is a state or provincial fire code that sets a minimum standard, and local jurisdictions have the option of adopting more stringent requirements.

Fire code requirements are often adopted or amended in reaction to fire disasters. This is known as the **catastrophic theory of reform**. Many egress code requirements can be traced back to disasters, such as the 1903 Iroquois Theatre (Chicago: 602 dead), the 1911 Triangle Shirtwaist Company (New York: 146 dead), the 1942 Cocoanut Grove Nightclub (Boston: 491 dead), the 1944 Ringling Brothers–Barnum and Bailey Circus (Hartford, Connecticut: 168 dead), the 1977 Beverly Hills Supper Club (Southland, Kentucky: 164 dead), the 1990 Happy Land Social Club (Bronx, New York: 87 dead), and the 2003 The Station nightclub (West Warwick, Rhode Island: 100 dead).

Authority having jurisdiction is a term used in NFPA documents to refer to "an organization, office, or individual responsible for enforcing the requirements of a code or standard, or for approving equipment, materials, an installation, or a procedure." The authority having jurisdiction for a state fire code is usually the state fire marshal. In the case of a provincial fire code, the authority having jurisdiction would be the provincial fire marshal or fire commissioner. The local fire chief, fire marshal, or code enforcement official would be the authority having jurisdiction for a local fire code.

The regulations contained in a fire code are enforced through code compliance inspections. These inspections could be conducted by the state fire marshal's office, by the local fire department, or by code enforcement officials who might or might not be a part of the fire department. The authority having jurisdiction delegates the power to enforce the code to the fire officers, inspectors, and other individuals who actually conduct inspections.

Building Code versus Fire Code

Building codes and fire prevention codes both establish legally enforceable minimum safety standards within a state, province, or local jurisdiction. A building code contains regulations that apply to the construction of a new building or to an extension or major renovation of an existing building, whereas a fire code applies to existing buildings and to situations that involve a potential fire risk or hazard. For example, the building code might require the installation of automatic sprinklers, a fire alarm system, and a minimum number of exits in a new building; the fire code would require the building owner to properly maintain the sprinkler and alarm systems and to keep the exits unlocked and unobstructed at all times when the building is occupied. Sometimes, a fire code includes certain requirements that apply to new buildings that are beyond the scope of the building code, such as a regulation requiring the installation of fire lanes and hydrants.

State Fire Codes

Most U.S. states and Canadian provinces have adopted a set of safety regulations that apply to all properties, without regard to

---

Fire Marks

Life Safety Code
NFPA 101, *Life Safety Code*, is a model code document that contains requirements specifically related to protecting the lives of building occupants, covering detailed information on exits. When NFPA 101 is adopted by a jurisdiction, it can be applied to both new and existing buildings.

local codes and ordinances. Where there is a state or provincial fire code, it is generally the minimum legal standard in all jurisdictions within that state or province. The state or provincial fire marshal usually delegates enforcement authority down to local fire officials.

Most states allow local authorities the option of adopting additional regulations or a more restrictive code. A few states have adopted **mini/max codes**, which mean that local jurisdictions do not have the option of adopting more restrictive regulations.

The NFPA *Fire Protection Handbook* identifies seven different organizational patterns for state fire marshal organizations in the United States. The state fire marshal may work in:

- The department of insurance
- The department of public safety
- A separate government department
- A regulatory agency
- The state police
- A cabinet-level office
- The state fire commission **Table 12-1 ▾**

Local Fire Codes

At the local level, fire and safety codes are enacted by adopting an **ordinance**, which is a law enacted by an authorized subdivision of a state, such as a city, county, or town. The local jurisdiction adopts an ordinance that establishes the fire code as a set of legally enforceable regulations and empowers the fire chief to conduct inspections and take enforcement actions. This authority can then be delegated to fire officers, fire inspectors, and other individuals within the fire department.

Model Codes

Model codes are documents developed by a standards developing organization, such as the NFPA, and made available for adoption by authorities having jurisdiction. The *National Building Code* was first published by the National Board of Fire Underwriters in 1905. A model code is developed through a consensus process using a network of technical committees. Most jurisdictions use model codes developed by the NFPA and the International Code Council.

States and local jurisdictions adopt a nationally recognized model code, with or without amendments, additions, and exclusions. A complete set of model codes includes a building code, electrical code, plumbing code, mechanical code, and fire prevention code. The primary advantages of a model code are that the same regulations apply in many jurisdictions, and all of the requirements are coordinated to work together without conflicts.

Table 12-1   State Fire Marshal Organizational Patterns

Under Department of Insurance
- Alabama
- Florida
- Georgia
- Idaho
- Mississippi
- New Mexico
- North Carolina
- Tennessee
- Texas

Under Department of Public Safety
- Alaska
- Colorado
- Connecticut
- District of Columbia
- Iowa
- Kentucky
- Maine
- Massachusetts
- Minnesota
- Missouri
- Nevada
- New Hampshire
- Oklahoma
- Utah
- West Virginia

Under a Separate Government Department
- Indiana (State Emergency Management)
- Kansas
- Louisiana (Department of State)

- Michigan (Department of Consumer and Industry Services)
- Montana (Justice, Attorney General)
- South Carolina (Department of Labor)
- South Dakota (Department of Community Affairs)
- Virginia (Department of Housing and Community Development)
- Wisconsin (Fire Investigation, Justice, Attorney General)
- Wyoming

Under a Regulatory Agency
- Arizona (Department of Building Fire Safety)
- Vermont
- Wisconsin (Other functions, Department of Commerce)

Under State Police
- Arkansas
- Maryland
- Michigan (Fire Investigation)
- Oregon
- Pennsylvania
- Washington

Under a Cabinet-Level Office
- California
- Illinois
- Nebraska
- New Jersey (Department of Community Affairs)
- New York (Secretary of State)
- North Dakota (Attorney General)
- Ohio (Department of Commerce)
- Rhode Island (Governor's Office)

Under State Fire Commission
- Delaware

*Courtesy of the National Fire Protection Association.

If a model code is adopted by a local jurisdiction, it may occur in one of two ways. **Adoption by reference** occurs when the jurisdiction passes an ordinance that adopts a specific edition of the model code. A local jurisdiction could adopt NFPA 1, *Fire Code* (2009) by reference. The requirements specified in NFPA 1 then become local requirements that can be enforced by designated local officials. A fire officer would need to obtain a copy of the 2009 edition of NFPA 1 to read the specific requirements.

Adoption by transcription occurs when the jurisdiction adopts the entire text of the model code and publishes it as part of the adopting ordinance. For example, a city would copy the language of the code and include it in its entirety within the ordinance. A fire officer would only need to read the city's ordinance to read the specific requirements rather than having to find the appropriate code book.

The fire officer must know which code and which annual edition is used by the local jurisdiction. Although the model code process updates the code every 2 to 5 years, the authority having jurisdiction must specifically adopt the new edition of a model code before it becomes legally enforceable. Different codes or different editions of the same code might apply to different occupancies. Some communities adopt selected portions of one or more model codes and defer to a state code or locally written ordinances to cover other issues. Some jurisdictions choose to maintain their own independent codes.

▮ Retroactive Code Requirements

Regulations that applied to a particular building at the time it was built remain in effect as long as it is occupied for the same purpose. The fire officer may have to determine the specific code document, title, and year used when the building was built to determine whether a building is still in compliance with the code requirements that apply. Most codes include provisions that can be applied to buildings that were constructed before a code was adopted.

If a building is remodeled or extensively renovated, or if the occupancy use changes, most codes specify that all of the current requirements of the code must be applied. New code requirements, adopted after a certificate of occupancy has been issued, do not apply unless specific language is included in the adopting ordinance. For example, Scottsdale, Arizona, established a mandatory requirement for sprinkler systems in all new residential construction in 1986. The ordinance did not require retroactive installation of sprinkler systems in residential structures built before that date.

On occasion, a state or local authority having jurisdiction passes a code revision that is specifically identified as applying retroactively to all affected occupancies. After the 1980 Las Vegas MGM Grand fire that killed 84 and injured 679, the state of Nevada required fire suppression systems to be installed in all existing casinos and hotels. Rhode Island and Massachusetts adopted retroactive sprinkler requirements for nightclubs in response to the 2003 fire in The Station nightclub that killed 100 and injured more than 200 patrons.

▮ Understanding Built-in Fire Protection Systems

Next to access and egress, the status of the built-in fire protection features is the second reason for a fire company to perform inspections. Built-in fire protection systems are designed as tools to assist fire fighters in fighting a fire. The officer should understand how systems work and how the codes are altered when fire protection systems are in place.

The codes often allow more flexibility in the design of a building because built-in fire protection systems are included. A building with a sprinkler system can be larger and taller, the travel distance to exits can be longer, and the access for fire apparatus could be restricted. All of these trade-offs depend on a properly functioning sprinkler system.

If there is a fire in the building, you are depending on the built-in fire protection systems to assist you. The dependability of these systems is of prime importance to fire fighters' safety. An inspection is the best method of ensuring the systems will work. Industrial Risk Insurers (IRI) is a large provider of coverage for industrial properties. The most frequent cause of large-loss industrial fires in IRI properties was because sprinkler systems had been shut down for repair or maintenance and not turned back on. As a result, Industrial Risk Insurers promotes the RSVP (Restore Shut Valve Promptly) program.

Los Angeles, Phoenix, and Fairfax County, Virginia, started a fire protection system testing program in the late 1980s. In the first year, they encountered significant failures in built-in fire protection systems. A periodic retesting program was put in place to improve the performance of built-in fire protection services.

▮ Water-Based Fire Protection Systems

Automatic sprinkler systems, standpipe systems, and fire pumps are the three primary components of water-based fire protection systems. An **automatic sprinkler system** consists of a series of pipes with small discharge nozzles (sprinklers) located throughout a building. When a fire occurs, heat rising from the fire causes one or more heads to open and release water onto the fire. When the water starts flowing, a water flow alarm is activated `Figure 12-4 ▾`. This alarm may be monitored by a central station alarm service and/or on-site safety/security service. Some systems are "local alarms" and sound a gong or bell only at the outside of the building `Figure 12-5 ▸`.

Depending on the usage and climate, automatic sprinkler systems may be wet pipe, dry pipe, deluge, or pre-action. In a

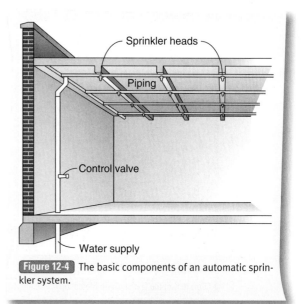

Sprinkler heads

Piping

Control valve

Water supply

`Figure 12-4` The basic components of an automatic sprinkler system.

Figure 12-5 Some systems are "local alarms" and sound a gong or bell only at the outside of the building.

wet-pipe system, there is water in all of the pipes throughout the system. When a sprinkler head opens, water is discharged immediately. In general, wet-pipe systems require less maintenance than dry-pipe systems. Because of the faster reaction, fewer sprinkler heads are activated to control most fires.

Dry-pipe systems are used in locations where a wet-pipe system would be likely to freeze, such as unheated storage facilities and parking garages. Instead of water, the pipes are filled with compressed air or nitrogen until a sprinkler head opens. When the air pressure drops, the dry-pipe valve opens and water is released into the system. Dry-pipe systems require higher maintenance because activation of the sprinkler system requires the entire sprinkler system to be drained. Sometimes, instead of installing a dry-pipe system, an antifreeze solution is added to the water in a wet-pipe system to protect an unheated area (e.g., a freezer or a loading dock).

Deluge systems are special versions of wet- or dry-pipe systems in which large quantities of water are needed to quickly control a fast-developing fire. Deluge systems are most often found in ordinance plants, aircraft hangers, and occupancies with flammable liquid hazards. All of the sprinkler heads are open and ready to discharge water as soon as the control valve opens. The system uses smoke, flame, or fire detectors to sense a fire and trigger the system.

Pre-action sprinkler systems are similar to dry-pipe systems, with a separate detection system to trigger the dry-pipe valve and fill the sprinkler pipes with water. At this point, it becomes equivalent to a wet-pipe system. A pre-action system is designed to reduce the risk of water damage due to accidental sprinkler discharge or a broken pipe.

Some sprinkler systems are designed to discharge protein or aqueous film-forming foam as the extinguishing agent. These are used in high hazard areas where the contents to be protected are flammable liquids, such as fuel storage and chemical process facilities. These are complex systems that require more frequent inspections and are custom designed for the storage area, processing

plant, or fueling facility that is protected. Often there is an on-site or nearby technician trained in the maintenance of these systems.

Standpipe systems provide the ability to connect fire hoses within a building. Standpipes are an arrangement of piping, valves, hose connections, and allied equipment that allow water to be discharged through hoses and nozzles to reach all parts of the building. Like automatic sprinkler systems, you may find dry standpipes in unheated areas. Standpipes are subdivided into three classes based on their expected use:

- Class I provides 2½" (64-mm) hose outlets, intended for use by fire department or fire brigade members trained in the use of large hose streams.
- Class II provides a 1½" (38-mm) hose coupling with a preconnected hose and nozzle in a hose station cabinet. The hose is designed for occupant use.
- Class III provides both 1½" (38-mm) and 2½" (64-mm) connections. The 1½" connection may have a preconnected hose line to be used by the occupants until the fire department arrives.

You should pay particular attention to the condition of the standpipe system; it is one of the few built-in fire protection devices specifically designed to help fire fighters. Fire officers should identify fire hose access and ensure proper calibration of pressure-regulating devices. A pressure-regulating device limits the discharge pressures from standpipe hose outlets. During a 1991 high-rise office fire at One Meridian Plaza in Philadelphia, fire fighters discovered that the pressure-regulating devices were improperly set, and only a weak flow could be obtained from each outlet. Usually found on the lower floors of high-rise standpipes, these were installed on floors 26 though 30 and impeded firefighting operations. Three fire fighters died while the fire consumed the 22nd through the 30th floor in 19 hours.

Fire Pumps

Fire pumps increase the water pressure in standpipe and automatic sprinkler systems Figure 12-6 . They are designed to start automatically when the water pressure drops in a system or a fire suppression system is activated. A field inspection generally is confined to a visual inspection to confirm that the pump appears to be in good condition and is free of physical damage. The fire officer should confirm that the fire pump has passed the annual performance test.

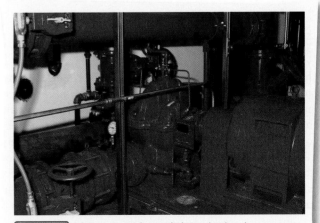

Figure 12-6 A fire pump may be needed to maintain or increase water pressure in standpipe and automatic sprinkler systems.

Fire Sprinkler Alarms

A fire alarm system is present in most buildings with automatic sprinklers. In many cases, smoke detectors and local fire alarm pull stations are also connected to the fire alarm system. Sophisticated alarm systems are often monitored by a central fire alarm service that calls in the fire alarm event to the local 9-1-1 center. Most of the notifications from fire alarm systems are nonfire events. Some of these events are caused by situations that appropriately trigger an alarm, such as cigarette smoke or burnt food. Other activations are due to defective or dirty smoke detectors. The fire officer must remain vigilant because the same notification process is used when a waterflow alarm occurs.

A waterflow alarm means that water has moved past a flow switch that is located inside the water supply pipe and set off the alarm. Nonfire events that set off the waterflow alarm often include public water supply surges, sprinkler heads broken accidentally, and burst pipes. Many older buildings have a single waterflow device for the entire structure, so the problem could be anywhere in the building.

A trouble alarm in a monitored fire alarm system is like a warning light in a car: All that it is telling you is something is not right. Old and inadequately maintained fire alarm systems that detect a fire or waterflow alarm may default into a trouble alarm after the reset button is pushed. Some communities require that a log book be provided in buildings with large and sophisticated alarm systems, usually in the fire control room. Both the building maintenance department and the fire department make entries in this log book with every activated fire alarm. Fire officers who work in districts with many monitored buildings develop a detailed knowledge of the features and problems in each fire alarm system. Although the code may specify a certain type of system, these fire officers appreciate that every building has unique characteristics.

If the fire alarm system resets without going back into alarm or trouble, a surge in the municipal water supply may have tripped the waterflow alarm. If the alarm system properly resets and does not reactivate within 1 or 2 minutes, you may not need to make a detailed floor-to-floor search.

However, if you respond back to a second waterflow activation within a couple of hours, assume that water is flowing somewhere. If you are making a return visit, make sure that a building representative responds.

Special Extinguishing Systems

There are four different types of special extinguishing systems: carbon dioxide, dry or wet chemical, Halon, or foam. Note these specialized systems in the pre-incident plan. In general, the ongoing code compliance inspection consists of visual inspection, looking for physical damage, and confirming that the safety seals are intact, as well as verification of the documentation for required tests and inspections. The safety seal is placed on any handle that would activate the system. If the safety seal is broken, the system could have been discharged.

Carbon Dioxide

Carbon dioxide systems are fixed systems that discharge carbon dioxide from either low- or high-pressure tanks, through a system of piping and nozzles, either to protect a specific device or process (e.g., a printing press) or to flood an enclosed space. Carbon dioxide extinguishes fire by displacing oxygen. The gas is heavier than air, so it settles in low spaces. Fixed systems are generally required to comply with NFPA 12, *Standard on Carbon Dioxide Extinguishing Systems*. Your pre-incident plan should require the use of a self-contained breathing apparatus (SCBA) if a fixed system has activated.

Dry or Wet Chemical

Fixed chemical extinguishing systems discharge a chemical extinguishing agent through a system of piping and nozzles. These systems can be found protecting commercial cooking devices and industrial processes where flammable or combustible liquids are used. Observe the condition of the nozzles. Many have a protective cap to protect the nozzle from becoming obstructed with cooking grease.

The wet chemical systems are preferred for protecting cooking equipment. The wet chemical agent reacts with hot grease to form a foam blanket, reducing the release of combustible vapors. The foam blanket cools the grill and reduces the possibility of a rekindle. The dry chemical systems leave a residue that is difficult to clean up.

Both systems are activated in one of two ways: (1) a fusible link that melts on flame contact or (2) a manual pull station. Activation of the system turns off the cooking device by closing the cooking fuel valve or turning off the electricity.

Halon

Halon 1301 was the extinguishing agent of choice for fire protection in computer rooms and to protect electronic equipment from the 1960s to the 1990s. Based on the weight of the agent, Halon 1301 is about 250 percent more efficient than carbon dioxide for extinguishing fires. Unfortunately, it also depletes the ozone layer in the atmosphere. Since 2000, Halon cannot be manufactured or imported into the United States, but legacy systems may still be recharged.

NFPA 2001, *Standard on Clean Agent Fire Extinguishing Systems*, covers the use of alternative agents that replace Halon systems protecting electrical and electronic telecommunications systems. Most of these systems are designed to flood a room or an enclosure and, like Halon, are toxic. The room or enclosure must be sealed, and the occupants have to leave before the agent is discharged. The system can be automatically or manually fired; both methods include a pre-alert warning for the occupants to leave the room before the agent discharges. The pre-incident survey should identify the chemical used, the duration of the discharge (10 seconds to 1 minute), the enclosure protected, the automatic activation sequence, and the location of the manual activation station.

Foam Systems

A low-expansion foam system is used to protect hazards involving flammable or combustible liquids, such as gasoline storage tanks. These systems discharge foam bubbles over a liquid surface to create a smothering blanket that extinguishes the fire and suppresses vapor production. High-expansion foam is used in areas where the goal is to fill a large space with foam in order to exclude air and smother the fire.

Fire Alarm and Detection Systems

A fire alarm system consists of devices that monitor for a fire and notify the appropriate personnel. Manual fire alarm boxes, smoke detectors, or heat detectors may activate the fire alarm system. The alarm may also be activated by the waterflow switch in a sprinkler system. The activated system notifies the appropriate personnel, including the building occupants, with audible and visual signals. These may be transmitted outside the building and to the fire department or a central station monitoring firm.

Understanding Fire Code Compliance Inspections

The objective of a fire code compliance inspection is to determine whether an existing property is in compliance with all of the applicable fire code requirements. Some codes call this a maintenance inspection. The goal is to observe the housekeeping to ensure that there are no fire hazards and confirm that all of the built-in fire protection features, such as fire exit doors and sprinkler systems, are in proper working order.

The fire department's authority and responsibilities for conducting code compliance inspections are usually included in the ordinance that adopts the local fire code. The responsibility for code enforcement is usually assigned to the fire chief or fire marshal. The fire chief can delegate the responsibility for code enforcement activities to different individuals or units within the department.

In many fire departments, fire officers and fire inspectors who are assigned to a fire prevention bureau or code enforcement division conduct inspections of specific types of properties. In departments where fire companies conduct code compliance inspections, there is usually an individual or a group assigned to coordinate their activities and provide technical assistance.

Fire Company Inspections

The purpose of conducting fire inspections is to identify hazards that exist and ensure that the violations are corrected. This process helps enable the fire department to be proactive in preventing fires. One method of maximizing the number of inspections is for them to be conducted by the fire companies.

Inspections conducted at the fire company level have been an activity advocated by the National Fire Protection Association since the first *NFPA Quarterly* was published more than a hundred years ago. Local fire companies have the ability to become familiar with their response areas and take actions to prevent the start of fires and, if a fire starts, minimize the fire spread and destruction of the neighborhood.

Before conducting an inspection, it is important to understand the scope of code enforcement authority that is delegated to a fire officer in your jurisdiction. There is wide diversity in the scope and authority of inspections conducted by fire companies. In some areas, fire suppression companies are authorized to enforce all fire code **regulations**; one metropolitan city authorizes fire officers to issue corrective orders that would require the

installation of automatic sprinklers in a business under renovation. In other fire departments, fire suppression companies are assigned to inspect only certain types of occupancies, whereas the fire marshal's office or fire prevention division inspects other types of occupancies. Some fire departments can only provide safety recommendations to citizens and property owners, whereas others can issue citations and compliance orders. The fire officer needs to determine the source and the scope of his or her code enforcement authority before conducting any fire safety inspection, as well as which particular codes are used by the jurisdiction.

Outside of fire emergencies, a fire officer generally cannot enter private property without the permission of the owner or occupant. Most fire codes include a section that authorizes code enforcement officials to enter private properties at any reasonable time to conduct fire and life-safety inspections; however, the scope of this authority usually depends on the type of property involved. In most cases, the permission of the owner or occupant is required for entrance into a dwelling unit, whereas access to public areas is less restricted. The fire code often contains a section that, if necessary, allows for the issuance of a court order requiring the owner or occupant to allow the fire department agent to enter the occupancy to conduct an inspection.

Classifying by Building or Occupancy

Many of the code requirements that apply to a particular building or occupancy are based on its classification. The codes classify a building by construction type, occupancy type, and use group. You need to know how to classify a building before you can determine the code requirements that apply to it. In addition to these three classifications, local zoning ordinances often regulate where certain types of occupancies or buildings are permitted.

Construction Type

The building itself is classified by **construction type**, which refers to the design and the materials used in construction. NFPA 220, *Standard on Types of Building Construction*, addresses building construction and firefighting. The type of building construction has a significant influence on firefighting strategy. Construction type is a fundamental size-up consideration when firefighting strategy for a burning structure is being determined.

The most commonly used model codes subdivide construction into five basic types:

- Type I: Fire resistive. The construction elements are noncombustible and are protected from the effects of fire by encasement, using concrete, gypsum, or spray-on coatings Figure 12-7 ▶ . Depending on the model code used, type I is divided into subtypes, based on the level of fire protection provided. The level of protection is described by the number of hours a building element can resist the effect of fire.
  - Type I is the most durable and lasting structure. Allow time for extensive fire suppression operations without collapse.

Voices of Experience

A few years ago while on vacation, I was involved in an incident that made me very aware of the importance of inspecting fire detection equipment or the lack thereof.

On the first day, we were sitting poolside and noticed a double duplex multi-family apartment house across the street. During the afternoon and continuing into the early evening there was a constant stream of vehicles arriving with many small children. This seemed a little unusual but not out of the ordinary since this was a resort community and people are constantly moving about.

In the late afternoon of the following day, as I looked toward the multi-family apartment building, I noticed puffs of light smoke coming from the area of the back deck. Because this was the usual time for people to cook on the grills, I thought nothing of it. My instincts were aroused when these puffs of smoke turned black and became more frequent.

I told my wife I was going across the street to investigate. Halfway across the street I realized that the smoke was coming from the open window in the back of the building and not from a gas grill. I yelled back to those at the motel to call 9-1-1.

Smoke was now constantly billowing from the window and I wanted to enter the building to do a search. Not having any PPE, I kicked in the door and entered the apartment, staying low to the floor. I noticed fire on my left and proceeded into the apartment to conduct the search. After two searches I was unable to locate any victims in that apartment.

Exiting that apartment, I went to the adjoining apartment and found the door open. The occupants, a mother and her two daughters, were in the front room and were totally oblivious that the rear of their building was on fire. Their only escape route was quickly being engulfed with smoke and fire. Within moments I had reluctantly led them out the back door to safety.

Being curious about the cause of the fire, I was allowed to stay with the local fire department during their investigation. What we found could have caused a catastrophe if the fire happened during the night.

The apartments were recently painted by a contractor who covered the smoke detectors with plastic bags when he spray painted the apartments. The problem came about when he finished the job and left the plastic bags on the smoke detectors.

By code, we have the authority to inspect multi-family buildings, but there are no rules or regulations that a painting contractor must obtain a permit before he can start a painting job. Once this contractor covered the smoke detectors, it was considered tampering with life safety equipment and he could face prosecution. Maybe we should have these contractors obtain a permit from the Fire Prevention Bureau with some do's and don'ts before being allowed to start. Following completion of the job and before occupancy is allowed, the Fire Prevention Bureau should inspect the complex. The question is "*When do we know when to inspect?*"

Richard Kosmoski
New Jersey Certified Fire Official
Fire Instructor, Middlesex County Fire Academy
Sayreville, New Jersey

Figure 12-7   A type I building.

Figure 12-9   A type III building.

- This type of construction often uses compartmentation instead of fire sprinklers to control fire spread, creating unsurvivable interior furnaces when fire fighters are trapped on the fire floor. Metal structural elements may be failing due to age and rust.
- Type II: Noncombustible. The structural elements can be made from either noncombustible or limited combustible materials **Figure 12-8 ▼**. Like type I, this type also has subdivisions, based on the level of fire resistance. Although the buildings are assembled from noncombustible components, the structural elements have limited or no fire resistance. A type IIA structural frame is expected to resist fire for 1 hour. The structural frame in a type IIB building is not expected to resist the effects of fire. A strip shopping center with cinderblock walls, unprotected steel columns, and steel bar joists supporting a steel roof deck is an example of a type IIB building.
  - Type II is a common 20th-century construction and the type of building Frances Brannigan referenced when describing the 20-minute interior firefighting rule of thumb in 1971.

- It is durable, but not a legacy building. It will require replacement in 30–40 years. This type of building is frequently updated with Type V structural elements.
- Type III: Limited combustible (ordinary). The exterior load-bearing walls of the building are noncombustible masonry **Figure 12-9 ▲**. A **masonry wall** may consist of brick, stone, concrete block, terra cotta, tile, adobe, or concrete. The interior structural elements may be combustible or a combination of combustible and noncombustible. Like types I and II, there are different levels of fire protection for type III buildings. The structural frame of a type IIIA building is protected, which means it is encased in concrete, gypsum, or spray-on coatings and is expected to have a fire-resistive rating of 1 or 2 hours. A type IIIB structural frame is unprotected and has no fire resistance rating.
  - Type III was used to build commercial, multiple-family, and mercantile types of buildings through the 1970s. Brannigan classifies these buildings as "Main Street USA."
  - This type of building is usually no higher than four stories; it was designed to preserve the load-bearing walls if fire consumed the building. The connection between the floor and the load-bearing wall is designed to pull out without damaging the wall.
- Type IV: Heavy timber. The exterior walls are noncombustible (masonry), and the interior structural elements are unprotected wood beams and columns with large cross-sectional dimensions **Figure 12-10 ▶**. Mill construction, which was used in many New England textile buildings built in the 1800s, is an example of heavy timber construction. Mill construction features massive wood columns and wood floors.
  - As durable as type I structures, most surviving buildings have been converted to residential, mercantile, or mixed use.
  - A well-seated fire in a nonsprinklered type IV building may exceed the municipal water supply capability.

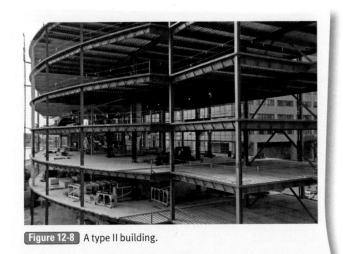

Figure 12-8   A type II building.

Figure 12-10   A type IV building.

- Type V: Wood frame. The entire structure may be constructed of wood or any other approved material . Sometimes, a masonry veneer is applied to the exterior, but the structural elements are wood frame.
  - The most common structure built since the 1980s, including single-family, multiple-family, mercantile, and low-rise commercial buildings.
  - This type of structure will maim or kill the first-arriving fire fighters if they enter the structure and fall through a fire-weakened floor.
  - Type V construction techniques and approach will be the foundation for future innovative building methods and materials.

It is often difficult to determine the type of construction from the exterior of a building. Many low-rise offices, apartment buildings, and residences appear to be ordinary construction; however, they are actually wood frame buildings with a brick, stone, or masonry veneer on the exterior walls. This type of wall covering does not contribute to the strength or fire resistance of the building.

Occupancy and Use Group

The **occupancy type** refers to the purpose for which a building or portion of a building is used or is intended to be used. The code requirements are determined by its **use group**. Occupancies are classified into use groups based on the characteristics of the occupants, the activities that are conducted, and the risk factors associated with the contents. Within each occupancy type are dozens of use groups.

Assembly

An assembly occupancy is used for the gathering of people for deliberation, worship, entertainment, eating, drinking, amusement, or awaiting transportation. The classification may be further divided into more specific types of assemblies. Examples of assembly-type occupancies include:
- Churches
- Taverns or bars
- Nightclubs
- Basketball arenas
- Restaurants
- Theaters

Business

A business occupancy is used for account and record keeping or transaction of business other than mercantile **Figure 12-12**. Examples of business occupancies include:
- Dental offices
- Banks
- Architects' offices
- Hair salons
- Colleges and universities
- Doctors' offices
- Investment offices
- Insurance offices
- Radio and television stations

Figure 12-11   A type V building.

Figure 12-12   Example of a business occupancy.

Figure 12-13   Example of an industrial occupancy.

Figure 12-14   Examples of mercantile occupancies.

Educational

An educational occupancy is used for educational purposes through the 12th grade. Typically, this designation refers to schools. Educational occupancy may also cover some day care centers for children older than 2 ½ years.

Industrial

An industrial occupancy is where products are manufactured or in which processing, assembling, mixing, packaging, finishing, decorating, or repair operations are conducted Figure 12-13 ▲. This would include occupancies such as:

- Automobile assembly plants
- Clothing manufacturers
- Food processing plants
- Cement plants
- Furniture production facilities

Health Care

A health care occupancy is used for purposes of medical or other treatment or care of four or more persons, where such occupants are mostly incapable of self-preservation due to age, physical or mental disability, or security measures not under the occupants' control. This would include hospitals and nursing homes.

Detention and Correctional

A detention and correctional occupancy is used to house four or more persons under varied degrees of restraint or security, where such occupants are mostly incapable of self-preservation because of security measures not under the occupants' control. This could include prisons, jails, and detention facilities.

Mercantile

A mercantile occupancy is used for the display and sale of merchandise Figure 12-14 ▶. This group includes the following:

- Retail stores
- Convenience stores
- Department stores
- Drug stores
- Shops

Residential

A residential occupancy provides sleeping accommodations for purposes other than health care, detention, or corrections. There are five subcategories:

- One- and two-family dwelling units: Buildings that contain no more than two dwelling units with independent cooking and bathroom facilities
- Lodging or rooming houses: Buildings that do not qualify as one- or two-family dwellings that provide sleeping accommodations for a total of 16 or fewer people on a transient or permanent basis, without personal care services, with or without meals, but without separate cooking facilities for individual occupants
- Hotels: Buildings under the same management in which there are sleeping accommodations for more than 16 persons and primarily used by transients for lodging with or without meals
- Dormitory: Buildings in which group sleeping accommodations are provided for more than 16 persons who are not members of the same family in one room, or a series of closely associated rooms, under joint occupancy and single management, with or without meals, but without individual cooking facilities

- Apartment building: Buildings containing three or more dwelling units with independent cooking and bathroom facilities

Storage

A storage occupancy is used primarily for the storage or sheltering of goods, merchandise, products, vehicles, or animals. Examples include:

- Cold storage plants
- Granaries
- Lumber yards
- Warehouses

Mixed

A mixed-use property has multiple types of occupancies within a single structure. An example would include an old commercial building that has been renovated to include multifamily residential loft apartments on the second and third floor and a bakery on the first floor.

Unusual

In addition to the previous occupancies, some do not fit neatly into the other categories. These occupancies can be placed in a miscellaneous category that represents unusual structures, such as towers, water tanks, and barns.

You can find the use group classifications in the building code or NFPA 101, *Life Safety Code* Table 12-2 ▶.

Once the use group classification is known, the allowable height, floor area, and construction type can be found in the building code. The building code (or NFPA 101) also includes the requirements for the number of exits, the maximum travel distance to an exit, and the minimum width of each exit. Requirements for the installation of fixed fire protection systems in new buildings are also included in the building code. Additional requirements for fire protection systems are sometimes included in the fire code, particularly for existing buildings.

Code enforcement starts by confirming that the occupancy is used for the approved purpose. For example, if a tenant space in a shopping center has been converted from a clothing store to a bar, the exit requirements and the built-in fire protection system requirements could be different.

NFPA 704 Marking System

Buildings with significant quantities of hazardous materials may be required to use a marking system. The most recognized standard is NFPA 704, *Standard System for the Identification of the Hazards of Materials for Emergency Response*. The marking system consists of a color-coded array of numbers or letters arranged in a diamond shape Figure 12-15 ▶.

Each color represents a specific type of hazard. Blue represents health hazards, red represents flammability hazards, and yellow represents the material's reactivity hazard. Within each color diamond, there is a number from 0 to 4 that represents the relative hazard of each. A 0 means that material poses essentially no hazard, whereas a 4 indicates extreme danger. For example, on the exterior of a building, there might be a marking with a 2 inside the blue area, a 4 inside the red area, and a 0 inside the yellow area. This means that inside the building, there are materials that pose a

Table 12-2   Classification of Use Groups

Major Use Classification	Occupancy Categories
Assembly	Theaters, auditoriums, and churches
	Arenas and stadiums
	Convention centers and meeting halls
	Bars and restaurants
Health care	Hospitals Nursing homes
Detention and correctional	Prisons
	Penitentiaries
Mercantile	Retail stores
Business	Offices
Industrial	Factories
Storage	Warehouses
	Parking garages
Educational	Schools
Residential	Homes
	Apartments
	Dormitories
	Hotels

Courtesy of the National Fire Protection Association.

moderate health hazard, a severe flammability hazard, and no real reactivity hazard. The last quadrant of the diamond is white and is used to represent special hazards. This could include letters or numbers. For example, it might include a placard that indicates to the responder that there is material inside that is water reactive.

This system also requires labels to be affixed to containers inside the structure to indicate the hazards of the substance. This might include labels such as "corrosive," "flammable," or "poison." In general, the system requires a 704 marker at each entrance to the building, on doorways to chemical storage areas, and on fixed storage tanks.

Preparing for an Inspection

Preparation is required before conducting a code enforcement inspection. The assumption in this section is that the fire officer has the responsibility, authorization, and training necessary to

Figure 12-15   Document any special hazard in the facility. © National Fire Protection Association.

conduct such an activity. This section covers how to prepare to conduct such an inspection.

The fire officer should consider the owner or occupant as a professional partner. This begins with preparing for the inspection by reviewing the applicable code provisions and any information on previous inspections that can be found in the occupancy file.

For the business owner or occupant, a visit by the local fire company is a major and disruptive event. Some jurisdictions send a pre-inspection or self-survey form to the business or property a few weeks before an official fire department inspection will occur.

Reviewing the Fire Code

Before conducting an inspection, you should review the sections of the code that apply to the specific property. In some cases, you may discover unanticipated situations during an inspection that require you to research an additional section of the code. The NFPA *Fire and Life Safety Inspection Manual* provides general information about a large variety of processes, equipment, and systems.

Review Prior Inspection Reports, Fire History, and Pre-Incident Plans

Review the inspection results from past activities. You may see a trend or a chronic problem. A facility that always requires two follow-up inspections to correct the violations that are noted during inspections might indicate that the fire officer needs to turn up the salesmanship. For example, when calling to schedule the inspection, the fire officer might take the opportunity to review the violations noted during the previous inspection and encourage the owner to check those items.

Look at the fire history for the occupancy. If a steakhouse restaurant has a history of a flue or hood fire every 4 to 6 months, you should plan to concentrate on the grill and hood system during your inspection. You should think about what you can recommend to reduce the frequency of accidental fires.

Bring a copy of the pre-incident plan with you. This is an excellent time for you to update contact information, such as names, email addresses, and phone numbers. You can also use the pre-incident plan to identify any modifications and additions that have been made to the occupancy.

Coordinate Activity with the Fire Prevention Division

Coordination is needed in departments where there is both a fire prevention division and a company-level inspection program. Some jurisdictions assign a list of occupancies to be inspected by each fire suppression company on a monthly or quarterly basis. In buildings where a process, storage, or occupancy is required to have an annual fire prevention permit, the local fire company's **ongoing compliance inspection** could be scheduled for 6 months after the fire prevention division issues the permit.

The authority having jurisdiction usually determines the frequency of inspections for each type of occupancy. For example, day care facilities might be required to have an annual inspection from the fire marshal's office, whereas business occupancies are inspected by a local fire company once every 2 or 3 years. The fire officer commanding the fire suppression company is expected to follow the inspection schedule and coordinate with the fire marshal's office.

Arrange a Visit

It is good practice to contact the owner or business representative to schedule a day and a time for the fire safety inspection. Many businesses are cyclical, and some days or months are more difficult to accommodate a fire department inspection than others. For example, early April may not be the best time for an inspection of an accounting firm's office because it is the peak of their annual work cycle.

In some cases, a time that is inconvenient for a business is an important time in terms of fire safety. Between October and December, retail stores are packed with extra stock for the holiday buying season. Some retail businesses earn 30 percent or more of their annual revenue at that time of year. A fire safety inspection of a store in an enclosed mall on December 1 may reveal boxes stored from floor to ceiling, obstructing the sprinkler heads. Empty cardboard boxes may be pushed up tight against the electrical panel, instead of maintaining a 30" (76-cm) clearance. The electrical panel may be hotter than usual because of all the extra power needed to run the holiday displays. Trash may have piled up in the storage room and have blocked the rear emergency exit.

Assemble Tools and References

Once you have reviewed the information on the occupancy to be inspected and the applicable fire prevention codes, you are ready to perform the inspection. In addition to your knowledge, you will need to bring some equipment to assist you during the visit. You should have the following tools with you:

- Pen, pencil, and eraser
- Inspection form
- Clipboard. Consider using the type of clipboard that law enforcement and EMS professionals use; it would allow you to store the completed forms in a safe place in case you need to respond to an emergency during a code enforcement period.
- Graph paper and ruler
- Handlight
- Digital camera with flash attachment
- Coveralls
- Measuring device, tape measure, or walking meter
- Fire department business cards
- Reference code books

Company Tips

Neatness Counts

Make sure that your uniform is neat and clean. Some areas will require wearing a safety helmet to complete a life safety inspection or pre-incident plan. Many members will utilize their fire helmet to meet this safety requirement. A "salty" helmet with a heavy buildup of carbon and other products of combustion can create a poor impression. If your hand gets dirty when you touch your helmet, it is time to clean it.

Plan to wear your safety shoes and bring your fire helmet, eye protection, and protective gloves if you will be inspecting an area that is under renovation or inspecting an occupancy that requires such protective equipment.

Conducting the Inspection

Some departments require that the entire fire company perform the fire inspection together. Business owners complain about the disruption that occurs when four to six fire fighters come into a business. Portable fire radios are blaring, and, sometimes, the fire fighters seem to be more interested in sightseeing or shopping than identifying fire safety hazards. Other departments deploy the fire company in two- or three-person teams to conduct inspections in adjacent occupancies. Make sure you know what procedure and practice your organization requires.

Conducting a fire inspection should be approached in a systematic manner. The fire officer should begin every inspection in the same manner. The first step should be to circle the area as you park the apparatus to get a general overview of the property. Second, meet the property owner or manager to let them know that you have arrived and will begin the inspection. Third, begin the inspection at the exterior of the building and work systematically throughout the inside of the building, beginning in the lowest level and working up. Fourth, conduct an exit interview with the contact person. Last, write a formal report on the inspection.

General Overview

Circle the property and observe all four sides of the building when you arrive. You are looking for any obvious access or storage problems and any new construction since your last visit. You are confirming the location of hydrants, sprinkler/standpipe hook-ups, and other outside features with your pre-incident survey sheet.

Park the fire apparatus in a location that does not disrupt the business and allows the fire company to respond if they are dispatched to an emergency. Some departments require that one fire fighter remain with the apparatus to both listen to the radio and protect the rig from vandals or thieves. The apparatus should not be parked in a fire lane. It is hard to convince the business owner to comply with the fire prevention code when the fire engine is violating the code by sitting in a fire lane when there is no emergency.

Meet with the Representative

Enter the business through the main door and make contact with the appropriate representative. If you have called ahead to schedule this visit, you already have a name and office location. Introduce your crew and briefly explain the goal of this visit. This is a great time to review and update all of the contact names, phone numbers, and information found in your pre-incident survey sheet. Ask to have a representative with the appropriate access cards and keys accompany the inspection team.

Company Tips

Conducting Inspections and Surveys
A managing fire officer has a four-step system that meets the organization's need for fire company inspections and pre-incident surveys and helps the company focus on what is important.

- Step 1: Schedule the inspections based on use group or occupancy. For example, all of the service stations and auto repair facilities are scheduled for November.
- Step 2: Have an in-station drill to review the fire prevention code sections that apply to the types of items that would be found during a typical code compliance inspection in that type of occupancy. In October, the fire officer schedules an in-station drill to review the fire code requirements for a service station or automotive repair shop and also delegates a senior fire fighter to schedule the inspections.
- Step 3: Break the fire company into two-person inspection teams. One of the team members would focus on the code enforcement, and the other would update the pre-incident survey form.
- Step 4: The company reviews their findings. This would be the time to discuss any changes that are required on incident action plans, based on the inspections. For example, if an auto repair shop has added a vehicle spray booth, the plan for that occupancy should be revised.

Inspecting from the Outside in, Bottom to Top

A fire company–level ongoing compliance inspection confirms the built-in fire protection systems are fully operational and makes sure the area is free of fire ignition sources. This begins with a walk around the exterior of the premises. Confirm that the address is present and properly identified on both the front and the rear of the building. If the building has a fire department connection, is it free from foreign objects inside the piping, and are the threads in workable condition? Verify that all means of access and egress are clear and in proper operating order; there is no more important reason for this type of inspection. Exit problems require immediate correction.

The fire officer should ensure that the location of dumpsters does not present a fire hazard. Outside storage buildings should be checked for compliance, particularly with the storage of hazardous materials. Are required markings, like NFPA 704, present?

After the outside walk-around, go to the basement, where you will find utility rooms, fire pumps, fire protection system control valves, back-up generators, and laundry rooms. These areas are susceptible to improper storage of combustibles near electrical panels, open junction boxes, improperly stored hazardous material, and blocked exits.

The fire officer should systematically work through the building, checking for fire safety issues. Like the exterior and the basement, the ability of occupants to quickly exit a building is a primary concern. Problems may include inventory stock that is blocking exits, locked doors, and exit and egress lights that are not working.

Look for conditions that are prone to starting fires: open electrical wiring, use of extension cords, and storage of combustibles too close to heat sources. Watch for storage of flammable liquids, as well as the use of candles, portable heating units, and fireplaces or wood stoves.

Last, consider items that would assist in extinguishing a fire once it has started. This would include the clearance between sprinkler heads and objects, the condition of smoke and heat detectors, and the operability of fire extinguishers. The fire officer should check to ensure that fire doors are not propped open and are in operable condition. If the structure is equipped with a commercial kitchen, the hood and duct system and fire suppression system should be checked for compliance.

▉ Exit Interview

It is important to wrap up your ongoing compliance inspection/pre-incident survey by meeting with the owner or designated representative to review what was found and issue any required correction orders. Remember that one of your roles is to be a partner with the business. A written report needs to be completed, with one copy going to the occupant.

▉ Writing the Inspection/Correction Report

There are different types of inspection/correction reports. Many use a check-off system allowing the officer to put a check next to the corresponding deficiency. This allows the citation to have the appropriate code without having to look up each violation. If violations are not corrected on a re-inspection, a notice of hazard is described in a narrative rather than noted in a check-off form.

The report needs to clearly describe any needed corrections and quote the appropriate sections of the code or ordinance. Some communities have developed report forms that list the most commonly occurring violations.

You should review this report with the owner or representative. If violations are found during the inspection, you have to explain to the responsible individual what needs to occur to correct the problem. You retain the original report for follow-up purposes. The owner or representative should sign the form and keep a copy of the report. The report should also be forwarded to the fire prevention division.

Life-threatening hazards, such as locked exits, must be corrected immediately. Less critical issues can be corrected within a reasonable time period, generally 30 to 90 days. You should arrange for any needed follow-up inspections to verify that the corrections have been made. Here is the procedure used by one fire department for non–life-threatening code enforcement issues:

- Fails first fire company–level inspection: Schedule follow-up inspection in 30 days.
- Fails second fire company–level inspection: Follow-up inspection in 15 to 30 days.
- Fails third fire company–level inspection: Issue sent to the fire prevention division for resolution. Depending on the nature of the issue and the history of this occupancy, the fire prevention division contacts the owner and makes an inspection within 2 to 30 days.
- Fails first fire prevention division inspection: Fire marshal issues a notice of violation or correction order. Time to comply with the notice or order is determined by the

situation and by local regulations. The owner can file for appeal or variance within 10 days of the first fire prevention division inspection.

- Fails second fire prevention division inspection: Second notice issued by fire marshal. Generally, time to comply is shorter. The owner is warned of more severe consequences if the issues remain unresolved.
- Fails third inspection: Follow-up inspection in days; building representative may be subject to a misdemeanor charge with potential fines, or jail, or both. The occupancy may be required to cease operations until the matter is resolved.
- Fails fourth inspection: Fines and legal action are initiated.

▉ General Inspection Requirements

The fire code includes general fire and life-safety requirements that apply to every type of occupancy. These items should be checked during every inspection. The general requirements include properly operating exit doors and unobstructed paths to fire exits. The general requirements also require built-in fire protection systems, such as automatic sprinklers and fire detection systems, as well as portable fire extinguishers, to be regularly inspected and properly maintained in operational condition. The fire code also includes general precautions that are required in the use or storage of hazardous, flammable, or combustible goods. Some fire codes include topics relating to emergency planning and conducting of fire drills, as well as features like rapid entry key boxes for fire department use.

▉ Access and Egress Issues

The fire department confirms that there is sufficient means of egress for the occupants. These provisions of the fire code are often violated because improper storage causes the exits to be obstructed.

Proper storage practices mean that the goods for sale or the items used by the business or industrial process are safely stored in accordance with the fire prevention code. In the example of a retail store during the holidays, the fire hazard is greatly increased when the store is jammed to the ceiling with stock. In some cases, a fire company may discover a very hazardous condition due to excessive, illegal, or dangerous storage practices. This situation requires immediate corrective action and, if available, assistance from the fire prevention division or the fire marshal's office.

▉ Exit Signs and Egress Lighting

Equally important in occupant evacuation are exit signs and egress lights. Exit signs are those that indicate the direction of exit to occupants during a fire. Egress lights light the path of the exit. These are particularly important when the occupants may be unfamiliar with the location of exits other than the main entrance, such as at theaters and nightclubs. Often, officers find exit lights that are burned out or the back-up battery no longer works. The same is true for egress lights. The occupant is required to maintain documentation of monthly checks performed on these systems.

▉ Portable Fire Extinguishers

Almost every building has portable fire extinguishers, which are provided for the occupants to control incipient fires. The fire prevention code describes requirements for the size, type, and locations of extinguishers needed in various occupancies, usually referring to NFPA 10, *Standard for Portable Fire Extinguishers*, for the details. Verify extinguishers are of the appropriate size and type. In addition to visually inspecting for physical damage, look to see that the instructions are visible, the safety or tamper seal is present, there is current inspection and testing documentation, and the pressure gauge is in the normal range.

Inspecting Built-in Fire Protection Systems

NFPA's *Fire Protection Systems: Inspection, Test and Maintenance Manual* provides guidance for inspection, testing, and maintenance. Each chapter includes inspection forms that can be used for 20 different types of built-in fire protection systems. The officer should check to see that all fire department connection caps are in place, and that they are unobstructed and accessible. For sprinkler systems, extra heads and a changing wrench should be readily available. The officer should ensure that all control valves are in the open position and are locked or have a supervisory alarm to protect the system from being accidentally shut down.

Only properly trained personnel can test fire protection systems. Los Angeles, Phoenix, and Fairfax County encountered embarrassing situations when they started a retesting program. Using a pumper to charge a standpipe system blew one standpipe right out of the building wall. Another pressure test filled a mercantile basement with hundreds of gallons of water. There are issues with liability and repair costs when the local fire company is operating built-in fire protection systems when there is no fire emergency. In some retesting programs, the fire prevention bureau requires a licensed fire protection contractor to perform the retest, and the contractor submits a certified report.

Electrical

See that no combustibles are stored around the electrical panels and that the panel covers have not been removed. The use of extension cords is often abused. Extension cords are designed to be used with items on a temporary basis, such as with a drill or a fan. They are not designed to be used as permanent wiring, such as with a refrigerator or lighting. Check for multiple extension cords being used in sequence and open electrical boxes for outlets, switches, or junctions. The use of open splices in electrical wiring is unacceptable.

Special Hazards

The nature of an industrial process by itself may be hazardous. For most processes with very high hazards, the occupancy may be required to have a **fire prevention division or hazardous use permit**. This is a local government permit that is renewed annually after the fire prevention division performs a code compliance inspection. The local fire company may be the first to discover a high-risk occupancy, like a Los Angeles fire company that discovered a fireworks warehouse during a routine inspection. This is especially true if your district includes "on spec" or "spec built" industrial parks. Owners construct buildings that are generic spaces and rent out space to a variety of tenants. This type of space has high occupant turnover, each with different fire risk factors. A space that was used for storing building materials last year may be used for assembling computer hardware equipment this year and be an auto repair shop next year.

Hazard Identification Signs

When required, visible hazard identification signs meeting NFPA 704 conditions should be placed on stationary containers and above-ground tanks and at entrances to locations where hazardous materials are stored, dispensed, used, or handled. Individual containers, cartons, or packages must be conspicuously marked or labeled. Material safety data sheets must be readily available on the premises for regulated material.

Selected Use Group–Specific Concerns

Each use group classification involves special concerns and considerations in addition to the general concerns. Some of the more significant considerations are described in the following sections.

Public Assembly

The goal for a code compliance inspection in public assembly occupancy is to ensure that all of the access and egress pathways are clear and in good order **Figure 12-16 ▾**. The number and the size of exits should have been approved before the notice of occupancy was given to the business.

Figure 12-16 All access and egress pathways must be clear.

Confirm that the exits are not blocked or locked and are in good working order, and that there has not been a change in the layout. The code specifies when panic hardware must be installed and the direction of swing of the doors. Like the number of exits, these are not usually an issue unless remodeling has occurred. Storage of combustibles under exit stairs is not permitted.

Exit lighting and egress lights are an important consideration for assembly-type occupancies. Typically, occupants are not familiar with exit locations, so methods to direct occupants to the exits are essential. Lighting requires battery back-up or a back-up generator. Evacuation plans may also be required.

Confirm use of flame-resistant material in the curtains, draperies, and other decorative materials. Watch for unacceptable storage of hazardous or flammable materials.

Commercial exhaust hoods and ducts also need to be inspected. The fire officer should check for unvented, fuel-fired heating equipment. These occupancies are also prone to propping open fire doors and smoke barriers.

Business

Make sure that access and egress pathways are clear and in good order. This includes exits that are blocked with office furnishings. Exit egress lights are problematic. Most occupants are familiar with building design, but fire protection equipment, such as fire extinguishers, often is not maintained properly.

These occupancies are notorious for inappropriate use of electrical cords. In older buildings, open electrical boxes and spliced wiring tend to be a problem. These occupancies also tend to store small amounts of flammable liquids and other hazardous materials inappropriately.

Educational

A single school fire produced the swiftest example of the catastrophic theory of reform. The Our Lady of Angels School burned on December 1, 1958. Despite heroic efforts by the Chicago Fire Department, 95 people died, most of them children. NFPA's Percy Bugbee reported that within days of the disaster, state and local officials ordered the immediate inspection of schools. The NFPA partnered with Los Angeles Fire Marshal Raymond Hill to conduct 172 fire tests to develop new fire safety regulations for schools. *Operation School Burning* provided smoke and heat data that guided the technical committee in drafting new standards. It was reported that major improvements had been made in 16,500 schools throughout the country within a year of publishing the new standards.

Like assembly-type occupancies, exit paths are essential for occupant safety. Fire officers should ensure that doors are not blocked or locked, preventing a safe exit. When inspecting these occupancies, the fire officer must also ensure that built-in fire protection systems are in working order and are properly maintained. Confirm that the evacuation plan is up-to-date and practiced. Staff should also be trained in fire emergency procedures.

Factory Industrial

Factory industrial occupancies include many of the same hazards as business ones. They also may have processes that are particular to that type of factory; these should be identified for specific fire hazards. Factories often have improperly stored combustibles. They are also prone to having fire protection systems that have been shut off or are improperly maintained and fire doors that are propped open.

Hazardous

Special consideration should be given to preventing fires from starting. Codes specify what items and quantities can be stored in a facility. They also specify the markings that are required. Signs should be posted that prohibit open flames and smoking. Fire doors and smoke barriers should be in working order. Fire protection systems should be regularly inspected and maintained. Fire emergency plans should be in place, along with evacuation plans, which should be practiced.

Health Care

Institutional occupancies include hospitals, nursing homes, and similar occupancies where the occupants are likely to require special assistance to evacuate. Built-in fixed fire protection systems, including automatic sprinklers, smoke detection systems, and sophisticated alarm systems, are required to provide additional evacuation time. Fire detection systems should also be checked for proper maintenance and operation. These facilities should have a fire evacuation plan that should be up-to-date and practiced.

Mercantile

Mercantile occupancies include retail shops and stores selling stocks of retail goods. Stand-alone 24-hour convenience markets, department stores, and big-box home improvement and discount stores are all mercantile occupancies. Mercantile fires experience a higher-than-average number of fire fighter line-of-duty death. Like in the assembly classification, the occupants of mercantile occupancies may not be aware of the location of exits. Ensuring that exits are properly marked and lit is a prime concern.

Housekeeping is also a concern in these occupancies. Frequently, exits and aisles are blocked with merchandise, sometimes to the point of covering sprinkler heads, standpipe connections, and fire extinguishers. Exits may also be locked. The storage of flammable liquids might also be noted.

Residential

When the fire officer is inspecting residential units, only common areas can be inspected unless otherwise requested by the occupant. Hallways, utility areas, entryways, common laundry rooms, and parking areas are all open for inspection. Residential concerns vary by specific occupancy type; however, in general, exits are a concern. Fire doors are routinely propped open and may have items stored in their path. Exit and egress lighting are often not maintained properly.

Fire protection systems also need to be checked for adequate maintenance and operation. Vandals often remove caps and place objects inside standpipes. Valves may be closed, preventing the system from operating. Smoke alarms may have been removed. Hose cabinets may have damaged threads.

Preventing fires from igniting is a prime concern. Poor housekeeping, such as bags of trash in hallways, are inviting for fire setters. Managers of garden apartments often store mowers and gasoline under stairways both inside and outside the structure.

Special Properties

This class has a wide variety of hazards. Before inspecting a special-use occupancy, the officer needs to review applicable code requirements that are specific to the occupancy type.

Detention

A working sprinkler system is essential because of the inability of occupants to protect themselves from fires. Because of the security measures of the facility, the fire department is often delayed in accessing the seat of the fire. Verify that the fire department hook-up is accessible and in working order. Confirm that the standpipe threads match those of the hose used by the department.

Many of these facilities have standpipe systems because of the travel distance to the exterior. Often, they have "house lines," which are preconnected 1 1/2" (38-mm) hose lines attached to the standpipe connection. These lines are used by workers on scene and may be in a locked metal cabinet. Like the standpipe connections, fire extinguishers may be locked in a metal cabinet; however, they should be properly marked.

Typically, these facilities do not have a great deal of combustibles, and the housekeeping is excellent.

Storage

Storage facilities often have sprinkler and standpipe systems, which should be checked for proper inspection and maintenance records. Frequently, this has not been completed as required. Storage facilities regularly stock material too close to the sprinkler heads. Fire extinguishers may have been removed and set on the ground or may be damaged from forklifts.

These facilities may also store hazardous materials, sometimes for brief periods. Determine the normal amount of on-site hazardous materials, as well as the maximum amount that could be present.

These facilities often block access and egress paths. The need to use space often causes the occupants to partially block exits with pallets of material. The side exit doors may also be locked, and the exit and egress lights may not work.

Mixed

Each mixed-use building is considered individually in its requirements, but the building must meet the most stringent requirement of all of the occupancies inside. This might include a building that has a fireworks storage area in one unit and a barbershop in another. Although each occupancy must meet the codes that apply to that specific occupancy, the building in general must meet the most stringent standard. In this case, it would be the fireworks storage.

Emergency Management and Business Continuity Plans

Natural and man-made catastrophes within the last decade resulted in significant changes for NFPA 1600, *Standard on Disaster/Emergency Management and Business Continuity Programs*. The standard includes five sections: prevention, mitigation, preparedness, response, and recovery.

The fire department pre-incident plan is a component of a business continuity program. Emergency management and business continuity plans are established for public, not-for-profit, and private entities, including risk management, security, and loss prevention practices. The fire department is a stakeholder in an organization's program. A business continuity program resembles an incident action plan within the National Response Framework.

Risk Assessment

The public, not-for-profit, and private entities conduct a **risk assessment** to identify hazards, monitor those hazards, determine the likelihood of their occurrence, and assess the vulnerability of people, property, the environment, and the entity itself to those hazards. These hazards are subdivided into natural hazards, like an earthquake or hurricane; accidental and intentional human-caused events; and technologically caused events.

Business continuity planning (BCP) practices expanded with the migration from paper to electronic records, with the 1989 San Francisco earthquake requiring the first large scale recovery of digital records after a disaster. When 15 million ft^2 (1.4 million m^2) of office space were destroyed at the World Trade Center in 2001, most of the financial organizations were able to operate from alternative centers minutes after the attack.

As stakeholders, local government agencies may help in providing an impact analysis on the health and safety of persons in the affected area. The fire department will also be involved in the analysis of properties, facilities, and infrastructure. If the property handles reportable levels of hazardous materials, this analysis may have been done as part of a Local Emergency Planning Council activity.

Senior officials of public, not-for-profit, and private entities will consider the regulatory and contractual obligations as well as the reputation or confidence of the organization. Each entity will evaluate the economic and financial conditions when developing the operations and delivery of services after a disaster or major emergency.

All parties will be assessing the impact on the environment. Some incidents will meet the federal government's definition of "incidents of national consequence" whereas others may have regional or international considerations.

Incident Prevention

A strategy is developed to prevent an incident that threatens people, property, and the environment, based on the risk assessment. The public, not-for-profit, and private entities are expected to evaluate and update the risk assessment regularly, adjusting the level of preventative measures to be commensurate with the risk.

Mitigation

Mitigation includes measures taken to limit or control the consequences, extent, or severity of an incident that cannot be reasonably prevented. The mitigation strategy is based on the results of hazard identification and risk assessment, impact analysis, program constraints, operational experience, and cost-benefits analysis.

Mitigation strategies include interim and long-term actions to reduce organizational vulnerability.

Resource Management and Logistics

The emergency/disaster management and business continuity plan document requires that the public, not-for-profit, and private entities provide the resources needed to run the program. They will need to locate, acquire, store, distribute, maintain, test, and account for services, personnel, resources, materials, and facilities used to support the program.

The fire department may provide subject matter experts or specialized resources as part of the program. It is the responsibility of the entity to develop a detailed resource storage, deployment, and recovery plan. This may include reimbursing for fire department time or resources. For example, who will pay for personal protective clothing that has to be replaced after a hazardous materials incident?

Publishing the Plan

The emergency/disaster management and business continuity plan document is published in six parts: strategic plan, emergency operations/response plan, prevention plan, mitigation plan, recovery plan, and continuity plan.

The emergency operations/response plan assigns responsibilities for carrying out specific actions in an emergency. This plan should link with the fire department incident action plan and be consistent with NFPA 1561, *Standard on Emergency Services Incident Management System*. The mitigation plan establishes interim and long-term actions to reduce the impact of hazards that cannot be eliminated. The recovery plan provides the short- and long-term priorities for restoration of functions, services, resources, facilities, and programs.

Training, Exercises, and Evaluation

The public, not-for-profit, and private entities are expected to develop a training program so that all members will obtain the skills to maintain and execute the program.

There are periodic reviews, testing, and exercises. There may be federal or industry requirements mandating the frequency and size of a simulation exercise. Postincident analysis and reports, performance evaluation, and lessons learned can be used to review the plan. As a stakeholder, the fire officer may be participating in or evaluating these activities.

Summary

Pre-incident planning and code enforcement require the fire officer to perform within the education, enforcement, and engineering areas of fire prevention. It requires additional training beyond the NFPA Fire Officer I and II that is covered in this textbook.

Significant change in this activity is the involvement of the fire department as a stakeholder in disaster/emergency management and business continuity programs, as outlined in NFPA 1600. This evolution will prepare business and industry to provide a coordinated response in the event of a disaster or emergency that is consistent with the fire department incident management plan as overseen by the federal National Response Framework.

You Are the Fire Officer: Conclusion

You start a reference file on the factory and company town, using a copy of the basic site plan and the preliminary information provided by the developer. You call the fire prevention division to consult with the person performing the plan review for the project.

You assign each fire fighter to research the special or unique characteristics of one type of occupancy that will be constructed. The driver-operator is assigned the task of identifying the fire flow requirements and evaluating the water supply and hydrant locations.

Wrap-Up

Chief Concepts

- NFPA 1620 provides a six-step method of developing a pre-incident plan.
  - Step 1: Evaluate physical elements and site considerations.
  - Step 2: Evaluate occupant considerations.
  - Step 3: Evaluate fire protection systems and water supply.
  - Step 4: Evaluate special hazards.
  - Step 5: Evaluate emergency operation considerations.
  - Step 6: Evaluate special or unusual characteristics of common occupancy.
- The state, commonwealth, or province determines the range and scope of local community fire code enforcement. The local community adopts an ordinance or a regulation to establish a fire code.
- Automatic sprinkler systems are wet-pipe or dry-pipe versions, with special deluge and pre-action systems to meet unique requirements.
- There are three classes of standpipes, based on the size of the hose coupling outlets and the intended use of the standpipe.
- There are five building construction classifications.
- Every inspection is completed when the fire officer conducts an exit interview with the owner or representative and leaves a copy of a fire inspection report.

Hot Terms

Adoption by reference Method of code adoption in which the specific edition of a model code is referred to within the adopting ordinance or regulation.

Adoption by transcription Method of code adoption in which the entire text of the code is published within the adopting ordinance or regulation.

Authority having jurisdiction An organization, office, or individual responsible for enforcing the requirements of a code or standard, or for approving equipment, materials, an installation, or a procedure.

Automatic sprinkler system A system of pipes with water under pressure that allows water to be discharged immediately when a sprinkler head operates.

Catastrophic theory of reform When fire prevention codes or firefighting procedures are changed in reaction to a fire disaster.

Construction type The combination of materials used in the construction of a building or structure, based on the varying degrees of fire resistance and combustibility.

Fire prevention division or hazardous use permit A local government permit renewed annually after the fire prevention division performs a code compliance inspection. A permit is required if the process, storage, or occupancy activity creates a life-safety hazard. Restaurants with more than 50 seats, flammable liquid storage, and printing shops that use ammonia are three examples of occupancies that may require a permit.

Floor plan View of a building's interior. Rooms, hallways, cabinets, and the like are drawn in the correct relationship to each other.

Fuel load The total quantity of all the combustible products that are within a room or space.

High-risk property Structure that contains the potential for a catastrophic property or life loss in the event of a fire.

High-value property Structure that contains equipment, materials, or items that have a high replacement value.

Masonry wall May consist of brick, stone, concrete block, terra cotta, tile, adobe, precast, or cast-in-place concrete.

Mini/max codes Codes developed and adopted at the state level for either mandatory or optional enforcement by local governments; these codes cannot be amended by local governments.

Mitigation Measures to be taken to limit or control the consequences, extent, or severity of an incident that cannot be reasonably prevented.

Model codes Codes generally developed through the consensus process with the use of technical committees developed by a code-making organization.

Occupancy type The purpose for which a building or a portion thereof is used or intended to be used.

Ongoing compliance inspection Inspection of an existing occupancy to observe the housekeeping and confirm that the built-in fire protection features, such as fire exit doors and sprinkler systems, are in working order.

Ordinance A law of an authorized subdivision of a state, such as a city, county, or town.

Plot plan A representation of the exterior of a structure, identifying doors, utilities access, and any special considerations or hazards.

Wrap-Up

Pre-incident plan A written document resulting from the gathering of general and detailed data to be used by responding personnel for determining the resources and actions necessary to mitigate anticipated emergencies at a specific facility.

Regulations Governmental orders written by a governmental agency in accordance with the statute or ordinance authorizing the agency to create the regulation. Regulations are not laws but have the force of law.

Risk assessment Involves identifying hazards, monitoring those hazards, determining the likelihood of their occurrence, and assessing the vulnerability of people, property, the environment, and the entity itself to those hazards.

Standpipe system An arrangement of piping, valves, hose connections, and allied equipment installed in a building or structure, with the hose connections located in such a manner that water can be discharged in streams or spray patterns through attached hose and nozzles, for the purpose of extinguishing a fire, thereby protecting a building or structure and its contents in addition to protecting the occupants. This is accomplished by means of connections to water supply systems or by means of pumps, tanks, and other equipment necessary to provide an adequate supply of water to the hose connections.

Use group Found in the building code classification system whereby buildings and structures are grouped together by their use and by the characteristics of their occupants.

Fire Officer *in Action*

Every year, about one-third of the tenants in the industrial park and the strip-style shopping center change owners. Updating the pre-incident plans, you discover that some of the newer tenants are living in their strip shopping center stores. Some industrial park tenants are very busy between 9 P.M. and dawn, but are closed up tight during the day. Other industrial park tenants have built mezzanines or internal spaces without permits, inspections, or updates of the built-in fire protection systems. You arrange for a meeting with the fire prevention division to see what can be done to ensure fire fighter safety and occupant survival.

This chapter describes two different roles for the company officer. In pre-incident planning the officer is considering what happens if the building burns, collapses, or expels hazardous materials. These considerations are developed into an emergency/disaster management and business continuity document that becomes part of an incident action plan in the event of an incident at that building.

The second role is to perform life safety inspections to ensure that the building is meeting the appropriate fire prevention code requirements, including the status of built-in fire protection features and general housekeeping.

1. What is a class III standpipe system?
   A. Provides both 1.5" (38-mm) and 2.5" (64-mm) fire hose connections
   B. Combination sprinkler system with 1.5" (38-mm) fire hose connections
   C. Provides a 2.5" (64-mm) connection with a reducer and 150' (48 meters) of 1.5" (38-mm) fire hose
   D. A class I system with pressure-reducing valves

2. A structure constructed entirely of wood is a _____ occupancy.
   A. type II
   B. type V
   C. type W
   D. type III

3. A nightclub falls into the _____ occupancy.
   A. mixed-use
   B. business
   C. entertainment
   D. assembly

4. _____ is an emergency management task that includes measures taken to limit or control the consequences, extent, or severity of an incident that cannot be reasonably prevented.
   A. A business continuity plan
   B. Mitigation
   C. Risk assessment
   D. Incident prevention

Budgeting and Organizational Change

NFPA 1021 Standard

Fire Officer I

4.1.1 **General Prerequisite Knowledge.** The organizational structure of the department; geographical configuration and characteristics of response districts; departmental operating procedures for administration, emergency operations, incident management system and safety; departmental budget process; information management and recordkeeping; the fire prevention and building safety codes and ordinances applicable to the jurisdiction; current trends, technologies, and socioeconomic and political factors that affect the fire service; cultural diversity; methods used by supervisors to obtain cooperation within a group of subordinates; the rights of management and members; agreements in force between the organization and members; generally accepted ethical practices, including a professional code of ethics; and policies and procedures regarding the operation of the department as they involve supervisors and members. [p 245–258]

4.3 **Community and Government Relations.** This duty involves dealing with inquiries of the community and communicating the role, image, and mission of the department to the public and delivering safety, injury, and fire prevention education programs, according to the following job performance requirements. [p 245–253]

4.3.1 Initiate action on a community need, given policies and procedures, so that the need is addressed. [p 245–253]

(A) Requisite Knowledge. Community demographics and service organizations, as well as verbal and nonverbal communication, and an understanding of the role and mission of the department. [p 245–253]

(B) Requisite Skills. Familiarity with public relations and the ability to communicate verbally. [p 245–260]

4.4.1 Recommend changes to existing departmental policies and/or implement a new departmental policy at the unit level, given a new departmental policy, so that the policy is communicated to and understood by unit members. [p 259–260]

(A) Requisite Knowledge. Written and oral communication. [p 259–260]

(B) Requisite Skills. The ability to relate interpersonally and to communicate change in a positive manner. [p 259–260]

4.4.3 Prepare a budget request, given a need and budget forms, so that the request is in the proper format and is supported with data. [p 256–258]

(A) Requisite Knowledge. Policies and procedures and the revenue sources and budget process. [p 245–258]

(B) Requisite Skill. The ability to communicate in writing. [p 245–258]

4.4.4 Explain the purpose of each management component of the organization, given an organization chart, so that the explanation is current and accurate and clearly identifies the purpose and mission of the organization. [p 245–258]

(A) Requisite Knowledge. Organizational structure of the department and functions of management. [p 245–258]

(B) Requisite Skills. The ability to communicate verbally in a clear and concise manner. [p 245–258]

Fire Officer II

5.1.1 **General Prerequisite Knowledge.** The organization of local government; enabling and regulatory legislation and the law-making process at the local, state/provincial, and federal levels; and the functions of other bureaus, divisions, agencies, and organizations and their roles and responsibilities that relate to the fire service. [p 246–249, 251–252, 259]

5.1.2 **General Prerequisite Skills.** Intergovernmental and interagency cooperation. [p 246–249, 251–252, 259]

5.3 **Community and Governmental Relations.** This duty involves dealing with inquiries of allied organizations in the community and projecting the role, mission, and image of the department to other organizations with similar goals and missions for the purpose of establishing strategic partnerships and delivering safety, injury, and fire prevention education programs, according to the following job performance requirements. [p 247–249, 251]

5.3.1 Explain the benefits to the organization of cooperating with allied organizations, given a specific problem or issue in the community, so that the purpose for establishing external agency relationships is clearly explained. [p 247–249, 251]

(A) Requisite Knowledge. Agency mission and goals and the types and functions of external agencies in the community. [p 247–249, 251]

(B) Requisite Skill. The ability to develop interpersonal relationships and to communicate orally and in writing. [p 247–249, 251]

5.4 **Administration.** This duty involves preparing a project or divisional budget, news releases, and policy changes, according to the following job performance requirements. [p 245–260]

5.4.1 Develop a policy or procedure, given an assignment, so that the recommended policy or procedure identifies the problem and proposes a solution. [p 245–260]

(A) Requisite Knowledge. Policies and procedures and problem identification. [p 245–260]

(B) Requisite Skill. The ability to communicate in writing and to solve problems.

5.4.2 Develop a project or divisional budget, given schedules and guidelines concerning its preparation, so that capital, operating, and personnel costs are determined and justified. [p 256–258]

(A) Requisite Knowledge. The supplies and equipment necessary for ongoing or new projects; repairs to existing facilities; new equipment, apparatus maintenance, and personnel costs; and appropriate budgeting system. [p 256–258]

(B) Requisite Skill. The ability to allocate finances, to relate interpersonally, and to communicate orally and in writing. [p 245–258]

5.4.3 Describe the process of purchasing, including soliciting and awarding bids, given established specifications, in order to ensure competitive bidding. [p 253, 255–256]

(A) Requisite Knowledge. Purchasing laws, policies, and procedures. [p 255–256]

(B) Requisite Skill. The ability to use evaluative methods and to communicate orally and in writing. [p 253, 255–256]

5.4.6 Develop a plan to accomplish change in the organization, given an agency's change of policy or procedures, so that effective change is implemented in a positive manner. [p 259–260]

(A) Requisite Knowledge. Planning and implementing change. [p 259–260]

(B) Requisite Skill. The ability to clearly communicate orally and in writing.

Introduction to Fire and Emergency Services Administration (FESHE) Course Outcomes

2. Explain the need for effective communication skills both written and verbal. [p 248–249, 255–256, 259–260]
6. Evaluate methods of managing available resources. [p 247–253]
7. Identify roles and responsibilities of leaders in organizations. [p 259]
9. Identify and assess safety needs for both emergency and nonemergency situations. [p 256–257]

Knowledge Objectives

After studying this chapter, you will be able to:

- List the supplies and equipment necessary for ongoing and new projects and repairs to existing facilities; new equipment, apparatus maintenance, and personnel costs; and an appropriate budgeting system.
- Understand purchasing laws, policies, and procedures.
- Describe the process of introducing organizational change.

Skills Objectives

After studying this chapter, you will be able to:

- Prepare a budget.
- Propose an organizational change

I t started with a wireless medical monitor, the size of a small Band-Aid, that transmits data on pulse, blood pressure, respirations, and level of oxygen in the blood. Another company constructed an integrated personal alert safety system (PASS) that amplifies the medical monitor signal, adding a unique identifier and three-dimensional global positioning system data. A relational database and mapping system automatically tracks every fire fighter on the incident by location and physical status. The fire department is applying for a grant to be one of the first beta-testers of this unique system.

The fire chief wants to establish a Fire Fighter Status and Location (FFSL) unit that will respond to every working incident in order to track members working on the fire ground. The system can improve incident scene fire fighter health and safety, both within the incident and at the rehab sector. There is an opportunity to pay for the start-up and the first 3 years of operation through a grant from a private organization.

You have been tasked to write up the budget submission and describe the implementation.

1. What do you have to consider when preparing a budget?
2. How do you prepare a budget proposal?
3. What procedures can you use to facilitate organizational change?

Introduction to Budgeting

A **budget** is an itemized summary of estimated or intended revenues and expenditures. **Revenues** are the income of an organization from all sources intended for the payment of expenses. **Expenditures** are the money spent for goods or services. Every fire department has a budget that defines the funds that are available to operate the organization for 1 year. The budget process is a cycle:

1. Identification of needs and required resources
2. Preparation of a budget request
3. Local government and public review of requested budget
4. Adoption of an approved budget
5. Administration of approved budget, with quarterly review and revision
6. Close out of budget year

Budget preparation is a technical and a political process. The technical part relates to calculating the funds that would be required to achieve different objectives, whereas the political part is related to elected officials making the decisions about which programs should be funded among numerous alternatives.

The fire department represents one of the larger sections of a municipal budget, along with schools, police, and public works. The elected officials determine tax rates to generate the revenue. They allocate the available funds among the different organizations and programs. Federal, state, and local regulations also influence the budget process.

The Budget Cycle

The budget document describes where the revenue comes from (input) and where it goes (output) in terms of personnel, operating, and capital expenditures. Annual budgets usually apply to a **fiscal year**, such as starting on July 1 and ending on June 30 of the following year. Using these dates, the budget for fiscal year 2013, referred to as FY13, would start on July 1, 2012, and end on June 30, 2013. The process for developing the FY13 budget would start in 2011, a full year before the beginning of the fiscal year **Table 13-1 ▶**.

Table 13-1   **Fiscal Year 2013 Timeline**

August–September 2011	Fire station commanders, section leaders, and program managers submit their FY13 requests to the fire chief. These are for proposed new programs, expensive new or replacement capital equipment, and physical plant repairs. This is when a fire officer would request funds for a new storage shed or a replacement dishwasher. Documentation to support replacing a fire truck or purchasing new equipment, such as an infrared camera or a hydraulic rescue tool, would be submitted, and a proposal for a special training program would be prepared. These documents are prepared by fire officers throughout the organization.
September 2011	The fire chief reviews the requests and develops a prioritized budget wish list for FY13. This is when department-wide initiatives or new programs are developed. Purchasing a new radio system would be on the wish list.
October 2011	The budget office distributes FY13 preparation packages. Included are the forms for personnel, operating, and capital budgets. The senior local administrative official (mayor, city manager, county executive) provides guidelines. For example, if the economic indicators predict that tax revenues will not increase during FY13, the guidelines could restrict increases in the operating budget to 0.5 percent or freeze the number of full-time equivalent positions.
November 2011	Deadline for agency heads to submit their FY13 budget requests to the budget director.
December 2011	The budget director assembles all of the proposals. Any request above the guidelines must be the result of a legally binding regulation, agreement, or settlement. Other submissions above the limit will need the support of elected officials.
January 2012	Local officials receive the preliminary budget from the budget director. There may be two or three alternative proposals that require a decision from the officials. Those decisions are made, and the proposal is completed. This is the "Proposed Fiscal Year 2013 Budget."
February or March 2012	The proposed budget is made available to the public for comment. Many cities make the proposed budget available on an Internet site, inviting public comment to the elected officials. In smaller communities, the proposed budget may show up in the local newspaper.
March or April 2012	Local officials hold a hearing or town meeting to receive input on the FY13 proposed budget. Based on the hearings and on additional information from staff and local government employees, the elected officials debate, revise, and amend the budget. During this process, the executive fire officer may be called on to make a presentation to explain the department's budget requests, particularly if large expenditures have been proposed. The department may be required to provide additional information or submit alternatives as a result of the public budget review process.
May 2012	Local leaders approve the amended budget. This becomes the "Approved" or "Adopted" FY 2013 budget for the municipality.
July 1, 2012	FY13 begins. Some fire departments immediately begin the process of ordering expensive and durable capital equipment, particularly items that have long lead times for delivery.
October 2012	Informal first-quarter budget review, looking for trends of expenditures to identify any problems. For example, if the cost of diesel fuel has increased and 60 percent of the motor fuel budget has been exhausted by the end of September, there will be no funds remaining to buy fuel in January. Amounts that have been approved for one purpose may have to be diverted to a higher priority.
January 2013	Formal midyear budget review. The approved budget may be revised to cover unplanned expenses or shortages. In some cases, this is in response to a decrease in revenue, such as an unanticipated decrease in sales tax revenue. Expenditures for the remainder of the year have to be reduced.
April 2013	Informal fourth-quarter review. Final adjustments in the budget are considered after 9 months of experience. Year-end figures can be projected with a high degree of accuracy. Some projects and activities may have to stop if they have exceeded their budget, unless unexpended funds from another account can be reallocated.
May 2013	The finance office begins closing out the budget year and reconciling the accounts. No additional purchases are allowed from the FY13 budget. Purchases are deferred until the beginning of FY14.

Consider the timeline for a replacement pumper. If the initial request is submitted in August 2011 and it is approved at every stage of the process, the funding is not authorized until July 2012. Table 13-1 provides a description of the process.

After the budget is authorized, the department issues a request for proposal to build the vehicle. It takes time to evaluate the submissions and award a contract. Construction does not start until the contract is awarded. The replacement pumper requested August 2011 may not be in the fire station before late 2013.

Base and Supplemental Budgets

Most municipal governments use a **base budget** in the financial planning process. The base budget is the level of funding that would be required to maintain all services at the currently authorized levels, including adjustments for inflation and salary increases. There is a built-in process that assumes that the current level of service has already been justified and that the starting point for any changes in the budget should be the cost of providing the same services next year.

Budgets can increase or decrease from the base level. Proposed increases in spending to provide additional services are classified as **supplemental budget** requests. When budgets have to be reduced, the change is calculated as a percentage of the base budget. A 15 percent budget cut means that only 85 percent of the base budget expenditures for the current year will be approved for the next fiscal year. The department would need to reduce its expenditures by more than 15 percent, however, due to inflation and already negotiated salary increases.

Increases in the fire department budget require early notification and support of elected officials. A new program that involves additional employees or a major capital expenditure would require funds that would have to be found by increasing revenues or decreasing expenditures from some other part of the budget. These decisions are made by the elected officials, and, in order to present a balanced proposed budget, the budget director usually has a good idea of what the elected officials are likely to approve.

Elected officials are both advocates and gatekeepers in developing the local budget. They want to keep taxes down and, at the same time, deliver the services that the voters expect. They do this through direct involvement in the budget preparation process and participation in public hearings. Elected officials often request private face-to-face meetings with fire department leadership to discuss issues that are expected to be controversial.

Revenue Sources

The revenue stream depends on the type of organization that operates the fire department and the formal relationship between the organization and the local community. There is a wide diversity in organizations and in revenue stream sources. Each type of organization has a different process for obtaining revenue and authorizing expenditures.

Most municipal fire departments operate as components of local governments, such as towns, cities, and counties. A fire district is a separate local government unit that is specifically organized to collect taxes in order to provide fire protection. In some areas, fire departments are operated as regional authorities or an equivalent structure.

Local Government Sources

The mix of revenues available to local governments varies considerably because state governments set the rules for local governments. The fire officer needs to know the rules that apply to his or her jurisdiction. General tax revenues can be spent for any purpose that is within the authorized powers of the local government.

Some funds are restricted and can only be used for certain purposes. A **fire tax district** is created to provide fire protection within a designated area. A special fire protection tax is charged to properties within the service area, in addition to any other county or municipal taxes. For example, the fire tax could be $0.05 per $100 of value on property within the defined area. All of the revenue from the fire tax is dedicated to pay for the provision of fire protection services.

Revenues can also be restricted; for example, a fuel tax can be used only to build or maintain roads. Taxes that are levied to repay capital improvement bonds can be used only for that purpose. However, when a severe recession occurs, all funding sources not restricted by federal legislation may be taken from one part of the budget to cover another part of the budget.

The United States Census summarizes the sources of state and local government tax revenues in nine general areas. The top three local tax revenue sources represent about two thirds

of the total revenues collected by local governments. These sources are:

- General sales and gross receipts taxes
- Property taxes
- Individual income taxes

The other six listed sources of revenues are:

- Corporation net income taxes
- Motor fuel sales taxes
- Motor vehicle and operators' licenses
- Tobacco product sales taxes
- Alcoholic beverage sales taxes
- Other fees and taxes (e.g., service fees, motel occupancy fees)

Many fire departments also obtain revenue through direct fees for service. Fire departments that operate ambulances often obtain sufficient revenue from charges for patient treatment and transportation to offset the operating costs of the service. Direct revenue might also come from fees for fire prevention permits, special inspections, and other services that are provided to individual property owners or contractors.

Volunteer Fire Departments

There are as many different ways to fund volunteer fire departments as there are paint schemes on fire apparatus. Some volunteer departments are operated by municipal governments and are completely supported by local tax revenues, with all capital and operating expenses being handled by the jurisdiction. The budget process treats the volunteer fire department as a local government agency.

Many volunteer fire departments are organized as independent 501(c) nonprofit corporations. Some of these volunteer organizations raise their own operating funds through public donations or subscriptions and are entirely independent of local government. In other areas, local tax revenues are allocated to the volunteer corporations. Their relationship with a local jurisdiction may be through a memorandum of understanding, a contract for services, or a regional association of independent volunteer fire departments. Some jurisdictions pass local legislation or approve a charter that details the relationship, duties, services, and compensation that the volunteer fire department will provide to the community in return for the tax revenues.

Tax revenues support the entire operation of some volunteer fire departments, whereas others supplement their tax allocation with fundraising activities. Sometimes, the volunteer corporations own and operate the fire stations and apparatus, whereas other volunteer organizations occupy buildings and operate vehicles that are owned by the local government.

Volunteer organizations use a wide variety of fundraising methods. Direct fundraising often includes door-to-door solicitations or direct mail campaigns. Many volunteer corporations sponsor activities to generate revenue, such as dinners, bake sales, car washes, raffles, and bingo nights. Some volunteer departments generate revenue by renting out social halls or meeting rooms for private events.

Bingo and Other Gaming Activities

A traditional volunteer fire department fundraising method is to sponsor bingo or other gaming activities. These activities are state or regionally regulated, with specific procedures for conducting the games and handling the money. Volunteer fire departments have a revenue advantage over other operators because they do not have to pay their members to run the games.

Some departments hold an annual carnival or other entertainment event that provides revenue to the department. Holding the event on fire department property and using unpaid staff allows the fire department to minimize operating costs. Many of these fundraisers are major events in the community, creating good will and strong citizen awareness and support.

Real Estate and Portfolio Management

As a corporation, some fire departments have invested in real estate to generate revenue for fire department operations. Others have developed a financial portfolio that provides interest income to the department.

▮ Grants

Municipal and volunteer fire departments can also obtain funds by applying for grants. One of the larger sources of grant funds is the Assistance to Firefighters Grant Program, which distributes federal funds to local jurisdictions.

Although the Assistance to Firefighters Grant Program is the largest program, there are hundreds of smaller grant programs offered by the federal government, state governments, private institutions, nonprofit organizations, and for-profit companies. Grants are competitive and require the local fire department to meet specific eligibility and documentation requirements. Many of the grant programs also require the local jurisdiction to provide a percentage of matching funds.

Some state agencies offer grants to local responders from specially designated funds, such as a fee on motor vehicle permits, a surcharge on traffic fines, or a tax on fire insurance premiums.

The U.S. Fire Administration provides a four-step method to develop a competitive grant proposal:

1. Conduct a community and fire department needs assessment.
2. Compare community needs to the priorities of the grant program.
3. Decide what to apply for.
4. Complete the application.

The process of following the four steps in developing a grant proposal requires painstaking compliance with the specific grant instructions. To qualify, applicants have to follow a specific set of procedures and provide a detailed level of financial and operational information. The requirements may seem excessive and intrusive; however, the grant administrators have to ensure that the money is spent wisely and within legislative guidelines.

The Assistance to Firefighters Grant Program applications are evaluated by a committee that ranks them on the basis of pre-established criteria. To be approved, the application must have a request that matches the goal of the grant program, show that it is cost efficient, and demonstrate a strong community or financial need.

To receive a grant, the first step is to determine whether a grant program exists that covers your problem area. Once you have identified the program, determine whether your organization is eligible under the rules of the grant program, as well as the process for applying for the grant.

A grant application has to describe how the department's needs fit the priorities of the grant program. The grant applicant needs to analyze the community, conduct a risk assessment, evaluate the existing capabilities of the fire department, and identify the department's needs.

There is strong competition for a limited amount of grant funds, so the applicant must provide a sound justification. Use the information gathered in the needs assessment in a narrative to describe the problem in a concise but detailed manner.

In addition to the needs assessment, the grant application must demonstrate the department's financial need and why it needs outside assistance to acquire these resources. The application should show that efforts have been made to obtain the funds from other sources. In describing the current financial situation, include factors such as an eroding tax base, expanding community growth, local/state legislation that restricts taxes, and any significant local economic problem, such as the loss of a major employer.

The application also needs to show that this solution provides a benefit at the lowest possible amount of funding. Factors include collaborating with other organizations to share expensive equipment. For the Fire Fighter Status and Location project described at the beginning of the chapter, the grant opportunity comes from a private source.

Nontraditional Revenue Sources

Fire departments have solicited donations from a targeted business or industry for a specific purpose. In one area, a resort hotel-motel association provided funding to purchase an additional aerial platform truck that serves the high-rise hotel district. Equipment for a hazardous materials team could be donated by a company in the community that produces hazardous materials. One fire department offered to sell advertising on its vehicles.

Cost Recovery

Local or state regulations may allow a fire department to recover the extraordinary expenses incurred in responding to a hazardous materials incident, particularly decontamination, spill clean-up, and recovery costs. The department can invoice a hazardous materials carrier for extraordinary charges, such as overtime and replacement of expendable supplies (e.g., absorbents and neutralizing agents). In addition, the carrier may be responsible for the expense of repair or replacement of durable equipment, such as contaminated monitoring equipment, chemical protective suits, fire fighter protective clothing, and fire suppression tools.

Cost recovery also applies to some special services supplied by fire departments. For example, if a movie production company is shooting a film that involves pyrotechnics, the fire department might provide an engine company to stand by for several hours. The cost of overtime to staff the unit and an hourly rate for the use of the apparatus could be charged to the production company.

Expenditures

The annual budget for the fire department describes how the funds that are available are spent during the year. In the case of a municipality, the fire department budget is usually a part of the overall budget for that governmental entity. The format of the budget should comply with recommendations of the **Governmental Accounting Standards Board (GASB)**, which is the standard-setting agency for governmental accounting.

The local government budget typically includes a system of accounts that classifies all expenditures within certain categories and complies with the generally accepted accounting principles as outlined by the GASB. The accounting system uses a database system to support a **line-item budget**. A line-item budget is a format in which expenditures are identified in a categorized line-by-line format. The accounting system may have a complex numbering system, such as 02-12301-876543, that includes a category for every type of expenditure. In this example, the line item describes the following:

- 02: The first two numbers identify the fund. The 02 indicates that this money is coming from the general fund.
- 12301: The second part (the five-digit number) identifies the department and the division or subdivision where the

money is being spent. Called the "object code" in some systems, this one says that the money is allocated to the Fire Department (12) and to the Suppression Division (301).
- 876543: The third part (the six-digit number) provides a detailed description of what the money is being used for. Called the "subobject code" in some systems, this one says that the money is to be used for replacement fire hose.

The accounting system allows budget analysts and financial managers to keep track of expenditures throughout a municipal government. For example, law enforcement, public works, schools, and fire departments all purchase work uniforms. The database system would allow a budget analyst to determine how much the municipality spent on all uniforms, as well as breaking expenditures down by department and type of uniform. This could be valuable information when one is considering the savings that could be obtained by purchasing all uniforms from one vendor.

Fire department expenditures are generally divided into three general areas: personnel costs, operating costs, and capital expenditures. The accounting system allows budget analysts and managers to keep track of how much money is spent on fire fighter salaries (personnel costs), fuel for vehicles (operating costs), and construction of new fire stations (capital expenditures).

Personnel Expenditures

More than 90 percent of the career fire department budget is allocated to salaries and benefits. This includes fringe benefits, such as pension fund contributions, workers' compensation, and life insurance. Civil service regulations, local administrative decisions, and labor contracts often determine the cost of these benefits. Fire fighters have one of the highest workers' compensation rates, based on the history of claims and payments.

Fringe benefits also reflect the estimated cost of providing sick and annual leave benefits to the employee. Many municipalities have a sliding scale, in which the amount of leave earned per pay period increases as the employee gains seniority. The base cost of this fringe benefit equals the amount of the employee's accrued leave. In many cases, the actual cost to the fire department also includes overtime to pay for another employee to fill the vacancy.

Operating Expenditures

The operating budget covers the basic expenditures that support the day-to-day delivery of municipal services. Uniforms, protective clothing, telephone charges, electricity for the fire stations, flashlight batteries, fire apparatus maintenance, and toilet paper are all examples of purchases that would be classified as operating expenses. These funds are allocated in categories that allow for some flexibility throughout the year. For example, a fire department could be authorized to spend $250,000 on protective clothing during the year. The actual numbers of coats, pants, helmets, boots, and gloves that would be purchased would depend on the needs and the cost per item at the time of purchase. When a budget needs to be trimmed down, operating expenditures are usually the first area to be considered. For example, employee training, travel, and consulting expenses are usually reduced when revenue falls below expectations.

The cost of operating a vehicle is calculated on a per-mile basis that includes fuel, scheduled maintenance, and anticipated repairs.

Maintenance costs for fire department vehicles are high because they are loaded with additional equipment, including emergency lighting and siren systems, two-way radios, and computers. Large fire apparatus, which is expensive to repair and accumulates low annual mileage, may have a rate of several dollars per mile. If the actual fleet management costs exceed the budget projections because of high fuel costs or unanticipated repairs, the additional amount is sometimes charged back to the fire department at the end of the fiscal year. That adjustment is included in the third-quarter adjustments.

Many municipalities include a per-mile vehicle replacement fee applied to automobiles, police cars, buses, ambulances, and light-duty vehicles. The amount comes from estimating the anticipated replacement cost and the projected number of years that the vehicle will be used. It is also used by fire departments that have dozens of pumpers and aerials and a fire apparatus replacement schedule. In most cases, the replacement of a pumper, aerial, or specialized suppression rig will be a capital item.

Replacement uniforms and protective clothing are another expense calculated in the operating budget   Figure 13-1 ▼ . For example, the fire department might budget a lump sum of $150 per fire fighter to cover the annual cost of cleaning and repairs to protective clothing and up to $400 per fire fighter for uniforms.

Mandated training, such as hazardous materials and cardio-pulmonary resuscitation recertification classes, is also factored into the operating costs. That figure includes the cost of the training per fire fighter per class. The cost to pay another fire fighter overtime to cover the position during the mandated training could also be included.

Capital Expenditures

Capital expenditures refer to the purchase of durable items that cost more than a predetermined amount and will last for more than one budget year. Local jurisdictions differ on the cost and service period. Items such as computers, hydraulic rescue tools, washing machines, and self-contained breathing apparatus (SCBA) are examples of equipment purchased as capital items   Figure 13-2 ▶ . In general, the municipality establishes a unique

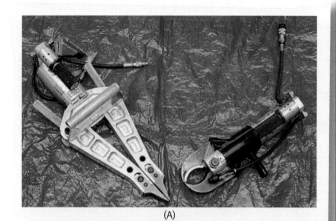
(A)

(B)

Figure 13-1   Replacement protective clothing is an operating expense calculated in the budget.

(C)

Figure 13-2   Three examples of equipment purchased as capital items. A. Tools. B. Computers. C. SCBA.

inventory record that documents the purchase price, source, assignment, and disposal of each capital item.

Sedans, pick-up trucks, and SUVs are usually part of the capital budget. Specialized vehicles, such as heavy fire apparatus,

Assessment Center Tips

Remember Annual Expenses

When discussing or responding to a budget proposal, remember to focus on the annual expenses in addition to the initial purchase cost. The annual costs fall into two areas: continuity and personnel.

Consider the purchase of semiautomatic cardiac defibrillators for three fire companies. What happens when one of the devices breaks down? Continuity considers the need for a spare unit to be used when one of the new devices is broken or unavailable, as well as the cost of maintenance and repairs. Additional costs would include the initial training of fire fighters to use the devices, plus annual recertification classes.

Personnel costs are the largest expense for career fire departments. A new ladder truck could cost $800,000 to purchase; however, the annual personnel cost to provide staffing for the vehicle could easily exceed $1 million. That cost could be difficult to justify for a company that will make fewer than 100 responses per year.

are also capital equipment, but they are often included in a special section of the capital budget because of their cost and complexity. In larger departments, replacement vehicles are purchased from a special set-aside fund, and only additional vehicles are included in the supplemental capital budget.

Capital improvement projects cover the construction, renovation, or expansion of municipal buildings or infrastructure. These are expensive projects that are often funded through long-term loans or bonds issued by the municipality.

Bond Referendums and Capital Projects

Large capital improvement projects, such as new fire station construction or major renovations, are funded through bond programs. A **bond** is a certificate of debt issued by a government or corporation; the bond guarantees payment of the original investment plus interest by a specified future date. The same type of funding mechanism is also used to build roads, water distribution systems, sewer systems, parks, libraries, and similar public facilities. The voters have to approve this type of expenditure through a referendum. If the voters approve, the municipality is authorized to borrow money from investors by issuing bonds up to a set amount. The authorization to sell the bonds is usually valid over a period of 5–10 years.

The bonds are repaid over a period of 10–30 years and return a fixed interest rate to investors. Bond funds are classified as a special revenue source and require compliance with a different set of accounting controls and reporting procedures. Annual status reports are required, and there are specific restrictions on the items that can be purchased with the bond revenue. If a city is issuing $100 million in bonds to be repaid over 20 years, a fraction of a percentage difference in the interest rate can have a significant impact to the taxpayers.

Bond funding allows the taxpayers to actually pay for the facility while it is being used, just as a mortgage allows a family to live in a home. The repayment period should be at least equal

to the anticipated life span of the facility. Bonds are repaid by collecting a special property tax or, if the facility will produce revenue, such as an airport or convention center, the bonds can be repaid from future income.

When bonds are used to build a new fire station, the authorized amount often includes the land, site improvements, building construction, permits, architectural and engineering fees, furniture and equipment that go into the building, as well as the apparatus and equipment that will be assigned to the new station.

Occasionally, a bond program is used to fund a large fire department apparatus purchase. The St. Louis Fire Department used a bond to cover the cost of an entire fleet of new apparatus in 1986. That large purchase allowed for the replacement of all existing pumpers and ladders with 30 new quints to establish the "Total Quint Concept." In fiscal year 2000, another bond program was authorized to purchase 30 replacement quints.

Lower Revenue Means Fewer Resources

Planned expenditures have to be balanced against anticipated revenues a year or more in advance. If the revenues do not meet expectations, adjustments have to be made to reduce spending. If revenues exceed expectations, some extra funds could be available during the year. Local revenue has a history of cyclical behavior, driven by external and internal forces that affect the delivery of services.

Changes in the local economy often result in major changes in the revenue collected by local government. If property values fall, property tax revenue decreases. Where local governments receive a percentage of sales taxes, a healthy economy provides increasing revenues and a weak economy results in a sudden revenue drop. A business that shuts down or reduces the number of employees can have a major impact on the local government budget.

Consider the impact of an empty office, retail, or manufacturing building. When the building is occupied and operating, it provides revenue to local government through property taxes and sales or gross receipts taxes. When the building is vacant, there are no sales or gross receipts to be taxed. An empty building generates dramatically lower revenues to local government. That is why local officials closely monitor the vacancy rate in office, business, and mercantile space.

If a new business moves into a community, it brings additional revenue to the local government. When a business moves to another jurisdiction, the sales tax revenue also moves to the new jurisdiction. The local jurisdiction could also lose the income tax from people who move with the business. If a property owner goes into bankruptcy, the building could be abandoned and the property taxes may not be paid.

Lower Revenue Options

Sometimes, the results of budget reductions drastically affect fire department operations. Fire departments have to make difficult choices when faced with declining revenues. Five such options are:

- Defer scheduled expenditures, such as apparatus replacement and station maintenance
- Regionalize or consolidate services

Fire Marks

The California Experience with Budget Calamities

In June 1978 California's Proposition 13 restricted the rate by which municipalities could raise property tax. This measure stopped the dramatic annual increases in property taxes to pay for municipal expenditures.

Although Proposition 13 did not have the predicted impact of devastating local government services, it significantly restricted future municipal growth. The 4700 special service districts, which include many fire organizations, lack the flexibility that local municipalities had to find other sources of revenue. Many of the smaller fire protection districts had to consolidate or contract out.

Orange County declared bankruptcy in 1994. During the 2-year restructuring, the Orange County Fire Authority was created. The new authority operates under finance rules that are different from those that apply to a county-operated fire department.

In 2008 the City of Vallejo filed for bankruptcy, citing the cost of police and fire labor contracts as a major factor. The initial filing for bankruptcy included a request by the city to void the four labor contracts that covered 400 employees.

- Privatize or contract out elements of the service provided by the department
- Reduce the career workforce
- Reduce the size of the department

Defer Scheduled Expenditures

Deferring scheduled expenditures means delaying the purchase of replacement fire apparatus or other expensive equipment. For example, a department that normally replaces pumpers on a 15-year schedule may decide to stretch the life of the rigs to 18 or 20 years. During the recession of the 1980s, Los Angeles and Chicago could not afford to replace aging ladder trucks. They obtained funding to rehabilitate existing rigs. The aerials were updated, rewired, repainted, and recertified as aerials. The Los Angeles Service Life Extension Program (SLEP) replaced gasoline motors and manual transmissions with diesel engines and automatic transmissions. This provided another 5–7 years of service for the rehabilitated aerials at less than half the cost of new replacement apparatus.

Privatize or Contract Out Elements of the Service Provided

Privatization means replacing municipal employees with contract employees. The concept is that the cost to the municipality to provide the service should be lower because a private company can operate more efficiently than a local government agency. Trash pickup, vehicle fleet maintenance, and school bus operations are three examples of services that are often contracted out by local government. The 1990s saw large, national emergency ambulance providers, such as American Medical Response and Rural-Metro, competing to provide 9-1-1 ambulance transportation in place of fire department–based services in many areas.

Some fire departments privatize or contract out services that are so specialized that a private company may be able to provide economic advantage to the municipality. Paramedic training, special operations classes, hazardous material site clean-up, ambulance billing, apparatus maintenance, and communications system maintenance are candidates for privatization.

Regionalize or Consolidate Services

Regional or consolidated fire departments are established to increase efficiency by reducing duplication in staff and services. Miami-Dade Fire Rescue Department in Florida started in 1935 as a fire patrol with one employee and one truck reporting to the Agriculture Department. The organization grew through numerous mergers with municipal fire departments and fire districts, as well as the expanding population in Dade County. In 1965, it became the Metropolitan Dade County Fire Department, and in 1997, it became the Miami-Dade Fire Rescue Department.

The Miami-Dade Fire Rescue Department provides fire suppression, emergency medical services, and specialized operations services in an 1883-square-mile area that includes the unincorporated portions of Dade County and 30 municipalities. The municipalities chose to have the county provide fire-rescue service instead of providing their own department. Miami-Dade has more than 2500 employees operating out of 66 fire stations and responds to an average of 603 calls per day.

Reducing the Career Workforce

Sometimes cities have to reduce their workforces. Fire departments try to protect staffing on emergency response vehicles by reducing positions in administrative and support areas. This has limited effectiveness because these are a small proportion of the workforce, and the individuals in these positions perform important tasks that support the emergency responders.

One option is to maintain the number of companies and reduce the staffing per vehicle. The other option is to limit

Safety Zone

Verify Your Sources

One of the best ways to ensure approval of a budget proposal is to identify a federal regulation, state law, or local ordinance that mandates equipment replacement or delivery of a program. The budget officials want to comply with all legally binding requirements but are not motivated to spend limited resources on recommended practices, nonbinding consensus standards, or suggested best practices. If the state requires replacement of Level A chemical suits after 5 years, the budget request should identify the specific regulation.

the number of units in service. Neither option is popular or attractive; however, budget realities sometimes leave no other choices.

Some fire departments temporarily close fire companies for one day or work shift at a time, reassigning the on-duty fire fighters to fill vacancies in other fire stations. This practice creates huge political repercussions if a civilian dies or suffers serious injury in an incident where the nearest fire company was closed for the day.

New York started fiscal year 1976 by laying off more than 40,000 city employees, including 1600 fire fighters. Although the city hired some back, 900 lost their Fire Department of New York (FDNY) jobs. Some of the laid-off became temporary employees under a federal Housing and Urban Development grant. They were assigned as the fourth or fifth member of an FDNY ladder company with the job of boarding up roofs and windows of fire-damaged buildings. It would take 2 years before the city could rehire all the laid-off fire fighters.

Reduce the Size of the Fire Department

Departments have been asked to reduce their career workforce in response to reduced local revenues. A fire department disbanded 17 engine and ladder companies between 1990 and 2009, reducing the fire fighter workforce by 23 percent. The seven fire companies that were closed in 2000 did not reduce the workforce, but shifted staff to reduce fire fighter overtime and staff additional ambulances. A truck, engine, and support vehicle were disbanded at the start of the FY2010 budget.

The recession that started in fiscal year 2008 has resulted in municipal revenue shortfalls as severe as the 1929 Depression. Cities are closing fire stations, disbanding fire companies, reducing on-duty staffing, and shrinking the range of services provided by the fire department. Some employees have been required to take furloughs, or unpaid days off.

Closing a fire station or eliminating a fire company often creates a significant and high-profile political issue that mobilizes the citizens as well as the fire fighters. No community wants to lose their neighborhood fire company.

The Purchasing Process

Most agencies will have a standardized method of making purchases. The fire officer must understand the policies and procedures of the organization. Failure to do so can have significant consequences for the organization. Because most fire departments are either political entities or nonprofit organizations, they are accountable to the public for the wise use of funds. Most fire departments are required to be audited, so purchasing violations are found sometimes during the auditing process.

Near Miss REPORT

Report Number: 09-0000254

Event Description: We responded to a morning trailer fire with a victim inside. First-due engine was running short (3) the captain had left and acted-up the senior fire fighter. The acted-up fire fighter was first on the scene and assumed command. His probationary fire fighter went inside the trailer alone for the rescue.

Once a company officer arrived and took command, conditions changed rapidly and the fire fighter inside was soon to be in a flashover setting with no radio communication.

Emergency traffic over the radio requested a "RIT" crew go in and get the fire fighter out while exterior crews ventilated the trailer.

Lesson Learned: Running short, slow down to get in with a hoseline.

Communications

Acting members: ability to fulfill the position.

Voices of Experience

Budgeting and operational change in the fire service are very challenging in today's economy. Diminishing budgets have a direct impact on the availability of personnel, equipment, and services fire departments are able to provide. Reductions in the fire department almost assuredly result in the reduction of suppression personnel and often the elimination of support personnel. Many departments are out-sourcing non-suppression personnel and services to try to cope with reduced budgets.

Historically, the budget process has been a guarded process that only a few officers understood or cared to understand. Today, officers at all levels need to know how the budget process works. This requires examination and research on the part of the officer to understand how the governmental agency obtains its funding, how they allocate that funding, and how to defend attacks against the budget line items. It is also incumbent upon all participants of the labor-management process to understand this process and to be able to speak intelligently with community leaders about the needs of the fire department.

Like most Midwestern fire-based EMS departments, 75 percent of our emergent call volume is EMS. Our budget is typical of medium-sized departments in the Midwest. Ninety-five percent of our $5.3 million budget is personnel costs. That leaves very little room in the operational line item to absorb a 5 percent or 6 percent reduction.

Recently, we were forced to make drastic operational changes due to a 17 percent reduction in daily minimum staffing from an average of 14 per shift to 12 per shift. This forced us to utilize non-traditional "jump companies" to keep outside stations open. Depending on the service need, three personnel would either respond an engine or a medic unit from our two outside stations. At Central Fire Headquarters the same "jump concept" is utilized to respond personnel on an engine, ladder, or medic unit.

It is conceivable that further cuts may be needed. These cuts will most likely come from the minimum staffing overtime line item, thus further reducing the available daily staffing. Any further reductions in daily staffing will likely result in the closure of a fire station. Closures in fire stations will result in extended response times, greater property damage due to fire, and greater care-to-door times for EMS calls.

Today's fire officer needs a thorough understanding of strategy, tactics, personnel management, operations, and the budget process. In the past we examined each of these components as individual pieces of the officer development process, but what we need to understand is that they are inter-related. The survival of the fire service is dependent upon the officer's ability to adapt to the changing budget process. Operational change, due to budget reduction, is inevitable. The focus of any operational change must continue to be the safety of our personnel.

Paul Ricci
Assistant Chief
Sandusky Fire Department
Sandusky, Ohio

> "Any further reductions in daily staffing will likely result in the closure of a fire station."

Petty Cash

The policies and procedures for purchasing typically vary by how much an item costs. For example, the fire department may allow any items under $100 to be purchased directly in a noncompetitive manner. Often this is done through a petty cash system. The petty cash system allows for a member of the fire department to be the custodian of an amount of cash that is provided by the organization. This individual is responsible for the cash that they are provided. As members of the fire department purchase small dollar value items, the petty cash custodian will reimburse them for the expense in exchange for the sales receipt.

Once the petty cash custodian is low on cash, the accumulated receipts are turned over to the finance department in exchange for a like amount of cash. In small organizations, the receipts are turned in to the governing body, such as the fire board that approves the purchases and then writes a check to the petty cash custodian. The check is then cashed and the funds are placed back in the petty cash box. The petty cash custodian will always have the authorized amount in either cash or receipts. This account is regularly audited.

Some personnel tend to think of the petty cash account as an endless fund, which is not true. On each receipt, the appropriate notation is made to identify which account, or part of the budget, is covering the expenditure. Petty cash allows for small purchases to be made without having to obtain purchase orders. Each time petty cash is spent it lowers the account balances.

Purchase Orders

Many items are purchased that exceed the petty cash limit that has been set by the organization. When this occurs, a purchase order system is typically used. A purchase order is a method of ensuring that there are sufficient funds available in a budget account to cover a purchase. Like petty cash, most organizations have a limit on the maximum amount of a purchase order; for example, an organization may allow purchase orders to be used for purchases up to $2000.

When a decision to purchase is made, the fire officer's role is to acquire the item at the most reasonable cost to the organization. To do this, most organizations allow phone bids for these items. For example, if a fire officer needs to purchase a pair of tires, he or she first determines what size is needed and then calls at least three places to get prices on the tires. He or she then takes the lowest price and completes a purchase order, which is commonly called a "PO." The purchase order typically requires an authorizing signature by an officer who has control over the budget area from which the funds are paid. In this example, it might be the chief. The purchase order is then entered into the purchasing system, which debits the maintenance account.

The fire officer then takes the purchase order to the location that has the tires and the sale is transacted. A copy of the purchase order is given to the vendor, who attaches the sales receipt to it and sends it to the fire department's finance department for reimbursement. In effect, the vendor allows the fire department to charge the merchandise until the purchase order is paid. Additionally, with the purchase order the fire department indicates to the vendor that there are funds available for the purchase and the purchase has been approved. Some vendors do not accept purchase orders.

Requisitions

For purchases that exceed a predetermined amount, such as $2000, a requisition is required rather than a purchase order. This occurs for large purchases, such as a new hydraulic extrication tool. A requisition has even more stringent requirements than a purchase order.

The requisition differs from a purchase order in that the exact amount of the purchase is not known. Instead, the department requests that a specific amount of funds be encumbered, or set aside from the budget account, which will more than cover the cost of the purchase. In the case of the rescue tool, the department might set aside $15,000. A bidding process will be followed and once complete, the requisition will allow for payment of the item.

The Bidding Process

Once funds have been encumbered, the fire department must go about the process of making the purchase. This is done in one of two ways. For smaller items, the fire department may develop specifications for bids. The second method, which is often used for very large or complex purchases, is a request for proposal (RFP).

To purchase based on specifications, commonly known as "specs," the fire department writes up exactly what it desires in the product. For example, a portable power unit for a fire truck must not exceed a size of 18" × 14" × 21" (46 × 36 × 53 cm). With a spec, every requirement must be met for the vendor to be considered. So for this example, if a vendor wanted to bid but the power unit was 19" × 13" × 20" (48 × 33 × 51 cm) its bid would be rejected. This may be important if the compartment that the power unit will be placed in will only accommodate the specified size. Some fire departments use this method so only one vendor or product can meet the specifications; however, this is an illegal and unethical practice. Specs should be used to specify what must be met for the purchase to be effective for the fire department.

Once developed, a bid sheet listing the specifications is sent to vendors for pricing. Most organizations have a bidding list of vendors that wish to do business with them. The vendors will place their name and address on the list so each time a bid comes up, a copy is sent to them for consideration. A notice of bid may also be required to be placed in the local paper. Other organizations place notice of bids in trade magazines and papers. The Internet also is becoming a method of getting information out to the public about upcoming bid processes.

Often, the use of specifications eliminates virtually every potential vendor or may greatly increase the price in order to meet the specifications. Sometimes, a bid specification limits the bidders. For example, the purchase of a mobile data computer (MDC) system for use in the fire apparatus would likely use the request for proposal (RFP) method rather than the bid specification method because of the complexity of the technology.

In an RFP, the fire department gives the general information about what is desired and allows the vendor to determine how it will meet the need. For example, the mobile data computer RFP might require that the system provide coverage to 99 percent of the city. One vendor might opt to place 10 antenna towers to obtain the coverage whereas another might opt for only 7 antenna towers but place them at a higher elevation. Both ideas meet the need of the fire department. The fire department also determines how the proposals will be evaluated. For example, the price might be worth 50 percent of the overall consideration, while delivery time might be worth 5 percent. Once the RFP is developed, it is sent out to potential vendors for bidding.

Whether the process uses an RFP or bid specifications, vendors are given a specific amount of time to reply. Their replies are placed in a sealed envelope indicating the bid number on the outside of the envelope. At a time that was outlined in the bid specifications or RFP, all bids or proposals are opened in public view. The bids are then awarded to the lowest bidder and the amount is attached to the requisition for payment when delivery is made.

When an RFP is used, each proposal is evaluated to determine how it meets the need. They are typically assessed based on performance as well as price. Once each performance requirement has been addressed and the price is considered, the RFP is awarded to the highest score. The price is then added to the requisition for payment when the item is delivered.

In recent years, there has been a trend to develop intergovernmental or interagency cooperation to streamline purchases and leverage greater discounts. The federal government issues many purchasing contracts for a wide range of equipment. Most vendors honor these rates for other governmental agencies. This streamlines the process because the bidding process has already occurred. In other cases, local organizations band together to purchase common items. For example, a group of fire departments may jointly develop an RFP to purchase SCBA. This ensures a lower bid because they are purchasing in greater quantity and it allows all agencies to use like equipment.

Navigating the Budgetary Process

Using the example presented in the opening case study, let's navigate through the budgetary process. In the opening case study, the department is preparing a budget to get funding from a private corporation to establish a Fire Fighter Safety and Location program.

Developing a Budget Proposal

Because this is a new unit providing a new service, the first step in the budget proposal process is to describe what the new unit will be doing and the impact if the unit is not funded.

Overview of Fire Fighter Safety and Location (FFSL) Program

The following is the narrative that is part of the budget submission. It describes what the Fire Fighter Safety and Location (FFSL) program will do, the impact of the FFSL program on current operations, and the consequences if the new program is not funded.

The FFSL is a two-person unit that responds to working incidents to track fire fighters operating in immediately dangerous to life and health (IDLH) environments and to monitor the medical status of members who rotate through the rehabilitation sector during an event. FFSL operates every hour of every day.

The FFSL has two responsibilities at a working incident. The first responsibility is to deploy the field antenna system to capture the three-dimensional global positioning data that are transmitting from the integrated personal alert safety system (PASS) units that are activated on the self-contained breathing apparatus. The managing or supervising fire officer will advise the Safety Sector of any individual who appears to be in distress or, upon activation of a PASS alarm, the location of the fire fighter within the incident.

The second responsibility is to track the vital signs of fire fighters who enter and exit the rehabilitation area during an incident. The primary goal is to identify those who may need immediate rehydration or medical attention. The secondary goal is to assure that the fire fighters are ready to leave the rehabilitation area by conforming with medical standards established by the operating medical director.

If not funded, the department will not reduce the identified risk factors that lead to fire fighter line-of-duty deaths and disability. The department will be unable to improve its ability to rescue a fire fighter in trouble within a burning building. The fire department will be unable to provide early detection of life-threatening health or environmental conditions that impact fire fighters.

FFSL Annual Personnel and Operating Expenditures

This part of the budget submission shows the annual expenses to operate the FFSL. There are personnel and operating expenses.

Personnel Expenses

The FFSL will have a managing fire officer, three supervising fire officers, and four technicians. In this example, the personnel cost is described in two areas. The first area is the direct salary for each position. Most civil service pay classification systems use a pay level that includes steps tied to seniority. Budget protocol requires that new positions reflect the average seniority step that the incumbents occupy. That means that the captain, the managing fire officer, is at seniority step 7; the lieutenant, the supervising fire officer, is at seniority step 5; and the technicians are at seniority step 3.

The second area is the calculation of the fringe benefits for each employee position. It is calculated as a percentage of the employee's salary. For the FFSL example, fire fighter fringe benefits account for 28.7 percent of the salary. That means that every $100 of salary costs the city $128.70.

Operating Expenses

The cost of operating the FFSL unit entails three general areas: vehicle cost, office operations, and personnel training. City fleet maintenance has determined the cost to operate and replace a small, all-wheel-drive emergency services hybrid SUV to be $0.84 a mile. This covers the fuel and scheduled maintenance of the vehicle. This also includes a per-mile charge that is designed to cover the replacement of the vehicle when it reaches the end of its scheduled service life. Vehicle service life is determined in three ways: years of service, miles accumulated, or specific vehicle cost. It is estimated that the unit will accumulate 17,000 miles a year. The city replaces small emergency service vehicles at 80,000 miles or after 10 years. The FFSL will need replacement after 4.7 years.

Office operations include rent, telecommunication charges, and equipment maintenance. In this example, the FFSL is operating out of a fire station, so rent is zero. The office has two hardwire lines to provide service for the phones, Internet access, and fax machine. The municipality's information technology (IT) office has an annual fee per hardwire line that covers telephones and Internet broadband access. IT also handles computer maintenance and repairs. As with the vehicle, the annual maintenance and operating charge covers the anticipated annual costs of maintenance, software upgrades, and repairs. It also includes a cost that will allow replacement of the computers on a 5-year cycle. The FFSL has two desktop computer workstations, one vehicle-based desktop workstation, and four laptop computers.

Operating expenses also include the expenses of mandated annual fire fighter training. The cost reflects the per-person cost of training, for both the tuition and the position replacement overtime to cover the FFSL person who is receiving the training. This covers training required by federal or state regulations, such as annual blood-borne pathogen and hazardous materials refresher training.

FFSL Capital Budget

The capital budget request would pay for the durable equipment needed to start the FFSL program. Many of the required items would be purchased from federal, state, or regional purchasing contracts. The budget summary shows the entire first-year capital expenditures to start the program, including the response vehicle, the administrative office, the monitoring equipment, and the impact of eight new fire department positions.

FFSL Vehicle

During the planning process with the administrative fire officer, you determine the appropriate type of vehicle. The state offers a range of vehicles for purchase. Here are the ones considered for the FFSL:

- Compact four-door sedan, $11,733
- Midsized hybrid four-door sedan, $15,105
- Full-sized sedan, police pursuit, $26,675
- Small all-wheel-drive hybrid SUV, emergency services, $24,765
- Large four-wheel-drive SUV, emergency services, $29,290

The administrative fire officer decides that the FFSL should be the same as other staff and command vehicles. In addition to the vehicle, equipment is needed to convert a small SUV into a fire department response vehicle, including emergency lights/siren, mobile fire radio and computer dispatch terminal, two portable fire radios, and semi-automatic cardiac defibrillator. Depending on the local budget process, these items may need to be identified in a line-item format consistent with budget documentation protocol **Figure 13-3 ▶**. More than half of the FFSL vehicle cost is for the specialized equipment that goes on or into the vehicle.

FFSL Administrative Office

Because the FFSL will be funded by a grant that requires incident reports and deployment research, the budget includes the cost of establishing two administrative workstations. The workstation will go into an available space at Fire Station 5. The capital equipment represents all of the durable items needed to establish the office: desks, chairs, computers, fax machine, and printer.

FFSL Uniforms and Protective Clothing

The expense of fire fighter protective clothing and the initial uniform issue are included in the initial year capital expenditures. Because the FFSL program creates the new positions, the expense shows up here.

FFSL Disposable Medical Monitors

The disposable wireless medical monitor "Band-Aid" is attached to the left or right side of the neck, over the carotid artery. Each device has enough power for 72 hours. The fire fighters need to put the monitor on their neck and synchronize the monitor with the PASS device during the morning equipment check. This allows the PASS device to identify the person using the SCBA to the FFSL monitoring and mapping programs. This will allow a quick identification of who is in trouble if a PASS alarm is activated or the fire fighter's vital signs indicate a problem.

Ask for Everything You Need

Fire departments try to do the most with the least resources. When submitting a proposal for a new project or service, assume that there are no existing resources and ask for everything. This includes asking for enough tools to ensure continuing operation. For example, the rehabilitation monitoring station computers, configured to work with the FFSL communications, cost $6,113 each. Only twice in the past 10 years were there two simultaneous rehabilitation stations operating during a disaster. However, ordering three properly configured laptops provides the redundancy to ensure continued operations.

Cost Recovery and Reduction

It will cost over $1 million a year to run the FFSL program in a city with 55 to 70 on-duty fire fighters. The cost recovery comes with a faster identification of a fire fighter in medical or situational distress on the fire ground. A declining oxygen level will alert much sooner than a fire fighter calling a may-day. This is important because carbon monoxide poisoning impairs perception and reaction times. Once a problem is detected, the FFSL mapping will immediately locate the fire fighter. This reduces the time needed to remove the distressed fire fighter from an IDLH environment. This may change a line-of-duty death to a near-miss.

Less dramatic, but equally important, is the tracking of vital signs when a fire fighter enters and leaves the rehabilitation area. Research is linking dehydration to cardiac-related fire-ground deaths. The cost recovery is reducing the severity level of a fire fighter in distress, reducing hospital time, and avoiding a lasting disability. The safety of the rescue team is improved because knowing the location of the fire fighter reduces the time spent searching within a hostile environment.

Fire Department - Firefighting Division
New Program: **Fire Fighter Safety and Location (FFSL) program**

Initial Year: Capital Expenditures	**Unit**	**Total**	**Source or comments**
FFSL Unit Vehicle:			
1 Small AWD SUV hybrid, emergency services	$24,765	$24,765	State vehicle contract
1 Mobile 800 mHz radio and computer terminal	$15,526	$15,526	Communications division quote
1 Fire emergency lighting system	$2,765	$2,765	Apparatus section quote
1 Safety striping and FD lettering	$411	$411	Apparatus section quote
1 FFSL monitoring and mapping computer (vehicle based)	$8,700	$8,700	Vendor quote
3 FFSL in-field antenna receiver system	$3,360	$10,080	Vendor quote
1 Semi-automatic cardiac defibrillator	$875	$875	Firefighting division quote
2 Portable 800 mHz radios	$3,135	$6,270	Communications division quote
Total Alarm Unit vehicle cost:		**$69,392**	
FFSL Personal Alert Safety System devices			
98 Integrated PASS device and FFSL transmitter	$677	$66,346	Vendor quote
55 Trade-in of existing PASS units	($175)	($9,625)	Vendor quote
Total FFSL PASS device deployment:		**$56,721**	
FFSL Rehabilitation Monitoring Stations			
3 Configured laptops with wireless communication	$6,113	$18,339	Vendor quote
Total FFSL Rehabilitation Monitoring Stations:		**$18,339**	Vendor quote
FFSL administrative office			
2 Cubicle work stations (desk and chair)	$747	$1,494	Facilities
2 File cabinet	$200	$400	Facilities
2 Computer workstation - Model B	$3,200	$6,400	Information Technology
2 Digital pagers/wireless voice	$400	$800	Information Technology
1 Fax machine	$344	$344	Information Technology
1 Workstation laser printer - Model B	$1,325	$1,325	Information Technology
Total Alarm Unit administrative office cost:		**$10,763**	
New Uniformed Fire Fighter Positions - maintaining a two-person team on duty every hour of every day			
8 Protective clothing ensembles	$1,876	$15,008	Firefighting division
4 Fire officer initial uniform issue	$865	$3,460	Firefighting division
4 Fire fighter initial uniform issue	$711	$2,844	Firefighting division
New Uniformed Fire Fighter Positions:		**$21,312**	
Initial Year Capital Expenditures:		**$158,188**	

Annual Operating Expenses	**Unit**	**Total**	**Source or comments**
Personnel Expenditures:			
1 Captain at Pay Step 28-7	$56,023	$56,023	Salary
3 Lieutenant at Pay Step 26-5	$46,785	$140,355	Salary
4 Technicians at Pay Step 22-3	$39,209	$156,836	Salary
Fringe benefits		$101,372	28.7% of fire fighter salary
Total annual personnel expenditures:		**$454,586**	
Operating Expenses:			
1 FFSL vehicle maintenance, operating and replacement	$14,280	$14,280	$.84 per mile, estimate 17,000 miles/year
1 Office rental	$0	$0	Existing Fire Station 5 space
2 Telecomunications fee per hard wire line	$1,800	$3,600	Internet and phone line charges
7 Computer maintenance, operating, and replacement	$500	$3,500	Information Technology
2 Paper, office supplies	$275	$550	Facilities
4 Mandated fire officer continuing training	$221	$884	Academy, per position cost
8 Mandated fire fighter continuing training	$176	$1,408	Academy, per position cost
FFSL office annual operating expenses:		**$24,222**	
FFSL Program Operating Expenses			
165 Disposable medical monitor devices, per case	$3,168	$522,720	Vendor quote
10 Replacement of destroyed or damaged PASS devices	$677	$6,770	Vendor quote (10% of inventory)
1 Replacement of FFSL in-field antenna receiver system	$3,360	$3,360	Telecommunications estimate
FFSL program annual operating expenses:		**$532,850**	
TOTAL FFSL annual operating expenses:		**$557,072**	
Annual Personnel and Operating Expenditures:		**$1,011,658**	

Figure 13-3   Summary budget submission sheet for the FFSL.

Organizational Change

Effective change occurs when preparation meets opportunity. Many fire department changes are because of federal and state administrative regulations, lawsuit settlements, arbitration, local municipal practices, and mandates from elected officials. Recent changes are due to diminished revenues and reimbursements.

The study of organizational change started in the mid-1970s, focusing on techniques: re-engineering, total quality management, the learning organization, six-sigma, and the balanced scorecard. Currently, emphasis is on organizational effectiveness, customer satisfaction, and organizational survival.

■ Preparing for a Change Opportunity

Fire codes are not the only item influenced by the catastrophic theory of reform. Many initiatives are funded in the wake of a disaster or tragedy. James O. Page provided this example about Los Angeles County Fire Chief Keith Klinger:

On the first night of the November 1961 Bel Air Fire he was driving Chief Petroff. Several times during the night they met on dusty fire roads with Chief Klinger and his driver. Throughout the night they could hear the boss on the radio, seeking updated information, arranging for meetings with other chiefs, inquiring about the welfare of personnel, and scolding food dispenser operators to "get over there and take care of those guys." He was 50 years old at the time but he didn't slow down or sneak off for a nap all night long.

The rest of the story was told to him by Kenny Hahn in 1977. According to Kenny, the Board of Supervisors was having its regular Tuesday morning meeting. "All of a sudden," Hahn said, "the wooden doors at the back of the meeting room swung open, and through them marched Chief Klinger. He was covered with soot and dust and I swear he must have had a fireman out in the lobby with a bellows full of smoke, puffing it through the doorway after the chief."

"He marched down that aisle like he'd just bought the Hall of Administration," Supervisor Hahn continued. "He wasn't on the agenda but he walked right up to the podium and took over the meeting. What could we do?" Hahn asked rhetorically. "It seemed like the whole county was on fire and the fire chief wanted to talk to us. Keith Klinger knew how and when to get attention."

"I noticed he had a folder in his hand," Kenny Hahn remembered. He then recalled how Klinger gave the board a blow-by-blow report on the battles that were under way in the mountains between Sepulveda Pass and Topanga. Then, Hahn recalled with a grin, Chief Klinger pulled from the folder a 10-year plan for improvement of fire protection in LA County. Again, he asked, "What could we do but vote yes on it?"

The Board of Supervisors adopted Chief Klinger's 10-year plan and provided the money for it, thus the fleet of brush rigs to be known as 400 engines, as well as several new stations. Obviously, Chief Klinger had the plan developed long before the Bel Air Fire and was just waiting for the best time to spring it on the Board of Supervisors.

Arlington County, Virginia, Fire Chief Edward P. Plaugher had a similar experience after the 2001 Pentagon attack. With dozens of military and federal government facilities in the county, Plaugher implemented a regional antiterrorism response plan in 1995. After the Pentagon attack, Plaugher provided the Board of Supervisors with a proposal to staff more units, establish four-employee minimum staffing for all fire companies, and improve the apparatus fleet. The proposal came from a constantly evolving plan to improve the department since he was appointed fire chief in 1993.

■ Budgeting for Change

Organizational change requires expenditures in time and resources. Even a "cost-neutral" change will require managerial time and employee training. The managing fire officer will identify the initial and continuing expenditures for every proposal.

Some changes are initially more expensive, but result in cheaper, faster, or better outcomes once implemented. The proposal may need to include an analysis on the return of investment or lifecycle costs. For example, the additional expense of purchasing a medium-duty, truck-style, diesel ambulance is offset by the reduced frequency in transmission and brake repairs and the longer service life in comparison to van-style light-duty ambulances.

■ The Process of Organizational Change

Efforts to implement a business model of organizational change have been unsuccessful within the local government environment. Although many of the tools change, like going from paper to digital records, the requirement for local governments to provide services that are too difficult or expensive to be performed by for-profit entities remains.

Change starts for the managing officer by considering the larger picture, identifying the internal and external groups or individuals affected by the proposal. Answer the "What's in it for me" (WIIFM) question for all of the internal and external players. For example, fire fighters may object to having their vital signs monitored continually. What readings will require the department to put someone on light duty? Who is qualified to make that call?

There may be resistance at the local government level, with a concern about what happens after the final year of the grant. Local leaders may be reluctant to add another $1 million per year to the fire department operating budget.

The interoperability task force may be concerned, because the other fire departments are not involved in this project but will be operating at the same incidents. Any player who perceives a loss of turf, power, or resources will resist change.

The managing fire officer has two techniques that can address the WIIFM and the concern about the loss of turf, power, or resources. Both techniques take time because they impact individual emotions and organizational culture.

Maintaining Communications

The most effective tool for facilitating change is providing communication in as transparent a way as possible. Some initiatives have had success using a website to describe the project, the status of action items, and scheduled meetings; celebrate successes; and provide frequently asked questions or an online reference.

Establishing Trust and Resolving Conflicts

Open communications and a transparent process start creating trust for the managing fire officer who is implementing the FFSL program. Many of the techniques described in Chapter 11 would be helpful with this project. Conflict often follows the Kübler-Ross grief and tragedy model:

- Anger: "How dare you say that we need to make changes!"
- Denial: "There are no problems and, therefore, no need to make changes."
- Bargaining: An opportunity to negotiate acceptance by tweaking the new change process.
- Depression: Common when the party suffers a real or perceived loss of authority, power, or resources.
- Acceptance: The change is occurring and it is not the end of the world.

Fire fighters are passionate about their job and obsessive about policies and practices. If you can show that the gain in implementing the change exceeds the pain of change, you can get to the acceptance step quicker.

Summary

Knowledge of the budget process provides a powerful set of tools for the fire officer. Although not part of most promotional exams, the ability to understand and contribute to the budget process is a professional development skill as identified by the International Association of Fire Chiefs (IAFC).

Fire departments are involved in complex revenue streams, using fees, grants, and other creative techniques to ensure enough resources are available to deliver effective services in the face of a downturn of traditional revenue sources.

All change requires resources. Considering change within the budget process is a realistic and effective approach. It prepares the managing fire officer to progress to administrative and executive fire officer positions.

You Are the Fire Officer: Conclusion

You submit the grant proposal on time. The challenge is just beginning. It is rare for a new procedure to get implemented without some tweaking or political maneuvering. You are getting push-back from some of the players that fear a loss of turf, power, or resources. The fire chief suggests that when the grant funding runs out the staffing will go from an officer and a fire fighter to a single civilian.

Wrap-Up

Chief Concepts

- A four-step method can be used to apply for a grant:
  - Conduct a community and fire department needs assessment.
  - Compare weakness to the priorities of the grant program.
  - Decide what to apply for.
  - Complete the application.
- The annual budget describes how the local government will spend revenue in operating, personnel, and capital expenditures.
- Local governments operate in a fiscal year that generally runs from July 1 to June 30.
- A proposed budget is submitted by the agency heads to the budget director. After negotiations, the proposed budget is released for public review, comment, and hearings.
- After the public comment period, elected officials revise the budget and issue the "Approved" or "Adopted" budget that is the financial document that will guide the jurisdiction.
- During the budget year, informal budget reviews occur during the third and ninth months. Minor revisions may occur.
- A more formal midyear budget review occurs at the 6-month period. The elected officials may reassign resources and revise objectives on the basis of tax revenues and local government issues.
- Lower revenue options:
  - Defer scheduled expenses.
  - Regionalize or consolidate services.
  - Reduce career workforce.
  - Privatize or contract out elements of the service provided by the department.
  - Reduce the size of the department.
- Change occurs when preparation meets opportunity. Emphasis should be on organizational effectiveness, customer satisfaction, and organizational survival.
- Opportunity follows the catastrophic theory of reform.
- Even "cost-neutral" change will require time and resources.
- Consider "What's in it for me" (WIIFM) for internal and external groups.
- Open communication and a transparent change process start the trust-building process.

Hot Terms

Base budget The level of funding that would be required to maintain all services at the currently authorized levels, including adjustments for inflation, salary increases, and other predictable cost changes.

Bond A certificate of debt issued by a government or corporation; a bond guarantees payment of the original investment plus interest by a specified future date.

Budget An itemized summary of estimated or intended expenditures for a given period, along with proposals for financing them.

Expenditures The act of spending money for goods or services.

Fire tax district Special service district created to finance the fire protection of a designated district.

Fiscal year Twelve-month period for which an organization plans to use its funds. Local governments' fiscal years are generally from July 1 to June 30.

Governmental Accounting Standards Board (GASB) The mission of the Governmental Accounting Standards Board is to establish and improve the standards of state and local governmental accounting and financial reporting that will result in useful information for users of financial reports and to guide and educate the public, including issuers, auditors, and users of those financial reports.

Line-item budget Budget format in which expenditures are identified in a categorized line-by-line format.

Revenues The income of a government from all sources appropriated for the payment of the public expenses.

Supplemental budget Proposed increases in spending to provide additional services.

Fire Officer *in Action*

The price of diesel fuel exceeds $15 a gallon. The fire chief is prohibiting the driving of fire apparatus except to and from emergencies. No more food or physical training runs, or driving around the neighborhood on street drills. The chief is proposing a dramatic reduction of suppression units sent to activated fire alarms or investigations, only sending a full first alarm with a confirmed report of a working structure fire. The administrative fire officer has asked every work location for suggestions on how to accomplish routine fire department tasks with the least amount of fuel use.

The budget process has not been a focus for promotional examinations. The recession that started in fiscal year 2008 has produced extreme challenges to the fire service. Understanding how your department handles the issues significantly improves your ability to provide suggestions when future cost reductions are required.

1. Which of the following represents a major local government revenue source?
- **A.** Fire lane parking tickets
- **B.** Property taxes
- **C.** State subsidy
- **D.** Federal grant

2. An early step in developing a grant proposal is to:
- **A.** inventory the community apparatus and specialized equipment.
- **B.** determine which other jurisdictions are applying for the grant.
- **C.** get a copy of the last successful grantee.
- **D.** determine what the grant is designed to cover.

3. Personal protective clothing falls under which category within the budget document?
- **A.** Personnel expenditures
- **B.** Capital expenditures
- **C.** Operating expenditures
- **D.** Facilities expenditures

4. When recommending a procedural or organizational change, a good practice is to:
- **A.** provide communication in as transparent a way as possible.
- **B.** wait until the plan is authorized by the local elected officials before revealing it to the fire fighters.
- **C.** use a press conference as a mechanism to inform the employees.
- **D.** accept input only from consultants, not from rank-and-file employees.

Fire Officer Communications

NFPA 1021 Standard

Fire Officer I

4.1.2 **General Prerequisite Skills.** The ability to effectively communicate in writing utilizing technology provided by the AHJ; write reports, letters, and memos utilizing word processing and spreadsheet programs; operate in an information management system; and effectively operate at all levels in the incident management system utilized by the AHJ. [p 267–269, 272–281]

4.2.1 Assign tasks or responsibilities to unit members, given an assignment at an emergency incident, so that the instructions are complete, clear, and concise; safety considerations are addressed; and the desired outcomes are conveyed. [p 271–272]

(A) Requisite Knowledge. Verbal communications during emergency incidents, techniques used to make assignments under stressful situations, and methods of confirming understanding. [p 271–272]

(B) Requisite Skills. The ability to condense instructions for frequently assigned unit tasks based on training and standard operating procedures. [p 271–272]

4.2.2 Assign tasks or responsibilities to unit members, given an assignment under nonemergency conditions at a station or other work location, so that the instructions are complete, clear, and concise; safety considerations are addressed; and the desired outcomes are conveyed. [p 267–271]

(A) Requisite Knowledge. Verbal communications under nonemergency situations, techniques used to make assignments under routine situations, and methods of confirming understanding. [p 267–271]

(B) Requisite Skills. The ability to issue instructions for frequently assigned unit tasks based on department policy. [p 267–271]

4.3 **Community and Government Relations.** This duty involves dealing with inquiries of the community and communicating the role, image, and mission of the department to the public and delivering safety, injury, and fire prevention education programs, according to the following job performance requirements. [p 279]

4.3.1 Initiate action on a community need, given policies and procedures, so that the need is addressed. [p 279]

(A) Requisite Knowledge. Community demographics and service organizations, as well as verbal and nonverbal communication, and an understanding of the role and mission of the department. [p 279]

(B) Requisite Skills. Familiarity with public relations and the ability to communicate verbally. [p 279]

4.4.5 Explain the needs and benefits of collecting incident response data, given the goals and mission of the organization, so that incident response reports are timely and accurate. [p 271–281]

(A) Requisite Knowledge. The agency's records management system. [p 278]

(B) Requisite Skills. The ability to communicate both orally and in writing. [p 267–281]

Fire Officer II

5.3 **Community and Governmental Relations.** This duty involves dealing with inquiries of allied organizations in the community and projecting the role, mission, and image of the department to other organizations with similar goals and missions for the purpose of establishing strategic partnerships and delivering safety, injury, and fire prevention education programs, according to the following job performance requirements. [p 273, 275–279]

5.4.4 Prepare a news release, given an event or topic, so that the information is accurate and formatted correctly. [p 279]

(A) Requisite Knowledge. Policies and procedures and the format used for news releases. [p 279]

(B) Requisite Skills. The ability to communicate orally and in writing. [267–281]

5.4.5 Prepare a concise report for transmittal to a supervisor, given fire department record(s) and a specific request for details such as trends, variances, or other related topics. [p 273, 275–278]

(A) Requisite Knowledge. The data processing system. [p 279–280]

(B) Requisite Skills. The ability to communicate in writing and to interpret data. [p 278–280]

5.6.3 Prepare a written report, given incident reporting data from the jurisdiction, so that the major causes for service demands are identified for various planning areas within the service area of the organization. [p 278]

(A) Requisite Knowledge. Analyzing data. [p 279–280]

(B) Requisite Skills. The ability to write clearly and to interpret response data correctly to identify the reasons for service demands. [p 278–280]

Introduction to Fire and Emergency Services Administration (FESHE) Course Outcomes

2. Explain the need for effective communication skills both written and verbal. [p 267–279, 281]

11. Identify the role of a company officer in Incident Command System (ICS). [p 271–272]

Knowledge Objectives

After studying this chapter, you will be able to:

- Discuss the communication cycle.
- Identify ways to improve listening skills.

- Understand the fire officer's responsibilities as an active listener to both subordinates and superiors.
- Describe the ways to counteract environmental noise.
- Identify the conditions that interfere with verbal communication.
- Explain the difference between formal and informal communications.
- Outline fire officer responsibilities for routine and infrequent reports.
- Explain what should be included in a news release.

Skills Objectives

After studying this chapter, you will be able to:

- Write a recommendation report using information about an issue or problem.
- Write a news release.

Your fire company is responding alone to a food-on-the-stove incident. You arrive and see fire leaping out of the first floor windows of a middle-unit townhouse. A soot-singed civilian is screaming that her children are trapped on the second floor. You pick up the microphone and report a working fire with occupants trapped.

While you do a quick 360-degree size-up, your crew deploys an attack line. They enter the townhouse before you complete your size-up and immediately fall through the floor. All you hear is one of your two members screaming into a portable radio.

At this point, dispatch calls you. "Is the fire out or do you need additional assistance?"

1. What should you do now?
2. Why is it important to complete the entire communications process when you are in an emergency situation?
3. What can you do in the future to improve information exchange?

▮ Introduction to Fire Officer Communication

Many fire officers wear a rank insignia that features bugles, representing the fire officer's speaking trumpet. The trumpet was used to shout orders on the fire ground at the turn of the 20th century. This symbol emphasizes the fire officer's requirement to communicate. Although the technology has advanced, communication skills are important. An officer must be able to communicate effectively in many different situations and contexts.

A fire officer must be able to process several different types of information to effectively supervise and support the fire company members. One characteristic of public safety services in the 21st century is the near overwhelming volume of information, particularly regarding newly identified hazards. A fire officer should regularly read a fire service trade magazine and visit websites that provide current information. The officer should share significant information with the company members.

Effective communication skills are essential for a fire officer. These skills are required to provide direction to the crew members, review new policies and procedures, and simply exchange information in a wide range of situations. Effectively transmitting radio reports requires a unique skill set. Communication skills are equally important when working with citizens, conducting tours, releasing public information, and preparing reports.

▮ The Communication Process

Communication is a repetitive circular process. Successful communication occurs when two people can exchange information and develop mutual understanding. When information flows from one person to another, the process is effective only when the person receiving the information is able to understand what the other person intended to transmit.

Effective communication does not occur unless the intended message has been received and understood. The message must make sense in the recipient's own terms, and it must convey the thought that the sender intended to communicate. The sender cannot be sure this has occurred unless the recipient sends some confirmation that the message arrived and was correctly interpreted.

▮ The Communication Cycle

The communication cycle contains five parts:
1. Message
2. Sender
3. Medium (with noise)
4. Receiver
5. Feedback

The Message

The message represents the text of the communication. In its purest form, the message contains only the information to be conveyed.

Messages do not have to be in the form of written or spoken words. For example, a stern facial expression with purposeful eye contact can convey a very clear message of disapproval, whereas a smile can convey approval. The messages are clear, yet no words accompany them.

The Sender
The sender is the person or entity who is sending the message. We think of the sender as a person, but it could be a sign, a sound, or an image. The message needs to be properly targeted to the right person and formulated so that the receiver will understand the meaning. The sender is responsible for the receiver properly understanding the message.

A fire officer must be skilled in order to transmit information and instructions to subordinates and coworkers. The tone of voice or the look that accompanies a spoken message can profoundly influence the receiver's interpretation. Body language, mannerisms, and other nonverbal cues impact the interpretation.

Senders convey messages that are not intended, not directed to anyone in particular, or not even meant to be messages. For example, a person can outwardly express personal disappointment; however, others may interpret the look as a message of disapproval aimed at them.

The Medium
The medium refers to the method used to convey the information from the sender to the receiver. The medium can be words that are spoken by the sender and heard directly by the receiver. Spoken words or sounds can also be transmitted as electromagnetic waves through a radio system. Written words, pictures, symbols, and gestures are all examples of messages transmitted through a visual medium. The

sender should consider the circumstances, the nature of the message, and the available methods before sending a message.

When information has to be transmitted to subordinates, a fire officer can post a notice on a bulletin board, announce it at a formal lineup, or mention it during a firehouse meal. The medium that is chosen influences the importance that is attached to the message. If it is really important, the fire officer might announce the key points at lineup, and then direct all members to read a written document and sign a sheet to acknowledge that they have read and understood it.

The Receiver
The receiver is the person who receives and interprets the message. There are many opportunities for error in the reception of a message. Although it is up to the sender to formulate and transmit the message in a form that should be clearly understandable to the receiver, it is the receiver's responsibility to capture and interpret the information. The same words can convey different meanings to different individuals, so the receiver does not automatically interpret a message as it was intended. In the fire service, the accuracy of the information that is received can be vital, so both the sender and the receiver have responsibilities to ensure that messages are properly expressed and interpreted. Many messages are directed to more than one person, so there can be multiple interpretations of the same message.

Feedback
The sender should never assume that information has been successfully transferred unless there is some confirmation that the message was received and understood. Feedback completes

Near Miss REPORT

Report Number: 08-0000332

Event Description: While relieving the off-going shift, a response for a commercial fire was dispatched. While pulling out of the station, a column of smoke could be seen. I immediately requested a second alarm assignment and advised units we were not on scene. Upon arrival, we had an extremely large warehouse with heavy smoke showing out of an occupancy that manufactured cardboard boxes. Crew of 4 stretched 200' 2 1/2" line on side C (pump operator remained outside) and there was zero visibility.

The line stretched was too short to reach the fire so I requested an additional 200'. We found origin of fire in the storage area. Before the nozzle was opened, we witnessed a large explosion. I yelled to hit the wall. I then tried to call emergency traffic and give information to command. Command had too much radio traffic (not priority information) with a unit not even assigned. We were unable to communicate and never were able to inform command nor receive acknowledgment. Afterward, we were able to see a 20 × 40 room which housed breaker panels. They blew up when impinged by fire.

Lesson Learned: Prioritize radio traffic, whether a command function or task assignment. Radio traffic should be limited from non-assigned units. Command should prioritize and units operating in the IDLH conditions have the highest priority. Protocol for calling emergency traffic was followed but was not successful. On the equipment front, the emergency call button was not activated. This would have alerted dispatch to command that a unit was in trouble. Radio has capacity to increase wattage when the emergency button is activated.

the communication cycle by confirming receipt and verifying the receiver's interpretation of the message. The importance that is attached to feedback depends on the nature of the message.

When relaying critical information during a stressful event, the sender should have the receiver repeat back the key points of the message in his or her own words. Without feedback, the sender cannot be confident that the message reached the receiver and was properly interpreted.

Effective communication should contain all five components of the communications process; if one or more is missing, communication does not occur.

The Basics of Effective Communication Skills

Almost every fire officer task depends on the ability to effectively communicate. An officer must be effective as both a sender and a receiver of information and, in many cases, as a processor of information that has to flow within the organization.

Active Listening

Success as a supervisor depends on how freely your subordinates talk to you, keep you informed, and tell you what is bothering them. Success also depends on your effectiveness in communicating with your superiors and keeping them informed.

The ability to listen becomes more important as one advances up the organization. A fire fighter must listen effectively to information that is coming from a higher level. A fire officer must also be able to listen to company members or other subordinates and accurately interpret their comments, concerns, and questions.

Listening is a skill that must be continually practiced to maintain proficiency. A typical listening situation for a fire officer could be a meeting with a fire fighter who is expressing a problem or a concern. Listening, in a face-to-face situation, is an active process that requires good eye contact, alert body posture, and frequent use of verbal engagement **Figure 14-1 ▶**. The purpose of active listening is to help the fire officer understand the fire fighter's viewpoint in order to solve an issue or a problem.

The following techniques may help to improve your listening skills:

- Do not assume anything. Do not anticipate what someone will say.
- Do not interrupt. Let the individual who is trying to express a point or position have a full say.
- Try to understand the need. Often, the initial complaint or problem is a symptom of the real underlying issue. Look for the real reason the person wants your attention.
- Do not react too quickly. Try not to jump to conclusions. Avoid becoming upset if the situation is poorly explained or if an inappropriate word is used. The goal is to understand the other person's viewpoint.

Stay Focused

It is easy to get sidetracked and bring unrelated issues into a conversation. Directed questioning is a good method to keep a conversation on the topic at hand. If the speaker starts to ramble, ask a specific question that moves the conversation back to the

Figure 14-1 Listening is an active process that requires good eye contact, alert body posture, and frequent use of verbal engagement.

appropriate subject. For example, if you are attempting to find out why a fire fighter failed to wear a uniform shirt to roll call and he starts talking about how he does not like the fire department patch, you could ask, "How does the patch affect whether or not you wear your uniform shirt to roll call?"

Ensure Accuracy

A fire officer needs to have up-to-date information on the fire department's standard operating guides, policies, and practices. An officer should also be familiar with the personnel regulations, the approved fire department budget, and, if applicable, the current union contract.

If a fire fighter is misinterpreting a factual point or a departmental policy, the fire officer is obligated to clarify or correct the information. Ignoring an inaccurate statement may only foster erroneous information on the grapevine. If necessary, obtain the accurate information from a chief officer or headquarters staff.

A fire officer sometimes has to exercise control over what is discussed in the work environment. Fire station discussions can easily encroach on subjects that are intensely personal, such as politics, religion, or social values, and can quickly escalate into confrontations. To proactively address potential problems, the fire officer needs to establish some ground rules about the topics and level of intensity when discussing issues.

Keep Your Supervisor Informed

The administrative fire officer depends on you to share information about what is happening at the fire station or in your work environment. Three areas where a fire officer needs to keep the chief officer informed are:

- Progress toward performance goals and project objectives: A fire officer needs to keep his or her chief apprised of work performance progress, such as training, inspections, smoke detector surveys, and target hazard documentation. It is especially important to let the chief know about anticipated problems early enough to get help and to keep the projects running on time.
- Matters that may cause controversy: The chief should be informed about conflicts with other fire officers or between shifts, or conflict that extends outside the organization. The chief also needs to know about any disciplinary issues or a controversial application of a departmental policy.
- Attitudes and morale: A fire officer spends the workday with a group of fire fighters at a single fire station, whereas the command officer spends much of the workday in meetings or on the road. The fire officer should communicate regularly with the chief about the general level of morale and fire fighter response to specific issues.

The Grapevine

Every organization has an informal communication system, often known as the "grapevine." The flow of informal and unofficial communications is inevitable in any organization that involves people. The grapevine flourishes in the vacuum created when the official organization does not provide the workforce with timely and accurate information about work-related issues. Much of the grapevine information is based on incomplete data, partial truths, and sometimes outright lies.

A fire officer can often get clues about what is going on but should never assume that grapevine information is accurate and should never use the grapevine to leak information or stir controversy. In addition, a fire officer needs to deal with grapevine rumors that are creating stress among the fire fighters by determining accurate information and sharing it with subordinates.

Overcoming Environmental Noise

Environmental noise is a physical or sociological condition that interferes with the message. Within this definition, "noise" includes anything that can clog or interfere with the medium that is delivering the message.

Physical noise includes background conversations, outside noises, or distracting sounds that make it difficult to hear. For example, siren noise makes it difficult to communicate over the radio. Digital radios and cell phones can suffer from poor reception or static. A rapid flow of incoming messages can overload the receiver's ability to deal with the information, even if the words come through clearly Figure 14-2 ▶. Darkness or bright flashing lights make it difficult to see clearly and interpret a visual message.

Another form of noise interference is when the receiver is distracted or thinking about something else and blocks out most or all of an incoming message. The human brain has difficulty processing

Figure 14-2 Trying to talk on a cellular phone with poor reception is one example of physical noise interfering with communication.

more than one input source at a time. For example, you could be having a conversation on the telephone with one individual when someone else calls your name; attention is diverted while you try to determine who called you and why. During that moment, you are likely to miss part of the telephone conversation. A similar situation occurs if the sender or receiver is unable to concentrate on or respond to the message due to fatigue, boredom, or fear.

Sociological environmental noise is a more subtle and difficult problem. Prejudice and bias are examples of sociological environmental noise. If the receiver does not believe that the sender is credible, the message is ignored or results in an inadequate response. The following list provides seven suggestions to improve communication:

- Do not struggle for power. Focus attention on the message, not on whether the sender has the authority to deliver it to the receiver. The situation, not the people involved, should drive the communication and the desired action.
- Avoid an offhand manner. If you want your information to be taken seriously, you must deliver it that way. Be clear and firm about matters that are important.

- Words have meaning. Select words that clearly convey your thoughts, and be mindful of the impact of the tone of voice.
- Do not assume that the receiver understands the message. Encourage the receiver to ask questions and seek clarification if the intent of the message is unclear. A good technique of confirming understanding is for the receiver to repeat the key points of the message back to the sender.
- Immediately seek feedback. If the receiver identifies an error or has a concern about the message, encourage that individual to make the statement sooner rather than later. It is always better to solve a problem or identify resistance while there still is time to make a change.
- Provide an appropriate level of detail. Think about the person who is receiving the message and how much information that individual needs. Consider a fire officer telling a fire fighter to set up the annual hose pressure test. The information needs are different for a fire fighter who has 22 years of experience as a pump operator versus a rookie who has 22 days on the job.
- Watch out for conflicting orders. Check to make sure that your message is consistent with information coming from other sources. A fire officer should develop a network of peers to consult with when dealing with unfamiliar or confusing situations.

These seven suggestions are all intended to improve communications dealing with administration and supervisory activities. The fire ground or emergency scene requires different communications practices.

Emergency Communications

A fire officer needs to communicate effectively during emergency incidents. There is no time to be elegant when communicating within the emergency environment. The direct approach entails asking precise questions, providing timely and accurate information, and giving clear and specific orders.

Under the time pressure of an emergency incident, management of the communications process can be as important as communicating effectively. The incident commander has to manage the exchange of information so that the most important messages go through and lower-priority or unnecessary communications do not get in the way.

The key points for emergency communications are:
- Be direct.
- Speak clearly.
- Use a normal tone of voice.
- If you are using a radio, hold the microphone about 2 inches (5 cm) from your mouth.
- If you are using a repeater system, allow for a time delay after keying the microphone.
- Use plain English rather than 10 codes.
- Try to avoid being in the proximity of other noise sources, such as running engines **Figure 14-3 ▶** .

Figure 14-3 Emergency communications require a direct approach.

"Unit Calling, Repeat . . ."

Radio messages must be accurate, brief, and clear. An officer should be as consistent as possible when sending verbal messages over the radio. The performance goal should be to sound the same and communicate just as effectively when reporting a minor incident as when communicating under intense stress.

Recordings of radio messages transmitted during emergency incidents are an effective training tool. Listening to others allows a fire officer to identify and emulate techniques that are clear and precise, leaving no doubt about the intent of the message. Listening to recordings of your own radio communications provides valuable feedback, allowing you to compare the output with your thought process at that moment.

Initial On-Scene Size-Up

The initial report should describe what you have, state what you are doing, and provide direction for other units that will be arriving. Routinely using the same terminology and format helps ensure that nothing is missed. An example might sound like this:

Engine 1: Dispatch, from Engine 1.

Dispatch: Go ahead Engine 1, this is dispatch.

Engine 1: Engine 1 has arrived at 2345 Central Avenue. We have a two-story, wood-frame, single-family dwelling with heavy smoke showing on side charlie and fire showing from the first floor on side Delta. Engine 1 has a supply line and will be making an offensive attack through side alpha. Engine 1 is establishing Central Avenue Command on side Alpha.

Dispatch: Copy Engine 1 at 2345 Central Avenue with a two-story, wood-frame, single-family dwelling, heavy smoke showing on side Charlie and fire showing on side Delta first floor. Engine 1 is making an offensive attack through side Alpha and establishing Central Avenue Command on side Alpha.

Through this simple exchange of information, everyone who needs to know is informed of the situation and the action being taken by the first-arriving unit, and the stage is set for all other units to take action based on standard operating procedures. The dispatcher's repeat of the key information provides solid confirmation that it was received and understood.

Using the Order Model

The order model is a standard method of transmitting an order to a unit or company at the incident scene. It is designed to ensure that the message is clearly stated, heard by the proper receiver, and properly understood. It also confirms that the receiver is complying with the instruction.

> Command: Ladder 2, from Command.
>
> Ladder 2: Ladder 2, go ahead Command.
>
> Command: Ladder 2, come in on side Charlie and conduct a primary search on the second floor. Also advise if there is any fire extension to that level.
>
> Ladder 2: Ladder 2 going to side Charlie to do a primary search on the second floor and check for fire extension.
>
> Command: Ladder 2, that is correct.

Radio Reports

Company fire officers frequently provide reports over the radio. Radio communications are essential for emergency operations because they provide an instantaneous connection and can link all of the individuals involved in the incident to share important information. During an emergency incident, both the sender and the receiver should strive to make their radio messages accurate, clear, and as brief as possible. Conditions are often stressful, with several parties competing for air time and the receiver's attention. There may be only a limited time to transmit an important message and ensure that it is received and understood.

The individual who is transmitting a radio report often feels an intense pressure to get the message out as quickly as possible. The realization that a large audience could be listening often adds to the sender's anxiety level.

When given the option, many fire officers prefer to use a telephone or face-to-face communications instead of a radio to transmit complicated or sensitive information. These options allow a more private exchange of information between two individuals, including the ability to discuss and clarify the information.

Written Communications

Written communication is needed to document routine and extraordinary fire department activities. It establishes institutional history and is the foundation of any activity that the fire department wishes to accomplish. As a senior chief explained, "If it is not written down, it did not happen."

Informal Communications

Informal communications include internal memos, emails, instant messages, and messages transmitted via mobile data terminals. Informal reports have a short life and are not archived as permanent records. They are used primarily to record or transmit information that will not be needed for reference in the future.

Some informal reports are retained for a period of time and may become a part of a formal report or investigation. For example, a written memo could document an informal conversation between a fire officer and a fire fighter regarding a performance issue. The memo might note that the employee was warned that corrective action is required by a particular date. If the performance is corrected, the warning memo is discarded after 12 months. If the performance remains uncorrected, the memo becomes part of the formal documentation of progressive discipline.

Formal Communications

A formal communication is an official fire department document printed on business stationery with the fire department letterhead. If it is a letter or report intended for someone outside the fire department, it is usually signed by the fire chief or a designated staff officer to establish that it is an official communication. Subordinates often prepare these documents and submit them for an administrative review to check the grammar and clarity. A fire chief or the staff officer who is designated to sign the document performs a final review before it is transmitted.

The fire department maintains a permanent copy of all formal reports and official correspondence from the fire chief and senior staff officers. Formal reports are usually archived.

Standard Operating Procedures

Standard operating procedures (SOPs) are written organizational directives that establish or prescribe specific operational or administrative methods to be followed routinely for the performance of designated operations or actions. SOPs are intended to provide a standard and consistent response to emergency incidents as well as personnel supervisory actions and administrative tasks. SOPs are a prime reference source for promotional exams and departmental training.

SOPs are formal, permanent documents that are published in a standard format, signed by the fire chief, and widely distributed. They remain in effect permanently or until they are rescinded or amended. Many fire departments conduct a periodic review of all SOPs so they can revise, update, or eliminate any that are outdated or are no longer applicable. Any changes in the SOPs must also be approved by the fire chief.

General Orders

General orders are formal documents that address a specific subject, policy, condition, or situation. They are usually signed by the fire chief and can be in effect for various periods, from a few days to permanently. Many departments use general orders to announce promotions and personnel transfers. Copies of the general orders should be available for reference at all fire stations.

Announcements

The formal organization may use announcements, information bulletins, newsletters, websites, or other methods to share additional information with fire department members. These methods are generally used to distribute short-term and nonessential information that is of interest.

Legal Correspondence

Fire departments are often called on to produce copies of documents or reports for legal purposes. Sometimes the fire department is directly involved in the legal action, whereas in other cases, the action involves other parties but relates to a situation in which the fire department responded or had some involvement. It is not unusual for a fire department to receive a legal order demanding all of the documentation that is on file pertaining to a particular incident. This is a task that can require extensive time and effort. In any situation for which reports or documentation is requested, consult with legal counsel.

The fire officer who prepared a report may be called on, sometimes years later, to sign an affidavit or appear in court to testify that the information provided in the report is complete and accurate. The fire officer might also have to respond to an **interrogatory**, which is a series of written questions asked by someone from an opposing party. The fire department must provide written answers to the interrogatories, under oath, and produce any associated documentation. If the initial incident report or other documents provide an accurate, factual, objective, complete, and clear presentation of the facts, the response to the interrogatory is often a simple affirmation of the written records. The task can become much more complex and embarrassing if the original documents are vague, incomplete, false, or missing.

Reporting

Some reports are prepared on a regular schedule, such as daily, weekly, monthly, or yearly. Other types of reports are prepared only in response to specific occurrences or when requested. To create a useful report, the fire officer must understand the specific information that is needed and provide it in a manner that is easily interpreted.

Types of Reports

Some reports are presented orally, and others are prepared electronically and are entered into computer systems. Some reports are formal, whereas others are informal. This chapter covers some of the common elements of reporting and provides examples of the most common types of reports that a fire officer might have to prepare. More experienced fire officers can also provide assistance in mastering the reporting requirements of a new position.

Verbal Reports

The most common form of reporting is verbal communication from one individual to another, either face-to-face or via a telephone or radio. In order to be effective, the transfer of information must be clear and concise, using terminology that is appropriate for the receiver.

Face-to-face conversation is the most effective means of conveying many types of information. The sender and the receiver can engage in a two-way exchange of information that incorporates body language, facial expressions, tone of voice, and inflection. When verbal communication must be conducted over radio or telephone, supplementary expressions are sacrificed.

Routine Reports

Routine reports provide information that is related to fire department personnel, programs, equipment, and facilities. Most fire departments also require company officers to maintain a **company journal** or log book **Figure 14-4 ▸**. A fixture in fire stations since the 19th century, the company journal provides an extemporaneous record of all of the emergency, routine, and special activities that occur at the fire station. Some fire departments have advanced from keeping the journal in a handwritten, hard-covered book to maintaining it as a database on the station computer.

The company journal serves as a permanent reference that can be consulted to determine what happened at that fire station at any particular time and date, as well as who was involved. The general orders usually list all of the different types of information that must be entered into the company journal. The company journal is the place to enter a record of any fire fighter injury, liability-creating event, and special visitors to the station.

Morning Report to the Administrative Fire Officer

Most career fire officers are required to provide some type of morning report to their superior officer. This report is made by telephone or on a simple form that is transmitted by fax or email.

One purpose of the morning report is to identify any personnel or resource shortage as soon as possible. For example, if the driver/operator calls in sick, the driver/operator from the off-going shift might have to remain at work to keep the engine company in service. The administrative fire officer needs to know that the company is short a driver/operator and has a person working involuntary overtime or holdover.

Monthly Activity and Training Report

The monthly activity and training report documents the company's activity during the preceding month. These reports typically include the number of emergency responses, training activities, inspections, public education events, and station visits that were

Voices of Experience

When fire officers discuss the topic of communication, the focus is often just on the incident scene. While this is critically important, we also need to remember the 90 percent of communication that takes place under nonemergency conditions—in the fire station, at the kitchen table, in the meeting room, etc. Just as an incomplete or wrongly communicated message can cause problems on the fire ground, it can also cause problems in these situations as well.

As I have progressed through my fire service career I have seen the importance of communications. Through my own experience as well as watching others, I have learned some important lessons. These were sometimes learned the hard way and each of them was valuable.

One lesson is knowing which method is most appropriate for communicating a given topic. In recent years, email has become a prevalent mode of communications for many fire departments. Unfortunately, what is sometimes gained in terms of *efficiency* and speed can be lost in terms of *effectiveness*. While email can be great for communicating "bullet points" or getting information to a large group, it can have negative consequences when it gets used to debate topics or carry out arguments or when the sender is tired, overloaded, or aggravated at the time they wrote it. Once the send button is hit, you don't have control over how the message is perceived and trying to correct misconceptions after the fact can be difficult.

Another lesson involves knowing the person or group you are communicating with. This ties in with knowing which methods work best for a given topic. It is important not to assume that what works best for you or what you prefer is what works best for the other person. Some individuals may have a preference that you "get right to the point," whereas others may need to hear more background information first. As we move up in rank, our words also have more impact and consequence and this must be considered as well.

Finally, there is the important lesson of seeking first to understand, then to be understood. When we allow our perceptions and what we *think* is happening or what we *think* the other person's motivation is to diminish how well we actually listen, it sets the stage for further misunderstanding. These are the times when we need to move away from the computer, let the office phone ring, and really focus on what the other person is trying to say.

Rudy Horist
Assistant Fire Chief
Elgin Fire Department
Elgin, Illinois

> **"What is sometimes gained in terms of efficiency and speed can be lost in terms of effectiveness."**

Fire Station 14 Company Journal

out	in	inc #	address/comments				sign
	0700		Thursday, Oct 11 - B shift on: Capt Smyth OIC				
			Engine	**Rescue**	**Medic**	**Chief 9**	
			Lt. Seth (OT)	Cpt Smyth	Lt. Willow	BC Grave	
			Tech Ayres	Tech Cegar	FF Villani		
			FF Robinson	FF Chase			
			FF Bartlow	FF Adams			
			Not here:				
			Reserve Medic 114 running as Medic 7				
			Lt. Johnson on Official Representation				
			Fire Fighter Tolliver on Annual Leave				
			Tech Madison detailed to Engine 10				
				Engine	Rescue	Medic	
			Radio	Ayres - 4	Cegar - 4	Villani - 2	
			Keys	Seth	Smyth	Willow	
			Drugs	Robinson	xxxxx	Villani	
	0745		Medic 14 disinfected				Vil
1030	0800		Rescue 14 to Station 22 for drill				Ceg
1011	1120	0345	7111 Hogarth St - injury		Medic		Vil
1134	1245	0375	10400 Richmond Hy - 2nd Rescue		Rescue		Ceg
1217	1244	0500	9103 Cross Chase Court - ALS		Medic		Vil
1600	1315		Facilities working on bathrooms				
1302	1430	0711	4110 Green Springs Dr #101 - sick		Medic		Vil
1533	1551	0987	I-95 N at Newington - Crash		Rescue		Ceg
1533	1607	0987	I-95 N at Newington - Crash		Engine		Sth
1533	1635	0987	I-95 N at Newington - Crash		Medic		Vil
	1840		12-lead monitor #1014 damaged on event #0987				Ceg
	1900		EMS 9 delivered loaner 12-lead #0733				
1836	1900	1135	9112 Chapel Oaks Way - auto fire		Engine		Sth
1845	1911	1270	8902 Harrivan Lane - chest pain		Medic		Wil
	2015		FF Sorce on sick leave for tomorrow				
2044	2340	1350	7401 Eastmoreland Ave - 2nd alarm		Rescue		Ceg
2109	2253	1350	7401 Eastmoreland Ave - 3rd alarm		Engine		Ceg
2315	2120		Quint 22 filling in at Station 14				MJ
			Friday, October 12				
0112	0201	0099	10614 Hampton Rd - trob breathing		Medic		Vil
0112	0144	0099	10614 Hampton Rd		Engine		Sth
0249	0305	0143	6600 Springfield Mall - Alarm bells		Engine		Sth

Figure 14-4 Example of a company journal.

conducted during the previous month. Some monthly reports include details such as a list of the number of feet of hose used and the number of ladders deployed during the month.

In many cases, the officer delegates the preparation of routine reports that do not involve personnel actions or supervisory responsibilities to subordinates; however, the officer is responsible for checking and signing the report before it is submitted.

Two other routine reports are the annual fire fighter performance appraisals and the fire safety inspection. These reports are covered in detail in Chapter 8, Evaluation and Discipline, and Chapter 12, Pre-Incident Planning and Code Enforcement.

Some fire companies or municipal agencies post a version of a monthly report on their public website. The reports often include digital pictures of the incidents and the people involved.

Special events or unusual situations that occurred during the month might also be included; for example, the monthly report might note that the company provided standby coverage for a presidential visit or participated in a local parade.

Incident Reports

An incident report is required for every emergency response. The nature and the complexity of the report depend on the situation: minor incidents generally require simple reports, whereas major incidents require extensive reports.

The **National Fire Incident Reporting System (NFIRS)** is a nationwide database managed by the U.S. Fire Administration that collects data related to fires and other incidents. The first-arriving officer completes an NFIRS data report for each response, including a narrative description of the situation and the action

Civilians Rescued From Burning Home

Twenty-eight fire fighters responded to an early morning house fire in the Grandview district. The first 9-1-1 call was at 10:47 pm on Monday December 10, 2012, reporting smoke coming from a three-story townhome.

Metro County fire fighters arrived at 11:07 pm, encountered smoke and flames coming from the first floor windows at 1928 Braniff Boulevard. Two elderly females were found on the third floor. They were taken out of the building through a window via an aerial tower, treated by paramedic/ fire fighters, and transported to University Hospital.

It took 17 minutes of aggressive fire suppression before Battalion Chief Devon Jones declared the fire "under control." The first floor of the townhouse was extensively damaged, with heat and smoke damage to the adjacent townhouses.

The cause of the fire remains under investigation. Preliminary results show that the fire started in the kitchen. Smoke detectors were in the home, but batteries were removed.

The monetary loss has not been calculated.

Submitted by T. L. Gaines, Metro County Fire Department

MCFDPIO@metrocounty.gov (111) 555-3473

Figure 14-5 Sample incident report.

that was taken. The full incident report includes a supplementary report from the officer in charge of each additional unit that responded.

Many departments use a narrative format when reporting on incidents **Figure 14-5 ▲**.

Some incidents require an **expanded incident report narrative**, in which all company members must submit a narrative description of their observations and activities during an incident. The fire officer should anticipate the need to provide expanded incident reports on incidents in which:

- The fire company was one of the first-arriving units at a fire with a civilian fatality.
- The fire company was one of the first-arriving units at an incident that has become a crime scene or an arson investigation.

- The fire company participated in an occupant rescue or other emergency scene activity that would qualify for official recognition, award, or bravery citation.
- The fire company was involved in an unusual, difficult, or high-profile activity that requires review by the fire chief or designated authority.
- Fire company activity occurred that may have contributed to a death or serious injury.
- Fire company activity occurred that may have created a liability.
- Fire company activity occurred that has initiated an internal investigation.

The author of a narrative should describe any observations that were made en route or on the scene and fully document his or her actions related to the incident. The narrative should provide a clear mental image of the situation and the actions that were taken.

Infrequent Reports

Infrequent reports usually require a fire officer's personal attention to ensure that the report's information is complete and concise. The special reports include:

- Fire fighter injury report
- Citizen complaint
- Property damage or liability-event report
- Vehicle accident report
- New equipment or procedure evaluation
- Suggestions to improve fire department operation
- Response to a grievance or complaint
- Fire fighter work improvement plans
- Research report

Some situations require two or more different report forms to cover the same event. For example, if a fire engine collides

Safety Zone

Knowing Reporting Procedures

A fire officer is responsible for the timely and proper completion of property damage and injury reports. That does not mean that the fire officer must personally fill out every form. All fire fighters should know how to fill out the required administrative paperwork for a minor injury, blood-borne pathogen exposure, or damage to fire department property.

The late notification of the duty officer on infectious disease encounters and the tardy or incomplete documentation of injuries or toxic exposures are continuing challenges to fire fighter health and safety. All fire fighters should know the reporting procedures to protect them throughout their careers.

SUPERVISOR'S ACCIDENT REPORT
WORKERS' COMPENSATION CLAIMS

Claims Management Inc.
PO Box 342
Sacramento, CA 95812-3042
(916) 631-1250
FAX (916) 635-6288

DATE & TIME
REPORTED:

OSHA CASE NO:

COMPANY Scotts Valley Fire Protection District	LOCATION 7 Erba Lane, Scotts Valley, CA, 95066	LOCATION CODE NO: 1100

A. EMPLOYEE	NAME		JOB TITLE	
	DEPARTMENT Scotts Valley Fire Protection District 1100	☒☒ LOST TIME ☒☒ NO LOST TIME	☒☒ FIRST AID	

B. TIME AND PLACE OF ACCIDENT	DATE	HOUR	DEPARTMENT 1100	IMMEDIATE SUPERVISOR
	IDENTIFY EXACT LOCATION WHERE ACCIDENT OCCURRED *(Be specific)*			
	JOB OR ACTIVITY AT TIME OF ACCIDENT *(Be specific)*			

C. WITNESSES - *List of Names and Addresses*

Name	Address

D. DESCRIBE THE ACCIDENT/ACCIDENT CAUSE - *Please be specific*

E. UNSAFE ACT/CORRECTIVE ACTION TAKEN - *Include both employee and supervisor corrective actions to prevent future occurrences.*

EMERGENCY - WENT TO THE DOCTOR ☒☒ YES ☒☒ NO	If yes, please fill out the following information: Name of Doctor: _____ Address of Doctor: _____

☒☒ NON-EMERGENCY, BUT PLAN ON SEEING A PHYSICIAN
Physician's Name: _____

Figure 14-6 Supervisor's accident report. Courtesy of Claims Management, Inc.

with a private vehicle at an intersection while responding to an emergency call, the fire officer will spend significant time completing a half-dozen forms, even if there are no injuries. There is a supervisor's report and the driver must complete an accident report. Another form identifies the damage to the fire apparatus and any repairs that are needed. The local jurisdiction's risk management division may also require a report. The insurance company that covers the fire apparatus usually requires the completion of another form.

A **supervisor's report** is required by state workers' compensation agencies whenever an employee is injured. This report is the control document that starts the state file relating to an injury or a disability claim **Figure 14-6**.

The supervisor's first report must be submitted within 24 to 72 hours of the incident, after performing an investigation into the facts. The report often identifies who is injured, the nature of the injury, how the incident occurred, what practices or conditions contributed to the incident, any potential loss, the typical frequency of occurrence, and what actions will be taken to prevent the same accident from occurring again.

Special Reports

A **chronological statement of events** is a detailed account of activities, such as a narrative report of the actions taken at an incident or accident. A **recommendation report** is a document that suggests a particular action or decision.

Some reports incorporate both a chronological section and a recommendation section. For example, the fire fighter line-of-duty death investigations produced by the National Institute for Occupational Safety and Health (NIOSH) include a chronological report of the event, followed by a series of recommendations that could prevent the occurrence of a similar situation.

The goal of a decision document is to provide enough information and persuasion so the intended individual or body accepts your recommendation. This type of report includes recommendations for employee recognition or formal discipline, for a new or improved procedure, or for adoption of a new device. A decision document usually includes:

- Statement of problem or issue: One or two sentences.
- Background: Brief description of how this became a problem or an issue.
- Restrictions: Outline of the restrictions affecting the decision. Factors such as a federal law, state regulations, local ordinances, budget or staff restrictions, or union contract are all restrictions that could affect a decision.
- Options: Where appropriate, provide more than one option and the rationale behind each option. In most cases, one option is to do nothing.
- Recommendation: Explain why the recommended option is the best decision. The recommendation should be based on considerations that would make sense to the decision maker. If the recommendation is going to a political body and involves a budget decision, it should be expressed in terms of lower cost, higher level of service, or reduced liability. The impact of the decision should be quantified as accurately as possible, with an explanation of how much money it will save, what new levels of service will be provided, or how much the liability will be reduced.
- Next action: The report should clearly state the action that should be taken to implement the recommendation. Some recommendations may require changes in departmental policy or budget. Others could involve an application for grant funds or a request for a change in state or federal legislation.

Writing a Report

Reports should be accurate and present the necessary information in an understandable format. The report could be to brief the reader, or it might provide a systematic analysis of an issue.

If the fire chief wants only to be briefed on a new piece of equipment and the fire officer delivers a fully researched presentation, both parties would be frustrated.

The fire officer must consider the intended audience. Consider a study that proposes replacing four engine and three ladder companies with three quints. If the report is internal, the technical information includes fire department jargon. If the report is for the city council, with copies going to the news media, many of the terms and concepts would have to be explained in detail.

Determine what will be the mode and format when preparing the report. This can range from an internal memo format with one or two pages of text to a lengthy formal report that includes photographs, charts, diagrams, and other supporting information.

Some reports require a recommendation. The fire officer will need to analyze and interpret data that are maintained in a database or a records management system. The fire department's records management system contains NFIRS run reports, training records, occupancy inspection records, pre-incident plans, hydrant testing records, staffing data, and company activity records.

A computer-aided dispatch system uses a combination of databases to verify addresses and determine the units that should be dispatched and maintains information that can be retrieved relating to individual addresses. All of the transactions performed by a computer-aided dispatch system are recorded in another database. This information can be retrieved to examine the history of a particular incident or to identify all of the incidents that have occurred at a particular location.

The fire officer might be looking at variances, such as differences between the projected budget and actual expenses in different accounts in a given year. When a variance is encountered, the analysis should consider both the cause and the effect. If the budget included an allocation of $10,000 for training and the expenditure records indicate that only $500 was spent, there would be a variance of $9500 that could be reallocated for some other purpose. This variance could also indicate that very little training was actually conducted.

Presenting a Report

A fire officer should be prepared to make an oral presentation of a decision document or to speak to a community group on a fire department issue. Using the written report as a guide, a verbal presentation should consist of four parts:

1. Get their attention: Have an opening that entices the audience to pay attention to your message.
2. Interest statement: Immediately and briefly explain why listeners should be interested in this topic.
3. Details: Organize the facts in a logical and systematic way that informs the listeners and supports the recommended decision.
4. Action: At the close of your presentation, ask the audience to take some specific action. The most effective closing statements are actions that relate directly to the interest statement at the beginning of the presentation.

Assessment Center Tips

Verbal Presentations

A promotional candidate should be prepared for two different types of verbal presentations. The first type is a meeting with a citizen or community group to explain a fire department practice or issue. This is equivalent to a verbal presentation of a decision document.

The second type of presentation could follow an emergency incident scenario. After the incident management portion of the exercise is completed, the candidate could be required to make a presentation of the incident as if he or she is speaking to the press or reporting to a senior command officer.

Preparing a News Release

Fire organizations need to be able to communicate effectively through the local mass media. A news release allows a fire department to reach a large audience at virtually no cost. The release could be structured to promote a public education message, draw attention to the community's risk problems, announce a new program or the opening of a new facility, or simply develop good public relations. A well-prepared news release encourages the media to cover your story.

The first step in preparing a news release is to formulate a plan. Consider what the release is to accomplish. Is it an urgent fire safety message or an invitation to the fire department picnic? Who is the target audience? Most messages are directed toward some particular group or audience. A safety message about a smoke alarm program is directed toward those residents who do not have smoke alarms.

During this first step, you should also identify what makes the story interesting and worthy of the time and energy the media source must expend to present it. The media may have hundreds of releases that they must prioritize in order to decide which ones to give attention. If the news release is about smoke alarms, the release should emphasize why this subject is important.

The second step is to develop the concept and write the release. The format should be clear, concise, and well organized. The first paragraph or two should cover the "who, what, when, where, and why" of the story. Successive paragraphs should provide further information, progressing from the most important information to the least important. Do not be wordy in a press release, and stick to the facts. The idea is to give the media enough information to convince them that the story warrants coverage. Each media outlet considers the release in the context of whether it would be of interest to their audience. If the release is on target, a reporter will follow up.

The use of the department's letterhead helps draw the attention of the news agency. Formatting can also help bring attention to the release. Use at least 1-inch (2.5-cm) margins, double space the text, and make the news release fit on one page. Draw attention by keeping it neat and clearly organized, and make sure that there are no typographical or grammatical errors.

At the top of the page, print "NEWS RELEASE" in all capital letters. The time and date should appear on the following line. Separate this heading from the text by a series of five to seven pound signs (#######). Pound signs should also be used to separate the text from the ending, which should include a contact name and number for further information.

The last step is to get the news release out to the media. It can be distributed through fax, email, or the postal service if there is sufficient time. It is important to distribute the release to all media outlets equally.

Using Computers

Computers are made up of physical components called hardware. Hardware includes the keyboard, monitor, printer, and central processing unit (CPU). The CPU houses the internal hardware components, such as the processor, memory, hard drive, and motherboard. The motherboard is a circuit board that provides the connections to all of the components of the computer. The processor is the component of the computer that actually does the calculations.

Memory is also attached to the motherboard. Memory is like the inbox on your desk. It holds all the items you are currently using and keeps them readily at hand. The hard drive is also attached to the motherboard. The hard drive stores information in a more permanent fashion, like the filing cabinet in your office. When the memory and hard drive are nearly full, it takes significantly longer to find the files.

Software is needed to instruct the computer on what to do. Software contains all of the instructions that tell the computer how to function.

Operating System

The operating system performs basic tasks, such as recognizing input from the keyboard, sending visual output to the monitor, keeping track of files, and controlling external devices, such as printers. Every computer needs an operating system to manage application programs. The application programs interface with the operating system to allow them to perform their tasks. Common operating systems are Linux, Mac OS, and Windows.

Word Processing

Word processing programs allow a user to type letters, memos, and reports. Most programs also have templates, which are premade layouts of the most common letters, memos, and reports, as well as fax cover sheets. Common word processing programs are Microsoft Word, GoogleDocs, and Corel WordPerfect.

Spreadsheets

Spreadsheet software is used to tabulate numbers and information in columns and rows. It is often used to analyze data and to develop charts. A spreadsheet resembles a ledger sheet, with the columns represented by letters and the rows identified by numbers. The intersection of a column and a row is called a cell. A cell can contain numbers, letters, or a formula.

A formula is used to make calculations. For example, a formula in cell A2 might read "=SUM(C3+D4)," which means that it adds the number in cells C3 and D4 and places the total in cell A2.

Spreadsheets have immense calculation capabilities. Each brand of software uses different methods of inputting formulas, so the user needs to learn the specific program being used. Some of the more common spreadsheet programs are Microsoft Excel, Corel Quattro Pro, and OpenOffice.org Calc.

Presentation Software

Presentation software is used to place information in a format that allows the audience to visually understand the message being presented. Presentation software has taken over the functions that used to be performed with slide shows, overhead transparencies, and video recorders. Today, a computer and a projection system allow the user to create slides and videos that can be projected onto a screen, produce handouts, and manage instructor notes. Common presentation software programs are Microsoft Power-Point, OpenOffice.org Impress, and Apple Keynote.

Communication

The most common form of electronic communication is email, which is equivalent to a memo or letter transmitted via the Internet. Email programs allow the user to type messages, attach files, and send them instantaneously to another email address. The most common email programs are Microsoft Outlook, Google Gmail, Hotmail, Yahoo!, and AOL.

Database

A database is composed of a series of interrelated tables. For example, the department might create a table that lists all of the addresses of every household in town, another table that lists all the addresses where fires have occurred, another that tracks the addresses where every smoke alarm has been installed, and yet another that lists the public education events by address. Using database software, reports can be generated to pull data from any of the various tables. For example, if the department wanted to know whether there has been a reduction in the number of fires in areas where public education has been delivered, this information could be extracted from the database very rapidly. Without a database, it would be very time consuming to search for this information. The most common database programs are Microsoft Access, Oracle, MySQL, and FileMaker.

Digital Mapping

Almost all of North America, topography, streets, utilities, and buildings have been translated into a spatial description that allows the creation of a digital representation of a neighborhood, street, or housing project. The process starts when a hard copy of a map or survey plan is transferred into a digital medium using a computer-aided design (CAD) program. The information is stored in a **geodatabase**, an object-relational database that uses geometry data types to represent the location of an object in the physical world.

The spatial description can be represented in a graphical form using programs like ESRI ArcGIS, MapInfo, MapServer, Microsoft MapPoint, and Google Earth. Two- and three-dimensional maps can be generated, appearing as directions in a vehicle navigation system or as an aerial view of a neighborhood. The latest systems have integrated satellite and street-level images to produce a virtual picture of a property.

Geographic Information System

A vehicle navigation system is one example of a **geographic information system (GIS)**. A GIS captures, stores, analyzes, manages, and presents data that is linked to location.

Tom Wieczorek, executive director for the Center for Public Safety Excellence, describes how intelligent GIS can support fire departments in developing deployment analysis and risk assessment. Adding incident history, building description, and population demographics to the GIS allows development of a map ranking the level of incident risk.

Going from retrospective to prospective, early event detection and situation analysis programs like FirstWatch provide real-time information to a dispatch center. Using operational research models developed by the RAND Corporation for the Fire Department of New York (FDNY) in the 1970s, fire companies can be moved to meet the anticipated workload during a dynamic event or high-demand period.

Information Management

Many organizations operate in paperless environments, where everything is stored and processed electronically. This requires an advanced management system to allow information to be filed appropriately so that data can be retrieved when they are needed in the future. A lost electronic file is no different from a lost paper file; it is the same as no file at all.

Many fire departments have local networks that link every computer within the organization. Each computer is connected to a server, where information can be stored, shared, and protected from loss. Each user is given a private storage location on the server that no one else can access. Certain users also have access to protected information. For example, the Fire Prevention Division may have a location on the server that holds all of the permit files. Only employees in this division can access these files.

The advantage of having all data stored on the network is that it allows all of the data to be regularly archived in a central location. If data are stored on an individual computer's hard drive, that information is usually not archived regularly by the user. This creates the potential for the permanent loss of files.

Fire Marks

Fire Organization Websites
In the 1990s, college student and Hyattsville, Maryland, volunteer Dave Iannone operated "The Metro DC Fire/Rescue Wire" and the Hyattsville VFD website. The Hyattsville website was one of the most sophisticated sites for its day. Now, most of the organizations within Prince George's County, Maryland, have active websites, with digital pictures, audio, and video clips. Iannone and Chris Herbert started Firehouse.com in 1998; within a decade they had 175,000 registered members and a website that receives a half-million visitors a month.

Blogs, Video Sharing, and Unofficial Websites

A blog is a website where extemporaneous entries are displayed in reverse chronological order. Some blogs provide commentary or news on a particular subject; others function as personal online diaries. A blog combines text, images, and links to other blogs, web pages, and other media. The ability for readers to leave comments in an interactive format is an important part of many blogs. At the time of publication, Technorati was tracking more than 133 million blogs.

Video sharing sites like YouTube have altered the news media. In January 2009, more than 100 million viewers watched over 14.8 billion videos; with more than 20 hours of video uploaded to YouTube every minute. A digital video capture of a 2 A.M. incident can be posted on YouTube within minutes and become the opening item in the morning news, linked to a national fire service discussion board or blog, and have hundreds of comments before 9 A.M.

There are tens of thousands of unofficial fire department websites that, like blogs, post pictures, videos, discussion forums, and incident information. For example, 30 of the 38 volunteer fire companies in Prince George's County, Maryland, have a website, as do the Volunteer Fire and Rescue Association, the Fire Commission, IAFF Local 1619, and the county fire department. There are also individual enthusiast sites. Not surprisingly, some interdepartment conflicts arise due to the pictures, videos, and commentary posted on individual company sites.

Fire investigators complain that the immediate posting of digital pictures and videos interferes with ongoing investigations. Investigators have asked website operators to remove digital images until an investigation is completed.

There is controversy about on-duty fire fighters taking pictures and videos during the initial emergency scene operations.

Getting It Done

Monitoring Through Internet Searches
Fire officers occasionally use Google to search for videos, images, or websites related to their department. Using names, nicknames, and slang related to the department, apparatus, or members, officers have found inappropriate, embarrassing, and potentially libelous digital images on the Internet.

It is unacceptable for the initial fire attack crew members to slow their actions in order to obtain good images. The use of images from helmet- or vehicle-mounted cameras remains unclear, as is the use of pictures taken by off-duty fire department members. Some departments require review of all emergency incident images by an administrative fire officer before they can be posted or shared.

There is also the issue of fire department members who have personal pages on virtual social networking sites like Facebook, MySpace, and LinkedIn. When off-duty fetish, criminal activity, or morally objectionable behavior is linked to a fire department member, the fire fighter's career is endangered. When the activity involves arson, child pornography, or other felony-level criminal activity it will probably result in termination and prosecution.

Summary

Effective communication is a core success skill for the fire officer. The skill becomes more important as the individual rises higher in the fire department rank structure. The 19th-century fire officer needed only the speaking trumpet to direct fire fighters at emergencies. Today's fire officer uses computers and wireless communication to function in the 21st-century environment.

You Are the Fire Officer: Conclusion

Your pulse is pounding and you realize that the dispatcher did not understand the first radio message. Fighting an urge to scream into the radio, you repeat that you have a fire within a townhouse with fire showing on the second floor. You have a report of children trapped on the second floor and you also have fire fighters who have fallen through a floor.

Using clear and concise words, you request a second, or greater, alarm and an EMS mass casualty response. You confirm that the dispatcher understands and is sending you help. You call the fire attack team that fell through the floor. One of the members responds and provides a location, unit, name, assignment, resource (LUNAR) report. They are on the first floor of the townhouse and appear to be in a laundry room. There is no fire on the first floor. The other fire fighter, who was screaming, has a broken leg.

Your 360-degree size-up noted that the structure was two stories in the front and three stories in the rear. The first floor rear includes a sliding glass door. You run back to the rear and force open the door. You assist in the removal of the injured fire fighter as a rising cacophony of sirens, air horns, and Jake brakes announce the impending arrival of help.

Wrap-Up

Chief Concepts

- There are five parts to the communication cycle: message, sender, medium, receiver, and feedback.
- To improve your listening skill, do not assume, do not interrupt, try to understand the need, and do not react too quickly.
- Fire officers who are active listeners are also up to date on the department's guides, policies, and procedures.
- Fire officers keep their bosses informed on progress toward goals and projects, potential controversial issues, fire fighter attitude, and morale.
- Environmental noise can be counteracted by avoiding a power struggle, communicating clearly and firmly, carefully choosing words, confirming that the receiver understands the message, and providing consistent and appropriately detailed messages.
- Informal reports are not considered official fire department permanent records.
- The company journal is the extemporaneous record of fire station activity.
- A written report will have formal, informal, and unanticipated readers.
- A generic decision document has a statement of problem/issue, background, restrictions, options, recommendation, and next action.
- Information technology has expanded the amount and quality of data available to the department, converting data into information that assists in decision making.
- Blogs, video sharing, and unofficial websites create new challenges in the control of digital images and departmental information.

Hot Terms

Chronological statement of events A detailed account of the fire company activities as related to an incident or accident.

Company journal Log book at the fire station that creates an extemporaneous record of the emergency, routine activities, and special activities that occurred at the fire station. The company journal also records any fire fighter injury, liability-creating event, and special visitors to the fire station.

Environmental noise A physical or sociological condition that interferes with the message in the communication process.

Expanded incident report narrative A report in which all company members submit a narrative on what they observed and activities they performed during an incident.

Formal communication Represents an official fire department communication. The letter or report is presented on stationery with the fire department letterhead and generally is signed by a chief officer or headquarters staff member.

General orders Short-term documents signed by the fire chief and lasting for a period of days to 1 year or more.

Geodatabase An object-relational database that uses geometry data types to represent the location of an object in the physical world.

Geographic information system (GIS) A GIS captures, stores, analyzes, manages, and presents data that are linked to a location.

Informal communications Internal memos, emails, instant messages, and computer-aided dispatch/mobile data terminal messages. Informal reports have a short life and are not archived as permanent records.

Interrogatory A series of formal written questions sent to the opposing side of a legal argument. The opposition must provide written answers under oath.

National Fire Incident Reporting System (NFIRS) A nationwide database at the National Fire Data Center under the U.S. Fire Administration that collects fire-related data in order to provide information on the national fire problem.

Recommendation report A decision document prepared by a fire officer for the senior staff. The goal is support for a decision or an action.

Standard operating procedures (SOPs) Written organizational directives that establish or prescribe specific operational or administrative methods to be followed routinely for the performance of designated operations or actions.

Supervisor's report A form that is required by most state workers' compensation agencies that is completed by the immediate supervisor after an injury or property damage accident.

Fire Officer *in Action*

The dispatch center provides the administrative fire officer with a copy of the radio transmissions and 9-1-1 conversation related to the townhouse fire. The first 9-1-1 call came from the occupant, reporting a fire on the stove. The telecommunicator asked if anyone else was in the house and the caller said no.

The initial on-scene report by the fire company was garbled. The dispatcher asked for a repeat of the message, but did not receive a reply. The next transmission was the screaming of the fire fighter who fell through the floor and broke a leg. The messages from the company officer on the portable radio were clear. A follow-up investigation showed damage to the mobile radio microphone cord that partially severed the wires.

Effective communication requires a coordination of verbal and written procedures, consideration of environmental factors, and awareness of information technology.

1. _____ is/are a series of formal written questions sent to the opposing side of a legal argument. The opposition must provide written answers under oath.
   A. General orders
   B. Arbitration
   C. Interrogatory
   D. Chronological statement of events

2. The townhouse fire resulted in a fire fighter may-day and a civilian death. In addition to the NFIRS report, the supervising or managing fire officer will also complete:
   A. an expanded incident report narrative.
   B. a PowerPoint presentation.
   C. an informal report to the administrative fire officer.
   D. a summary from a counseling session.

3. What is the first step in presenting a verbal presentation?
   A. Provide the presenter's professional background.
   B. Paint a picture of what happens if nothing is done.
   C. Introduce your proposed action.
   D. Get the audience's attention.

4. _____ is not part of the communication cycle.
   A. Medium
   B. Translation
   C. Feedback
   D. Receiver

Managing Incidents

NFPA 1021 Standard

Fire Officer I

4.6* **Emergency Service Delivery.**
This duty involves supervising emergency operations, conducting pre-incident planning, and deploying assigned resources in accordance with the local emergency plan and according to the following job performance requirements. [p 287–303]

4.6.1 Develop an initial action plan, given size-up information for an incident and assigned emergency response resources, so that resources are deployed to control the emergency. [p 292–293, 295–297]

(A) Requisite Knowledge*. Elements of a size-up, standard operating procedures for emergency operations, and fire behavior. [p 292]

(B) Requisite Skills. The ability to analyze emergency scene conditions; to activate the local emergency plan, including localized evacuation procedures; to allocate resources; and to communicate orally. [p 289–300]

4.6.2* Implement an action plan at an emergency operation, given assigned resources, type of incident, and a preliminary plan, so that resources are deployed to mitigate the situation. [p 292–293, 295–297]

(A) Requisite Knowledge. Standard operating procedures, resources available for the mitigation of fire and other emergency incidents, an incident management system, scene safety, and a personnel accountability system. [p 290–295]

(B) Requisite Skills. The ability to implement an incident management system, to communicate orally, to manage scene safety, and to supervise and account for assigned personnel under emergency conditions. [p 287–295]

4.6.3 Develop and conduct a post-incident analysis, given a single unit incident and post-incident analysis policies, procedures, and forms, so that all required critical elements are identified and communicated, and the approved forms are completed and processed in accordance with policies and procedures. [p 300, 302–303]

(A) Requisite Knowledge. Elements of a post-incident analysis, basic building construction, basic fire protection systems and features, basic water supply, basic fuel loading, fire growth and development, and departmental procedures relating to dispatch response tactics and operations and customer service. [p 300, 302–303]

(B) Requisite Skills. The ability to write reports, to communicate orally, and to evaluate skills. [p 300, 302–303]

Fire Officer II

5.6 **Emergency Service Delivery.**
This duty involves supervising multi-unit emergency operations, conducting pre-incident planning, and deploying assigned resources, according to the following job requirements. [p 290–292]

5.6.1 Produce operational plans, given an emergency incident requiring multi-unit operations, the current edition of NFPA 1600, and AHJ-approved safety procedures, so that required resources and their assignments are obtained and plans are carried out in compliance with NFPA 1600 and approved safety procedures resulting in the mitigation of the incident. [p 289, 292–293, 295–297]

(A) Requisite Knowledge. Standard operating procedures; national, state/provincial, and local information resources available for the mitigation of emergency incidents; an incident management system; and a personnel accountability system. [p 288–300]

(B) Requisite Skills. The ability to implement an incident management system, to communicate orally, to supervise and account for assigned personnel under emergency conditions, and to serve in command staff and unit supervision positions within the Incident Management System. [p 288–300]

5.6.2 Develop and conduct a post-incident analysis, given multi-unit incident and post-incident analysis policies, procedures, and forms, so that all required critical elements are identified and communicated and the approved forms are completed and processed. [p 300, 302–303]

(A) Requisite Knowledge. Elements of a post-incident analysis, basic building construction, basic fire protection systems and features, basic water supply, basic fuel loading, fire growth and development, and departmental procedures relating to dispatch response, strategy tactics and operations, and customer service. [p 300, 302–303]

(B) Requisite Skills. The ability to write reports, to communicate orally, and to evaluate skills. [p 300, 302–303]

Additional NFPA Standards

NFPA 1500 *Standard on Fire Department Occupational Safety and Health Program*

NFPA 1521 *Standard for Fire Department Safety Officer*

NFPA 1561 *Standard on Emergency Services Incident Management System*

NFPA 1600 *Standard on Disaster/Emergency Management and Business Continuity Programs*

Introduction to Fire and Emergency Services Administration (FESHE) Course Outcomes

2. Explain the need for effective communication skills both written and verbal. [p 291, 296, 298, 300, 302–303]

9. Identify and assess safety needs for both emergency and nonemergency situations. [p 291, 300]

11. Identify the role of a company officer in Incident Command System (ICS). [p 288–300]

12. Describe the benefits of documentation. [p 300, 302–303]

Knowledge Objectives

After studying this chapter, you will be able to:

- Explain how the Incident Command System was created.
- Describe the National Incident Management System.
- Describe the role and elements of the Federal Response Plan.
- Explain the importance of fire fighter safety and accountability within the incident management system.
- Describe the "two-in-two-out" rule.
- Explain how to effectively accomplish a transfer of command.
- Describe the fire officer's role in incident management.
- Explain the use of divisions and groups within the incident management system.
- Describe the task level of incident management.
- Describe the post-incident review.

Skills Objectives

After studying this chapter, you will be able to:

- Demonstrate making an initial radio report.
- Demonstrate conducting a face-to-face transfer of command.

A single fire company is dispatched for a vehicle collision with injury in an apartment complex. On arrival, you see a vehicle has hit the side of a five-story, type V, multi-family building. Walking to the vehicle, you briefly hear a hissing sound, followed by an overwhelming explosion. You observe a pillar of fire that is igniting the vinyl siding of the building.

1. As the first-arriving officer, what is your responsibility for an incident management system?
2. What mode of command will you assume on this fire?
3. What components of the incident management system would you use?

Introduction to Managing Incidents

Every fire officer should be prepared to function in a variety of roles in the incident management system. A fire officer must be prepared to perform the duties of a first-arriving officer at any incident, including assuming initial command of the incident, establishing the basic management structure, and following standard operating procedures. A fire officer must also be fully competent at working within the Incident Command System (ICS) at every incident. This chapter introduces the model procedures for incident management. Chapter 16, Fire Attack, focuses on the fire attack and on applying the incident management procedures to this type of situation.

Incident Management in the "New Normal"

Fire service incident management was a local activity, using unique terms and practices in each community. Two divergent programs, the Southern California FIRESCOPE and the Phoenix, Arizona, Fire Ground Commander programs, provide the foundation for the National Incident Management System.

FIRESCOPE and Fire Ground Commander

FIrefighting REsources of Southern California Organized for Potential Emergencies (FIRESCOPE) was created in the wake of massive southern California wildfires. These fast-moving fires crossed boundary lines and burned for days within multiple jurisdictions.

The community-based command and control systems were ineffective at managing operations of this scale and complexity.

Resources were often deployed to combat a fire in one jurisdiction that could have been better used in another jurisdiction. Units from different agencies had problems communicating and coordinating their actions. Deficiencies in radio system compatibility, standardized terminology, and equipment compatibility led to a disconnected and inefficient use of scarce and critical resources.

FIRESCOPE set out to resolve the jurisdictional, interoperability, and standardization issues. The program managers developed a standardized method of setting up an incident management structure, coordinating strategy and tactics, managing resources, and disseminating information.

In 1982, it was adopted as a cornerstone of the National Interagency Incident Management System (NIIMS), and a year later, the National Fire Academy adopted it as the model system for emergency management.

In Arizona, Phoenix executive fire officers Alan Brunacini, Bruce Varner, and Chuck Kime were developing Fire Ground Commander (FGC) to meet the needs of an all-hazards metro city fire department. The notes and handouts from the weekend seminars became the 1983 NFPA *Fire Command* textbook.

The Fire Ground Commander focused on small and medium-sized urban emergencies, such as structural fires, mass casualty events, and hazardous material events. FIRESCOPE, in contrast, handled the challenges at major large-scale, wildland fires. Multiple jurisdiction events, campaign operations, and significant interagency coordination were significant issues in southern California.

According to Brunacini, "The California response-and-assistance network uses the Incident Command System as the basis for an incredible, statewide automatic-aid system. California's unique mutual response system serves as the large-scale, multi-agency organizational gold standard for our entire business."

Developing One System

The first edition of NFPA 1561, *Standard on Emergency Services Incident Management System*, was issued in 1990. The National Fire Service Incident Management System Consortium was developing model procedure guides for various incidents. After the 2001 terrorist attacks, the federal government expanded the 1992 Federal Response Plan to engage state and local responders.

Hurricanes Katrina and Rita identified gaps in national, state, and local emergency preparedness. In October 2006 the *Post-Katrina Emergency Management Reform Act* established structure, funding, and authority to improve national preparedness. This includes a National Response Framework and a National Incident Management System that supersedes earlier incident management practices.

National Response Framework

The federal government established the **National Response Framework (NRF)** in March 2008. NRF is a comprehensive, national, all-hazards approach to domestic incident response by describing specific authorities and best practices for managing incidents. It builds upon the National Incident Management System (NIMS) (discussed in the next section), which provides a consistent template for managing incidents.

The Homeland Security Act assigns the DHS Administrator to build a comprehensive National Incident Management System with federal, state, and local government personnel, agencies, and authorities to respond to attacks and disasters; consolidate existing federal emergency response plans into a single, coordinated national response plan; and administer and ensure the implementation of the NRF, including coordinating and ensuring the readiness of each Emergency Support Function under the NRF.

Federal, state, and some private sector and nongovernmental entities organize their resources and capabilities under 15 Emergency Support Functions (ESFs). ESFs align categories of resources and provide strategic objectives for their use. The ESF annexes that are of interest to a fire officer are listed in **Table 15-1 ▼**.

National Incident Management System

The 1974 Robert T. Stafford Disaster Relief and Emergency Assistance Act was amended in 2007 to provide federal government disaster and emergency assistance to state and local governments, tribal nations, eligible private nonprofit organizations, and individuals affected by a declared major disaster or emergency **Figure 15-1 ▶**. The Stafford Act covers all hazards, including natural disasters and terrorist events. To be eligible for Stafford Act funding, as laid out in Homeland Security Presidential Directive 5, Management of Domestic Incidents, requires adoption and implementation of the NIMS.

The **National Incident Management System (NIMS)** is a core set of doctrine, concepts, principles, terminology, and organizational processes. It allows for effective, efficient, and collaborative incident management across all emergency management and incident response organizations. Because it is used consistently nationwide, NIMS makes it possible for federal, state, tribal, and local governments; the private sector; and nongovernmental organizations to work together more easily. This facilitates preparing for, preventing, responding to, recovering from, and mitigating the effects of a variety of incidents. By tying disaster reimbursement with compliance, the federal government is compelling NIMS implementation by hospitals, industries, and law enforcement.

The NIMS has five components: preparedness, communications and information management, resource management, command and management, and ongoing management and maintenance. The FY08 to FY12 NIMS training plan includes updates of existing NIMS training, completing additional courses, delivery of training to all stakeholders, delivery of exercises, and evaluations of incidents and missions.

Incident Command System

The **Incident Command System (ICS)** is within the NIMS Command and Management component. Local emergency response agencies were required to adopt ICS by October 2004 to remain eligible for federal disaster assistance. This included training in the core NIMS curriculum:

- IS 700: National Incident Management System (NIMS) an Introduction
- ICS 100: Introduction to the Incident Management System
- ICS 200: ICS for Single Resources and Initial Action Incidents

Table 15-1   **ESF Annexes Applicable to Fire Officers**

Annex	Primary Agencies
ESF 4: Firefighting	Department of Agriculture/Forest Service
ESF 5: Emergency Management	Department of Homeland Security (DHS)/Federal Emergency Management Agency (FEMA)
ESF 6: Mass Care, Emergency Assistance, Housing and Human Services	DHS/FEMA
ESF 8: Public Health and Medical Services	Department of Health and Human Services
ESF 9: Search and Rescue	DHS/FEMA
	DHS/U.S. Coast Guard (USCG)
	Department of the Interior (DOI)/National Park Service (NPS)
	Department of Defense (DOD)
ESF 10: Oil and Hazardous Materials Response	Environmental Protection Agency, DHS/USCG
ESF 13: Public Safety and Security	Department of Justice (DOJ)

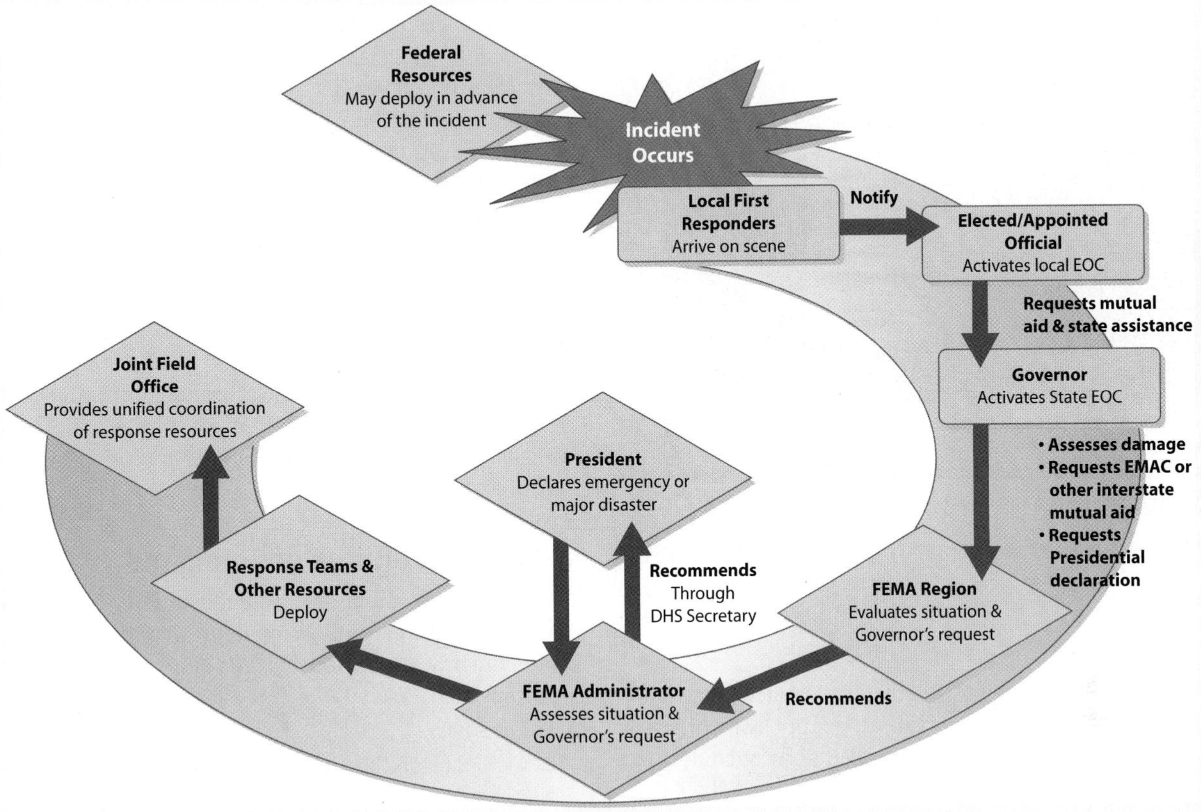

Figure 15-1 Overview of Stafford Act support to the states. Courtesy of the National Response Plan (NRP)/United States Army Combined Arms Center.

- ICS 300: Intermediate ICS
- ICS 400: Advanced ICS
- ICS 701: NIMS Multiagency Coordination System
- ICS 702: NIMS Public Information System

Students attending courses at the National Fire Academy are required to have completed ICS 100 and 200 before arriving on campus Figure 15-2.

Components of NIMS			Levels of Training: Awareness		Levels of Training: Advanced	Levels of Training: Practicum
Preparedness		IS-700	IS 800 IS 705			
Communications & Info Management		IS-700	IS 704			
Resource Management		IS-700	IS 703 IS 706 IS 707			
Command & Management	ICS	IS-700	ICS 100 ICS 200		ICS 300 ICS 400	Position-specific courses
Command & Management	MACS	IS-700	ICS 701			
Command & Management	Public Info	IS-700	ICS 702			
Ongoing Management & Maintenance						

Figure 15-2 Core curriculum aligned with NIMS components and by level of training. Courtesy of NIMS/FEMA.

The Fire Officer's Role in Incident Management

Every fire officer is expected to function as an initial incident commander, as well as a company-level supervisor within the ICS. A supervising fire officer functioning as the incident commander must supervise the work of a group of fire fighters, report to a managing or administrative fire officer, and work within a structured plan at the scene of an incident. A supervising fire officer may be required to initiate the procedure to assemble a multiple-agency response.

The first-arriving fire officer has the responsibility to establish command and manage the incident until relieved by a higher-ranking officer. This task is accomplished in addition to supervising the members of his or her company. Managing an incident requires the fire officer to develop strategies and tactics, determine required resources, and decide how those resources will be used.

ICS is flexible and can be incrementally implemented. The command structure for an incident should only be as large as the incident requires. The goal is to use the model ICS structure to assign all of the functions that have to be performed at that incident Figure 15-3. Most fire department tactical and task activities fall under the Operations section within ICS.

ICS allows the company officer to maintain a manageable span of control. One officer can provide effective supervision to a

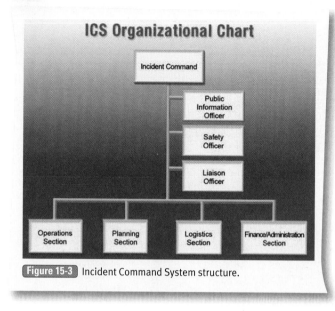

Figure 15-3 Incident Command System structure.

limited number of subordinates, whether those subordinates are individual workers or supervisors directing groups of workers. For emergency operations, a recommended span of control is three to five individuals reporting to one supervisor. The span of control is maintained by adding levels of management as an officer's effective span of control is exceeded.

Levels of Command

There are three levels of command in the ICS, and a set of responsibilities is assigned to each level. At small-scale incidents, or in the early stages of a larger incident, one company officer covers all three levels simultaneously. When the incident expands, the management responsibilities are subdivided. A fire officer may have a role at any level.

The overall direction and goals are set at the **strategic level**. The incident commander always functions at the strategic level. A strategic goal, in a defensive situation, could be to stop the extension of a fire to any adjacent structure.

Tactical-level objectives define the actions that are necessary to achieve the strategic goals. A tactical-level supervisor would manage a group of resources to accomplish the tactical objective. In medium- to large-scale incidents, the tactical-level components would be called divisions, groups, or units, and each of these components could include several companies. A company-level officer would be in charge of each company, and all of the company-level officers would report to the tactical-level supervisor.

Tactical assignments are usually defined by a geographical area (e.g., one part of a building) or a functional responsibility (e.g., ventilation), or sometimes by a combination of the two (e.g., ventilation in a particular part of the building).

In large incidents, an additional level of management may be added to maintain a reasonable span of control. Branches are established to group tactical components. An officer assigned to a branch would oversee some combination of divisions, groups,

or units. It is unlikely that a company officer would be required to fill a branch-level position.

Task-level assignments are the actions required to achieve the tactical objectives. This is where the physical work is accomplished. Individual fire companies or teams of fire fighters perform task-level activities, such as searching for victims, operating hose lines, or opening ceilings.

Strategic-Level Incident Management

The first-arriving managing and supervising fire officers are required to focus on the strategic level as they arrive at an emergency. Depending on the size and complexity of the incident, the officer may quickly move to the strategic or task level as the administrative fire officer and other departmental resources arrive.

Responsibilities of Command

The **incident commander** is the individual who is responsible for the management of all incident operations. The incident commander is responsible for the completion of three strategic priorities:

1. Life safety
2. Incident stabilization
3. Property conservation

The incident commander is also responsible for:

- Building a command structure that matches the organizational needs of the incident
- Translating the strategic priorities into tactical objectives
- Assigning the resources that are required to perform the tactical assignments

Establishing Command

The first fire officer or fire department member to arrive at the scene is always required to assume command of the incident. The initial incident commander remains in charge of the incident until command is transferred or the situation is stabilized and terminated. If the incident expands, a higher-ranking officer arrives and assumes command, and the initial incident commander's role changes.

Command must be established and ICS must be used at every event. Vehicle fires and EMS responses are handled by a single company under the supervision of a company-level officer. These incidents do not require a highly structured command presence; nevertheless, the basic functions of incident command must still be performed. For single-company incidents, establishing command may only require notification to the dispatcher that the company is on the scene and is handling the situation.

For incidents requiring two or more companies, the first fire department member or company-level officer on the scene must establish command and initiate an incident management structure that is appropriate for the incident.

Activating the command process includes providing an initial radio report and announcing that command has been established. The initial radio report should provide an accurate description of the situation for units that are still en route. The report should also identify the actions that arriving units will take. In that report, the company officer should include:

- Identification of the company or unit arriving at the scene.
- A brief description of the incident situation. That may include the building size, height, and occupancy, or the magnitude of a multiple-vehicle collision.
- Obvious conditions, such as a working fire, multiple patients, or a hazardous materials spill, or a dangerous situation, like a man with a gun.
- Brief description of the action to be taken (e.g., "Engine 2 is advancing an attack line into the first floor.").
- Declaration of the strategy to be used (offensive or defensive).
- Any obvious safety concerns.
- Assumption, identification, and location of command.
- Request for additional resources (or release of resources), if required.

Command Options

The first-arriving company-level officer has three options when arriving at the incident and assuming command: investigation, fast attack, or command mode. The decision is based on the situation that is presented to the fire officer.

Investigation Mode

There may be nothing showing or may appear to be a very minor situation. The first-arriving company will conduct an investigation. The other units assigned to the event will stage and remain uncommitted. The first-arriving company-level officer performs the role of initial incident commander as well as supervising the company performing the investigation.

Fast-Attack Mode

Some situations require immediate action by the first-arriving fire company to save a life. An example is when a single company arrives at a fire with occupants in imminent danger at upper-floor windows. The company-level officer performs the initial incident command responsibilities through a portable radio while engaged in a fast attack.

The fast-attack mode ends when one of the following occurs:

1. The situation is stabilized.
2. The situation is not stabilized and the company officer must withdraw to the exterior and establish a command post.
3. Command is transferred to another officer.

Command Mode

Some events are so large, complex, or dangerous that they require the immediate establishment of command by the first-arriving company-level officer. The company-level officer's personal involvement in tactical operations is less important than the command responsibility. The company-level officer should

Getting It Done

Examples of Initial Radio Reports

For a single-company incident: "Engine 2 is on the scene of a fully involved auto fire with no exposures. Engine 2 can handle."

For an EMS incident: "Quint 618 is on the scene of a four-vehicle crash with multiple patients and one trapped. Transmit a special alarm for a heavy rescue company, a second suppression company, and an EMS task force. Quint 618 will be Palisade Command."

For an offensive structure fire: "Engine 1034 has a two-story motel with a working fire in a second-floor room on side Charlie. Engine 1034 has a hydrant and is going in with a hand line to start primary search. This will be an offensive fire attack. Engine 1034 will be Cedar Avenue Command."

For a defensive structural fire: "Engine 422 is on the scene of a one-story strip shopping center with an end unit heavily involved. Engine 422 is laying a LDH supply line and will be using a master stream to cover the exposure on side Delta. This is a defensive fire. Engine 422 will be Loisdale Command."

Safety Zone

"Nothing Showing" Calls Can Be Dangerous

Many fire fighter fatalities and serious injuries occur at incidents that started as "nothing showing" events. After entering the building, the first-in fire company would suddenly encounter a rapidly deteriorating situation. Vital firefighting equipment, such as self-contained breathing apparatus (SCBA) and tools, was left behind on the apparatus. Fire fighters were unprepared for the situation they encountered.

When incidents, particularly in commercial and industrial buildings, are investigated, consider the following factors:

- Assume that there is a fire in the building and bring the appropriate personal protective equipment.
- Use thermal imaging devices.
- Go to the location of a monitored fire alarm activation. Fires in newer buildings are often detected by the fire alarm system while they are still in the incipient stage. Do not assume that the alarm is false if nothing is visible from the exterior. Check the fire alarm panel for the location of the activated device and be prepared for action when you go there to investigate.
- Use two or three companies to perform the initial investigation in larger buildings.
- Hold the entire first-alarm assignment until the investigation is completed. Many fire departments have the assisting companies reduce their response from emergency to nonemergency when the first-arriving company reports no indication of fire. This policy reduces the risk of accidents during emergency response.

FIRE AND RESCUE DEPARTMENTS OF NORTHERN VIRGINIA
***INITIAL* INCIDENT COMMAND BOARD**

ADDRESS/COMMAND:					INITIAL RIT ENGINE:	
FIRST ALARM		Floor, Group, Division, Branch	Floor, Group, Division, Branch	Floor, Group, Division, Branch	**OPERATIONS CHANNEL:**	
ENGINE					**TYPE OF OCCUPANCY:**	
ENGINE					**TASKS**	**REQUESTS**
ENGINE					WATER SUPPLY	GAS COMPANY
ENGINE					PRIMARY SEARCH	POWER COMPANY
TRUCK					LADDERS	FIRE INVESTIGATOR
TRUCK					VENTILATION	POLICE
RESCUE					UTILITY CONTROL	LIGHT/AIR UNITS
MEDIC					**NOTES**	
AMB						
BFC						
EMS CAPT						
SECOND ALARM		Floor, Group, Division, Branch	Floor, Group, Division, Branch	Floor, Group, Division, Branch		
ENGINE						
ENGINE						
ENGINE						
ENGINE						
TRUCK						
TRUCK						
RESCUE						
MEDIC						
AMB					**CHECK ALL OPERATING UNITS AIR SUPPLY. CREWS**	
BFC					**WITH *2000 PSI OR LESS* NEED TO BE REPLACED**	
EMS CAPT					**AND SENT TO THE *MEDICAL UNIT*.**	

RECEO-VS	CHECK EXPOSURES ⇒	B3	B2	B1	FIRE UNIT	D1	D2	D3

Figure 15-4 Tactical worksheet. Courtesy of the Northern Virginia Regional Commission.

establish a command position in a safe and effective location and initiate a **tactical worksheet** **Figure 15-4 ▲**. The role of the initial incident commander at this type of situation is to direct incoming units to take effective action. While the initial incident commander remains outside, the rest of the company members should do one of the following:

- Initiate fire suppression with one of the members assigned as the acting company officer. The acting company officer must be equipped with a portable radio, and the crew must be capable of performing safely without the initial incident commander.
- After the initial incident commander assigns them, the remaining company members work under another company officer.
- Stay with the initial incident commander to perform staff functions that assist command.

Functions of Command

There are nine functions of command.
- Determining strategy
- Selecting incident tactics
- Setting the action plan
- Developing the ICS organization
- Managing resources

- Coordinating resource activities
- Providing for scene safety
- Releasing information about the incident
- Coordinating with outside agencies

Immediate command functions include determining strategy, selecting incident tactics, and setting an action plan. These three functions must be completed as part of the first-arriving officer's size-up and initial actions. The initial radio report as described earlier will cover these three functions.

The initial fire-ground assignments and company-level tasks may be predetermined by departmental SOP or practice. The company officer will ensure that the SOP is followed and will look for conditions that may require a deviation from the standard initial actions.

Once the initial actions are under way, the incident commander will work on the next four functions: developing the ICS organization, managing resources, coordinating resource activities, and providing for scene safety. Designate a safety officer as well as unit, group, or division commander. Request additional resources and manage them through staging and assignments within the incident management system.

Once the incident action plan is fully operating, the incident commander will work on the last two functions: releasing information about the incident and coordinating with outside agencies.

Transfer of Command

An administrative fire officer assumes command of significant incidents. At a large incident, command could be transferred more than once, depending on the situation and the chain of command, as successively higher-ranking or more qualified officers arrive at the incident scene.

Command should be transferred only to improve the quality of the command organization. It should not be transferred simply because a higher-ranking officer has arrived, and it should not be transferred more times than are necessary to effectively manage the incident.

There is a structure to a transfer of command:

1. The officer assuming command communicates with the initial incident commander. This can occur over the radio, but a face-to-face meeting is preferred.
2. The initial incident commander briefs the new incident commander, including:
   - Incident conditions: the location and extent of fire, the number of patients, status of the hazardous materials spill or leak, etc.
   - Tactical worksheet and incident action plan
   - Progress toward completion of the tactical objectives
   - Safety considerations
   - Deployment and assignment of operating companies and personnel
   - Need for additional resources
3. Command is officially transferred only when the new incident commander has been briefed.
4. The fact that command has been transferred is communicated to the dispatch center and all units operating on the fire scene.

The initial incident commander should always review the tactical worksheet with the new incident commander. The worksheet outlines the status and location of personnel and resources in a standard form. This is especially important on those events where the first-arriving company initially started in the command mode.

After the transfer of command has occurred, the new incident commander determines the most appropriate assignment for the previous incident commander. A company-level officer may be assigned as a division or **group supervisor** or may remain with the new incident commander at the command post. The size and the complexity of the incident determine how much the management structure will need to be expanded.

Fire Fighter Accountability

Concurrent with the evolution of incident management systems was accounting for fire fighters at the incident. Although the concept had been used in the United Kingdom for several years, it was required by the first edition of NFPA 1500 in 1987. Legal sanctions accelerated adoption of fire fighter accountability practices.

The Seattle Fire Department lost six fire fighters between 1987 and 1995. The Washington State Department of Labor and Industry found the department was negligent in SCBA training and tracking of fire crews at large-scale operations. At one incident a significant amount of time elapsed before the incident commander was aware that a fire fighter was missing. A judge allowed the spouses and estates of the deceased to sue Seattle executive and

administrative fire officers directly for $55 million worth of personal liability. As part of the court-mandated corrective action, the department purchased personal alert safety system (PASS) devices and implemented a passport-style accountability system.

In September 1994, the National Institute for Occupational Safety and Health (NIOSH) released a report, *A Request for Assistance in Preventing Injuries and Deaths of Fire Fighters*. This alert bulletin made the following statement:

A recent NIOSH investigation identified four factors essential to protecting fire fighters from injury and death: (1) following established firefighting policies and procedures, (2) implementing an adequate respirator maintenance program, (3) establishing fire fighter accountability at the fire scene, and (4) using personal alert safety system devices at the fire scene. Deficiencies in any of these factors can create a life-threatening situation for fire fighters.

Additional pressure to improved accountability came from a request by the International Association of Fire Fighters (IAFF) for a clarification of Occupational Safety and Health Administration (OSHA) respiratory protection regulation 29 CFR 1910.134. This section of the Code of Federal Regulations covered industrial employees operating in confined spaces, toxic environments, or oxygen-deficient atmospheres. These work areas are classified as immediately dangerous to life and health (IDLH).

OSHA ruled in 1996 that fire fighters working within a structure fire were operating in an IDLH atmosphere. Fire departments must comply with 29 CFR 1910.134 while self-contained breathing apparatus is being used. A minimum of two fire fighters enter the IDLH area together and remain in visual or

Near Miss REPORT

Report Number: 08-0000396

Event Description: I was on the first-arriving engine company of a structure fire in a two-story abandoned house. Heavy flame and thick smoke were coming from the Delta side of the structure. Other companies arrived before my crew deployed into the structure. I took one back-end fire fighter with six years of experience and also a rookie with two weeks of experience.

The six-year fire fighter and I deployed to the interior Delta side, first floor, for attack, but only after assigning the two week fire fighter to the exterior doorway to pull hose. After making an initial attack, we bucked out of the Delta side to the exterior. I recovered the two-week fire fighter and we went to the truck to exchange air bottles.

We re-entered the Alpha side interior where I once again left the two-week fire fighter at the exterior doorway to pull hose and advanced the line with the six-year fire fighter. We were only 6-8' inside the structure, Alpha side, and I decided to back out just enough to direct the two-week fire fighter on the hoseline behind me. We re-entered the interior to resume the attack.

Shortly after re-entry, Command called for a PAR. I tapped the six-year fire fighter on the shoulder as he operated the nozzle. I then turned and tapped who I thought was my two-week fire fighter on the helmet. I then radioed Scene Command with PAR. Command asked me to verify PAR and I complied. When retreating to the exterior with empty air, I found the fire fighter I thought to be my two-week fire fighter.

My two-week fire fighter was a 22-year veteran that had been freelancing. Another crew directed my two-week fire fighter to advance with them into the second-story interior. When I questioned the two-week fire fighter why he followed them, he told me he thought the other crew was me and the six-year fire fighter because we all look the same. He could not tell us from the other crews because our turnout gear is all alike.

Lesson Learned: Just because you give an assignment, it doesn't mean the assignment will be carried out. Freelancing is dangerous. Officers should be distinguishable on the fire scene.

---

voice contact with one another at all times. In addition, at least two properly equipped and trained fire fighters must:

- Be positioned outside the IDLH atmosphere
- Account for the interior teams
- Remain capable of rescue of the interior team or teams

This interpretation became known as the **two-in-two-out rule** and evolved into the Rapid Intervention Team (RIT) concept that was incorporated into NFPA 1500. NFPA 1407, *Standard for Fire Service Rapid Entry Teams* was issued in Fall 2009 describing basic training procedures and information.

After the Transfer of Command: Building the Incident Management System

Many incidents can be handled by the first-arriving fire officer. The incident management system will expand to handle the larger and more complex incidents. Managing and supervising fire officers may be assigned to these incident management system assignments.

Command Staff

After command has been transferred, the company-level officer may be assigned to the **command staff**. Individuals on the command staff perform functions that are reported directly to the incident commander. The safety officer, liaison officer, and information officer are always part of the command staff; these duties cannot be delegated to other sections of the incident organization.

Aides, assistants, and advisors may be assigned to work directly for the incident commander. An aide is a fire fighter (sometimes a fire officer) who serves as a direct assistant to a command officer. In many fire departments, operational-level command officers have regularly assigned aides who drive the command vehicle and perform administrative support duties, as well as functioning as aides at incident scenes. In other departments, aides are assigned only as needed at incident scenes. The incident commander may assign more than one aide to perform support functions at a major incident. Aides can also be assigned to other officers in the command structure.

Safety Officer

The **safety officer** is responsible for ensuring that safety issues are managed effectively at the incident scene. The safety officer is the eyes and ears of the incident commander for identifying and evaluating hazardous conditions, watching out for unsafe practices, and ensuring that safety procedures are followed. Normally, the safety officer is appointed early during an incident. As the incident becomes more complex, additional qualified personnel can be assigned as assistant safety officers to subdivide the responsibilities.

The safety officer is an advisor to the incident commander but has the authority to stop or suspend operations when unsafe situations occur. This authority is clearly stated in national standards, including NFPA 1500, *Standard on Fire Department Occupational Safety and Health Program*; NFPA 1521, *Standard for Fire Department Safety Officer*; and NFPA 1561, *Standard on Emergency Services Incident Management System*. Several state and federal regulations require the assignment of a safety officer at hazardous materials incidents and certain technical rescue incidents.

The safety officer should be a qualified individual who is knowledgeable in fire behavior, building construction and collapse potential, firefighting strategy and tactics, hazardous materials, rescue practices, and departmental safety rules and regulations. A safety officer should also have considerable experience in incident response and specialized training in occupational safety and health. Many fire departments have full-time safety officers who perform administrative functions relating to health and safety when they are not responding to emergency incidents.

Liaison Officer

The **liaison officer** is the incident commander's point of contact for representatives from outside agencies and is responsible for exchanging information with representatives from those agencies. During an active incident, the incident commander may not have time to meet directly with everyone who comes to the command post. The liaison officer position takes the incident commander's place, obtaining and providing information, or directing people to the proper location or authority. The liaison area should be adjacent to, but not inside, the command post.

The incident commander can also assign a liaison officer to directly represent the fire department with another agency. At a complex incident involving extensive interaction between the police and fire departments, the fire department could assign a liaison officer to work with the police, or the police commander could assign a liaison officer to the fire department command post.

Public Information Officer

The **public information officer** is responsible for gathering and releasing incident information to the news media and other appropriate agencies. At a major incident, the public wants to know what is being done. The public information officer serves as the contact person for media requests so that the incident commander can concentrate on managing the incident. A media briefing location should be established that is separate from the command post.

■ General Staff Functions

When an incident is too large or too complex for just one person to manage effectively, the incident commander may appoint officers to oversee major components of the operation. Four standard components are defined in the ICS model. Everything that occurs at an emergency incident can be divided among these four major functional components:

1. Operations
2. Planning
3. Logistics
4. Finance/administration

The incident commander decides which (if any) of these four components need to be activated, when to activate them, and who should be placed in each position. Remember that the blocks on the ICS organization chart refer to functional areas or job descriptions, not to positions that must always be staffed. The positions are assigned only when they are needed and, if a position is not assigned, the incident commander is responsible for managing that function.

The chiefs in charge of the four major sections are known as the **ICS general staff**. The four chiefs on the ICS general staff, when they are assigned, may conduct their operations from the main command post or from a different location. At a large incident, the four functional organizations may operate from different locations, but the general staff chiefs are always in direct contact with the incident commander Figure 15-5 ▶.

Operations

The **operations section** is responsible for the management of all actions that are directly related to controlling the incident. The operations section fights the fire, rescues any trapped individuals, treats the patients, and does whatever else is necessary to deal with the emergency situation. This is the part of the organization that produces the most visible results.

For most structure fires, the incident commander directly supervises the functions of the operations section. A separate **operations section chief** is used at complex incidents so that the incident commander can focus on the overall situation while the operations section chief focuses on the strategy and tactics that are required to get the job done.

Planning

The **planning section** is responsible for the collection, evaluation, dissemination, and use of information relevant to the incident. The planning section works with status boards and pre-incident plans, as well as building construction drawings, maps, aerial photographs, diagrams, and reference materials.

The planning section is also responsible for developing and updating the incident action plan (IAP). The planning section is similar to the scheduling department at a company. It plans what needs to be done by whom and what resources are needed. To perform this role, the planning section must be in close and regular contact with all of the other sections.

The incident commander activates the planning section when information needs to be obtained, managed, and analyzed. The **planning section chief** reports directly to the incident commander. Individuals assigned to planning functions examine the current situation, review available information, predict the probable course of events, and prepare recommendations for strategies and tactics. The planning section also keeps track of resources at large-scale incidents and provides the incident commander with regular situation and resource status reports.

The planning functions may be delegated to subunits. These include resources unit, situation unit, documentation unit, demobilization unit, and technical specialists.

Incident Action Plan

The **incident action plan (IAP)** is a basic component of ICS; all incidents require an action plan. The IAP outlines the strategic objectives and states how emergency operations will be conducted. At most incidents, the IAP is relatively simple and can be expressed by the incident commander in a few words or phrases.

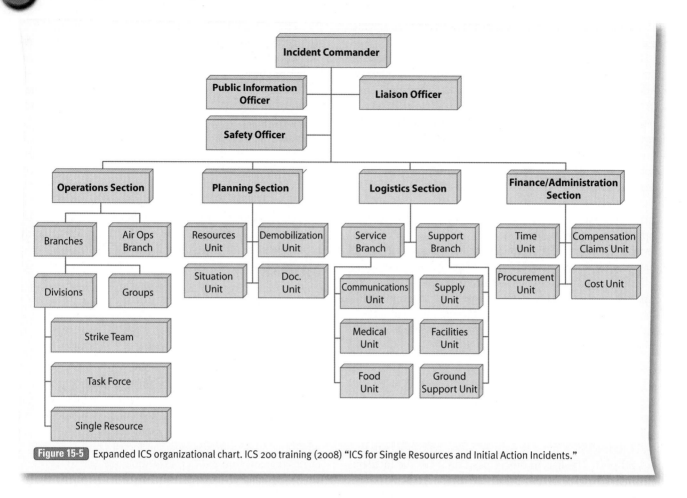

Figure 15-5 Expanded ICS organizational chart. ICS 200 training (2008) "ICS for Single Resources and Initial Action Incidents."

A written IAP is required for large or complex incidents that have an extended duration. The IAP for a large-scale incident can be a lengthy document that is regularly updated and used for daily briefings of the command staff.

Logistics

The <u>logistics section</u> is responsible for providing supplies, services, facilities, and materials during the incident. The <u>logistics section chief</u> reports directly to the incident commander. Among the responsibilities of this section would be keeping apparatus fueled, providing food and refreshments for the fire fighters, obtaining the foam concentrate needed to fight a flammable liquids fire, and arranging for a bulldozer to remove a large pile of debris. The logistics section is similar to the purchasing and support services departments of a large company. It ensures that adequate resources are always available and functional.

Logistical functions are routinely performed by personnel assigned to a support services division. These groups work in the background to ensure that the members of the operations section have whatever they need to get the job done. Resource-intensive or long-duration situations may require assignment of a logistics section chief because service and support requirements are so complex or extensive that they need their own management component.

The logistics section may use subunits to provide the necessary support for large incidents. These units may include a supply unit, facilities unit, ground support unit, communications unit, food unit, and medical unit.

Finance/Administration

The <u>finance/administration section</u> is the fourth major ICS component under the incident commander. This section is responsible for the administrative, accounting, and financial aspects of an incident, as well as any legal issues that may arise. This function is not activated at most incidents, because cost and accounting issues are usually addressed before or after the incident. A finance/administration section may be needed at large-scale and long-term incidents that require immediate fiscal management, particularly when outside resources must be procured quickly.

Fire Marks

Murrah Federal Building Bombing
The bombing of the Murrah Federal building in Oklahoma City on April 19, 1995 added a new element to the incident management process. It was the largest response of local, regional, and state agencies to a single incident in the southwest. It also was the largest mobilization of the federally funded Urban Search and Rescue Task Forces, with 11 of the 28 teams responding to the incident. The federal after-action report identified the value of establishing a written incident action plan (IAP) for every 12-hour work period. Each IAP identified the strategic goals, tactical objectives, and support requirements for the work period. While written IAPs were used in wildland fires, the rescue and recovery operations at Oklahoma City established the practice of using IAPs for all types of emergency incidents.

A finance/administration section is usually established during a natural disaster, when state or federal expense reimbursements are expected, or during a hazardous materials incident in which reimbursement may come from the shipper, carrier, chemical manufacturer, or insurance company. The finance section is equivalent to the finance department in a company; it accounts for all activities of the company, pays the bills, ensures that there is enough money to keep the company running, and keeps track of the costs.

The finance/administration section may incorporate subunits to efficiently handle its duties. These include a time unit, procurement unit, compensation and claims unit, and cost unit.

Tactical Level of Incident Management

Divisions, groups, and units are where the strategy developed in the incident action plan is converted into action. Supervising and managing fire officers will be commanding these incident management assignments.

Divisions, Groups, and Units

Divisions, groups, and units are tactical-level management elements that are used to assemble companies and resources for a common purpose. The flexibility of the ICS enables organizational units to be created as needed, depending on the size and the scope of the incident. In the early stages of an incident, individual companies are often assigned to work in different areas or perform different tasks. As the incident grows and more companies are assigned to areas or functions, the incident commander can establish tactical-level units to place multiple resources under one supervisor. This keeps the incident commander's span of control within desired limits and provides direct coordination of common efforts.

Divisions represent geographical operations, such as one floor or one side of a building. **Groups** represent functional operations, such as the ventilation group. **Unit** is a generic term that can be applied to either a geographical or a functional component. An officer is assigned to supervise each division, group, or unit as it is created **Figure 15-6 ◄**.

These organizational elements are particularly useful when several resources are working near each other, such as on the same floor inside a building, or are doing similar tasks, such as providing ventilation. The assigned supervisor can directly observe and coordinate the actions of several crews.

Divisions

A division is composed of the resources responsible for operations within a defined geographical area. This area could be a floor inside a building, the rear of the fire building, or a geographical area at a brush fire. Divisions are most often used during routine fire department emergency operations. By assigning all of the units in one area to one supervisor, the incident commander has to communicate with only that one individual. The assigned supervisor coordinates the activities of all resources that are working within that area, in accordance with the incident commander's strategic plan.

The division structure provides effective coordination of the tactics being used by different companies working in the same area. For example, the **division supervisor** would coordinate the actions of a crew that is advancing a hose line into the fire area, a crew that is conducting search-and-rescue operations in the same area, and a crew that is performing horizontal ventilation.

Groups

An alternative way of organizing resources is by function rather than by location. A group is composed of resources assigned to a specific function, such as ventilation, search and rescue, or water supply. The officer assigned to supervise the ventilation group uses the radio designation "ventilation group" and is required to coordinate activities with other division, group, or unit supervisors.

Groups are responsible for performing an assignment, wherever it may be required, and often work within more than one division. Sometimes, groups are established with both functional and geographical designations, such as a west wing search-and-rescue group and an east wing search-and-rescue group.

Units

A unit is an organizational element with functional responsibility for a specific incident activity, such as planning, logistics, or a specific geographic assignment. The officer assigned to supervise the Air Resupply unit uses the radio designation "Air Resupply" and is required to coordinate activities with other division, group, or unit supervisors.

The unit is the smallest organizational element within the incident management system and is more frequently used in large or complex incidents where there are many specialized groups operating.

Division/Group/Unit Supervisor Responsibilities

In many cases, a company-level officer is assigned by the incident commander to function as a division/group/unit supervisor. When an incident is rapidly escalating, the company-level officers on the first alarm are often assigned to geographical or functional

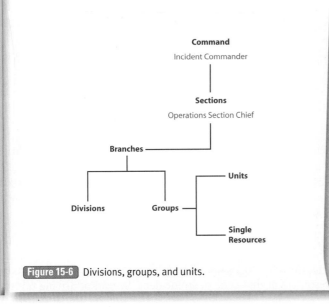

Figure 15-6 Divisions, groups, and units.

areas and designated as the division/group/unit supervisor. For example, the company officer in charge of the first-arriving ladder company could be assigned as the "ventilation group" supervisor. Additional companies assigned to ventilation would become part of this group. When additional command officers arrive, the ventilation group supervisor might be relieved by a chief officer and revert to functioning as a company-level officer.

As a division/group/unit supervisor, the company-level officer directly supervises and monitors the operations of assigned resources and provides updates to command. Duties of a division/group/unit supervisor include:

- Using an appropriate radio designation, such as roof division, division A, or rescue group
- Completing the objectives assigned by command
- Accounting for all assigned companies and personnel
- Ensuring that operations are conducted safely
- Monitoring work progress
- Redirecting activities as necessary
- Coordinating actions with related activities and adjacent division/group/unit supervisors
- Monitoring the welfare of assigned personnel
- Requesting additional resources as needed
- Providing the incident commander with essential and frequent progress reports
- Reallocating or releasing assigned resources

Division, group, and unit supervisors all function at the same level within ICS, regardless of the rank of the individual who is assigned. Divisions do not report to groups, and groups do not report to divisions; the supervisors are required to coordinate their actions and activities with each other. A division supervisor must be aware of everything that is happening within the assigned geographical area, so a group supervisor must coordinate with the division supervisor when the group enters the division's geographical area, particularly if the group's assignment will affect the division's personnel, operations, or safety. Effective communication among divisions and groups is critical during emergency operations.

Branches

A **branch** is a supervisory level established in either the operations or logistics function to provide a span of control. At a major incident, several different activities may occur in separate geographical areas or several distinct, but related functions may be performed. The span of control might still be a problem, even after the incident commander has established divisions/groups/units. In these situations, the incident commander can establish branches to place a **branch director** in charge of a number of divisions/groups/units. For example, after a tornado, if the destruction occurred in two separate areas across town, the incident commander could establish two branches. If there are numerous injuries, the incident commander could establish a medical branch to coordinate all aspects of caring for the victims, including triage, treatment, and transportation.

Location Designators

The exterior sides of a building are sides A (Alpha), B (Baker), C (Charlie), and D (Delta). The front of the building is side A,

with sides B, C, and D following in a clockwise direction around the building. The companies working in front of the building are assigned to division A, and the radio designation for their supervisor is "Division A." Similar terminology is used for the sides and rear of the building.

The areas adjacent to a burning building are called exposures. Exposures take the same letter as the adjacent side of the building. A fire fighter facing side A can see the adjacent building on the left (Exposure B) and the building to the right (Exposure D). If the burning building is in a row of buildings, the buildings to the left are called Exposures B, B1, B2, and so on. The buildings to the right are Exposures D, D1, D2, and so on.

Within a building, divisions commonly take the number of the floor on which they are working. For example, fire fighters working on the fifth floor would be in division 5, and the radio designation for the chief assigned to that area would be "Division 5." Crews performing different tasks on the fifth floor would all be part of this division. Common simple terminology should be used for designations whenever possible, such as "roof division" and "basement division."

Fire Officer Greater Alarm Responsibilities

The incident commander calls for additional resources or greater alarms when more resources are needed at an incident. In many cases, the incident commander calls for a greater alarm to have stand-by resources immediately available, in case they are needed or in anticipation of a future need. There are many duties that a company-level officer may be called on to perform when responding as an additional resource. These include:

- Reinforcing a fire attack strategy by adding resources.
- Relieving an exhausted crew from the first alarm and performing the fire-ground task that they were performing.
- Performing support activities, such as salvage, overhaul, moving equipment to a forward staging area, going door-to-door to evacuate a neighborhood, assisting occupants in recovering personal items, or providing runners to support a face-to-face communications system.
- Maintaining a ready reserve in a staging area.
- Performing additional related duties. This catchall phrase means that the officer and/or company may be performing tasks that are not traditional fire company tasks, but are time or mission critical. For instance, they could be assigned to an evacuation center to provide assurance and assessment of the occupants who fled from an apartment fire.

Sometimes, the additional resources are dispatched as strike teams or task forces, as opposed to individual companies. This means that the companies are directed to combine forces to respond and operate as a group. They could be assigned to meet at a designated location and travel to the scene together or to assemble at the staging area.

Staging

Staging refers to a standard procedure to manage uncommitted resources at the scene of an incident. Instead of driving up to

Assessment Center Tips

Rapidly Organize Your Greater Alarm

You will probably need to call for a special, additional, or greater alarm when handling a simulated emergency incident during a promotional examination. Practice identifying the staging location as you call for additional help. For example: "5th Street Command to Communications, transmit a second alarm. Staging for the second alarm units will be at the shopping center parking lot at 5th Street and Avenue D. Designate the officer on the first-arriving engine to assume the staging officer assignment."

Make sure that the terms you use and the assignments you make comply with the ICS used in your jurisdiction. You are evaluated on how well you implement the local system.

the incident scene and committing apparatus and crews without direction from the incident commander, staged resources stand by uncommitted and wait for instructions. These resources are immediately available for use where and when the incident commander needs them. This allows the incident commander to determine the most appropriate assignment for each unit.

Many fire departments have policies that automatically commit only one or two first-arriving companies and direct all others to level I staging. In level I staging, only the predesignated units respond directly to the scene, and later-arriving units remain uncommitted and wait for instructions. For example, the second-arriving aerial ladder would stop about a block short of the fire building, where it would have the option of going to whichever side it is needed. Standard operating procedures guide when units stage, where they stage, and how they communicate with the incident commander to indicate they are available for an assignment.

Level II staging, which is generally used for greater alarm incidents, directs responding companies to a designated stand-by location away from the immediate incident scene. A designated officer supervises the level II staging area with the "staging" radio identification and assigns units from the staging area as they are requested by the incident commander. The incident commander might direct the staging officer to maintain a minimum level of resources in the staging area.

Task Level of Incident Management

Individual companies, sometimes referred to as single resources, operate at the task level. A company is considered as a single resource because it is the basic work unit assigned to perform most emergency scene tasks. Certain types of tasks are routinely assigned to engine companies, ladder companies, rescue companies, and other types of companies. The normal role of a company-level officer is to supervise a group of fire fighters who are operating as a company at the task level.

At a simple incident, the company-level officers report directly to the incident commander. When divisions/groups/units are established, company-level officers report to their assigned division/group/unit supervisor and continue to supervise the operations of their respective companies in performing assigned tasks. Sometimes, the assigned supervisor is one of the company-level officers. Company-level officers advise the division/group/unit supervisor of work progress, preferably face-to-face. All company-level officer requests for additional resources or assistance must go through the division/group/unit supervisor.

Task Forces and Strike Teams

Task forces and strike teams are groups of single resources that have been assigned to work together for a specific purpose or for a period of time. Grouping resources reduces the span of control by placing several units under a single supervisor.

A **task force** includes two to five single resources that are assembled to accomplish a specific task. For example, a task force could be composed of two engines and one truck company, or two engines and two brush units, or one rescue company and four ambulances. A task force operates under the supervision of a **task force leader**. In many cases, the designated task force leader is one of the company-level officers. All communications for the separate units within the task force are directed to the task force leader.

Task forces are often part of a fire department's standard dispatch philosophy. In the Los Angeles City Fire Department, a task force consists of two engines and one ladder truck, staffed by a total of 10 fire fighters. Some departments create task forces of one engine and one brush unit for response during wildland fire season. The brush unit responds with the engine company wherever it goes.

A **strike team** is five units of the same type with an assigned leader. A strike team could be five engines (engine strike team), five trucks (truck strike team), or five ambulances (EMS strike team) **Figure 15-7 ▼**. A strike team operates under the supervision of a **strike team leader**. In most cases, the strike team leader is a command officer.

Strike teams are commonly used for wildland fires, where dozens or hundreds of companies may respond. During wildland fire seasons, many departments establish strike teams of five engine companies that are dispatched and work together on major wildland fires. The assigned companies are assigned to rendezvous at a designated location and then respond to the

Figure 15-7 A strike team is five units of the same type under one leader.

scene together. Each engine has an officer and a crew of fire fighters, and one officer, typically a battalion chief, is designated as the strike team leader. All communications for the strike team are directed to the strike team leader.

EMS strike teams, consisting of five ambulances and a supervisor, are often organized to respond to multicasualty incidents or disasters. Rather than requesting 15 ambulances and establishing an organizational structure to supervise 15 single resources, the incident commander can request three EMS strike teams and coordinate their operations through the three strike team leaders.

Greater Alarm Infrastructure

Small fire departments also tend to have limited infrastructure support. When they are working at a major incident, there could be a considerable delay before a rehabilitation unit is established with rest and recovery supplies. Fire officers should prepare for this by making sure that their apparatus carries enough water and food to support the fire company for a reasonable period of time. In wildland areas, many companies plan for a 2-day excursion, carrying 5 gallons of water and six energy bars per fire fighter. In the summer, 10 gallons of drinking water per fire fighter per day could be required.

Post-incident Review

Some form of review should be conducted at the company level after every call in which the company performs emergency operations. In most cases, this can be an informal discussion conducted by the company officer to review the incident, discuss the situation, and evaluate team performance. Every situation should be viewed as a potential learning experience, and the company officer should provide feedback to the crew members to reinforce positive performance and identify areas where there is room for improvement. When a deficiency is noted, a plan for addressing the problem should be developed at the same time.

Situations that involve multiple-company operations should be reviewed from the perspective of how well the overall team performed, in addition to individual fire officers conducting company-level reviews. The format of the multicompany review necessarily depends on the nature and the magnitude of the incident, whether it was a one-room house fire involving three or four companies or a five-alarm blaze that involved dozens of companies

and activated every component of the fire department. It is relatively easy to gather the companies that were involved in a one-alarm fire to conduct a basic review of the incident soon after the event. The critique of a large-scale incident can be a major event involving extensive planning, scheduling, and preparations.

Some organizations conduct critiques only for multiple-alarm or unusually large-scale incidents or for situations where something obviously went wrong. This creates the impression that there is nothing to be learned from the more ordinary-sized incidents, unless someone made a mistake. Whether it is conducted on a large or small scale, the primary purpose of a review process should always be to serve as an educational and training tool, not to place blame for improper or deficient actions. Approach every post-incident review in a nonthreatening manner and provide an open format for discussion as an aid to future training.

The review process should be conducted routinely and in a systematic manner. The same basic factors should be examined and discussed. Many fire departments have a standard critique format and procedure.

Preparing Information for an Incident Review

Multiple-company incident review is conducted by the officer who was the incident commander; however, the preparatory work is often delegated to one of the lower-ranking officers. An administrative fire officer might assign the officer in charge of the first-due company to prepare the information and make the necessary arrangements for the review of a one-alarm incident. The battalion chief could be responsible for preparing the information for a critique of a multiple-alarm situation that will be presented by a higher-ranking chief officer; however, the work of gathering and assembling the information might still be delegated to a company officer or staff officer.

Before the actual review occurs, information about the situation leading up to the incident and whatever occurred before the arrival of the first fire department unit should be obtained. This sets the stage for a discussion of the actual event from an operational perspective and often reveals important factors that were unknown before the incident occurred. The presentation of this information allows individuals to understand what happened and what they observed at the time but did not have the opportunity to interpret.

In the case of a fire involving a building, this process should begin with a review of the building, including its size and arrangement, construction type, date of construction, known modifications or renovations, and any known history that could be significant. All built-in fire protection features, including firewalls, automatic sprinklers, standpipes, and alarm and detection systems, should be identified, providing as much detail as possible. Any structural factors that could affect fire spread, such as unsealed openings around pipes, open shafts, and concealed spaces, should also be identified.

Was the building occupied, unoccupied, vacant, abandoned, or undergoing renovation? If it was a business, what did they do, make, or store on the premises? What was the fire load? Did the fire grow and spread more quickly because of the nature of the contents?

Company Tips

Addressing Poor Performance

Reviews should always focus on lessons and positive experiences; however, there will always be situations where errors must be addressed. The serious discussion between the fire officer and his or her immediate supervisor should occur in private, not in front of a room full of observers. At the company level, poor performance by one particular crew member should be discussed one-on-one in the office, not at the kitchen table. Areas where the whole crew needs to make improvements should be discussed with the entire crew.

FIRE OFFICER II

Voices of Experience

As a new lieutenant in a large metropolitan fire department, I was assigned to a heavy rescue company where I was fortunate to find myself working alongside several senior fire fighters, one of whom began his fire service career in the year I was born! I learned a lot from these folks—sometimes on purpose, and sometimes by accident.

One night we were dispatched as part of a first-alarm assignment (four engines, one ladder truck, one heavy rescue, one medic unit, and one battalion chief) to a one-story single-family dwelling fire in an older neighborhood. Since we were returning from another call, we arrived on-scene before any other units to find heavy fire and smoke showing from two ground-floor windows.

Because of their long experience and our time training together, I rarely had to give my crew detailed instructions at incidents. As the parking brake engaged, I released my seatbelt, grabbed my tools, and ran to the front door while quickly scanning the building and surrounding area as part of my initial size-up; I didn't see any sign of the occupants and noted a car parked in the driveway. While kneeling down at the front door to mask-up, I looked back for my crew and was frustrated to see they were still gathering tools and equipment from the side compartments. I couldn't understand why they were moving so "slowly" when we needed to get inside and do a search for victims!

> "If we had made immediate entry, according to my original plan, we probably would have fallen through the floor."

As I cinched my facepiece straps I realized that I didn't hear any sirens presaging the imminent arrival of the first-due engine and ladder; it appeared that we were going to be on our own for a few minutes. I was suddenly aware, thanks to my crew giving me a moment to think, that our usual assignment as part of the overall response had changed; instead of immediately forcing entry and getting inside to perform search and rescue, I needed to establish command, do a 360-degree survey of the building, and communicate more information to the next-arriving companies so they could effectively deploy their hoselines and ladders.

During my 360-degree lap of the building, I discovered the basement was almost completely involved in fire; if we had made immediate entry, according to my original plan, we probably would have fallen through the floor. I reported this information over the radio and we began setting up for a defensive operation. While we never talked about it, I implicitly understood the few moments of clarity provided by my "slow" fire fighters had actually caused me to step back, do the right things, and probably save our own lives.

One of the basic tenets of incident management is that command must be established as early as possible in the incident, often by the first-arriving unit to the scene; I was fortunate to have this lesson reinforced early in my career by an outstanding crew.

Adam K. Thiel
Fire Chief
Alexandria Fire Department
Alexandria, Virginia

Do we know where, when, and how the fire started? Do we know what was burning initially and how the fire spread? Did the fire involve the contents or the structure or both? Do we know who discovered the fire, what they observed, and when the fire department was called?

If the desired information about the building and occupancy is not readily available, the officer who is gathering this information can refer to several sources of information, such as the inspection file at the fire prevention bureau or the city department that issues building permits and keeps copies of the plans on file. Valuable information can often be obtained from pre-incident plans, and it is an excellent quality control technique to compare the prefire plan information with the situation that was actually encountered. This is also a good opportunity to determine whether the information on file was useful to the incident commander.

The fire investigator can usually provide information about the fire's cause and origin, and checking with the communications center provides information about the source of the alarm and a complete listing of the resources that responded. This should include a list of apparatus as well as the staffing level on each vehicle, if this is a variable. A recording of the incident radio communications should be obtained, as should photographs or videos of the incident.

If water supply was a significant factor in the incident, additional information about the water system should be obtained and evaluated. The locations of hydrants and available flow data should be obtained. The main sizes and whether the area is within a distribution grid or at the end of a dead-end main could be important.

All of this information should set the stage for the analysis of the situation that occurred from the time the first unit arrived at the scene. This analysis of the situation before arrival often reveals factors or circumstances that were unknown to the companies that were involved in the incident. The officer who is preparing for the incident review should be prepared to present all of this information in a manner that is factual and informative.

■ Conducting a Critique

The actual critique should be scheduled as soon as convenient after the event, allowing time to gather and prepare the necessary information. It is best to capture the information while it is still fresh in the minds of the participants. If it was a small-scale incident, it may be possible to have all of the involved companies in attendance so that all of the crew members can be involved in the discussion and learning experience. The attendance at a critique of a larger-scale incident could be limited to company officers and command officers; however, it is a good idea to invite the entire crew of any company that played a significant role or was involved in some unusual occurrence.

Start the critique with an overview presentation of the background and basic information about the incident, including the timeline and the units that were dispatched. The first-arriving officer should then be asked to describe the situation as it was presented on arrival and the actions that were taken. This description should include the first officer's role as the initial incident commander as well as the tactical operations performed by that company.

Each successive company should then take a turn explaining what they saw and what they did. An effective visual method is to draw a plot layout of the area on a dry erase board and to have each company-level officer locate where they positioned their apparatus, where they operated, and what actions they took. This should gradually produce a complete diagram of the incident. Photographs, videos, and radio recordings can be inserted at pertinent points in the discussion. If a large number of companies were involved in the incident, the presentation could be made by division/group/unit officers instead of individual company officers.

During this process, the moderator should keep the analysis directed at key factors, including the initial strategy and any changes in strategy that occurred, how the command structure was developed, how resources were allocated, and what special or unusual problems were encountered. The discussion should also focus on standard operating procedures, including whether or not they were followed and how well they worked in relation to the situation that occurred. The officer acting as moderator should help the participants differentiate between problems that occurred because procedures were not followed and areas where procedures should be changed or updated Table 15-2 ▾. The best way to evaluate the effectiveness of procedures is to determine whether following them actually produced the anticipated results. If there was a deviation from standard procedures, the reasons and the impact should be discussed.

Once every company-level officer has had a chance to make a presentation, the officer directing the critique should provide his or her perspective and assessment of the operation, including both the positive and the negative factors. The critique officer should make a point of evaluating the outcome in terms of where and when the fire was stopped and the extent of damage that occurred. If the outcome was positive and everything met expectations, praise should be widely distributed. Where there is room for improvement, it should be noted in terms of valuable lessons learned.

Have the attendees identify the positive and negative points and list the actions that should be taken to improve future

Table 15-2   **Incident Review Questions**
• Did the pre-incident plan provide accurate and useful information?
• Are there factors that could have or should have been addressed by fire prevention before the incident?
• Were the appropriate units dispatched based on procedures and the information that was received?
• Were the units dispatched in a timely manner?
• Was the appropriate information obtained and transmitted to the responding units?
• What was the situation on arrival?
• What was the initial strategy as determined by the initial incident commander?
• How did the strategy change during the incident?
• How was the incident command structure developed?
• Were there adequate resources for the situation?
• How were the resources allocated and assigned?
• Were standard operating procedures followed?
• Do any standard operating procedures need to be changed?
• What unusual circumstances were encountered, and how were they addressed?
• Is additional training needed?
• Did all support systems function effectively?

performance. These points should be listed on the board for everyone to see as the final product of the critique.

Documentation and Follow-up

The last step in conducting an incident review is to write a summary of the incident for departmental records. The results of this review should be documented in a standard format; many fire departments use a form to document the essential findings. Use of a form allows the information to be stored in a consistent manner and ensures that all relevant information is collected. The form should also list recommendations for changes in procedures, if necessary.

The completed package should then be forwarded to the appropriate section or sections within the department for review and follow-up. The training division should receive a copy of every incident review to determine training deficiencies and needs. Recommendations for policy changes should also be forwarded to the appropriate personnel for consideration.

The completed report should include all of the basic incident report information included from the National Fire Incident Reporting System (NFIRS) as well as the operational analysis information.

Summary

The "new normal" has changed the focus and role of fire department incident management. Fire incidents used to be a local activity, with unique command features that were based on tradition and organizational history.

Using presidential Homeland Security declarations, alterations to the Stafford Act, and the Post-Katrina Emergency Management Reform Act, the federal government has tied financial assistance and response to disasters with local compliance to the 2008 National Response Framework and the National Incident Management System.

This means that the fire officer on a first-arriving unit is expected to operate within the federal National Incident Management System, regardless of whether the incident requires the services of only a single fire unit or is the start of a week-long, multiple-jurisdiction "Incident of National Significance."

You Are the Fire Officer: Conclusion

What started as a single unit incident has immediately become a major event with multiple priorities. You have gone from a single unit offensive mode to fast attack mode. You need to call for enough help to handle a working fire in a large apartment building where the exterior wall is burning and the fire is entering apartments through windows broken by the explosion. The driver of the car that struck the natural gas distribution meter may still be in the car.

The incident will be divided into search and fire suppression branches within the apartment building, control of the burning natural gas pipe, and rescue of any occupants at the crash site. You will need a safety officer and unit officers early in the incident. This will be an event that requires liaison with other agencies and will generate media attention. Remember the rehab unit.

Wrap-Up

Chief Concepts

- The incident management system evolved from the southern California FIRESCOPE and the Phoenix Fire Ground Command System.
- Since the 2006 Katrina hurricane, the federal government has established a National Incident Management System that applies to federal, state, and some private sector and nongovernmental entities.
- The National Response Framework outlines the federal government response to natural catastrophe, fire, flood, explosion, or any instance in which the president determines that federal assistance is needed on a local operation.
- Negligence lawsuits in Seattle and an OSHA ruling on respiratory protection drove the emphasis on fire fighter safety and accountability at an emergency incident.
- The OSHA ruling on the OSHA Respiratory Protection regulations applied an industrial standard to structural firefighting for operations in confined spaces or in toxic or oxygen-deficient atmospheres that became the two-in-two-out rule.
- The first-arriving fire officer is responsible for taking command of an incident.
- Establishing command includes providing an initial radio report.
- Strategic-level command covers the overall direction and goals of the incident.
- Tactical-level command objectives achieve the strategic-level command goals through the resources assigned to the division or group supervisor.
- Task-level command is where the tactical-level objectives are achieved by the work performed by individual companies or individuals. This is the work that is actually performed at an incident.
- The first-arriving fire officer has three options when arriving at the incident: investigation, fast attack, or command.
- If the first-arriving fire officer selects the command mode, the officer needs to remain outside and initiate a tactical worksheet.
- Transfer of command is generally a face-to-face meeting in which the tactical worksheet is reviewed.
- Divisions represent geographical operations, whereas groups represent functional operations. Unit is a generic term that can be applied to either a geographical or a functional component.
- One of the most important responsibilities of a fire officer is to complete a post-incident review of significant operations.
- Situations that involve multiple-company operations should be reviewed from the perspective of how well the overall team performed and how individual companies performed.

Hot Terms

Branch A supervisory level established in either the operations or logistics function to provide a span of control.

Branch director A supervisory position in charge of a number of divisions and/or groups. This position reports to a section chief or the incident commander.

Command staff Positions that are established to assume responsibility for key activities in the incident management system that are not a part of the line organization; these include safety officer, public information officer, and liaison officer.

Division A supervisory level established to divide an incident into geographical areas of operations.

Division supervisor A supervisory position in charge of a geographical operation at the tactical level.

Finance/administration section A section of the incident management system that is responsible for the accounting and financial aspects of an incident, as well as any legal issues that may arise.

Group A supervisory level established to divide the incident into functional areas of operation.

Group supervisor A supervisory position in charge of a functional operation at the tactical level. This position reports to a branch director, the operations section chief, or the incident commander.

ICS general staff The group of incident managers composed of the operations section chief, planning section chief, logistics section chief, and finance/administration section chief.

Incident action plan (IAP) The objectives reflecting the overall incident strategy, tactics, risk management, and member safety that are developed by the incident commander. Incident action plans are updated throughout the incident.

Incident Command System (ICS) A system that defines the roles and responsibilities to be assumed by personnel and the operating procedures to be used in the management and direction of emergency operations.

Incident commander The person who is responsible for all decisions relating to the management of the incident and is in charge of the incident site.

Wrap-Up

Liaison officer The incident commander's representative or a point of contact for representatives from outside agencies.

Logistics section Responsible for providing facilities, services, and materials for the incident. Includes the communications, medical, and food units within the service branch, as well as the supply, facilities, and ground support units within the support branch. The logistics section chief is part of the general staff.

Logistics section chief A supervisory position that is responsible for providing supplies, services, facilities, and materials during the incident. The person in this position reports directly to the incident commander.

National Incident Management System (NIMS) Provides a consistent nationwide template to enable federal, state, tribal, and local governments; the private sector; and nongovernmental organizations to work together to prepare for, prevent, respond to, recover from, and mitigate the effects of incidents, regardless of cause, size, location, or complexity, in order to reduce the loss of life, property, and harm to the environment.

National Response Framework (NRF) A comprehensive, national, all-hazards approach to domestic incident response by describing specific authorities and best practices for managing incidents that builds upon the National Incident Management System (NIMS), which provides a consistent template for managing incidents.

Operations section Responsible for all tactical operations at the incident. In the national model, the operations section can be as large as 5 branches, 25 divisions/groups or units, or 125 single resources, task forces, or strike teams.

Operations section chief A supervisory position that is responsible for the management of all actions that are directly related to controlling the incident. This position reports directly to the incident commander.

Planning section Responsible for the collection, evaluation, dissemination, and use of information about the development of the incident and the status of resources. Includes the situation status, resource status, and documentation units as well as technical specialists. The planning section chief is part of the general staff.

Planning section chief A supervisory position that is responsible for the collection, evaluation, dissemination, and use of information relevant to the incident. This position reports directly to the incident commander.

Public information officer A command staff position that is responsible for gathering and releasing incident information to the news media and other appropriate agencies. This position reports directly to the incident commander.

Safety officer The person who is responsible for all decisions relating to the management of the incident and is in charge of the incident site as well as safety issues.

Staging A specific function in which resources are assembled in an area at or near the incident scene to await instructions or assignments.

Strategic level Command level that entails the overall direction and goals of the incident.

Strike team Specified combinations of the same kind and type of resources, with common communications and a leader.

Strike team leader A supervisory position that is in charge of a group of similar resources.

Tactical level Command level in which objectives must be achieved to meet the strategic goals. The tactical-level supervisor or officer is responsible for completing assigned objectives.

Tactical worksheet A form that allows the incident commander to ensure all tactical issues are addressed and to diagram an incident with the location of resources on the diagram.

Task force Any combination of single resources assembled for a particular tactical need, with common communications and a leader.

Task force leader A supervisory position that is in charge of a group of dissimilar resources.

Task level Command level in which specific tasks are assigned to companies; these tasks are geared toward meeting tactical-level requirements.

Two-in-two-out rule OSHA Respiratory Regulation (29 CFR 1910.134) that requires a two-person team to operate within an environment that is immediately dangerous to life and health (IDLH) and a minimum of a two-person team to be available outside the IDLH atmosphere to remain capable of rapid rescue of the interior team.

Unit Either a geographical or a functional assignment.

Fire Officer *in Action*

Although the federal government has established a National Incident Management System, it is the local fire department with standard operating procedures that organizes the resources to control the incident. For example, the vehicle collision in the apartment parking lot received a standard local response of an engine company and a paramedic ambulance.

After the natural gas explosion, the supervising fire officer reported the change in conditions and called for a first alarm assignment for an apartment fire. Local procedures dispatched four engines, two aerials, a heavy rescue company, and two command officers. The type V constructed building had combustible vinyl siding and flames entering second- and third-floor apartments through windows that were broken by the explosion or fire.

The first-arriving administrative fire officer called for a second alarm, bringing another six fire companies to the incident. As part of local protocol, a half-dozen support units were dispatched as part of the second alarm, including a rehabilitation unit, safety officer, fire investigator, paramedic supervisor, public information officer, and the on-duty executive fire officer.

Attack lines were deployed into six apartments to keep the fire from expanding into the apartment building. The Incident Command System included a branch for the relocation of displaced residents. The incident commander assigned the hazardous materials team to a sector to assist the gas company when there were difficulties in isolating and closing the gas service.

The incident management system provides a framework to handle all types of incidents at any level of complexity, from an investigation of an odor in a building to a major incident.

1. Which of the following is NOT a responsibility of the incident commander?
   A. Stabilize the incident
   B. Stop property loss
   C. Maintain the public order
   D. Protect life

2. Documenting the incident strategy, tactics, and risk management during a 12-hour period at an incident is found in the:
   A. after action report.
   B. incident action plan.
   C. public information officer's press release.
   D. Stafford Act reimbursement log.

3. A group of five pumpers operating together under an assigned leader is called a:
   A. wildland greater alarm assignment.
   B. task force.
   C. mutual aid disaster team.
   D. strike team.

4. The fast attack mode ends when:
   A. the company officer must withdraw and establish a command post.
   B. level I staging is filled.
   C. the first administrative fire officer arrives at the incident scene.
   D. a greater alarm is requested by the company officer.

NFPA 1021 Standard

Fire Officer I

4.6* **Emergency Service Delivery.**
This duty involves supervising emergency operations, conducting pre-incident planning, and deploying assigned resources in accordance with the local emergency plan and according to the following job performance requirements. [p 310–312, 314, 319–321]

4.6.1 Develop an initial action plan, given size-up information for an incident and assigned emergency response resources, so that resources are deployed to control the emergency. [p 312–314]

(A) *Requisite Knowledge. Elements of a size-up, standard operating procedures for emergency operations, and fire behavior. [p 312–314]

(B) Requisite Skills. The ability to analyze emergency scene conditions; to activate the local emergency plan, including localized evacuation procedures; to allocate resources; and to communicate orally. [p 310–326]

4.6.2* Implement an action plan at an emergency operation, given assigned resources, type of incident, and a preliminary plan, so that resources are deployed to mitigate the situation. [p 310–326]

(A) Requisite Knowledge. Standard operating procedures, resources available for the mitigation of fire and other emergency incidents, an incident management system, scene safety, and a personnel accountability system. [p 312–315, 319–324]

(B) Requisite Skills. The ability to implement an incident management system, to communicate orally, to manage scene safety, and to supervise and account for assigned personnel under emergency conditions. [p 310–326]

Fire Officer II

5.6 **Emergency Service Delivery.**
This duty involves supervising multi-unit emergency operations, conducting pre-incident planning, and deploying assigned resources, according to the following job requirements. [p 311–315, 319–324, 326]

5.6.1 Produce operational plans, given an emergency incident requiring multi-unit operations, the current edition of NFPA 1600, and AHJ-approved safety procedures, so that required resources and their assignments are obtained and plans are carried out in compliance with NFPA 1600 and approved safety procedures resulting in the mitigation of the incident. [p 311–315, 321–323]

(A) Requisite Knowledge. Standard operating procedures; national, state/provincial, and local information resources available for the mitigation of emergency incidents; an incident management system; and a personnel accountability system. [p 311–315, 321–323]

(B) Requisite Skills. The ability to implement an incident management system, to communicate orally, to supervise and account for assigned personnel under emergency conditions, and to serve in command staff and unit supervision positions within the Incident Management System. [p 310–326]

Additional NFPA Standards

NFPA 1561 *Standard on Emergency Services Incident Management System*

NFPA 1600 *Standard on Disaster/Emergency Management and Business Continuity Programs*

NFPA 1710 *Standard for the Organization and Deployment of Fire Suppression Operations, Emergency Medical Operations, and Special Operations to the Public by Career Fire Departments*

NFPA 1720 *Standard for the Organization and Deployment of Fire Suppression Operations, Emergency Medical Operations, and Special Operations to the Public by Volunteer Fire Departments*

Introduction to Fire and Emergency Services Administration (FESHE) Course Outcomes

2. Explain the need for effective communication skills both written and verbal. [p 310]

3. Articulate the concepts of span and control, effective delegation, and division of labor. [p 321–322]

7. Identify roles and responsibilities of leaders in organizations. [p 310–311]

8. Compare and contrast the traits of effective versus ineffective supervision and management styles. [p 311]

9. Identify and assess safety needs for both emergency and nonemergency situations. [p 311, 314–317, 322–326]

11. Identify the role of a company officer in Incident Command System (ICS). [p 311–315, 317–321, 323–324, 326]

Knowledge Objectives

After studying this chapter, you will be able to:

- Describe how to supervise a single company.
- Describe how to supervise multiple companies.
- Describe how to size up the incident.
- Describe Lloyd Layman's five-step size-up process.
- Describe the National Fire Academy size-up process.
- Describe risk-benefit analysis.
- Describe how to determine fire flow.
- Describe how to develop an incident action plan.
- Describe how to assign resources.
- Describe the tactical safety considerations.
- Describe the general structure fire considerations.

Skills Objectives

After studying this chapter, you will be able to:

- Perform a scene size-up.
- Assign resources.

I t is 1:24 A.M. and your fire company is responding alone to investigate an "odor of smoke" incident. You arrive at the incident address and find a three-story garden-style apartment with flames shooting out of a first-floor window and occupants screaming for help above the fire on the second floor.

1. What information would you consider in order to make an accurate size-up?
2. What are your strategic priorities?
3. What tactics might you use to meet those strategies?

Introduction to Fire Attack

Individuals at the managing or supervising fire officer levels directly supervise a crew who operate at fire and other emergency incidents **Figure 16-1 ▸**. The fire officer is a vital player with important responsibilities for the safety of fire fighters and civilians, as well as ensuring that the incident is mitigated.

This chapter explains how to perform an incident size-up, how to calculate fire flow requirements, how to determine incident priorities, and how to develop tactical assignments to control a fire. It also covers establishing of command, command transfer procedures, multiple alarm incidents, and special considerations.

Supervising a Single Company

A fire company is a basic tactical unit for emergency operations, and the fire officer is a working supervisor. When the company is assigned to advance an attack line into a structure, the fire officer is with the crew members, directing and leading them as well as continually evaluating the environment for hazards such as backdraft, flashover, or structural collapse. The officer's personal and physical involvement must never compromise his or her supervisory responsibilities.

In addition to leading and participating in company-level operations, the fire officer is also evaluating their effectiveness. The engine company officer must determine whether the fire stream is making progress or losing ground against the fire. The truck company officer must judge whether the ventilation opening is of sufficient size so that the attack line can advance. It is up to

the officer to figure out how to resolve fire-ground problems. For example, if the fire attack is not making progress, the officer must determine whether an additional line or a larger line is needed or if the attack should be made from a different direction.

The fire officer is expected to relay relevant information to his or her supervisor. At a structure fire, this would include informing command when the company's assignment has been completed or when an assignment will be delayed or cannot be completed. This also includes informing the supervisor immediately when problems or hazards are identified, such as signs of collapse. The fire officer serves as eyes and ears for the incident commander or for an assigned supervisor in the incident management structure.

Closeness of Supervision

The inherent risks associated with emergency operations demand close supervision at all times. Fire fighters who have been assigned specific tasks to complete, such as advancing a hose line to the seat of the fire or performing vertical ventilation, often focus on the task and ignore everything else. Aggressive, competent fire fighters are good at completing their tasks. While focusing on their assignment, however, they are not paying attention to their surroundings. The role of the fire officer is to look at the big picture: monitoring progress, coordinating activities with other companies, and looking out for hazards.

The level of supervision should be balanced with the experience of the company members and the nature of the assignment that is being performed. An inexperienced crew performing a high-risk task requires more direct supervision than a highly

Figure 16-1 Supervising emergency operations is a core fire officer task.

experienced crew performing a more routine task. A company that is making an interior attack on a fire in a large warehouse requires more direct supervision than a company operating a master stream to protect an exposure. The fire officer must know where the crew members are located and what they are doing. Within an interior attack, the fire officer should be able to see all crew members and communicate directly with them at all times.

Form of Leadership

Emergency scene supervision requires a more authoritarian form of leadership than in nonemergency activities. During nonemergency situations, the fire officer should use a participative style of leadership in order to promote productivity and group cohesiveness. There is time for discussion and feedback before decisions are made, and the consequences seldom involve an imminent risk of injury or death. Emergency operations present a very different environment for leadership and decision making.

Time is often of the essence at emergency incidents. Decisions are needed quickly, and actions must be implemented in a timely manner. These situations do not provide the luxury of time to gather the team, discuss what needs to be done, and reach a group consensus. The fire officer must frequently make decisions with little or no input from subordinates, and every member of the team must be prepared to perform a specific set of tasks with a minimum of instruction.

When the fire officer gives the order to lay a line, the fire fighter who is assigned to catch the hydrant must know exactly what to do. While that fire fighter is dragging the hose to the hydrant, the fire officer is gathering information to develop the next step in the action plan. The fire fighter who will be advancing the attack line can anticipate what comes next and does not need a lengthy explanation from the officer.

The key to using an authoritative style of leadership effectively is to develop the trust and confidence of the subordinates before the incident. A fire officer who has demonstrated sound

knowledge of the technical aspects of firefighting and has developed a trusting relationship with the fire fighters is ready to lead them into action. Fire fighters are less likely to trust the judgment of an officer who does not appear to fully understand the job and has not made it clear that their safety is foremost in his or her mind.

Fire fighters must be prepared to follow the officer's directions; however, they should also feel confident giving relevant information to the officer. For example, a fire fighter who has been given orders to perform vertical ventilation should not hesitate to inform the officer of safety conditions that the officer might not know about.

■ Supervising Multiple Companies

The first-arriving officer at a fire incident assumes the role of incident commander. When operating as the incident commander, the fire officer has an even greater level of responsibility; the incident commander is responsible for every company on the scene and for management of the overall operation.

As covered in Chapter 15, the initial incident commander's responsibilities include conducting a size-up, developing an action plan to mitigate the situation, assigning the resources to execute the plan, developing a command structure to manage the plan, and ensuring that the plan is completed safely. The fire officer who is functioning as the incident commander is responsible for all of these functions.

The fire officer might also be assigned as a division/group/unit leader, supervising multiple units, or even as a branch director within the Incident Command System (ICS). In each of these situations, the officer serves as a relay point within the command structure. Direction comes down through the system, from the incident commander, through the intermediate levels, to the individual companies or units. At the same time, information is transmitted upward from subordinates to the officer, who either acts on the information directly or relays the information up to the next level. For example, a division supervisor could be assigned to oversee operations directed toward keeping the fire out of an exposure. The division supervisor would be expected to manage that operation and supervise the assigned companies, giving regular progress reports to the incident commander.

Standardized Actions

A fire officer has great latitude in deciding how work is accomplished during nonemergency activities; however, the opposite is true at emergency scenes. Emergency operations must be conducted in a very structured and consistent manner. This is accomplished by placing a strong emphasis on standard operating procedures (SOPs).

SOPs provide a framework to allow activities at an emergency scene to be completed in an efficient manner and with all components of the organization working in concert with each other. The SOPs are the equivalent of a football team playbook; they explain the standard approach that should be followed in a particular situation. This allows the incident commander to develop an incident action plan. The use of SOPs also promotes a standard approach to safety.

The use of an incident command system (ICS) allows for a consistent approach to developing a command structure that

FIRE OFFICER II

establishes efficient management and effective control at incidents. The consistency allows everyone operating at the scene to know precisely to whom they report and from whom they receive their orders. The ICS incorporates a process that identifies a strategic goal for the incident, followed by an incident action plan that is translated into specific objectives for each tactical unit.

Command Staff Assignments

The command staff positions include safety officer, liaison officer, and public information officer. While working in one of these positions, the fire officer reports directly to the incident commander. The incident commander could assign any available and qualified officer to perform one of these roles at an incident scene.

The safety officer is responsible for overseeing the incident from a safety perspective, keeping the incident commander informed of safety concerns and taking preventive action when an immediate hazard is identified. When assigned as liaison officer, the fire officer functions as the link between the incident commander and representatives from various agencies. At a hazardous materials incident, this could be the property manager or a chemical manufacturer. When operating as the public information officer, the fire officer is responsible for gathering information that is to be released to the general public, developing news releases, and giving interviews or press conferences. The person in this position acts as the public spokesperson for the incident commander.

Sizing Up the Incident

Size-up is a systematic process of gathering and processing information to evaluate the situation and then translating that information into a plan to deal with the situation. The art of sizing up an incident requires a diverse knowledge about emergency incidents. The fire officer could be faced with a house fire, a hazardous materials release, and a major automobile collision in one shift. Each situation requires the ability to recognize and analyze the important factors, develop a plan of action, and then implement the plan. Each type of incident involves specialized knowledge.

Inexperienced fire officers may think that size-up begins when they get their first glimpse of the incident scene and ends when they have decided on a plan of action. In reality, size-up begins long before arrival and continues until the incident is stabilized. Deciding on an initial plan of action after arriving on the scene is only one step in a continuous process. As the plan is executed, new information must be gathered and processed to determine whether it is working as anticipated and to make any adjustments that are required.

The initial size-up at the scene of an incident must often be conducted under intense pressure to "do something." This pressure could be coming from the public, from fire companies that are arriving and anxious to get into action, and from within the fire officer who feels the anxiety of being responsible for deciding what to do quickly. The end result of a good size-up is an **incident action plan (IAP)** that considers all the pertinent information, defines strategies and tactics, and assigns resources to complete those tactics.

A comprehensive evaluation of a situation requires information that often is not available. An officer must look carefully and assess what can be seen, make reasonable assumptions about what cannot be seen, and anticipate what is likely to happen. An experienced officer can develop an initial plan, and then adjust the plan as more information becomes available.

The additional information could require a major change in the plan. The incident commander must consciously differentiate between what is known, what is assumed, and what is anticipated in processing size-up information.

The first phase of size-up begins long before the incident occurs. Everything the fire officer learns, observes, and experiences goes into a memory bank. Those memories are used to make size-up decisions. Knowledge of the types of building construction used in the area, the available water sources, and the available resources are factors when making a size-up. Whenever a pre-incident plan is updated, including a familiarization visit, the fire officer is performing part of a future incident size-up.

Pre-arrival Information

The specific size-up for an incident begins with the dispatch. The name, location, and reported nature of the incident all help the fire officer begin anticipating what is going on at the scene.

On-Scene Observations

The ability to quickly size-up a fire situation requires both a systematic approach and a solid foundation of reference information. SOPs list the essential size-up factors. These include building size and arrangement, type of construction, occupancy, fire and smoke conditions, and other factors, such as weather and time of day. The SOPs should guide the officer's systematic thinking to ensure that all of the important factors are considered.

A fire officer must understand and recognize basic fire behavior. It is easy enough to identify fire coming out of a window and smoke pushing out from the eaves of a house. An understanding of conduction, convection, and radiation is needed to predict where the fire is burning and where it will spread. An officer who has studied and observed fire behavior is able to process this information very quickly, without having to spend an excessive amount of time looking at the building and thinking about what is going on inside.

Visualization is one of the most significant factors in size-up. Every previous situation that the fire officer has experienced or observed, including some that might have been observed only in training sessions, photographs, or videos, is stored in the individual's memory. When a new situation is observed, the mind subconsciously looks for a matching image to create a template for this new observation. Instead of methodically processing the information, the brain can instinctively jump to a similar observation and apply the stored experience to the new set of circumstances. The same process works for other human senses, particularly smell, taste, and touch, fostering rapid recognition and interpretation of many situations.

Experienced fire officers know vigorous dark smoke churning means higher heat flux, indicating flashover conditions. Darker smoke is generally closer to the seat of the fire than lighter-colored smoke. By evaluating where the smoke is coming from in the building, the fire officer may be able to predict where the fire will be traveling, unless there are mitigating circumstances.

The volume and color of the smoke aid the officer in determining the need for ventilation. Because heat and smoke travel together, large amounts of smoke coming from a structure normally indicate that there is a need to ventilate in order to let the heat out of the building. Dark smoke is also an indication that there are more carbon particles suspended within it and the possibility of a lack of oxygen.

Understanding fire behavior is also needed to develop action plans. Fires in the ignition phase could be extinguished with a portable fire extinguisher. Fires in the growth phase may require a small handline. A fully developed fire requires multiple large handlines or master streams.

Fuel loading is also an important factor during size-up. A fire in class A combustibles normally dictates a direct attack with water. Those involving Class B combustibles require the use of foams. When using foams, the officer must determine what type and how much foam will be needed to complete the job before beginning the application.

■ Lloyd Layman's Five-Step Size-Up Process

In 1940, Chief Lloyd Layman authored *Fundamentals of Fire Tactics*, which used the military training model with an emphasis on general principles. Layman presented a five-step process for analyzing emergency situations:

1. Facts
2. Probabilities
3. Situation
4. Decision
5. Plan of operation

Facts

Facts are the things that are known about the situation. Chief Layman was an early advocate of pre-incident planning. This allowed the first-arriving officer to quickly determine the type of construction, whether fire protection systems are in place, fire flow requirements, water supply sources, special hazards, and other significant factors. Additional facts will be known on dispatch, such as time of day, location of incident, weather, and specific resources that the fire department is sending. Visible smoke and fire conditions are additional facts. The more facts the fire officer can secure, the more accurately the situation can be sized up.

Probabilities

Probabilities are things that are likely to happen or can be anticipated based on the known facts. For example, we know that the structure has a fire sprinkler system but we cannot be sure that it is going to work. It may have been turned off for maintenance or it could have been damaged by an explosion. Fire officers consider the most likely possibilities based on the known facts.

The fire officer has to anticipate where the fire is likely to spread. If the fire is burning freely in the attic, what impact is it likely to have on the stability of the building? This skill requires an in-depth knowledge about factors such as building construction, fire behavior, hazardous materials, sprinkler system performance, and human behavior in fires.

Another aspect of considering probabilities is the amount of time it takes to accomplish fire-ground tasks. How far is the fire likely to advance before the hose lines are in place to attack?

How long will it take to get ground ladders to Charlie side? How fast can a crew with a back-up hose line get to your location? How large will the fire be in 10 minutes?

Situation

The situation assessment involves three considerations. The first consideration is whether the resources on scene and en route will be sufficient to handle the incident. The incident commander will be able to determine whether additional companies will be needed to control the problem. The incident commander has to anticipate the time lag between requesting resources and having them on the scene available to use.

Many fire departments have policies that require an incident commander to request additional resources whenever all of the units will be committed, in order to provide an on-scene reserve. Some departments require the first-arriving company to immediately call for a second alarm if any smoke or fire is showing from a high-rise, hospital, or big box mercantile building.

The second consideration is the specific capabilities and limitations of the responding resources in relation to the problem. Consider the amount of water carried, pump capacity, amount and diameter of hose, and length and type of aerial devices. Staffing is an important factor. A company staffed with two fire fighters cannot perform as much work as a six fire fighter company. The incident commander might have to call for additional resources or adjust performance expectations downward.

The third consideration consists of the capabilities and limitations of the personnel, based on training and experience. Consider your personal experience with this type of incident, as well as the experience and capabilities of your crew members. If two fire fighters out of a four-person crew are rookies, the company requires more direct supervision.

Decision

The fourth step in Layman's size-up procedure is making fire attack decisions. This step requires the fire officer to make specific judgment decisions based on the known facts and probabilities, as well as the situation evaluation. The officer needs to answer four questions:

1. Are there enough resources responding to and on the scene to extinguish the fire or mitigate the situation?
2. Are sufficient resources available and do conditions allow for an interior attack?
3. What is the most effective assignment of on-scene resources?
4. What is the most effective assignment of responding resources?

The answers to these questions allow the fire officer to develop the overall strategy and the underlying tactics to mitigate the incident.

Plan of Operation

The final step in Layman's decision process is to develop the actual plan that will be used to mitigate the incident. Most other methods of size-up do not include this step as part of the size-up; they refer to the information that is obtained during size-up being used to develop a plan of action.

National Fire Academy Size-Up Process

The National Fire Academy (NFA) has developed a size-up system that includes three phases. The process has been widely distributed through the Managing Company Tactical Operations courses provided by the NFA. The three phases of size-up are:

- Phase one: Pre-incident information
- Phase two: Initial size-up
- Phase three: Ongoing size-up

Phase One: Pre-Incident Information

Phase one considers what you know before the incident occurs. This consideration closely mirrors the facts step identified by Layman. Pre-incident plans often provide valuable information for this phase **Figure 16-2 ▾**. Information about the building and the occupancy, such as the building layout and construction type, built-in protection systems, type of business, and nature of the contents, are all needed for an accurate size-up. Pre-incident information should also identify water supply sources, including their location, accessibility, and capacity.

If pre-incident plans are not available, the information must be determined through on-scene observation and research. The fire officer's intuition and experience may have to be relied on until this information can be obtained.

Fire officers are preloaded with a large amount of useful information that figures into the size-up process. Environmental information includes factors such as heat and cold extremes, humidity, wind, and snow or ice accumulations. The time of day can also be a critical piece of information; responding to fire in

a multifamily residential building at 2:00 A.M. involves different considerations from the same fire at 2:00 P.M. A fire officer also needs to know departmental and mutual aid resources.

Phase Two: Initial Size-Up

The second phase of the size-up begins on receipt of an alarm. There are three questions that need to be answered:

1. What do I have?
2. Where is it going?
3. How do I control it?

Start with the information that was established in phase one and build upon it. Phase two considers the specific conditions that are present at the incident. This could require a walk around a fire building to observe conditions, such as the location and the size of the fire, as well as the volume, color, movement, and location of the smoke. The size-up should also look at the size and construction of the building, identify any exposures that are present, and look for indications of special concerns, such as the possibility that lightweight construction could be involved.

To determine where the fire is going, the fire officer must consider the current location and stage of development of the fire, in what direction it is likely to spread, and whether there is a risk of flashover or backdraft. The accuracy of these predictions is heavily dependent on the fire officer's knowledge and experience. This stage is closely associated with the probabilities step noted by Layman.

The third question is, "How do I control it?" This is where a fire officer has to consider different alternatives in relation to the available resources, including apparatus, equipment, and personnel. The development of strategy and tactics must be based on the resources that are available and, if additional resources are needed, the time it will take for them to arrive. This stage is closely associated with the situation step noted by Layman.

Phase Three: Ongoing Size-Up

The third phase addresses the need to continually size up the situation as it evolves. This includes an ongoing analysis of the situation and an ongoing evaluation of the effectiveness of the plan being executed. The incident commander should always be prepared to modify the plan if the situation changes, including switching between offensive and defensive strategy. Additional resources could be required to address unanticipated needs or to replace tired crews. The ongoing evaluation could also indicate that resources can be redeployed from one task to another where they would be more effective.

This ongoing evaluation requires a constant flow of feedback information to the incident commander. The incident commander needs to know when:

- An assignment is completed
- An assignment cannot be completed
- Additional resources are needed
- Resources can be released
- Conditions have changed
- Additional problems have been identified
- Emergency conditions exist

Risk/Benefit Analysis

Risk/benefit analysis is a key factor of size-up when selecting the appropriate strategic mode. The strategic mode regulates the degree

Figure 16-2 Pre-incident plans provide information the fire officer needs during the information phase.

Figure 16-3 Defensive fire attack in a vacant apartment building.

of risk to fire fighters that is acceptable in a given situation. The degree of risk that is acceptable is determined by the realistic benefits that can be anticipated by taking a particular course of action.

The potential benefits include the possibility of saving lives or preventing injury to persons who are in danger, preventing property damage, and protecting the environment from harm. The major risk factor is the possibility of death, disability, or injury to fire fighters. Potential occurrences, such as flashover, backdraft, or structural collapse, have to be evaluated. Each alternative action plan involves a different combination of risks and benefits.

The fire officer can manage these risks. Training, experience, protective clothing and equipment, communications equipment, SOPs, accountability systems, and the rapid intervention team allow crews to work with a reasonable level of safety in a hazardous environment.

Fire fighters accept a higher personal risk when it is necessary to save a life; however, there is no justification for risking the lives of fire fighters when no property of value can be saved, such as in fires in vacant or abandoned buildings or situations where significant damage is already done. Building construction features, such as lightweight trusses, can also cause the risks of offensive operations to outweigh the potential benefits. When defensive operations are conducted, the risks to fire fighters are significantly reduced.

The risk analysis determines the appropriate strategy for an incident: offensive, defensive, or transitional. An **offensive attack** is an advance into the fire building by fire fighters with hose lines or other extinguishing agents to overpower the fire. Offensive attack is the activity that drives most fire department training, operations, and organizational structures.

When this mode is selected, the incident commander believes that the benefits associated with controlling and extinguishing the fire outweigh the risks to the fire fighters. This necessarily requires the resources to mount an attack that is powerful enough to overwhelm the fire within the time that is available to operate safely inside the burning building. The risks of an offensive attack can be justified only when there are

realistic benefits. There must be lives or valuable property that can be saved in order to justify taking this risk.

The **defensive attack** is used when the risks outweigh the expected benefits. In this mode, fire fighters are not to be allowed to enter the structure or to operate from positions that involve avoidable risks. Defensive operations are conducted from the exterior, using large streams to contain and, if possible, overwhelm a fire **Figure 16-3**. Defensive attacks are used when there is a risk of a structural collapse or the lack of adequate resources to conduct an adequate interior attack to control the fire. Defensive strategy is also the appropriate choice in a situation where the building and contents would be a total loss even if an aggressive interior attack could control the fire.

A **transitional attack** refers to a situation in which an operation is changing or preparing to change. Transitional applies where an offensive attack is initiated, with the recognition that it could be unsuccessful or the situation could deteriorate quickly. The offensive attack is conducted cautiously and in a manner that allows interior attack crews to be withdrawn quickly. At the same time, back-up resources are positioned in defensive positions in case they are needed. This strategy could also be used to quickly conduct a search and rescue operation ahead of a fire, knowing there is only a limited time for crews to get in and get out.

A transitional attack could also apply to a situation in which an initial attack is made with an exterior master stream to knock down a large body of fire, while crews prepare to conduct an offensive interior attack. The switch to offensive attack should occur only if there are still lives or property to be saved after the heavy streams have been shut down.

Smoke, Wind, Size, and Fire Flow

After a presentation from the National Institute of Standards and Technology on Fire Dynamics for the Fire Service, Bobby Halton, editor of *Fire Engineering* magazine, said "The evolution of our tactical capability will only be enhanced by the evolution of our mental agility." This section will look at how issues with smoke, wind, size, and fire flow will affect the risk/benefit analysis.

Company Tips

How Aggressive Suppression?

Discussion of fire suppression risk assessment can polarize into two extremes:

- Running into a collapsing and burning Type V building to rescue a checkbook.
- Never entering a building with smoke showing until you get independent confirmation that there is a savable life requiring rescue, more than 16 fire fighters are on the scene, and all command vests are properly deployed.

The sweet spot for effective interior fire operations is somewhere between these two extremes. It depends on the built environment as well as fire department resources, experience, and training.

What is appropriate for a department that can deliver 36 fire fighters, 12 company officers, and 3 command officers in 15 minutes is not appropriate for a rural fire department that can assemble 6 fire fighters and 1 company officer within the same time.

Smoke Is Fuel

The time-and-temperature curve for fire development within a structure was established by the NFPA and the American Society for Testing and Materials (ASTM) in 1918. This curve was published after work sessions with Underwriter Laboratories, Factory Mutual, National Board of Fire Underwriters, and National Bureau of Standards in order to establish the procedure now called ASTM E119-08a *Standard Test Methods for Fire Tests of Building Construction and Materials*.

Sofas, beds, chairs, and tables have evolved from the wood and wool used in the 1920s to plastic and resin-based composites. Fire instructors identify the 21st-century items as being constructed from "solidified gasoline" to describe the increase in the rate of heat release during a room-and-contents fire. In the mid-2000s David Dodson, a battalion chief and safety officer, started a program to teach company officers how to read smoke. Chief Dodson found another by-product of 21st-century room-and-contents fire; here is his description:

> Consider a primary search in a smoke-filled structure. The hallway is filled with turbulent smoke so thick you can barely see your helmet or boxlight. Products of combustion are coating your facepiece and the velocity of the smoke is accelerating.
>
> You are seconds away from burning to death. You are surrounded with the vaporizing contents of the room that is about to be heated to flashover. You are getting coated in a combustible glaze that will flash-fry as soon as one of the half dozen combustible gases reaches ignition. Once the first gas starts to burn, the rest of the gases will ignite.

After reviewing hundreds of structure fire videos, Dodson shows incidents where the black boiling smoke seems to lighten up moments before flashover. United Kingdom fire fighter Paul Grimwood, author of *3D Firefighting*, speculates that this lightening is the sublimation of vaporized carbon into ash. Ash is the dense and hot product of carbon combustion.

Wind

Since the 1999 line-of-duty death of District of Columbia fire fighters Anthony Phillips and Louis Matthews, staff from the Building and Fire Research Laboratory (BFRL) of the National Institute of Standards and Technology (NIST) Fire Safety Engineering Division work to identify fire fighter survival factors. Starting with the Fire Dynamics Simulator, NIST has steadily increased the quality of fire development simulation with each analysis of a significant fire or full-size test burns. Fire protection engineer Dan Madrzykowski, who has 15+ years with NIST, points out that a simulation or a live-fire video presents a more effective demonstration than " . . . a tall stack of computer print-outs."

Steve Kerber, another fire protection engineer, has been the primary investigator on wind-driven fires. The goal of the work was to provide guidance on the safe use of positive pressure ventilation (PPV) and validating a modeling of PPV with a computational fluid dynamics model. Field experiments were conducted in Toledo, Ohio, and Chicago, Illinois. The research concluded with 14 experiments conducted at a seven-story apartment building on Governor's Island in New York City in February 2008. Initial results of these studies were presented at the 2008 Fire Department Instructor's Conference and a few regional presentations.

NIST Technical Note 1629, *Fire Fighting Tactics Under Wind Driven Fire Conditions: 7-Story Building Experiments*, was published in April 2009 and provides the details. The 593-page report also summarizes earlier laboratory research, incident analysis, and field experiments on fire development, ventilation, and wind-control devices. The report also provides fire dynamic analysis from line-of-duty death investigations.

Tactical Considerations

Sometimes fire fighters making initial entry are facing underventilated structure fires when the fire is beginning to die down or the atmosphere is too rich for a flashover. The process of making their entry results in a flashover that envelopes the initial attack crew. Often, fire fighters who are operating within a burning structure are between the fire and the point where fire exits the structure. They will not survive the heat flux as the flame front travels over them.

In wind-driven fires, the flame front becomes a blowtorch. The failure of a window or door can render a hallway from ambient to a nonsurvivable furnace in seconds.

Thermal imaging cameras (TICs) do not show the heat within the "white" areas of the screen, potentially leading the viewer to think that the area is tenable. This situation is documented by the postfire statement of "we felt a lot of heat but saw no flame." In addition, TICs cannot show imminent collapse of a floor during a basement fire. Madrzykowski used this example to urge that "we need to understand the structure better" as part of the initial attack size-up.

The NIST observation on TIC performance was confirmed by research conducted by Underwriters Laboratories with the Chicago Fire Department. Supported by the IAFC and funded by a grant from the Department of Homeland Security, the online course "Fire Behavior in a Single Family Occupancy" included research that confirmed the inability of a TIC to detect a fire below a floor, such as a fire in a basement. The components of the floor are slow to heat up. When the TIC identified hot spots, the floor components already were failing.

Kerber shared the results of positive pressure ventilation (PPV) tests in large and high-rise structures. He suggests that a coordinated fire attack with PPV requires a 60-second wait after starting the fans for the fire to react to the introduction of oxygen before starting fire attack. In Toledo and New York City it was determined that the carbon monoxide (CO) created by the PPV fan is less than the CO generated by the fire or the running of power saws inside the structure.

The take-away messages from the NIST research include:

- Wind is a factor. It must be included in the incident size-up.
- Smoke is fuel. Oxygen-depleted combustion products, containing carbon dioxide, carbon monoxide, and unburned hydrocarbons, will fill the rooms of a structure. Once fresh air is introduced, it provides the oxygen needed to sustain the transition through flashover.
- Venting does not always equal cooling. In the NIST experiments, the cool air forced into the broken upwind window resulted in an initial period of cooling in the room of origin, typically followed by a transition to flashover, if a flow path was available.
- Fire-induced flows. The NIST experiments usually had a flow from an outside window, through the fire room, and out to a bulkhead door. Fire growth resulted in a flow of 11 miles per hour within the structure.
- Avoid the fire flow path. Thermal conditions within the fire flow path were higher than 752°F (400°C), an unsurvivable environment for fire fighters in PPE gear.
- Control fire flow path. Use of PPV, wind control devices, and externally applied water can control the fire flow path and reduce the interior environment to tenable conditions for fire fighters wearing PPE.

Size

Technician I Kyle Wilson died April 16, 2007, while operating as a member of a Prince William County, Virginia, initial fire attack team. The investigation pointed out that NFPA 1710, *Standard for the Organization and Deployment of Fire Suppression Operations, Emergency Medical Operations, and Special Operations to the Public by Career Fire Departments*, requires an initial force of 14 fire fighters to conduct a basic fire attack on a single-family dwelling. The standard states that these resources should be assembled on the scene within 8 minutes after dispatch.

The minimum staffing is required to operate two handlines and provide a search team, a ventilation team, and a rapid intervention crew, plus apparatus operators, an incident commander, and a safety officer. The standard calls for an uninterrupted water supply capable of delivering 400 gpm and a minimum initial flow of 300 gpm through two handlines. This is based on a 2000 ft² (610 m²), single family occupancy, two-story detached home without a basement.

Technician Wilson died in a 6000 ft² (1829 m²), single family occupancy, two-story detached home with a basement. Wilson was part of a team conducting a primary search while a large volume

of fire from outside the building was extending into the attic, pushed by sustained 25- to 35-mph winds with 50-mph gusts. The blowtorch effect resulted in failure of the Type V wood components and generated a partial structural collapse within minutes of the fire department arrival, trapping Technician Wilson.

The use of a small residential home as a tactical reference was identified as a problem in the aftermath of the March 14, 2001, death of Phoenix fire fighter/paramedic Bret Tarver. Tarver was operating within a 20,132 ft² (6136 m²) supermarket. Tarver and his partner became disoriented when they tried to leave the fire floor after their SCBA low air alarms activated. Traditional task-level training assumes that fire fighters have a short travel distance to get out of an immediately dangerous to life and health (IDLH) environment. Chapter 6, Safety and Risk Management, describes the results of Phoenix's rapid intervention research in large buildings.

Big box retail stores will exceed 100,000 ft² (30,480 m²), with 40' (12-m) ceilings. Merchandise may be stacked as high as 30' ft (9 m) with automatic fire suppression sprinklers that comply with the high-piled combustible storage provisions. The combined water supply requirements, for both the built-in and municipal fire suppression needs, may be four times greater than in a typical retail store.

Fire Flow

The volume of applied water must be sufficient to absorb the heat that is being released. The estimated fire flow is an approximation of the rate of water application (usually measured in gallons per minute [gpm]) that would be required to control a fire in a particular building or section of a building. There have been many formulas, including the Kimball Rule of Thumb and the Royer/Nelson Iowa Rate of Flow.

The incident commander can use this information to estimate the number of hose lines that will be required, the number of engine companies that will be required to operate those hose lines, and the water supply that will be needed to mount an effective fire attack. The Los Angeles Fire Department sends a heavier "Category B" dispatch to structures with a required fire flow of 4500 gpm or higher.

If the incident commander does not have the resources to extinguish the fire with an overwhelming fire flow, the available resources will be deployed in a defensive fire attack to contain the fire.

(A)

(B)

(C)

(D)

Figure 16-4 Wind-driven fire development.

NFA Fire Flow Formula

The <u>National Fire Academy fire flow formula</u> is a simplified version of calculations developed by the insurance industry, with a nod to the Kimball and Royer/Nelson approaches. The NFA version uses a four-part calculation and is based on square footage rather than cubic footage:

Example

Estimate the fire flow for a two-story building that is 30' wide by 40' long (9 m by 12 m), is 25 percent involved in fire, and has an exposure building 25' (8 m) away on each side:

Step 1: Determine the fire flow requirement per floor. For each floor of a building, estimate the length and width of the building and divide by three.

> Needed Fire Flow (NFF) = (Length × Width)/3
> NFF = (30 × 40)/3
> NFF = (1200)/3
> Needed Fire Flow per floor = 400 gpm

Step 2: Determine the fire flow requirements for the total building. Multiply the NFF per floor times the number of floors.

> NFF = 400 gpm × 2 floors
> Needed Fire Flow for entire building = 800 gpm

Step 3: Determine the NFF based on the percentage of fire involvement for the building. Multiply the NFF by the percentage involvement.

> NFF = [(L × W)/3] × [# of floors] × [% involved]
> NFF = 800 gpm × 25%
> Needed Fire Flow with 25% fire involvement
>   = 200 gpm

Step 4: Increase the NFF by 25 percent for each exposure. An exposure is any building within 50' (15 m). (Buildings more than 50' but less than 100' [30 m] can also be counted.) An exposure is also any floor in the building that is five floors or less above the fire floor.

> For two exposures add 25% + 25% = 50%
> NFF = 200 gpm + 50% (100)
> Needed Fire Flow with 25% involvement and two exposures = 300 gpm

■ "When the Science Hits the Streets"

The science of fire suppression is advanced in leaps. It was at the 1950 Fire Department Instructor's Conference when Lloyd Layman's presentation, "Little Drops of Water," electrified the audience. The indirect method of fire attack was developed when Layman was commanding the Coast Guard Fire Fighting School during World War II. This presentation changed the design of automatic fire sprinklers. Once confirmed by Miami and other fire departments, there was a decade of using high-pressure fog for indirect structural fire attack.

Purple-K, an effective potassium bicarbonate powder for extinguishing flammable liquid fires, was developed by the U.S. Naval Research Laboratory. Walter M. Haessler was a fire protection engineer who researched the effectiveness of Purple-K while certifying portable fire extinguisher performance. His 1962 NFPA-published monograph, *Uninhibited Chain Reactions*, established the tetrahedron of fire concept.

Dan Madrzykowski and Steven Kerber shared results of the Building and Fire Research Laboratory experiments with wind-driven fires at the 2008 Fire Department Instructor's Conference. Former Fire Commissioner Raymond E. Orozco used the phrase "When Science Meets the Street" to describe the impact of the research conducted by the Underwriters Laboratories Inc. with the Chicago Fire Department.

Suppression strategy, tactics, and techniques in 2015 will reflect the research by the National Institute of Standards and Technology, Underwriters Laboratories Inc., and others to improve firefighting effectiveness and safety.

Developing an Incident Action Plan

After size-up, the incident commander develops an incident action plan, based on the incident priorities. There are two major components to the incident action plan:

1. The determination of the appropriate strategy to mitigate an incident
2. The development of tactics to execute the strategy

Strategies are general, whereas tactics are specific and measurable. In nonfire terms, strategies are equivalent to goals, whereas tactics are the objectives used to meet the goals. The incident commander identifies the strategic goal and then identifies the tactical objectives that are required to reach the goal. Resources (companies) are assigned to perform individual tasks that allow the tactical objectives to be achieved.

SOPs are used to provide a consistent structure to the process of establishing strategies, tactics, and tasks. Without SOPs, every situation might be managed differently, depending on the individual who was in command and his or her interpretation of the situation, as well as the decisions made by individual company officers. SOPs guide the decision-making process and ensure consistency between officers and events. On arrival at a house fire with smoke showing, the SOP might require the first-arriving officer to assume command, the first-arriving engine company to begin a fire attack, and the second-arriving engine to provide a supply line for the first engine. The SOPs assist everyone involved in the process by outlining what actions to take when presented with a given situation. In effect, the SOPs are pre-established components of an incident action plan.

■ Incident Priorities

The three basic priorities for an incident action plan are:

1. Life safety
2. Incident stabilization
3. Property conservation

Life safety always takes precedence over stabilizing the incident. Incident stabilization and property conservation are often addressed simultaneously, although property conservation is lower on the priority list. If the incident commander does not have sufficient resources to perform both, property conservation measures might have to be delayed until the incident has been stabilized. In reality, many of the actions that are taken to stabilize an incident also work toward preserving property.

The life-safety priority refers to all people who are at risk due to the incident, including the general public as well as fire department personnel. The incident stabilization priority is directed toward keeping the incident from getting any worse. If one structure is fully involved, protecting the exposures is part of incident stabilization. The burning structure cannot be saved, but the fire will not spread beyond that building.

Property conservation is directed toward preventing any additional damage from occurring. This could include minimizing water damage by covering building contents with salvage covers or removing valuable property from harm's way. Property can also be protected by venting smoke and heat from a building and then covering the openings to keep wind and rain from causing additional damage.

■ Tactical Priorities

The incident commander is faced with a number of issues that require attention. If an abundance of resources arrived simultaneously, it would be simple to assign them in a manner that would get everything done immediately. Because this rarely happens, fire officers need a simple method to address the most critical concerns first. Tactical priorities provide a list and an order of priority for dealing with them. The tactical priorities and the information obtained in the size-up are used to develop the incident action plan.

Lloyd Layman developed a list of seven factors to assist in developing an action plan. Although there are other techniques, Layman's method remains one of the most popular, 50 years after it was first published.

RECEO VS is an acronym that covers the critical factors in developing a strategy. The first five factors are in a priority order, whereas the last two may be used at any point to support the first five. The acronym represents:

- Rescue
- Exposures
- Confinement
- Extinguishment
- Overhaul
- Ventilation
- Salvage

Figure 16-5   Rescue is the most important tactical priority.

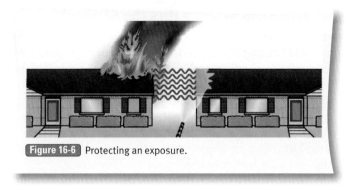

Figure 16-6   Protecting an exposure.

Rescue

Because life safety is our highest incident priority, rescue is at the top of the list Figure 16-5 ▲ . The rescue category includes all functions related to searching for potential victims and removing them from danger to a location of safety, as well as any other activity that is necessary to reduce the risk of death or injury. Medical care after the occupants have been removed is a rescue priority.

Because rescue is the first priority, there could be a situation in which the first-arriving company is committed to rescue a trapped occupant while the fire is allowed to burn. In many cases, the best method of protecting the occupants from harm is to quickly extinguish the fire. In other cases, the fire has to be attacked to give the rescuers time to enter the structure and search for occupants.

Exposures

The next priority is to keep the incident from getting bigger. Exposure protection is most important in a situation in which the incident commander does not immediately have the resources that would be needed to fully control or extinguish the fire. The best method of protecting an exposed property is to make an aggressive attack and extinguish the fire before it can spread. In the same manner that quickly extinguishing the fire reduces the rescue problem, rapid extinguishment also alleviates the exposure problem.

If the fire is beyond the capabilities of an initial attack, the best decision a fire officer can make is to protect the adjacent properties rather than waste resources attempting an inadequate attack on the original fire building. If there are insufficient resources to save the fire building, use the available resources to protect adjacent buildings Figure 16-6 ▶ .

Confinement

The third priority is to prevent the fire from spreading to the uninvolved areas of the same property. If the fire is in one room, the objective should be to confine the fire to that room of origin. If confining the fire to one room is not possible, then the objective could be to confine it to the floor or area of origin.

Confinement can also be directed toward supporting rescue efforts. Confining the fire to the room of origin can provide time for other crews to perform rescue. Confinement could also be directed toward protecting exit stairways and corridors in order to safely remove occupants.

Extinguishment

The next priority is to extinguish the fire or mitigate the incident. Ultimately, the goal is to extinguish all fires, and in many cases, rapid extinguishment takes care of the rescue, exposure, and confinement priorities. The priority order becomes important when rapid extinguishment is not feasible. The higher priorities must always be considered ahead of extinguishment. There are many tactical options for accomplishing extinguishment, using either offensive or defensive strategies.

Overhaul

The fifth and last of the tactical priorities is overhaul, the activity that makes sure that the fire is completely out. After all of the visible fire has been extinguished, the area must be checked for residual fire, including any fire that could have extended into areas not originally involved, such as walls, ceilings, and attic spaces. Incomplete overhaul may allow the fire to rekindle.

Ventilation

Ventilation is designed to remove heat, smoke, and the products of combustion from a fire area and allow cool, fresh air to enter. Effective ventilation can improve the survival chances for occupants and also can improve conditions to allow fire fighters to enter and operate inside a building. Releasing smoke and heat can also reduce the risk of backdraft or fire spread within a building and reduce the damage to contents.

Ventilation was not included in Layman's prioritized list because he advocated an indirect attack, applying high-pressure fog stream into a sealed structure. Any ventilation decreased the effectiveness of this attack method. Research by NIST into wind-controlled fires requires reconsideration of ventilation practices. We may return to a version of the Layman delayed ventilation for certain types of fires.

Salvage

Salvage is the other tactical activity that does not have a specific place on the priority list. Salvage includes protecting or removing property that could be damaged by the fire, smoke, water, or fire-fighting operations. Salvage also includes securing the building and protecting it from weather. Salvage can be placed wherever it is appropriate in a particular situation and is often performed in parallel with the other priorities, if resources are available. Salvage often involves placing salvage covers to protect property during overhaul. It can also include actions such as only using enough water to extinguish the fire in order to limit water damage.

■ Determining Task Assignments

Tactical priorities are subdivided into tasks and assigned to companies. Tasks are specific assignments that are typically performed by one company or a small number of companies working together.

During the early stages of an incident there are more tasks to be performed than there are companies available to do the work. The incident commander makes assignments based on tactical priorities and available resources. The incident commander must prioritize assignments and distribute them to companies as they arrive or become available. Companies that have completed one task assignment may have to be reassigned to another.

The incident commander should use the tactical priorities to determine the relative importance of each task that needs to be performed, in the context of the specific situation. SOPs often guide these decisions, as do the staffing and standard equipment on each piece of apparatus.

The normal function of the company should also be considered, when possible. In most cases, an engine company would be assigned to attack the fire, while a paramedic ambulance provides medical care and a truck company is assigned to ventilate. At times, the incident commander has to assign companies to perform tasks that might not fit their normal role, simply because of the importance of the task and the limited resources that are available. For example, if no medic unit is on scene, the incident commander might have to assign an engine company to provide medical care. If the truck company will be delayed, an engine company could be assigned to perform ventilation.

The rescue priority could be assigned to different companies to perform primary and secondary search of the structure, raise ladders, remove trapped occupants, provide medical care and transport, and establish a rapid intervention crew **Figure 16-7 ▶**.

Task assignments for the exposure priority typically include establishing a water supply and setting up master streams for use on the fire building or on the exterior of the exposure. Placing a handline on the unburned side of a firewall or in the cockloft of an exposure building could also be a task assignment. Removing combustible material from the windows of the exposed buildings is another example.

Task assignments for fire confinement typically include advancing handlines to the room of origin, into stairways, and into the attic or the floor above the fire. Ventilation tasks could be performed to support confinement efforts.

Task assignments for extinguishing a fire typically include establishing a water supply, advancing a handline to the seat

Figure 16-7 Task assignments for a rescue strategy may include performing a primary and secondary search of the structure.

of the fire, and applying water or other extinguishing agents. When making assignments for hose lines, the fire officer should indicate the line size as well as placement, unless predetermined by departmental SOPs.

Task assignments for overhauling a fire typically include pulling ceilings and walls in the burned areas, removing door and floor trim where charred, checking the attic and basement for extension, checking the floors above and below the fire, and removing or wetting all burned material.

Task assignments for ventilating a fire typically include vertical ventilation, horizontal ventilation, positive pressure ventilation, negative pressure ventilation, and natural ventilation. When making assignments for ventilation, the officer should indicate the location of entry points, the potential victims, and the location of the fire to ensure that proper techniques are used.

Task assignments for salvage typically include throwing salvage covers over large, valuable items; removing lingering smoke; soaking up water from floors; deactivating sprinklers; and removing important documents or memorabilia.

■ Assigning Resources

The resources assigned to an incident vary greatly and are influenced by history, tradition, and budgets. One department may respond to a structure fire with a single engine and tanker, whereas another might send four engines, two aerials, a rescue or squad company, and two command officers.

Some situations exceed the capabilities of any organization and require assistance from other agencies or jurisdictions. In most cases, the first level of assistance for fires is mutual aid from surrounding fire departments. Most fire departments also have working relationships with other local agencies, such as law enforcement and public works, to obtain the resources that are likely to be needed in most situations.

When the need for resources exceeds the normal capabilities of the fire department and involves numerous other agencies, a fire officer may have to activate the local emergency plan. This plan defines the responsibilities of each responding agency and outlines the basic steps that must be taken for a particular situation. For example, firefighting is normally the responsibility of the fire department, along with technical rescue and hazardous materials incidents. Traffic control, crowd control, perimeter security, and terrorist incidents are usually the responsibility of the police or sheriff's department. Every fire officer should be familiar with the local plan and the role of the fire department within it.

The most common method a fire officer can use to activate the local emergency plan is to notify the dispatch center, which should have the information and procedures to either activate the local emergency plan or notify the emergency management office. The emergency management office normally is responsible for maintaining and coordinating the plan, as well as facilitating the response to emergency situations.

The local emergency plan usually includes an evacuation component that can be used for any situation where the residents or occupants of an area have to be protected from a dangerous situation. A hazardous materials release or major fire event could require the evacuation of a local area. In many cases, the evacuation plan can be activated on a limited scale, without activating the overall local emergency plan.

The police department is the primary agency for notifying people within an evacuation area of what to do. The use of the public address feature on vehicle sirens, the emergency broadcast system, or a reverse 9-1-1 telephone system can aid in the process. For large-scale or long-term evacuations, the Red Cross may need to be contacted to establish emergency shelters for the residents. Depending on the situation, fire crews could be available to assist evacuating residents; however, they might be occupied with fighting a fire or other mitigation actions. Mutual aid companies or other agencies could be needed to assist with evacuation.

Consider the nature of the event when establishing an evacuation area. A natural gas leak in a street generally requires short-term evacuation of residents in the immediate area. An apartment building fire could result in several residents being displaced for a long period of time, but it affects only those who live in the building or complex. A major hazardous materials release could result in the evacuation of a large area and thousands of people. The fire officer should consult the *Emergency Response Guidebook*, the local hazardous materials team, or CHEMTREC for guidance in determining evacuation distances, based on the product, quantity, and environmental conditions.

Tactical Safety Considerations

Fighting fires is an inherently dangerous activity that exposes fire fighters to a wide variety of risks and hazards. Many advances have been made, including improved protective clothing and equipment, more effective methods of fighting fires, and better procedures for managing fire incidents. The hazards still exist, however, and new threats appear.

Protective clothing and self-contained breathing apparatus provide better protection than was available to earlier generations. In order to benefit from this protection, the full ensemble of personal protective clothing and equipment must be worn whenever fire fighters are exposed to hazardous conditions.

Personal protective equipment (PPE) allows fire fighters to enter deeper into burning buildings and stay longer, often without sensing the level of heat in that environment. PPE has saved lives and prevented disabling injuries. It also exposes fire fighters to an increased risk of being trapped by a sudden flashover, becoming disoriented inside a burning building, or running out of air in a highly toxic atmosphere.

The PPE's weight, bulk, and thermal properties must be considered during extreme weather conditions. Fire fighters need sufficient rehabilitation between work periods.

Scene Safety

Many fires occur at night, a fire scene full of tripping hazards and obstacles. The use of lights to adequately illuminate the incident scene is an important safety consideration. During cold weather, the fire officer may need to have abrasive materials spread about the scene to improve traction on ice.

Getting It Done

Taking Care of Yourself

The fire officer should encourage fire fighters to stash a spare pair of socks, gloves, a towel, and a watch cap on the fire apparatus. Getting soaked in a winter fire is a common occupational hazard. If they add a t-shirt, shorts, and coveralls to the stash, fire fighters can have dry clothes after a soaking or a technical decontamination.

Exposure to asbestos building components, drug laboratories, and chemical/biological weapons are three situations in which a fire crew may have to disrobe at the incident scene. A technical decontamination may result in the fire fighters losing their uniforms in the warm zone. Without the stash, the only option may be a disposable Tyvek gown.

It is important that fire fighters change out of their duty uniforms and clean up before they leave the fire station, even after a slow duty day. Infectious diseases, asbestos, industrial chemicals, and products of combustion are hazards that should stay at the fire station and not be brought home to expose family members.

The fire officer should also encourage the crew to carry an on-duty wallet. The on-duty wallet can be lost during decontamination without much disruption to the fire fighter's personal life. The on-duty wallet includes a duplicate driver's license, copies of any state- or department-mandated certification cards (like EMT-paramedic), a phone calling card, and, if desired, a credit card. The fire fighter should include enough money to get through the day and should leave the regular wallet locked in the car or the fire department locker.

The same goes for a cellular phone or PDA. Consider using a disposable or spare phone while on duty. Set up call-forwarding so incoming calls are forwarded to your disposable phone.

Create a set of duty keys. Make duplicates of the keys you need to get in the fire station, your locker, and your car. Like the on-duty wallet, you can lose your duty keys and not have your life too disrupted. Do not forget to put a durable identification tag on your duty keys so that you can get them back after they are sanitized! Lock up your regular keys, smart/cell phone, wedding ring, and other valuable jewelry before you report for duty. It may be a good idea to use a combination lock to secure your storage space.

Many emergency operations are conducted in the street, exposing fire fighters to traffic hazards. Apparatus should be positioned to protect the scene, and the officer should request traffic control to reduce the risk of fire fighters being struck by vehicles.

Hazardous areas at the incident scene, such as holes and potential collapse zones, should be clearly identified. Some fire departments use barrier tape to establish exclusion zones around hazards. Many departments utilize hot/warm/cold zones to identify the level of hazard:

- Hot zone: Highest hazard, minimum number of crew allowed in area. This could be an area of structural weakness, a hazardous environment, or an area requiring use of full PPE including SCBA. A hot zone is under direct supervision of a unit, branch, or group leader with an identified entry/exit point.
- Warm zone: Moderate hazard, restricted number of emergency service members allowed in area. This is an area that remains part of an active incident and requires the use of some protective equipment (helmet, boots, or respirator) or special vigilance.

- Cold zone: Minimum hazard, no restrictions on number of crew or nonemergency service individuals allowed in area.

Rapid Intervention Crews

The **rapid intervention crew (RIC)** is "a dedicated crew of firefighters who are assigned for rapid deployment to rescue lost or trapped members." NFPA 1407, *Standard for Training Fire Service Rapid Intervention Crews* describes required capabilities. An **initial rapid intervention crew (IRIC)** is two members from the initial attack crew, as defined in NFPA 1710, *Standard for the Organization and Deployment of Fire Suppression Operations, Emergency Medical Operations and Special Operations to the Public by Career Fire Departments*. A rapid intervention crew (RIC) is four members. This dedicated crew is not to be confused with the IRIC. Most fire departments add an additional fire company to the first alarm assignment for a structure fire, either as part of the initial dispatch or as soon as a working fire is reported to function as the RIC.

Near Miss REPORT

Report Number: 08-0000334

Event Description: I was captain on Engine [number deleted]. We responded to an apartment fire, [location deleted] with persons trapped. Our engine arrived first on scene, parked just west of A side, close to a hydrant. Engine [A] and Engine [B] misunderstood the address and proceeded north on [location deleted]. Command did not know the other units missed the address. I thought we were going to be fire attack.

Command assigned Engine [A] fire attack, even though they were not on the scene. My engine was assigned search and rescue for a woman trapped on the second floor. The brick, 2 story apartment was showing heavy fire and smoke on the first floor. We entered the apartment on the A side and proceeded to the second floor. We encountered heavy smoke and heat. I advised the crew to go back and get a 1 3/4" line. I advised the nozzleman to stay at the door as my other fire fighter and I searched the second floor. The conditions started to deteriorate. I advised the fire fighter with me to take out as many windows as he came across.

I continued to search toward the back of the apartment when the room exploded in flames. I was blown approximately 20' across the room. I was unconscious for maybe a few minutes. When I came to, I began searching for the other fire fighters in my crew. I found the fire fighter that was searching in the other direction. Together, we found the nozzleman. He was under a door and some debris.

I was told that Command radioed that he believed we just had a backdraft. Two fire fighters outside the front door were blown into the front yard. Fire attack found heavy fire on the first floor, did not realize that they extinguished the fire inside the apartment and started spraying water on what they thought was inside, but was actually outside. This heated the upper floors. I believe that this caused the smoke explosion. Command tried to contact my crew and I did not respond for a few minutes. We remained on the second floor and assisted with fire attack until we ran out of air.

[Brackets denote information changed by the reviewer]

Lesson Learned: Coordinating fire attack is critical to avoiding injury. Companies should be aware of where other companies are operating so that hoselines can be used to protect search teams. Search teams should be aware of which direction the fire attack is being directed. Fire streams should not be directed into structures from outside when crews are operating inside. Be sure to take a hoseline when searching. May-day procedures should be in place for companies that do not respond to calls from Command.

Many departments have developed RIC training and deployment procedures. The standard radio term "may-day" is used to indicate that a fire fighter is lost, missing, or in life-threatening danger. The incident commander's standard response to a may-day is to activate the RIC. When this occurs, the RIC operation has priority over all other tactical functions.

The RIC is generally positioned outside the building, near an entrance, ready for immediate action. The SOP generally specifies a list of equipment that will be prepared for use, including forcible entry tools, a thermal imaging camera, a search rope, medical equipment, and a spare SCBA or full replacement cylinder. Every RIC team member should have a portable radio and should be wearing full PPE. While standing by, the RIC members should closely monitor fire-ground radio communications to keep track of activities and listen for a may-day or any indication of a problem.

Personnel Accountability Report

A **personnel accountability report (PAR)** is a systematic method of accounting for all personnel at an emergency incident. When the incident commander requests a PAR, each fire officer physically verifies that all assigned members are present and confirms this information to the incident commander. This should occur at regular, predetermined intervals throughout the incident, generally every 10 minutes. A PAR should also be requested at tactical benchmarks, such as a change from offensive strategy to defensive strategy.

The fire officer must be in visual or physical contact with all company members in order to verify their status. The fire officer then communicates with the incident commander by radio to report a PAR. Any time a fire fighter cannot be accounted for, he or she is considered missing until proven otherwise. A report of a missing fire fighter always becomes the highest priority at the incident scene.

If unusual or unplanned events occur at an incident, a PAR should always be performed. For example, if there is a report of an explosion, a structural collapse, or a fire fighter missing or in need of assistance, a PAR should immediately take place so that it can be determined how many personnel might be missing, the missing individuals can be identified, and their last known location and assignment can be established. This last location would be the starting point for any search and rescue crews.

General Structure Fire Considerations

Chapter 12, Pre-incident Planning and Code Enforcement, divided buildings into use groups and briefly discussed common firefighting hazards that are associated with some of those groups. This section provides an overview of typical firefighting hazards, listed by traditional fire-ground classifications.

Single-Family Dwellings

More civilian fire deaths occur in one- and two-family dwellings than any other type of occupancy in the United States. These dwellings are not required to undergo a regular safety inspection. Structures range from a single-story frame house of less than 1000 ft² (305 m²) to a mansion with more than 15,000 ft²

(4572 m²) on a private estate in a nonhydrant area. Balloon frame construction, which was common until the 1930s, incorporated void spaces in the exterior walls, where a fire could spread rapidly from the basement to the attic. Single-family dwellings constructed within the past 20 years have lightweight structural components supporting floors and roofs. Rapid fire spread will occur in both modern and balloon frame buildings.

An extensive variety of hazards can be found, including improper storage of gasoline, faulty electrical wiring, and poor housekeeping habits. Each of these conditions increases the probability that a fire will occur at that home. Many homes contain storage for a retail business in the basement, attic, garage, or spare bedroom. Other home-based businesses, such as vehicle repair businesses or home-cleaning services, store and use chemical products.

A trend in older communities is to convert buildings that were originally designed as single-family homes into apartments. Another trend is multiple families living together in one crowded house. This means that a fire company could arrive at a fire in a single-family dwelling and encounter a dozen trapped residents.

Low-Rise Multiple-Family Dwellings

Low-rise multiple-family dwellings that have been built since the mid-1980s are often Type V (wood frame) construction, with lightweight wood truss component in the floors and roof. Creative building designs often create voids as well as architectural features like cathedral ceilings that limit the ability of fire companies to open up ceilings to check for fire extension. Like newer single-family dwellings, newer or renovated multiple-family dwellings usually have extensive insulation.

Often, the first-arriving fire companies responding to an activated fire alarm in a garden-style apartment report nothing showing on arrival. When they force the door of the apartment that has the activated smoke detector, they encounter thick smoke boiling out of the apartment. Sometimes, a room has already flashed over, and the fire has died out due to oxygen starvation.

"Solidified gasoline" furniture and tighter insulation means that the room retains much more heat than older buildings. Internal exposures will suffer more heat damage and there is a higher chance of fire spread. Lightweight wood trusses require

Company Tips

Zero-Clearance Chimneys
Constructed decades ago, a zero-clearance chimney has an inside steel flue and an outside flue, separated by 1" (2.5 cm) of air. The owner may have added glass doors to the fireplace, increasing the temperature inside the flue beyond its design parameters. To further complicate this scenario, an artificial log may be used in place of a traditional log. The artificial log plus fireplace doors plus an aging flue cause a failure of the flue that manifests itself by heating up the structural wood components around the flue. In some cases, the homeowner discovers the problem when someone knocks on their front door on a bitter cold winter night to tell them that their attic is on fire.

Voices of Experience

Garden apartment fires present unique issues when comparing them to other structure fires. Knowledge of building construction and a good size-up are key when fighting fire in these structures. Typical units are stacked on top of one another to take advantage of common soffit spaces so water and vent pipes can run from the top to the bottom of the structure.

Several years ago I was dispatched on the first-due ladder truck with a five-person crew to a fire in a garden apartment. My ladder truck arrived on the scene with the first-due engine to find a four-story garden apartment with fire on the lower level. The engine company pulled a hose line and began to knock the fire. As I approached the building I noticed smoke coming from the roof line. While my driver placed ground ladders to the upper floors, I split my attack crew in half. I instructed my hook-and-can fire fighter and my outside-vent fire fighter to report to the floor above the fire to perform search and rescue and check for fire extension, then report back to me via radio. My forcible-entry fire fighter and I went to the fire apartment for search and rescue. The engine made quick work of the fire, but my two fire fighters on the second floor reported fire in the walls. The command officer ordered a second hose line to the floor above, and instructed my two fire fighters on the floor above to check the cockloft and give a report. They reported to command that the fire had entered the cockloft and was running the roof line. A third line went to the top floor and my two fire fighters, along with the second-due truck, pulled the ceilings to expose a large volume of fire. That fire was also quickly knocked. It was determined that the fire had entered the soffit space in the kitchen and had run the pipe chases to the cockloft.

> **"A third line went to the top floor and my two fire fighters, along with the second-due truck, pulled the ceilings to expose a large volume of fire."**

Understanding how these structures are built along with a good size-up can aid in stopping an advancing fire. Getting crews and resources to all levels of a garden apartment as well as checking the cockloft can make the difference in containment of a fire and preventing large property loss.

David Polikoff
Captain
Montgomery County Fire and Rescue
Montgomery County, Maryland

minimal fire damage to collapse, which often occurs suddenly when the first arriving company starts their operation.

Multiple-family dwellings with NFPA-compliant 13-R automatic sprinkler systems provide greater protection to the occupants but may increase the degree of difficulty for the fire fighters. Buildings with an R-13 system will be taller with longer hallways. Apparatus access may be restricted due to creative landscaping. Even with these features, a fire within the 13-R protected area will be small with less damage.

The pre-incident plan should identify the best apparatus access positions and the hose lays required to get to every space in the building. Make sure that the fire apparatus can actually get into gated communities and make provisions to allow prompt access to the property every hour of every day. The R-13 system does not protect the attic area or the void spaces. Fires that start in these areas, or start on the outside combustible vinyl siding, spread rapidly.

General High-Rise Considerations

A working fire within a high-rise structure requires more fire fighters and an expanded ICS. The incident commander is responsible for the overall management of the incident. Because of the complexity of a high-rise fire, the incident commander often expands ICS to include planning, operations, and logistics sections. The incident commander also usually appoints staff to fill the positions of safety officer, liaison officer, and public information officer on the command staff.

High-rise fires require a tremendous amount of personnel deployed in a carefully concerted effort to achieve the strategic objectives. This requires dividing the incident into manageable units (i.e., branches), each with a supervisor. The last operations section position that a fire officer might be called on to fill at a high-rise incident is the staging area supervisor. The staging area supervisor is responsible for managing all activities within the staging area, including layout, check-in functions, and traffic control within the area. All fire officers must understand their relationship within the ICS. The large-scale operation of a high-rise fire necessitates a large ICS structure to maintain a manageable span of control and a margin of safety and accountability.

The fire officer may also be called on to fill a position within the logistics section. Like the operations section, the logistics section is divided into branches. Typically, there is a support branch and a service branch. The **service branch** is headed by the service branch director and is responsible for communications and fire fighter rehabilitation. The **support branch** is headed by the support branch director and is responsible for ensuring adequate supplies, personnel, and equipment. A high-rise incident requires large amounts of personnel and support equipment, so a support branch director must be established as an early part of the command staff.

The support branch director is often located at a base. A **base** is an area where the primary logistics functions are coordinated and administered. The incident command post may be colocated with the base, and there is only one base per incident. Spare SCBA cylinders are the first priority for the support branch.

It is important to quickly gain control of the lobby. The **lobby control officer** controls the entering and exiting of both civilians and fire fighters in the lobby. In addition, the lobby control officer oversees the use of the elevators, operates the local building communication system, and assists in the control of the heating, ventilating, and air conditioning systems. The lobby control officer reports to the logistics section chief or, if that position is not established, the incident commander.

A labor-intensive task is the **stairwell support group**. These are the folks who move equipment and water supply hose lines up and down the stairwells. The stairwell support unit leader reports to the support branch director or the logistics section chief. One practice is to position fire fighters on every third floor. Equipment moves up the high rise in an assembly-line fashion, as the third-floor fire fighter moves equipment up to the sixth-floor fire fighter.

Residential high-rise buildings present extra challenges to fire officers; three wind-driven fires have killed members of the first-arriving fire company since 1983. Recommended practices include:

- Comply with your organization's SOPs.
- Consider bringing the big attack line first.
- Beware of weather conditions.
- Assemble an adequate crew.

Summary

Commanding a fire attack team is a core fire officer requirement. It is an activity that differentiates the fire officer from other emergency service members and is rich in history, tradition, and lore. Fire attack competence is a requirement for fire service leaders.

While some fire attack practices date back to the 19th century, and the general concepts of "modern" fire attack start in 1940, the art and science of skilled fire attack evolve with each new item of understanding about human behavior, organizational change, the built environment, and the science of extinguishment. Information from the earlier chapters provided the concepts, principles, and practices that guide the fire officer to becoming a competent fire commander.

This chapter focused on performing an incident size-up, determining incident priorities, and developing tactical assignments to control a fire. It also covered establishing command, command transfer, multiple-alarm incidents, and special considerations.

You Are the Fire Officer: Conclusion

The response starts with a clear, concise radio report to Communications, describing what you see and asking for more resources.

Your size-up includes what you know about the building:

- Nonsprinklered three-story apartment complex.
- Forty-year-old ordinary construction building.
- Scissor stairs in the center of the building that serve 11 apartments.
- Stairway area is enclosed with roof and is glassed-in.
- Access way to the roof in the stairway atrium.
- Hollow core fire doors into each apartment.
- Natural gas heat.

These buildings experience rapid fire spread through the vertical pipe chases, often leading to well-developed attic fires. This complex has a history of few fire incidents and many elderly occupants. You estimate that there are one or two occupants in each apartment. That means you have one or two dozen endangered residents.

On arrival you see fire coming out of a first-floor bedroom window. The sliding glass door to the fire apartment has not crazed over, and you can see that smoke is in the top third of the living room. The stairway is clear. The second-floor occupants calling for help are standing on their balcony; you can see light smoke coming from the second-floor apartment. A mnemonic such as WALLACE WAS HOT, along with a lap around the building, could help you complete your size-up.

Life safety is your highest incident action plan priority. It appears that your crew can quickly remove the second-story residents using the interior stairs. The incident action plan will be offensive once the second-floor occupants are removed. You also have to provide medical care for the occupants after they are removed. After the occupants have been removed and the fire has been controlled, you can turn your attention to property conservation.

This incident exceeds the capabilities of a single fire company. As the initial incident commander, you are responsible for calling for additional help and assigning all of the first-alarm companies as they arrive.

Wrap-Up

Chief Concepts

- The primary responsibility of a fire officer is to supervise the company while assigned tasks are being performed.
- The initial responsibilities of the incident commander include:
  - Conducting a scene size-up
  - Developing an action plan
  - Assigning resources
  - Developing a command structure
  - Ensuring that the plan is safely completed
- Size-up is a systematic process of gathering and processing information to evaluate the situation and then translating that information into an action plan.
- Lloyd Layman's five steps for analyzing emergency situations are:
  - Facts
  - Probabilities
  - Situation
  - Decisions
  - Plan of operations
- The NFA's three basic phases of size-up are:
  - Phase one: preincident information
  - Phase two: initial size-up
  - Phase three: ongoing size-up
- A risk/benefit analysis must be incorporated into the size-up process.
- Once an accurate size-up has been performed, the incident commander must develop an incident action plan.
- Single-family dwellings are the most frequent source of civilian fire deaths and may have the widest variety of occupant activity and construction features.

Hot Terms

Base The location at which the primary logistics functions are coordinated and administered. The incident command post may be colocated with the base. There is only one base per incident.

Defensive attack When suppression operations are conducted outside the fire structure and feature the use of large-capacity fire streams placed between the fire and the exposures to prevent fire extension.

Incident action plan (IAP) The objectives reflecting the overall incident strategy, tactics, risk management, and member safety that are developed by the incident commander. Incident action plans are updated throughout the incident.

Initial Rapid Intervention Crew (IRIC) Two members from the initial attack crew who are assigned for rapid deployment to rescue lost or trapped members.

Lobby control officer The lobby control officer controls the entering and exiting of both civilians and fire fighters in the lobby. Also oversees the use of the elevators, operates the local building communication system, and assists in the control of the heating, ventilating, and air conditioning systems.

National Fire Academy fire flow formula Used to allow a first-arriving fire officer to quickly determine the needed fire flow. Determined on a per-floor calculation, the length is multiplied by the width, and the result is divided by three.

Offensive attack An advance into the fire building by fire fighters with hose lines or other extinguishing agents to overpower the fire.

Personnel accountability report (PAR) Systematic method of accounting for all personnel at an emergency incident.

Rapid intervention crew (RIC) A dedicated crew of four fire fighters who are assigned for rapid deployment to rescue lost or trapped members.

Service branch A major division within the logistics section of the ICS. Oversees the communications, medical, and food units.

Size-up A systematic process of gathering and processing information to evaluate the situation and then translating that information into a plan to deal with the situation.

Stairwell support group This group moves equipment and water supply hose lines up and down the stairwells. The stairwell support unit leader reports to the support branch director or the logistics section chief.

Support branch A major division within the logistics section of the ICS. Oversees the supply, facilities, and ground support units.

Transitional attack A situation in which an operation is changing or preparing to change.

Fire Officer *in Action*

After calling for the balance of a first-alarm assignment for the apartment fire described at the start of the chapter, you remain the incident commander. One of the two occupants removed from the second-floor apartment is suffering from an asthma attack. There appears to be a working fire in the attic and in the pipe chases on the second and third floors. A greater alarm is struck.

The fire attack "toolbox" of techniques, practices, and procedures is designed to handle operations from a small trashcan on fire to a high-rise fire with flames showing from 12–15 windows. The techniques, practices, and procedures can expand and contract based on the current incident situation.

1. An *indirect* fire attack should be considered when:
   A. the primary goal of the incident action plan is to remove citizens in distress.
   B. the greater alarm can deploy attack lines to all floors of the building and the attic.
   C. the fire in the attic cannot be extinguished by crews operating inside the building.
   D. the amount of fire loss exceeds the insured value of the building.

2. A fire officer uses _____ supervision during incident size-up.
   A. authoritarian
   B. bureaucratic
   C. inclusive
   D. participative

3. _____ ensure(s) consistent and repeatable procedures at a local emergency incident.
   A. The IFSTA Model Incident Command System
   B. NFPA 1710
   C. The National Response Framework
   D. Standard operating procedures

4. Knowing that this type of garden apartment building has frequent vertical fire extension through pipe chases, with the result of well-developed attic fires, is an example of:
   A. a Layman probability.
   B. defective building codes.
   C. a pre-incident walk-through discovery.
   D. an NIMS unit assignment.

Fire Cause Determination

NFPA 1021 Standard

Fire Officer I

4.5* **Inspection and Investigation.**
This duty involves conducting inspections to identify hazards and address violations, performing a fire investigation to determine preliminary cause, securing the incident scene, and preserving evidence, according to the following job performance requirements. [p 332–334, 343–344]

4.5.1 Describe the procedures of the AHJ for conducting fire inspections, given any of the following occupancies, so that all hazards, including hazardous materials, are identified, approved forms are completed, and approved action is initiated.

    (1) Assembly
    (2) Educational
    (3) Health care
    (4) Detention and correctional
    (5) Residential
    (6) Mercantile
    (7) Business
    (8) Industrial
    (9) Storage
    (10) Unusual structures
    (11) Mixed occupancies [p 341–343]

4.5.3 Secure an incident scene, given rope or barrier tape, so that unauthorized persons can recognize the perimeters of the scene and are kept from restricted areas, and all evidence or potential evidence is protected from damage or destruction. [p 343–344]

(A) Requisite Knowledge. Types of evidence, the importance of fire scene security, and evidence preservation. [p 344]

(B) Requisite Skills. The ability to establish perimeters at an incident scene. [p 343–344]

Fire Officer II

5.5 **Inspection and Investigation.**
This duty involves conducting fire investigations to determine origin and preliminary cause, according to the following job performance requirements. [p 332–343]

5.5.1 Determine the point of origin and preliminary cause of a fire, given a fire scene, photographs, diagrams, pertinent data, and/or sketches, to determine if arson is suspected. [p 333–343]

(A) Requisite Knowledge. Methods used by arsonists, common causes of fire, basic cause and origin determination, fire growth and development, and documentation of preliminary fire investigative procedures. [p 342–343]

(B) Requisite Skills. The ability to communicate orally and in writing and to apply knowledge using deductive skills. [p 336, 345–346]

Additional NFPA Standards

NFPA 921 *Guide for Fire and Explosion Investigations*
NFPA 1033 *Standard for Professional Qualifications for Fire Investigator*

Introduction to Fire and Emergency Services Administration (FESHE) Course Outcomes

2. Explain the need for effective communication skills both written and verbal. [p 336–337, 344–345]
7. Identify roles and responsibilities of leaders in organizations. [p 332–333, 335, 337, 343–344]
9. Identify and assess safety needs for both emergency and nonemergency situations. [p 342, 345]
12. Describe the benefits of documentation. [p 345–346]

Knowledge Objectives

After studying this chapter, you will be able to:

- Describe the role of the fire officer in determining the cause of a fire.
- List the common causes of fire.
- Explain when to request a fire investigator.
- Describe how to find the point of origin.
- Describe fire patterns.
- Describe how to determine the cause of the fire.
- Describe how to conduct interviews.
- Describe how to determine the cause of a vehicle fire.
- Describe how to determine the cause of a wildland fire.
- Describe the fire cause classifications.
- Describe the indicators of incendiary fires.
- Discuss arson.
- Discuss the legal considerations of fire cause determination.
- Discuss how to write an investigation report.

Skills Objectives

After studying this chapter, you will be able to:

- Determine the point of origin and the preliminary cause of a fire.
- Demonstrate how to secure the scene using rope or barrier tape to prevent unauthorized persons from entering the incident scene.
- Complete a fire incident report.

*I*t is your first commercial fire as a managing fire officer. The fire was in an electronics store in an older strip shopping mall. The back of the shop is used to install automobile electronics and sound systems. It appears that part of the shop was also used to modify vehicles because you see acetylene torches, flammable liquid cleaning tanks, and large tool chests in one of the six vehicle installation bays. The point of origin seems to be in one of these bays, below the charred remains of a sports car that is on a portable hydraulic lift.

There is extensive damage to three service bays, two vehicles are destroyed, and two vehicles are damaged. Three employees suffered from minor burns and smoke inhalation.

The fire investigator is tied up on another call and asks you to look it over and determine whether it will be necessary to call in another investigator on overtime. You have filled out dozens of National Fire Incident Reporting System (NFIRS) reports for automobile fires, food-on-the-stove fires, and other routine events, but you have never written up an industrial commercial fire. This will be a new experience.

You have some recollection of the occupancy from a familiarization visit 9 months earlier. You do not remember seeing all of the automotive repair and modification equipment from that visit. A closer inspection shows three charred 55-gallon drums of racing fuel. You wonder whether the business had applied for a fire prevention permit for flammable liquid storage.

1. How will you determine the origin of this fire?
2. How will you determine the cause of this fire?
3. How will you secure the scene?

Introduction to Fire Cause Determination

A preliminary investigation must be conducted to determine how a fire started. This is done once the fire is extinguished and before the property is turned back over to the owner. One of the best starting points for efforts to prevent future fires is to understand the causes of fires that have occurred in the past. In addition, the investigation is also important to establish the responsibility for fires that could have been caused by criminal acts, negligence, or fire code violation. The results of an investigation may lead to legal actions, ranging from prosecution for arson or related crimes to civil litigation over deaths, injuries, or property damage.

The incident commander is responsible for conducting the preliminary investigation, as well as completing the National Fire Incident Reporting System (NFIRS) documents or a local equivalent report. Sometimes, the incident commander delegates this responsibility to another fire officer, typically the officer in charge of the first-arriving company.

The first goal is to determine whether a formal fire investigation is needed. The authority having jurisdiction (AHJ) provides criteria for requesting an investigator. The fire officer has to consider all of the circumstances as well as the probable cause of the fire. Any fire that results in a serious injury or fatality meets the criteria for a formal investigation. Any fire that appears to be arson or related to a criminal act also meets the criteria. Some agencies automatically dispatch a fire investigator to all working structure fires.

The legal responsibility for conducting fire investigations is defined by state legislation or regulations. The fire officer should determine the laws that apply to fire investigations and which agency is responsible for conducting investigations in different circumstances. This information is available from the fire marshal's office or from an equivalent agency.

In situations where there is no formal investigation and a fire investigator does not respond to the scene to determine the cause and origin, the incident commander is responsible for determining and reporting the fire cause.

Common Causes of Fires

A list of all possible fire causes would be exhaustive, but a few causes are responsible for a large number of fires **Table 17-1 ▾**. Although these data do not help in determining the cause of a particular fire, it is important to understand why fires occur.

Requesting an Investigator

The fire officer should be able to determine a point of origin and a cause, or probable cause, of most fires. On small or routine incidents, this is the only investigation that is conducted, and in those cases, the fire officer must carefully document the findings.

A qualified fire investigator has specialized training in determining the cause and the origin of fires and, in most cases, is certified in accordance with NFPA 1033, *Standard for Professional Qualifications for Fire Investigator*. The investigator could be a fire department member, an employee of a county or state fire marshal's office, or a law enforcement officer. In some cases, a fire department investigator is responsible for determining the cause and the origin of fires, and a law enforcement agency is responsible for any subsequent criminal investigation. State and federal law enforcement agencies become involved in some investigations, particularly where arson is suspected.

The agency or organization that is responsible for conducting fire investigations has a set of guidelines on when to request an investigator. The investigator has more expertise in determining and documenting origin, gathering evidence, and interviewing. A fire investigator's primary responsibility is to develop a properly documented case to be forwarded to the prosecutor. The investigator must have credibility and experience in courtroom procedures to ensure that the facts are presented accurately.

A fire investigator should also be called when there is a death or serious burn injury. If the cause of the fire was deliberate, a crime has occurred and must be fully investigated. If the cause was accidental, the information gathered in the investigation is important as a means to identify and evaluate fire prevention strategies. The coroner or medical examiner's office also must be contacted.

Fire investigators often respond to large loss fires, even if the cause is known. A fire investigator might also be called for situations that could have caused great harm, such as a fire in a hospital or college dormitory, even if the fire was quickly controlled.

The fire officer has to evaluate the circumstances of the situation in relation to local guidelines and policies; however, it is a good practice to request an investigator whenever the facts do not seem to make sense or there is a compelling reason to know the exact cause of the fire.

In many cases, the fire officer is able to determine the exact cause of a fire. Sometimes, the cause cannot be established with certainty; however, a probable cause can be identified. As noted in the statistics, the causes of many fires are never determined or at least are not reported.

Finding the Point of Origin

The first step is to determine the point of origin. The **point of origin** is the exact physical location where a heat source and a fuel come in contact with each other. The point of origin is usually determined by examination of fire damage and fire pattern evidence at the fire scene. Flames, heat, and smoke leave distinct patterns that can often be traced back in order to identify the area or the specific location where the heat ignited the fuel and the burning first occurred. A fire investigator usually starts in the area where the least amount of damage occurred and follows the patterns back toward the area of greatest fire damage.

In many cases, eyewitnesses can provide valuable information that can assist in determining the point of origin. Sometimes, a witness actually saw what happened and can describe where it occurred. In other cases, witness accounts can identify the area where the fire was first observed or where indications such as visible smoke or a burning odor were noted. Sometimes, the witnesses can describe what was in the suspected area of origin before the fire and what normally occurred in that area.

Determining the point of origin requires the analysis of information from four sources:

1. The physical marks, or **fire patterns**, left by the fire.
2. The observations reported by persons who witnessed the fire or were aware of conditions present at the time of the fire.
3. Analysis of the physics and chemistry of fire initiation, development, and growth as an instrument to related known or hypothesized fire conditions capable of producing those conditions.

Table 17-1 **Causes of Fire in Residential Structures and Buildings (2005)**	
Cooking	33.4%
Unknown	19.3%
Heating	10.8%
Electrical malfunction	6.2%
Other unintentional, careless	5.1%
Open flame	5.0%
Other heat	3.9%
Intentional	3.9%
Equipment misoperation or failure	3.5%
Smoking	1.9%
Appliances	1.8%
Exposure	1.6%
Natural	1.4%

Source: United States Fire Administration. Residential Structure and Building Fires. National Fire Data Center, October 2008.

FIRE OFFICER II

4. Noting the location where electrical arcing has caused damage, as well as the electrical circuit involved. An electrical arc is a luminous discharge of electricity from one object to another, typically leaving a blackening of objects in the immediate area.

Fire Growth and Development

To determine the point of origin, the fire officer must understand fire growth and development. Fire fighters learn the basic concepts of fire behavior in Fire Fighter I and II courses. The fire officer should also understand the three methods of heat transfer: conduction, convection, and radiation. The fire officer must be able to apply these basic concepts in order to understand fire growth and interpret the spread of a fire.

Consider a fire in a sofa. The heat and products of combustion rise until they reach the ceiling. They then spread out laterally until they reach the walls. The heat and smoke then begin to bank down lower and lower into the room until the room is filled or until an open doorway or window is encountered. At that point, the heat and smoke tend to flow outward through the opening into the adjoining room or space, filling that area from the ceiling down. This process continues until the entire structure is filled with heat and smoke or until an opening to the outside is found, allowing the smoke and heat to escape from the structure.

While this process is occurring, the fire is continuing to develop and grow in the area of origin. The fire follows the same growth pattern as the smoke and heat, rising toward the ceiling, spreading out, and banking down **Figure 17-1 ▾**. When a room fire reaches the point that flames are spreading across the ceiling, tremendous heat energy is radiated down onto the combustible contents within the room, such as sofas, chairs, and tables. As these items are heated, they begin to release combustible fuels into the atmosphere and eventually reach their ignition temperatures. As each item ignites, the rate of heat release increases dramatically. The temperature increases even faster. Within a few seconds, the entire room flashes over, or becomes fully involved in fire. The fire then quickly spreads to adjoining rooms. The process often causes windows to break or holes to burn through to the outside of the structure, releasing heat and smoke and making the fire visible on the exterior of the structure.

If the fire does not have a fresh supply of oxygen, the fire slowly dies down to a smoldering phase. This may occur before or after a flashover has occurred. At this point, the fire sometimes completely dies out and self-extinguishes. In other cases, a fresh supply of oxygen is introduced, and the fire resumes the fully developed phase.

Fire Patterns

The point of origin can often be identified by interpreting fire patterns. The fire pattern provides the fire officer with a history of the fire. A flaming fire produces a plume of smoke, heat, and flame. As a fire burns up against a wall, it spreads up and out, creating a V- or U-shaped pattern. The origin of the fire is typically at the base of the V- or U-shaped pattern **Figure 17-2 ▾**. This type of pattern is also known as a movement pattern because it allows the fire officer to trace the fire and smoke patterns back to the origin.

The second type of pattern indicates the intensity of the fire. The intensity pattern indicates how much heat (energy) was transferred to the surrounding area and objects. The intensity is indicated by the response of various materials to the fire's rate of heat release and heat flux. The intensity may produce a line of demarcation, which can indicate the area closest to the point where the greatest amount of heat was produced.

The analysis of **char** is closely related to the fire intensity pattern. Char is the blackened remains of a carbon-based material after it has been burned. For the fire officer, the depth of char can assist in determining the direction of fire spread; generally, the deeper the char, the longer the fire burned and therefore the closer to the area of origin **Figure 17-3 ▸**.

Depth of char is only one indicator of the apparent duration and intensity of a fire. Depth of char can be influenced by several factors, including different types of wood and the fire's intensity, so deep charring can occur in locations that are remote from the area of origin. For example, there can be deep charring at a location where a fire vents through a window or doorway because of the intensity of the fire as it combines with the fresh air in that area.

Figure 17-1 The 2003 West Warwick, Rhode Island, nightclub fire killed 100 people.

Figure 17-2 Often, the point of a V-shaped pattern is near or at the point of origin.

Figure 17-3 Depth of char.

Determining the Cause of the Fire

The cause refers to the particular set of circumstances and factors that were necessary for the fire to have occurred. The cause determination can be approached as a three-step process:

1. The first step is to determine the source of ignition. This generally includes some device or piece of equipment that was involved in the ignition. A competent ignition source must be present to ignite the fuel.

2. The second step is to determine the fuel that was first ignited. Both the type of material and the form of the material should be identified.

3. The third step is to determine the circumstances or human actions that allowed the ignition source and the fuel to come together, resulting in a fire.

This three-step process only describes the factors that must be determined in order to conclusively establish the cause of a fire. NFPA 921 provides a systematic and scientific process. It is not sufficient to identify and focus on a possible cause that fits the circumstances that were noted. The fire officer or fire investigator must eliminate any alternative theories or explanations.

Before declaring a fire cause as incendiary or intentional, all accidental causes must be considered and eliminated. Potential causes should be ruled out only if there is definite evidence that they could not have caused the fire. For example, an electric heater can be ruled out if it was unplugged and had not been used for several weeks before the fire occurred.

The fire officer does not have to consider every possible fire cause in relation to every fire that occurs. The facts of the situation can be considered in relation to previous knowledge to identify a list of potential causes for the fire. The fire officer must then apply deductive reasoning, considering each possible cause one by one and sequentially eliminating each cause that is not supported by the evidence.

For example, when investigating a house fire where an outside wall was ignited from the exterior in proximity to the electrical service, the fire officer must consider any potential source of ignition that could start a fire in that location. The fire officer would identify lightning as a potential cause of the fire.

The officer would then determine whether there was lightning in the area at the time of the fire. If it can be established that there was no lightning, then lightning can be eliminated from the list of possible causes. The officer may consider that the fire could have been caused by a short circuit. If careful examination of the electrical equipment indicates that no short circuit occurred, that possibility can be eliminated.

The cause of a fire cannot be established until all potential causes have been identified and considered and only one cannot be eliminated. After all investigative possibilities have been exhausted, if there are still two or more potential causes, the cause of the fire is considered undetermined.

Source and Form of Heat Ignition

The **source of ignition** is the energy source that caused the material to ignite. The source of ignition must have been located at or near the point of origin. The fire officer may be able only to infer a probable ignition source; some sources of ignition remain at the point of origin in recognizable form, whereas others may be altered, destroyed, or removed.

If the equipment that provided the source of ignition was a cigarette lighter, for example, the form of the heat of ignition would be an open flame. The person who started the fire could have removed the source of ignition by placing it back in his or her pocket and leaving the scene. If the pilot light flame from a gas-fueled water heater ignited the fire, however, the source of ignition is likely to be found unaltered at the point of origin.

The source of ignition must have enough energy and must remain in contact with the fuel long enough to cause it to ignite. A competent ignition source has three components:

1. Generation: The ignition source must produce sufficient heat energy to raise the fuel to its ignition temperature.

2. Transmission: Sufficient heat energy must be transmitted from the source to the fuel to raise the fuel to its ignition temperature. Heat can be transferred through conduction, convection, or radiation.

3. Heating: The heat transfer from the source to the fuel must continue long enough for the fuel to be heated to its ignition temperature.

Material First Ignited

The **type of material** first ignited refers to the nature of the material itself. For example, the type of material might be cotton. The **form of material** tells how that material is used. For example, cotton could be in the form of cotton plants in a field, baled cotton fibers, rolls of cotton thread, woven cloth, or clothing.

The physical configuration of the fuel is an important characteristic in ignition. A 12" by 12" (30- by 30-cm) beam and a pile of shavings are both wood; however, the beam is much more difficult to ignite. This information is significant in determining what type of heat source could have ignited the material. A pile of wood shavings could be ignited by a dropped match, whereas the wooden beam would have to be exposed to a more powerful source of heat.

Ignition Factor or Cause

The third important factor in determining the cause of a fire is the sequence of events that brought together the source of

Near Miss REPORT

Report Number: 08-0000398

Event Description: At 1321 hours, our fire department was dispatched for a "lighting strike" of a dwelling. When the fire department arrived on location, they found an outlet in the laundry room had shorted-out and scorched the wall. There was also water on the floor. The circuit breaker box was not properly labeled to identify which circuit breaker controlled the damaged outlet. A circuit breaker believed to be the correct one was shut off.

The ladder company crew came in to check for extension. A member of the ladder crew came in with a "hot stick" to determine if the outlet was still energized. The fire fighter checked the outlet and reported that the outlet was de-energized. In the process of removing the outlet cover, the wires in the outlet arced again. The power to the outlet was still on.

The fire fighter could have received a fatal electrical shock as he was kneeling on the floor in the water while checking the outlet. The fire fighter had failed to verify that the "hot stick" was properly set by checking it against a known functioning energized power source.

Lesson Learned: Fire fighter should not have assumed the power was secured. Prior to checking the outlet, the fire fighter should have verified that the "hot stick" was properly set and functioning. All fire fighters need to be familiar with the SOP for using specific equipment. Fire fighters need periodic review and retraining for equipment that may not be frequently used.

---

ignition and the fuel. The cause of a fire could be a human act that was either accidental or deliberate. In the case of an accidental cause, negligence could be a factor; this is the failure to exercise appropriate care to avoid an accident. The fire cause could also be related to a mechanical failure; a poor design or improper assembly of a device; a worn-out piece of equipment; or a natural force, such as lightning.

<u>Failure analysis</u> is a logical, systematic examination of an item, component, assembly, or structure and its place and function within a system, conducted in order to identify and analyze the probability, causes, and consequences of potential and real failures. If the cause of a fire is determined to be related to some malfunction that occurred within a device or system, failure analysis is performed to identify what happened and why. The failure analysis could identify a design error, malfunctioning component, inadequate maintenance, operator error, or some other factor.

Fire Analysis

<u>Fire analysis</u> is the scientific process of examining a fire occurrence to determine all of the relevant facts, including the origin, cause, and subsequent development of the fire, as well as determining the responsibility for whatever occurred. Fire analysis brings together all of the available information that can be obtained to examine what happened. The fire officer may need to construct a timeline of events to establish the sequence of events that led up to the fire **Figure 17-4 ▶**.

Conducting Interviews

The fire officer may have to interview fire victims, witnesses, fire fighters, and suspected perpetrators. Make every effort to conduct separate interviews to provide the most accurate representation of what each person observed or experienced. When interviews are done in a group, it is human nature for individuals to change their story to match what others have said.

Each individual may have relevant information that will help to identify the cause and origin of a fire. For example, a witness might indicate that she saw lightning strike the building a few seconds before seeing smoke and flames. In another case, a witness might report that a can of gasoline was spilled accidentally and a few minutes later the basement erupted in flames.

The first-arriving fire team might report that the front door was broken open when they arrived. They also have valuable information about the color of the smoke or flames or the area where the fire was burning when they arrived.

The fire officer should review the facts that are already known or believed to be known before conducting an interview. During the interview, determine whether the witness is corroborating that information or providing conflicting information. The fact that the witness has a different version of the story does not mean that the person is lying; the person could have made a different observation or have an inaccurate memory. It could be that the original information was inaccurate and the witness is providing good information. The presence of conflicting information means that further investigation is required for the actual facts to be determined.

A witness statement should be disregarded only if it can be established with certainty that the information is incorrect; for example, if a witness indicates that he saw a man running from the back of a house before the fire started, yet there are no footprints in the snow.

Open-ended questions allow a witness to tell what they saw or know, whereas questions that seek a "yes" or "no" answer limit

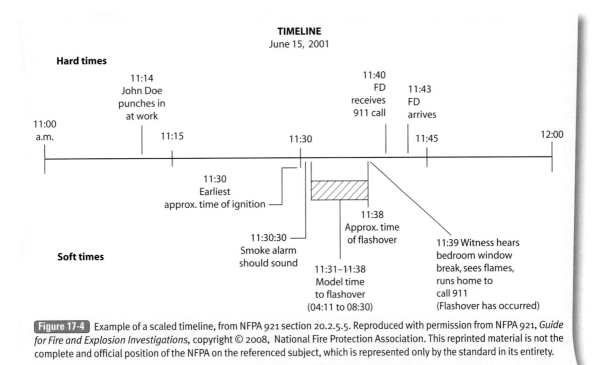

Figure 17-4 Example of a scaled timeline, from NFPA 921 section 20.2.5.5. Reproduced with permission from NFPA 921, *Guide for Fire and Explosion Investigations,* copyright © 2008, National Fire Protection Association. This reprinted material is not the complete and official position of the NFPA on the referenced subject, which is represented only by the standard in its entirety.

the exchange of information. For example, the fire officer should ask the first-arriving crew to describe what they saw when they arrived, not ask whether there were flames showing through the front windows. After the witness provides an overall description, more direct questions can be asked to clarify the facts.

Cause and origin interviews are typically conducted at the fire scene. The officer should introduce himself or herself to the witness and note the person's name and contact information in case a follow-up interview is needed. Interviews are normally conducted with only one individual at a time so that the comments made by one person do not influence the responses of another. The fire officer should not interrupt the witness unless it is essential for clarification. After the interview is complete, the officer should thank the individual for providing assistance and provide his or her contact information in case the witness thinks of anything else.

A witness can be reluctant to make a full disclosure of information. A special type of interview, called an interrogation, is used when an individual who is attempting to conceal information is being questioned. Interrogations require special knowledge and skills that are beyond the scope and duties of this text. Only a trained fire or police investigator should conduct interrogations.

Interview information must be documented. The documentation could be in the form of the interviewer's notes or a handwritten statement from the witness. An alternative is to record the witness statement with a tape recorder or video camera.

Vehicle Fire Cause Determination

Fire departments respond to more vehicle fires than structure fires. Chapter 25 of NFPA 921, *Guide for Fire and Explosion Investigations*, provides a standardized procedure to conduct a vehicle fire investigation.

According to the U.S. Fire Administration, 47 percent of vehicle fires were caused by mechanical failure or malfunction.

Some cars are burned to defraud insurance companies, whereas others are burned for revenge or to conceal other crimes.

The fire officer should trace the fire's development back to the point of origin, looking for the area where the most fire damage occurred and the areas of low burn. Like structural fires, there is often a distinctive burn pattern, such as a "V," that helps identify the area of origin; however, this should be viewed with caution because of the characteristics of the materials that are present within most vehicles. Fires involving the fuel or tires tend to be particularly intense.

Once the point of origin is established, the specific cause must still be determined. If the cause is related to a mechanical malfunction, the exact cause may be difficult to establish without an automotive expert. Additional factors should be considered to determine whether more extensive investigative techniques are required. Most fire departments do not conduct extensive investigations on vehicle fires unless arson is suspected because of the time and expense involved.

The first objective in most vehicle fire investigations is to look for indications of arson. Because some expensive vehicles are burned when the owners are unable to keep up with the high payments, insurance companies often conduct a more thorough investigation if they suspect that the fire was not accidental. The vehicle should be examined for its general condition. Is it a new, expensive vehicle burning at the end of a dead-end street in an unpopulated area? Have items of value, such as an expensive stereo, been removed? Are there indications of accelerant use inside the vehicle? Fuel lines should be checked to see whether they have been loosened. The condition of the tires might indicate that new tires were removed and replaced with old tires before the fire.

Several potential sources of accidental ignition must be considered for a vehicle fire. The mechanical sources include the electrical system, exhaust system, catalytic converter, and turbocharger. In older-model cars, a carburetor backfire can cause a fire within the engine compartment. Smoking materials cause passenger compartment fires.

Note the make, model, and year, as well as the vehicle identification number (VIN). This information allows for a review of previous fires that have occurred in the same make, model, and year. Some makes and models have a history of fires, and the reporting of this information is useful in identifying product defects. A diagram of the scene should be drawn. When possible, photographs should be taken before the vehicle is removed.

Interview the driver/owner to determine when the vehicle was last driven and how far, total vehicle mileage, operating abnormalities, date of last service and when it was last fueled, how the vehicle was equipped (e.g., sound system, custom wheels), and what personal items were in the vehicle.

If the vehicle was being driven, determine the speed and any loads that were being pulled, any abnormal indications before the fire, when and where the smoke and/or fire was observed, what actions were taken by the driver, and how much time elapsed before the fire department arrived.

Wildland Fire Cause Determination

According to the National Interagency Fire Center, there were an average of 69,000 wildland fires, consuming 6.3 million acres a year, between 1999 and 2008. The characteristics of wildland fires are quite different from those of structural fires.

Wildland fires are influenced by environmental conditions, including topography, fuel load, wind, and weather. Wildfires tend to spread vertically through convection, from lower vegetation to taller vegetation, and horizontally through radiation. The rate of spread varies, depending on the type of material burning and its density, the wind speed and direction, the humidity and fuel moisture content, the slope of the terrain, and natural features, such as valleys.

When a fire burn pattern on the side of a hill is investigated, the point of origin is most likely on the lower part of the slope, but not necessarily at the lowest point. The fire burns up the hill very rapidly, but it also burns down the hill at a slower rate. The wind also dramatically affects fire progression, on flat ground or on a slope.

Evaluating the degree of burn on the fuels helps establish the direction of travel. Ash residue also can be evaluated to establish

the direction in which it was blown, which indicates the wind direction. When a tree burns and falls, the remaining trunk is usually burned at an angle, creating a point. The point is generally on the side of the stump opposite the direction of fire approach.

Once the area of origin has been determined, the investigator can look for clues to help determine the cause of the fire. The following are examples of evidence that might be found at the point of origin:

- Campfire remains
- Time-delay devices
- Cigarette remains
- Lighters
- Multiple ignition points
- Splintered trees (indicating lightning strikes)
- Fulgurites (glassy, rootlike residue resulting from lightning strikes)
- Piles of fuel that are prone to spontaneous combustion
- Barrels used to burn trash
- Fallen electrical wires
- Trees on power lines
- Railroad tracks

Although none of these observations necessarily determines the exact cause of a fire, they are helpful in identifying potential causes. The National Wildfire Coordinating Group has developed professional qualification standards for wildland fire investigators.

Fire Cause Classifications

Once the facts are known, the circumstances of the fire are generally divided into four classifications according to NFPA 921. They are:

1. **Accidental fire cause:** All of those fires for which the proven cause does not involve a deliberate human act to ignite a fire or spread fire into an area where it should not go.
2. **Natural fire cause:** Fires caused by lightning, earthquakes, wind, and other natural forces without human intervention.
3. **Incendiary fire cause:** Any fire that is deliberately ignited under circumstances in which the person knows that the fire should not be ignited.
4. **Undetermined fire cause:** Any fire in which the cause cannot be proved is classified as undetermined. The fire may still be under investigation.

Accidental Fire Causes

Many accidental fires result from human activities that are not intended to start or spread a fire. Finding evidence of smoking, such as a lighter, ashtray, or cigarettes, near the point of origin may suggest that careless smoking could be involved as an ignition factor.

The most frequent ignition cause in residential fires is unattended cooking. When the fire originates on the stove, a pan is found on one of the burners, and the burner knob is on, the likely cause of the fire is accidental. Heating is the third leading cause of residential fires, with the two primary ignition causes being improper maintenance and combustibles located too close to the heating device. Other accidental fire causes

Getting It Done

Motor Vehicle Fire Cause Checklist
Motor vehicles are complex machinery that benefits from a checklist:

- Fuel source
  - Ignitable liquids
  - Gaseous fuels
  - Solid fuels
- Ignition sources
  - Open flames
  - Electrical sources
- Hot surfaces
  - Mechanical sparks
  - Smoking materials

include refueling of gasoline-powered equipment near an ignition source, placement of fireplace ashes in a combustible container, and sparks from welding.

When wood is continuously subjected to a moderate level of heat, below its normal ignition temperature, over a long period, it starts to break down into carbon. This chemical decomposition, called **pyrolysis**, results in a gradual lowering of the ignition temperature of the wood until autoignition occurs. Pyrolysis should be considered if the area of origin includes steam pipes, fluorescent light ballasts, flue pipes for a fireplace, or a wood-burning stove. The use of zero-clearance fireplace flues has caused an increase in this type of ignition.

The most common electrical fire scenario is misuse by the occupant, such as overloading electrical circuits, using lightweight extension cords for major appliances, or operating too many devices for the electrical service. Any of these errors can overload the wiring or the device, causing heat and eventually igniting a fire.

Large appliances, like dishwashers and clothes dryers, may fail if their high-temperature controls or timing mechanisms fail. Excessive lint buildup that impedes air exchange and overheats the dryers is a frequent accidental scenario at multiple-family buildings with community laundry rooms. Dishwashers are susceptible to plastic dishes and utensils falling onto the element. Always check for an object on the heating element when the dishwasher is the point of origin.

Electrical devices and appliances that start fires usually produce evidence of electrical damage on their power supply cord relatively close to the device or appliance. The wire insulation near the appliance often melts because of the heat produced by the burning appliance or other materials ignited by it. The melted insulation usually results in energized conductors contacting each other, resulting in a short circuit in the cord. An exception would be an appliance or device malfunction that results in the circuit breaker or fuse tripping so that the power cord is de-energized before the insulation melts.

Natural Fire Causes

A lightning strike in a forest can ignite a tree, and a strike on a building can have a tremendous destructive effect, sending several thousand volts through the electrical wiring system. This surge can destroy televisions, computers, telephone systems, and anything else that is plugged into the electrical system. Lightning can also follow antenna wires, telephone wires, metal plumbing, or the steel structure of a building and ignite combustibles that come in contact with them.

When lightning is suspected, the fire officer should look for a contact point, usually near the top of the structure, concentrating on roof peaks with metal edging, antennas, or large metal objects, such as air-handling units. If the building's electrical system is involved, the surge usually fuses or melts the main fuses or circuit breakers. The surge can travel through television cables, phone lines, power lines, or even natural gas or propane lines, so the point of entry for each utility should be examined for scorching. The joints of aluminum rain gutters may show the same signs.

If there are indications that lightning could have caused a fire, the fire officer should check with weather services to determine whether any lighting strikes were recorded in the area.

Some fire investigators commonly include a notation, such as, "It was a clear and calm night with no lightning observed," in the introduction to their reports to establish that this potential source of ignition was eliminated.

Earthquakes, tornadoes, floods, and hurricanes can all cause fires. Natural forces often cause power lines to contact trees or fall onto structures or cause sheets of metal to be blown across power lines. Earthquakes may also cause gas mains to break and release flammable gases into structures, where they find an ignition source. Extremely hot lava from a volcano can easily start a forest fire or incinerate a community.

Incendiary Fire Causes

Because an incendiary fire is an intentional occurrence, the direct cause is a person; however, the methods and the reasons for starting fires vary tremendously. Because of the complexity in determining an incendiary fire, a later section of this chapter is devoted to explaining why people intentionally set fires and what evidence might indicate that a fire was intentional.

An incendiary fire is one that is intentionally started when the person knows it should not be started. An incendiary fire is not necessarily arson. **Arson** is the crime of maliciously and intentionally or recklessly starting a fire or causing an explosion. The legal definition of arson as a criminal act is uniquely defined within different jurisdictions. The fire officer may be involved in making the determination of a fire's cause and origin and classifying it as incendiary, but ultimately, a prosecuting attorney or grand jury decides whether it will result in arson charges. The fire officer should be aware of the local or state judicial system that governs his or her jurisdiction.

Undetermined Fire Causes

No matter how much training and experience a fire investigator has acquired, how many resources are available, or how much effort is expended, sometimes the cause of a fire cannot be determined. This may be because of the extensive damage caused by the fire or the firefighting activities. This may occur when there are multiple reasons that the fire could have started that cannot be ruled out. The evidence may not exist to make a definitive determination of the cause.

For example, if a gas can is found next to the water heater, it may be impossible to determine whether the vapors were accidentally ignited from the pilot light or intentionally ignited by an arsonist. Without additional evidence, either situation is possible, so the cause cannot be determined with certainty.

The absence of any logical cause is another reason for classifying the cause of a fire as undetermined. Something caused the fire, but the lack of evidence or knowledge precludes the fire investigator from determining the cause.

Voices of Experience

The investigation of fires is a critical part of the operations side of a fire department. The cause is always the first question asked by the jurisdiction leaders and the media. As an investigator I have arrived on many scenes and been asked what caused the fire by one of the above just as I exit my vehicle. The fire officer must be careful because this first explanation as to the cause will be set in concrete and the investigator will be molded inside of it for the life of the investigation. The fire officer must understand the need to gather initial information about the cause and pass this on to the investigator, but many times this is also shared with the person who may not need to know this information.

A fire at a single-family residence on New Year's Eve caused major damage; the media was quick to arrive and to ask questions as to the origin and cause of the fire. The fire officer proceeded to tell the media the fire started in the garage and bedroom end of the basement rancher. He also proceeded to tell about the family's loss of their pet and possessions. The homeowner told him of a problem in the garage and that he believed it caused the fire. This was passed to the media. The cause was published as an accident by the local newspaper on New Year's day.

When the investigation started, it was found that the owner was under investigation by his employer for fraud and embezzlement of funds. As the investigation progressed it was determined that the fire was arson to destroy the records of the owner that would be used by the company to prove the charges. The owner quickly brought up the media report of the fire being accidental and the media made the investigation unit look like the bad guys for harassing this person by saying he had set his house on fire and destroyed all his family's possessions and the beloved pet. During the court proceedings, the fire officer was grilled by the defense about his reporting the fire was accidental.

Fire officers need to release only enough information about the event that they have trained knowledge of; the origin and cause should be left to the investigators. The best answer is to report the fire as undetermined so that the investigation remains unbiased and open for the investigator. The key point should be that the fire officer leaves the announcement as to the cause to the investigation unit.

Mike Dalton
Fire Investigator
Knox County Fire Investigation Unit
Knox County, Tennessee

> **"The media made the investigation unit look like the bad guys for harassing this person."**

Assessment Center Tips

When to Call for a Fire Investigator
One of the goals of an assessment center is to see whether the candidate understands the roles and responsibilities of other agencies that interact with the fire department. On either the incident scenario or a problem-solving scenario, the candidate may be faced with a situation in which one part is to identify a condition that requires action by the fire investigation unit. The fire officer should be able to recite, from memory, the situations and conditions for which a fire investigator must be notified. Be prepared to explain the actions that the candidate will take while waiting for the investigator, such as scene security and evidence preservation.

Indicators of Incendiary Fires

The fire officer needs to eliminate both accidental and natural causes before making a determination of an incendiary cause. Failure to eliminate accidental and natural causes makes it impossible to prove an incendiary cause.

Once the other causes have been eliminated, there are many conditions or factors that may indicate an intentional fire. These typically fall under five general categories:

1. Disabled built-in fire protection
2. Delayed notification and/or difficulty in getting to the fire
3. Accelerants and trailers
4. Multiple points of origin
5. Tampered or altered equipment

Disabled Built-in Fire Protection

A disabled sprinkler system may be encountered in fires involving large industrial or commercial occupancies. If a building was protected by a central station monitoring company, that service should be one of the first entities to call 9-1-1. If the call came from another source, particularly witnesses outside the property, the system might have been disabled or impaired. Part of the investigation is identifying the status of fire suppression and alarm systems at the time the fire started.

Look for damaged or vandalized sprinkler hook-ups, hose cabinets, hard-wired smoke detectors, and high-rise communication systems. At one high rise under construction, a tractor-trailer–sized dumpster was placed in front of the standpipe/sprinkler connection that prevented the fire department from being able to use the built-in fire protection system.

Delayed Notification and/or Difficulty in Getting to the Fire

One of the reasons businesses use a central station monitoring service is to ensure a prompt notification when a smoke detector, water-flow, or manual pull station is activated. In an urban environment, an interior fire could have been burning for 20 to 25 minutes before a passerby observed flames coming from a window. If the business is in a rural setting or an isolated industrial park, the first notification may be after the roof collapses and there are flames shooting up into the sky.

Be alert for conditions or situations that delay the fire department's ability to get to the fire. Malfunctioning keys and key cards, vandalized doors, and materials blocking access are conditions to note. In addition, points of origin that are in the attic, basement, or closet should receive special consideration. Professional arsonists prefer these places to start a fire because they are prone to going unnoticed for longer periods of time. Sometimes, an arsonist starts another fire to divert resources to a different location, delaying response to the real target property.

Accelerants and Trailers

Accelerants are agents used to initiate a fire or increase the rate of fire growth. Trained canines or survey instruments can detect ignitable liquid accelerants. In addition, ignitable-liquid–fueled fires often leave distinct burn and char patterns **Figure 17-5 ▼**.

Trailers are materials used to spread a fire from one area of a structure to another, causing a fire to grow more quickly. Materials most often used as trailers are:

- Paper towels
- Black gunpowder
- Film wrapped in paper rags
- Kerosene or other combustible liquids
- Gasoline or other flammable liquids
- Decorative streamers
- Cotton batting
- Paper
- Sheets or rolls of fabric softener

Figure 17-5 Ignitable-liquid–fueled fires leave distinct burn and char patterns.

- Newspapers
- Combinations of these items

Trailers usually leave a distinct fire pattern that resembles the material's shape and often runs from one room to the next. Liquids typically have irregular edges and areas that are burned where the liquids pooled because of low spots. Sometimes, a trailer of ignitable liquid is poured from one electrical outlet to another to simulate an electrical fire in an attempt to mislead the investigator.

Multiple Points of Origin

Professional arsonists frequently set multiple ignition points inside a building. This is done in case one burns out prematurely as well as to maximize the amount of fire growth before the fire department can respond. In other cases, the arsonist is attempting to cause confusion or to trap the occupants by blocking access to all of the exits.

Igniting fires in multiple buildings at the same time presents the fire department with a complicated problem. The responding units may be sufficient to manage one fire, but two or three buildings on fire create a challenge.

Multiple points of origin do not mean that a fire was intentional. In some cases, material falling from the ceiling and burning at the floor level can create a secondary "U" or "V" pattern, resembling an additional point of origin. There may be nothing at that location that could have caused the fire to ignite.

Multiple points of origin occur when an electrical surge causes ignitions at different locations in a building's electrical system. In one instance, an overpressurized natural gas system started fires simultaneously in several kitchens and water heater closets, where each pilot light became a torch.

Tampered or Altered Equipment

Document unusual conditions noticed at the fire scene. Look for indications of a forcible entry before the fire department arrived. If the fire origin was electrical, look for electrical devices that could have been altered or appear unusual. A professional arsonist wants to try to make a fire appear accidental.

Safety Zone

Gasoline and the Amateur Arsonist
Gasoline is the flammable liquid of choice of most amateur arsonists. Many have been burned while trying to ignite poured gasoline.

Amateur arsonists use excessive fuel, creating a dangerous situation for the first-arriving fire fighters. The fire sometimes flashes but then dies down due to insufficient oxygen, leaving only parts of the fire area burning. Because of the fuel-rich conditions, there are still unburned flammable liquid puddles in the fire area that can reignite or flash over when additional air comes into the fire area. When the fire attack team makes entry and begins to advance, the mixture could be ripe for re-ignition.

Fire fighters in full protective clothing and SCBA cannot smell the flammable liquid. They must be alert for explosive fire development, stay with the hose line, and be prepared to make a hasty exit.

Company Tips

NFPA Statistical Snapshot
1. An estimated 323,900 intentional fires of all types were set in 2005.
2. These fires resulted in 493 deaths and $1.1 billion in property damage.
3. An estimated 55,000 structure fires were deliberately set or suspected of having been deliberately set.
4. Three fire fighters died and 7600 fire fighters were injured while operating at intentional fires.
5. The number of intentional fires and associated losses has been on a steady decline since 1980.

Arson

Arson is the crime of maliciously and intentionally, or recklessly, starting a fire or causing an explosion. National statistics indicate that one in every four fires is of incendiary origin. Arson consistently has the highest rate of juvenile involvement when compared with all other Federal Bureau of Investigation index of the most serious felonies.

Arson Motives

There are six basic motives for arson:
1. Profit
2. Crime concealment
3. Excitement
4. Spite/revenge
5. Extremism
6. Vandalism

Profit

Most often, the plan is to collect insurance money. Indicators of insurance fraud include inability to meet payments, failure to complete business contracts, poor sales volume, or lack of supplies and inventory. Investigators verify the insurance coverage amounts and look for recent changes in the policy. In some cases, the insured makes claims for burned inventory that was actually removed before the fire or that never existed.

Elaborate schemes have been conducted to manipulate supply and demand for various products by destroying inventory or manufacturing facilities in order to cause a shortage and drive up prices. Arson has also been used for extortion or to eliminate competition. Sometimes, a contractor or individual starts a fire and then offers to repair the damaged property. Many vacant buildings have been burned to save the cost of demolition.

Crime Concealment

A business owner or employee may elect to burn the business in order to destroy records that show embezzlement of cash, supplies, or inventory. Burglars might set fire to a building to destroy evidence of their entry, eliminate fingerprints, or even conceal the fact that a theft occurred before the fire. Arson is also used to destroy evidence of other crimes, including murder, or to create a distraction while a crime is taking place in another location.

Excitement

Sometimes, a fire is started for the excitement of the arsonist, who could be seeking thrills, attention, or recognition. The arsonist could be planning to make a dramatic rescue to gain praise or to obtain recognition for discovering or extinguishing the fire. The arsonist may simply enjoy the excitement and spectacle that are generated by a fire.

Spite/Revenge

Spite and revenge fires involve intense emotions. The arsonist may intentionally set a fire in the exit stairway of an occupied building in the middle of the night. Frequently, the process is initiated by hatred, jealousy, or other uncontrollable emotions following a lover's quarrel, divorce, or bar fight. In many cases, the individual has consumed alcohol or drugs before starting the fire.

Extremism

Extremism for a variety of causes has been the reasoning for setting fires for many centuries. Abortion clinics, religious institutions, businesses that were ecologically damaging, and businesses undergoing labor disputes have all been arson targets. An extremist may want to cause a monetary loss to the person or business or to bring attention to a cause. Radical activists know that a large loss of life focuses attention on their cause. Incendiary devices are frequently used in these crimes.

Vandalism

The motive for vandalism is simply to cause damage for its own sake. Vandalism is most often directed toward schools, abandoned structures, vegetation, and trash containers. In most cases, the fire setter is within walking distance of his or her home.

Legal Considerations

Although the law provides that there is a public interest in determining the cause and origin of a fire, there are also the competing interests of a citizen's rights to privacy and due process. The fire officer who investigated the fire is called to testify in court and may be challenged on issues of proper procedure.

■ Searches

When a fire has occurred and the fire department has been called, the fire department has the right to determine the cause and origin of the fire. This process must be accomplished in accordance with the law. In *Michigan v. Tyler* (1978), the U.S. Supreme Court held: "Fire officials are charged not only with extinguishing fires, but with finding their cause. Prompt determination of the fire's origin may be necessary to prevent its recurrence, as through the detection of continuing dangers such as faulty wiring or a defective furnace. Immediate investigation may also be necessary to preserve evidence from intentional or accidental destruction."

The fire officer must take care to avoid an unlawful search and seizure, which is prohibited by the Fourth Amendment of the U.S. Constitution. Typically, no search warrant is needed to enter a fire scene and collect evidence when the fire department remains on scene for a reasonable length of time to determine the cause of the fire and as long as the evidence is in plain view of the investigator. This principle has been reaffirmed by the U.S. Supreme Court in *Michigan v. Clifford* (1984).

The aftermath of a fire often presents exigencies that will not tolerate the delay necessary to obtain a warrant or to secure the owner's consent to inspect fire-damaged premises. Because determining the cause and origin of a fire serves a compelling public interest, the warrant requirement does not apply to such cases.

The plain view doctrine allows for potential evidence to be seized during the processing of a fire scene, if the fire investigator had a legal right to be there and the evidence is in plain view. Under *Michigan v. Clifford*, the court held that as fire fighters remove rubble or search other areas where the cause of fires is likely to be found, an object that comes into view during such process may be preserved. In the same decision, the court also found that once the investigator has determined where the fire started, the scope of the search authority is limited to that area. After the cause and the origin have both been determined, a search warrant or consent is required for any further search.

If re-entry is needed after the fire department leaves the scene or to conduct a search for evidence of a crime after the cause and origin have been determined, the investigator must obtain a search warrant or receive permission from the occupant. The general provisions were clarified in *Michigan v. Tyler*:

- No search warrant is needed when fighting a fire or remaining on scene for a reasonable period of time to determine the cause of a fire and any evidence is admissible under the plain view doctrine.
- Administrative search warrants are needed for re-entry that is not a continuation of a valid search when the purpose is to determine the cause of the fire.
- A criminal search warrant is needed when re-entry is not a continuation of a valid search and the purpose is to gain evidence for prosecution.

■ Securing the Scene

A fire officer who conducts a preliminary fire cause investigation and suspects that a crime has occurred should immediately request the response of a fire investigator. When this occurs, the scene must be secured in order to protect any evidence that exists. If the fire department leaves the scene unsecured, any evidence that is collected after that point could be called into question. The fire officer must ensure that fire department personnel maintain custody of the scene until the investigator arrives.

Protecting the scene includes preventing unauthorized personnel from entering into the scene. To create a security perimeter, the fire officer can use fire line tape or police crime scene tape secured to objects, such as trees and fence posts. Natural barriers can also be used to aid in securing the area. Objects such as fences or hedges can be used along with barrier tape to completely surround an area.

All access to and from the area must be controlled. To be certain that no unauthorized personnel enter the area, a fire fighter or law enforcement officer may need to be posted to limit access. Posting a guard preserves the chain of custody over the scene and any evidence that is present until the fire investigator arrives. Failure to do so may require the fire department to get a warrant to return to the fire scene.

Barrier tape may also be placed across a doorway to prevent unauthorized entry into a room or building. To secure and protect smaller areas of evidence, the fire officer may decide to cover them with a plastic sheet or tarp.

Limit the number of fire personnel that are allowed into the secured area. The area should be treated as a crime scene, and only activities that are essential to control the emergency or protect the scene should be conducted. Fire fighters should not collect artifacts as souvenirs of their firefighting adventure, particularly when a scene is being secured.

Evidence

<u>Evidence</u> includes material objects as well as documentary or oral statements that are admissible as testimony in a court of law. Evidence proves or disproves a fact or issue. The fire officer must consider three types of evidence.

1. <u>Demonstrative evidence</u>: Tangible items that can be identified by witnesses, such as incendiary devices and fire scene debris
2. <u>Documentary evidence</u>: Evidence in written form, such as reports, records, photographs, sketches, and witness statements
3. <u>Testimonial evidence</u>: Witnesses speaking under oath

If the fire officer has determined that the fire requires a formal investigation, then every effort should be made to protect and preserve the fire scene evidence. The structure, contents, fixtures, and furnishings should remain in their prefire locations, as intact and undisturbed as possible.

Evidence plays a vital role in the successful prosecution of arson cases. In order to prove arson, the fire investigator must rule out all potential accidental and natural causes of the fire. The investigator must consider all possible circumstances, conditions, or agencies that could have brought together a fuel, an ignition source, and an oxidizer, resulting in a fire or combustion explosion.

<u>Artifacts</u>, in the context of fire evidence, could include the remains of the material first ignited, the ignition source, or other items or components that are in some way related to the fire ignition, development, or spread. An artifact could also be an item on which fire patterns are present, in which case preservation of the artifact is not for the item itself, but for the fire pattern that appears on the item.

Protecting Evidence

One part of the fire investigator's job is to dig out the fire scene. Like an archaeologist, the fire investigator removes each layer of fire debris. The investigator's ultimate goal is to identify the point of origin and the cause of the fire. Because fire follows the rules of science, an analysis of how the fire spread assists in determining where it originated and whether the cause was accidental or intentional.

<u>Fire scene reconstruction</u> is the process of re-creating the physical scene before the fire occurred, either physically or theoretically. As debris is removed, the contents and structural elements are replaced in their prefire positions, as much as possible.

Figure 17-6 The fire officer must protect the fire scene evidence from the public and from excessive overhaul and salvage.

Where the damage and destruction are too extensive to physically restore the scene, whatever information is available is used to fill in the blanks. The investigator interprets the fire scene and documents the fire development by examining the damage to objects, devices, and surfaces. Throughout the process, the investigator must concentrate on locating, examining, and preserving evidence.

The fire officer has to determine when to stop fire suppression or overhaul operations in order to preserve evidence for the investigator. From the fire investigator's viewpoint, the less the fire fighters disturb, the more intact the scene remains.

The worst case occurs when the fire investigator arrives at an apartment fire and discovers that the fire company has removed all of the fire debris, including the fire-damaged ceilings, walls, and doors. This prevents the investigator from being able to evaluate the evidence in the context of the fire. This would be akin to the police trying to reconstruct a motor vehicle collision from which both cars have been removed to the junkyard and no witnesses remain on the scene: It is possible, but very difficult.

The fire officer is responsible for protecting the fire scene evidence from the public and from excessive overhaul and salvage **Figure 17-6**. In addition, the fire officer is the first step in the chain of evidence that is vital to successful prosecution of arson cases. Once the fire department arrives, it is responsible for preventing evidence contamination; this requires fire fighters to stay until the fire investigator arrives. The chain of evidence requires that evidence remain secured and documented, from the fire scene to the courtroom. Most investigators document all physical evidence before collecting it by taking high-resolution photographs.

Legal Proceedings

A company officer may be called on to testify as a witness. Whereas a witness can provide the court with testimony based only on his or her personal knowledge and observations, an expert witness has scientific, technical, or other specialized knowledge that can be relied on to interpret the facts. The role of an expert witness is to assist the judge and jurors to understand the evidence or to determine the true facts in an issue. An expert witness is allowed to give an opinion based on the facts or the data of the case.

The first step is to fully prepare. This includes reviewing all reports, photographs, and diagrams of the incident. Also, review any previous depositions or testimony that you have given on the incident. The prosecutor will review your testimony and qualifications.

When testifying:

- Dress appropriately.
- Follow the prosecutor's directions.
- Sit up with both feet on the floor.
- Avoid gesturing.
- Keep answers short and to the point.
- Use language a jury can understand.
- Be courteous and patient.
- Be honest.
- Do not hesitate or avoid answering questions.
- Speak clearly and loudly.
- If you do not remember, do not guess.

Remember that the defense counsel's job is to try to discredit you and your statements. It is essential to remain calm and cool, even when the lawyer is asking questions or making statements that are designed to make you look bad. Your role is to present factual information that you can support. Every fire officer should be familiar with NFPA 921, *Guide for Fire and Explosion Investigations*.

Getting It Done

Coordinated Overhaul and Salvage

Although it is important not to destroy evidence during fire suppression operations, the fire officer's first responsibility is to maintain a safe fire ground. That means promptly controlling and extinguishing the fire. After the searches are completed and the fire is declared under control, the fire officer has to consider evidence preservation. Overhaul and salvage operations should be coordinated with the fire investigator, if the investigator is on the scene. For example, a fire investigator may wish to photograph or document the conditions of a ceiling before it is removed during overhaul operations. If the investigator's response is delayed, the fire officer should attempt to identify and protect the area of origin, limiting overhaul in that area to the absolute minimum required to ensure that the fire does not rekindle.

NFPA 921 makes the following recommendations:

- Use caution with straight or solid-stream water patterns; they can move, damage, or destroy physical evidence.
- Restrict the use of water for washing down, and try to avoid possible areas of origin.
- Refrain from moving any knobs or switches.
- Use caution with power tools in the fire scene. Refuel away from the fire scene.
- Limit the number of fire fighters performing overhaul and salvage until the fire investigator is finished documenting the scene.

Documentation and Reports

All fires must be properly documented and reported according to the fire department's standard procedures. Most fire departments use the NFIRS reporting system or a variation of it. The basic report includes the incident number, alarm time and date, location of the incident, property ownership, building construction and occupancy type, weather conditions, responding units and personnel, and numerous other factors that are required for full documentation of an incident.

The data collected through NFIRS provides important information for the fire department to identify risk factors and trends and to plan the most efficient utilization of resources to prevent fires and respond to emergencies. This information also flows into the state and national database systems to provide a better understanding of the overall fire problem. These data are used to assist the fire service and elected officials in determining where resources should be directed and when laws should be changed.

Preliminary Investigation Documentation

In addition to the basic incident report, the fire officer often writes up a special narrative report if the cause of the fire is incendiary or if unusual circumstances are involved. The narrative report is particularly valuable if the fire officer is called to court in the future. Chapter 14 discusses communications in more detail, but fire officers could use the following format to document activities and observations for the fire investigator:

A. Receipt of the alarm
  1. Who reported the fire?
  2. What time was the alarm received?
  3. Who discovered the fire?
B. Response to the incident
  1. Did the fire company encounter any suspicious activity while responding to the fire?
  2. How much time elapsed from dispatch to arrival?
  3. What were the observed weather conditions?
C. Accessibility at the scene
  1. What was the general condition of the fire?
    a. What were the extent and the intensity of the fire?
    b. What was the location of the fire or fires?
    c. Did anyone meet the fire fighters on arrival?
    d. Were any familiar spectators at the scene?
  2. Circumstances on arrival
    a. Was anything unusual, considering the fire load?
    b. Where was smoke and fire coming from?
    c. What were the rapidity and the spread of the fire?
  3. Gaining access to building
    a. How did the fire fighters get into the structure?
    b. If entry was forced, how did this occur and who forced entry?
D. Fire suppression
  1. What were your immediate observations on entry?
    a. Where was the fire centered?
    b. Were there any unusual flame, smoke, or odors?

   2. What were the conditions of fire extinguishment?
   - **a.** Did the fire flash when it was hit by water?
   - **b.** Was the fire difficult to extinguish?
   3. Were there obstructions to fire suppression?
   - **a.** Were fire protection systems tampered with?
   4. Was the alarm system functioning properly?
- **E.** Civilian contacts
   1. Did witnesses make statements to fire fighters?
   2. Was the owner at the scene?
   3. What were the names of persons allowed into the fire scene?
- **F.** Scene integrity
   1. Were any physical evidence or artifacts removed from scene?
   2. Were any photos/videos taken during fire suppression?
   3. Did any fire department make holes in walls or ceilings?

The report must be clear, complete, and factual. The fire officer must make no assumptions or speculations. The best report is a narrative that accurately and completely describes what the fire officer observed and what the fire company did.

Figure 17-7 The fire investigator is interested in determining fire cause and origin to help prevent future fires and prosecute criminal actions.

Investigation Report

The fire officer who conducts the investigation will also submit a narrative report. The information is provided in chronological order, beginning with a description of the structure before the event occurred. This description would include the building height and dimensions, construction type, structural condition, occupancy, and utility services. It would then describe the alarm notification information, including the time of the call, the name of the caller, and what the caller said.

The report should fully describe the results of the fire scene examination, beginning with a description of the exterior damage. This is followed by a description of the interior damage, including the determination of fire origin and cause, as well as the examination and elimination of any other possible fire causes. This section is closed with the officer's opinion and conclusion as to the cause and origin of the fire.

Attached to this report would be the information obtained from interviews and witnesses as well as statements from responders. Attachments should also include statements of evidence that were collected, warrants, and sketches.

After the Fire Officials Are Gone

Many fire investigations continue long after the fire department has cleared the event. There is a continuing investigation by the insurance company. This is not to undermine or question the fire investigator's conclusions but rather to look at the fire from the insurance company's point of view.

The role of the fire investigator and the insurance company investigator varies. The fire investigator could be interested in determining the fire cause and origin to help prevent future fires or to help prosecute criminal actions **Figure 17-7**. The insurance investigator might also be looking at the factors that contributed to the loss, such as the absence or inadequacy of fixed fire protection systems and whether the applicable codes were followed.

The insurance company has a financial interest in any fire loss. The insurance company is often interested in determining other parties that could have liability for the loss, even if there is no criminal responsibility. Although a criminal prosecution requires proof beyond a reasonable doubt, a civil action involving the insurance company is decided by a preponderance of the evidence. Many insurance cases are not settled for several years after the fire is extinguished.

Summary

Determining the initial origin and cause of fires is a responsibility of the company officer. This determination may lead to requesting a fire investigator, or it may conclude with the officer's report. Either way, it is essential for a fire officer to be able to make this determination.

The origin is the point where the fire began. After it is determined, the fire officer must determine what material was first ignited, how it was ignited, and why this occurred. This allows the fire to be categorized into one of four classifications: intentional, accidental, natural, and undetermined.

If the fire is intentional, the fire officer needs to gather the information that is used by a prosecutor to determine whether it rises to the level of arson. This usually includes writing a detailed narrative with all relevant information about the cause and origin of the fire.

You Are the Fire Officer: Conclusion

Working from the unburned area toward the burned area, it is clear that the point of origin was below the charred sports car that is still a few feet above the garage floor on a hydraulic lift. The technician's story matches what you see: a quarter-full gasoline tank was dropped or fell from the sports car and the fumes found an ignition source, perhaps the electric-powered space heater that is near where the tank fell.

Although you are ready to classify the fire as accidental, you consult with the on-duty fire investigator who is still tied up at another fire. The investigator tells you to call for the overtime investigator because of the injured civilians, the dollar loss, and what appears to be an unpermitted operation of a vehicle repair shop within an electronic installation store. The investigator also suggests that you have the hazardous materials team assist with handling the heat-damaged racing fuel drums, the acetylene torch, and the other hazardous materials found in the store.

Wrap-Up

Chief Concepts

- Know your jurisdiction's criteria on when to call a fire investigator.
- The point of origin is where a heat source and fuel come in contact with each other and start a fire.
- U- and V-shaped fire patterns can lead the fire officer to the point of origin.
- Char can be used to estimate fire intensity and direction.
- The four fire cause classifications are accidental, natural, incendiary, and undetermined.
- When considering intentional fires, look for disabled built-in fire protection, delayed notification/difficulty in getting to the fire, accelerants and trailers, multiple points of origin, and tampered/altered equipment.
- Arson is a crime of maliciously and intentionally, or recklessly, starting a fire or causing an explosion.
- There are six basic motives for arson: profit, crime concealment, excitement, spite/revenge, extremism, and vandalism.
- Evidence is the documentary or oral statements and material objects admissible as testimony in a court of law.
- Artifacts are the remains of the first material ignited, ignition source, or related components; they should be left where they are found.
- Fire investigators systematically remove each layer of fire debris as part of the fire scene reconstruction.
- Delay the overhaul and salvage until the fire investigator arrives and completes the initial scene survey.
- All fires must be properly documented and reported according to the fire department's standard operating procedures.

Hot Terms

Accelerant Agents, often ignitable liquids used to initiate a fire or increase the rate of growth or spread of fire.

Arson The crime of maliciously and intentionally, or recklessly, starting a fire or causing an explosion.

Artifacts The remains of the material first ignited, the ignition source, or other items or components in some way related to the fire ignition, development, or spread. An artifact may also be an item on which fire patterns are present, in which case the preservation of the artifact is not for the item itself but for the fire pattern that appears on the item.

Char Carbonaceous material that has been burned and has a blackened appearance.

Demonstrative evidence Tangible items that can be identified by witnesses, such as incendiary devices and fire scene debris.

Documentary evidence Evidence in written form, such as reports, records, photographs, sketches, and witness statements.

Evidence The documentary or oral statements and the material objects admissible as testimony in a court of law.

Failure analysis A logical, systematic examination of an item, component, assembly, or structure and its place and function within a system, conducted in order to identify and analyze the probability, causes, and consequences of potential and real failures.

Fire analysis The process of determining the origin, cause, development, and responsibility, as well as the failure analysis of a fire or explosion.

Fire patterns Physical marks left on an object by the fire.

Fire scene reconstruction The process of re-creating the physical scene during fire scene analysis through the removal of debris and the replacement of contents or structural elements in their prefire position.

Form of material What the ignited material is being used for; for example, the form of cotton material might be clothing or bales of cotton.

Point of origin The exact physical location where a heat source and a fuel come in contact with each other and a fire begins.

Pyrolysis The destructive distillation of organic compounds in an oxygen-free environment that converts the organic matter into gases, liquids, and char.

Source of ignition Devices or equipment that, because of their intended modes of use or operation, are capable of providing sufficient thermal energy to ignite flammable gas–air mixtures.

Testimonial evidence Witnesses speaking under oath.

Trailers Materials used to spread fire from one area of a structure to another.

Type of material What the ignited material is made of. For example, the type of material might be cotton.

Fire Officer *in Action*

As a supervising fire officer, you responded to and commanded a fire at an apartment where food was left unattended on a stove. The fire spread to materials next to the burning pot and extended up into the cabinets and exhaust ductwork. There were a 10-year-old, an 8-year-old, and a 6-year-old in the apartment with no adult supervision. The apartment was a mess and there was no battery in the smoke detector. You called for police assistance because of the unsupervised children.

The managing or supervising fire officer who handled the fire is in the best position to ensure that evidence is preserved and an initial investigation into the cause of the fire is started. An accurate fire cause investigation benefits all.

1. Which of the following is NOT a fire cause as defined by NFPA 921?

A. Accidental

B. Arson

C. Undetermined

D. Natural

2. You have started a cause and origin investigation and have determined that you need a fire investigator. It will take 4 hours for the investigator to get to the fire scene. You will:

A. make sure that the incident scene perimeter is completely wrapped in fire-line tape before leaving. Post a "Stay out" sign and notify the owner or representative when the fire investigator will arrive.

B. secure the perimeter of the incident scene with fire-line tape and arrange for a continuous and uninterrupted fire department on-scene presence until the fire investigator arrives.

C. cancel the fire investigator and make sure that you completely document the incident scene with a digital camera.

D. obtain a contact number from the owner or representative before leaving the incident scene, scheduling a follow-up time when the fire investigator arrives.

3. The first objective of a cause and origin investigation is to:

A. determine the source of ignition.

B. find the point of origin.

C. determine the fuel that was first ignited.

D. Deduce how the ignition source and fuel came together.

4. The most common cause of a fire in a residential structure is:

A. children playing with matches.

B. careless discard of smoking materials.

C. cooking.

D. congested dryer duct.

Crew Resource Management

NFPA 1021 Standard

Fire Officer I

4.4.1 Recommend changes to existing departmental policies and/or implement a new departmental policy at the unit level, given a new departmental policy, so that the policy is communicated to and understood by unit members. [p 353–362]

(A) Requisite Knowledge. Written and oral communication. [p 360–362]

(B) Requisite Skills. The ability to relate interpersonally and to communicate change in a positive manner. [p 360–362]

Fire Officer II

5.4.6 Develop a plan to accomplish change in the organization, given an agency's change of policy or procedures, so that effective change is implemented in a positive manner. [p 353–362]

(A) Requisite Knowledge. Planning and implementing change. [p 352–360]

(B) Requisite Skill. The ability to clearly communicate orally and in writing. [p 354, 356]

Introduction to Fire and Emergency Services Administration (FESHE) Course Outcomes

2. Explain the need for effective communication skills both written and verbal. [p 354, 356]

4. Recognize appropriate appraising and disciplinary actions and the impact on employee behavior. [p 353]

6. Evaluate methods of managing available resources. [p 357–358]

7. Identify roles and responsibilities of leaders in organizations. [p 356–357, 362]

9. Identify and assess safety needs for both emergency and nonemergency situations. [p 352–354, 360–361]

11. Identify the role of a company officer in Incident Command System (ICS). [p 358–360]

13. Identify and analyze the major causes involved in line of duty firefighter deaths related to health, wellness, fitness and vehicle operations. [p 353]

Knowledge Objectives

After studying this chapter, you will be able to:

- Discuss the origins of crew resource management (CRM).
- List Dupont's "dirty dozen" human factors that contribute to tragedy.
- Describe the six-point CRM model that can be used in the fire service.
- Describe the five steps in a successful debriefing.

Skills Objectives

There are no skills objectives for this chapter.

I t looks like a war zone. A gasoline tank truck has overturned, ruptured, and ignited. The burning petroleum has splashed into the lobby of an eight-story community hospital. The black boiling smoke is entering the open windows of a nursing home. Most of the other fire companies dispatched to this incident are held up at a railroad crossing. As you pull to a stop, the apparatus operator tells you that the pumper's transmission has failed. There will be no attack lines from your rig. Behind your rig, your see smoke and flames coming from manhole covers.

1. When faced with multiple tasks, how can you best allocate resources to ensure that the emergency situation is mitigated and everyone returns home safely afterward?

Origins of Crew Resource Management

On December 28, 1978, United Airlines Flight 173, a McDonnell Douglas DC8, was making a routine flight from Denver to Portland with 189 passengers and crew onboard. During final approach to Portland International Airport, an unfamiliar "thump" was felt as the landing gear deployed, and the "gear down and locked" light on the cockpit instrument panel did not illuminate. The captain decided to circle the airport while he, the first officer, and the flight engineer attempted to figure out the problem. The flight engineer told the captain that "15 minutes is really gonna run us low on fuel here." The captain looked at the fuel gauge and decided that the plane could circle while the crew troubleshooted the landing gear issue.

One hour passed as Flight 173 circled the Portland area while the three flight deck crew members sought confirmation that the gear was down and flight attendants prepared the passenger cabin for an emergency evacuation. Aviation traditions of the day held that the captain was the "infallible head of the ship" and was therefore never questioned. While the crew continued to fuss with the lights and the pilot awaited word on cabin preparations, the plane's engines coughed, sputtered, and finally went quiet, one at a time. The captain realized that the plane had run out of fuel moments before it slammed into the Oregon pines. Ten people were killed, including the flight engineer, and 23 were critically injured.

The DC-8 used by Flight 173 was a fully functional, mechanically sound airframe that crashed because the humans flying the machine became over-engrossed in a burned-out light bulb. The pilot became so absorbed in the burned-out bulb that he forgot to fly the plane. A new behavioral modification training system known as **crew resource management (CRM)** was developed in a 1979 National Aeronautics and Space Administration (NASA) workshop that examined the role of human error in aviation accidents.

Over the objections of senior pilots, CRM became mandatory training. Resistance continued until a July 1989 incident when United Airlines Flight 232 was bound for Los Angeles from Chicago. The plane experienced a catastrophic failure of the center engine. All three hydraulic lines necessary for controlling flaps, rudders, and other flight controls were severed. This in-flight disaster robbed the crew of primary and redundant safety features that are built into every airframe. When these conditions were presented within a flight simulation exercise, it always resulted in an unrecoverable spin with all lives lost.

The United 232 flight crew and a check ride pilot, using engine controls alone, managed to bring the crippled plane into the Sioux City, Iowa, airport. The plane made a spectacular crash landing that was captured on film by news crews. One hundred eighty-four of the 295 people on board survived. The crew attributed their success to CRM training as they initiated behaviors to overcome the five factors that contribute to human error. United 232 was the landmark event that validated CRM's worth.

Researching and Validating CRM Concepts

Part of the 80 percent reduction in the aviation industry's accident rate is attributed to the development, refinement, and system-wide adoption of CRM. Lieutenant Colonel Tony Kern (retired

U.S. Air Force), writing in *Controlling Pilot Error: Culture, Environment and CRM*, states: "CRM is designed to train team members how to achieve maximum mission effectiveness in a time-constrained environment under stress. That is a concept with nearly universal utility and timeless applicability."

Doctor Kern received the Distinguished Leadership Award from *Aviation Week and Space Technology* in 2003 for leadership in controversial decisions to ground nine air tankers during the 2002 U.S. wildland fire season after fatal accidents involving a Lockheed C-130A and a Consolidated PB4Y2; to bar future aerial firefighting contracts for these two airplane types in the Forest Service of the U.S. Department of Agriculture and the U.S. Bureau of Land Management; and to restrict aircraft operations in cooperation with other firefighting agencies. Kern is the national aviation director of the U.S. Forest Service.

Professor Robert Helmreich and his staff at the University of Texas Human Factors Research Project (HFRP) studied CRM and its application to aerospace, aviation, the military, maritime, and the medical profession. Much of the peer-reviewed publications and industry practices on this subject have come from the HFRP. The project started in the early 1980s as the Aerospace Crew Research Project, exploring the relationship among personality, group culture, and performance. HFRP closed in December 2008.

Human Error

Gordon Dupont has a background in aviation maintenance procedures and accident investigation and spent 7 years reviewing aviation accidents as a technical investigator for the Canadian Aviation Safety Board. In the 1990s, Dupont was the Special Programs Coordinator for Transport Canada, responsible for coordinating with the aviation industry in the development of programs that would reduce maintenance error.

While developing the Human Performance in Maintenance program, Dupont considered the similarities between errors that occur in the cockpit and those that occur in the maintenance hangar. Dupont's "dirty dozen" are considered a comprehensive list of reasons and ways that humans make mistakes:

- Lack of communication
- Complacency
- Lack of knowledge
- Distraction
- Lack of teamwork
- Fatigue
- Lack of resources
- Pressure
- Lack of assertiveness
- Stress
- Lack of awareness
- Norms

Whereas Dupont considered the human factor, Doctor James Reason looked at the systems approach to human error management. In "Human Errors: Models and Management" in the *British Journal of Medicine* (2000), Reason stated:

> High technology systems have many defensive layers: some are engineered (alarms, physical barriers, automatic shutdowns, etc.), others rely on people (surgeons, anaesthetists, pilots, control room operators, etc.), and yet others depend on procedures and administrative controls. Their function is to protect potential victims and assets from local hazards. Mostly they do this very effectively, but there are always weaknesses.

He points out that each layer of defense is more like a slice of Swiss cheese than a solid barrier. The presence of a hole in one defensive layer does not create a bad outcome event. But when the holes in all of the levels of defense align is when there is a bad or catastrophic outcome **Figure 18-1 ▶**.

Active Failures and Latent Conditions

Reason provides two reasons why holes appear in the layers of defense.

1. **Active failures** are the unsafe acts committed by people who are in direct contact with the situation or system. They have direct and short-lived effects on the integrity of the defenses. Not wearing a seat belt while in a moving vehicle is an example of an active failure.
2. **Latent conditions** are the inevitable "resident pathogens" within the system. Latent conditions have two kinds of adverse effect. First, they can translate into error-provoking conditions within the local workplace. Examples include time pressure, understaffing, inadequate equipment, fatigue, and inexperience. They also can create long-lasting holes or weaknesses in the defenses. Examples include untrustworthy alarms, unworkable procedures, and design and construction deficiencies.

These conditions may lie dormant within the system for many years before they combine with active failures and local triggers to create an accident opportunity. The ancient planes used for wildland firefighting that were grounded by Dr. Kern in 2002 provide an example of latent conditions.

Error Management Model

CRM is an error management model with three activities: avoidance, entrapment, and mitigating consequences. Error avoidance provides the greatest opportunity for trapping and preventing

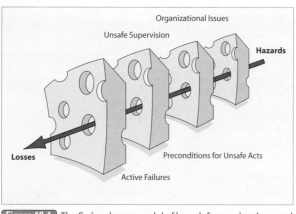

Figure 18-1 The Swiss cheese model of how defenses, barriers, and safeguards may be penetrated by an accident trajectory. Modified from Reason (2000), "Human Errors: Models and Management."

Reproduced from *Br Med J*, J. Reason, vol. 320. pp. 768–770, © 2000 with permission from *BMJ* Publishing Group Ltd.

errors from becoming a catastrophe. Errors that are not avoided are trapped at the second level. Errors that slip through the first two levels require mitigation. Mitigation is the action taken by emergency responders to minimize the effect of an emergency on a community **Figure 18-2 ▾**.

The CRM Model

There are several variations of CRM models. The fire service model covers six areas: communication skills, teamwork, task allocation, critical decision making, situational awareness, and debriefing. Regardless of which CRM model is used, the following conditions are assumed as accurate. Everyone within the department needs to recognize that:

- *No one* is infallible.
- Humans create technology; therefore, technology is fallible.

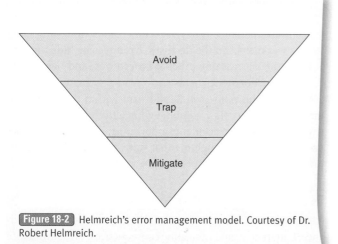

Figure 18-2 Helmreich's error management model. Courtesy of Dr. Robert Helmreich.

- Catastrophes are the result of a chain of events.
- Everyone has an obligation to speak up when seeing something wrong.
- People who work together effectively are less likely to have accidents.
- In order for the team to become more effective, every member of the team must participate.

■ Communication Skills

Communication is the successful transfer and understanding of a thought from one person to another. Chapter 14 provides detailed information about how to communicate successfully.

Review of airline disaster cockpit recordings shows that flight crews engaged in several forms of miscommunication that contributed to catastrophic events. These communication failures included misinterpretations of take-off, altitude, and landing instructions; airlines that fostered and condoned "fighter pilot" mentalities in their captains; a lack of assertiveness on the part of crew members who recognized problems before the captain; and distractions in the cockpit that inhibited focused communication, such as idle chitchat during preflight checks.

Developing a standard language and teaching appropriate assertive behavior are the keys to reducing errors resulting from miscommunication. In the aviation application, CRM advocates maintaining a "sterile cockpit," a cockpit environment where all communication and focus is on flight operations.

When a fire apparatus is en route to an alarm, the crew members are sitting in an enclosed box, surrounded by switches, gauges, wheels, pedals, and noise and moving down the road. In order for the machines to move efficiently and safely, the people inside need to focus on the mission and communicate effectively. The entire crew should be exchanging only information that is pertinent to responding to and arriving safely at the scene of the alarm.

The CRM-enriched environment generates a climate where the freedom to question is encouraged. Crew members are encouraged to respectfully speak up when they see something that causes concern by using clear, concise questions and observations. A discrepancy between what is going on and what should be occurring is often the first indication of an error.

The freedom for all crew members to question something that appears unusual or is not understood is a delicate subject for many organizations. Members need to speak directly in a manner that does not challenge the authority of a superior.

The British aviation world initially termed CRM "charm school" because of the philosophy's promotion of specific steps subordinates should take to question authority and authority's specific direction to listen to subordinates. Cockpit voice recordings of a number of air disasters reveal first officers and flight engineers attempting to bring an issue to the captain's attention by speaking indirectly about the subject.

Inquiry and advocacy are discrete, learnable skills that promote synergy between the mechanical element and human players. Inquiry is the process of questioning a situation that causes concern. Advocacy is the statement of opinion that recommends what one believes is the proper course of action under a specific set of circumstances. Using the two skills effectively, in concert

Voices of Experience

As we worked our way through summer afternoon traffic in Engine 66, sirens blaring, my Apparatus Operator Doug interrupted my concentrated study of the building pre-plan. "Wow, that's quite a column of smoke," he said. Quickly looking up, I saw a large, black roiling cloud pushing into the clear sky a half mile distant. We were responding to a popular fast food restaurant, and it was apparent the fire had gained significant ground in just a few short minutes.

Pulling into the lot, with Ladder 67 close behind, we were approached by the store manager, who yelled into my open window that he wasn't sure if everyone was out of the building. Heavy fire poured from an open rear kitchen doorway and from a gambrel-style roof with an attached tile-roofed facade. After providing a size-up for incoming units, I passed command to the arriving battalion chief and my crew of four assumed the role of fire attack.

Dragging a line from the engine, we hit the fire hard through the open doorway, then directed the stream into the roof overhang, darkening the fire. As we prepared for entry, I could hear loud "pops" as clay tiles slid off the roof onto the pavement. Suddenly, I felt a hand on my shoulder and looked over to see Fire Fighter Mark from Truck 67, a seasoned veteran. "What's the plan?" he asked. "We're going to make entry, Engine 51's pulling a back-up line, and you can search," I said. Mark took a quick look overhead, leaned into my ear and said, "I think we should back out." I briefly struggled to reconcile my feelings. I was the officer, I had given an order, and we seemed to be making progress. I turned and asked Mark "Why's that?"

"I'm concerned about that roof; it looks like its peeling back," he said. Bowing to his experience, and wanting to take a second look at the roof, I reached forward and grabbed the coat of my lead fire fighter, and we pulled back. Seconds later, the entire overhang fell to the ground in a large cloud of dust and debris, completely obscuring the place where two crews had just been preparing for entry.

Astonished, my crew and I immediately credited Mark with recognizing something none of us had seen. The roof was made of real ceramic tile, not a lightweight composite, and Mark had seen from underneath that it appeared to be simply nailed to the façade. I'd learned a powerful lesson as a new lieutenant. Conflicting views present you with a chance to be *curious* instead of *defensive*.

From that moment forward, I realized that curiosity is a key component of good leadership. If you remain open to input, and respond to queries with curiosity, you will demonstrate to your team that you are interested in their input. This will allow your team to sustain a shared understanding of the risks, resources, and plans during dynamic incidents, and build your crew resource management skills.

Paul LeSage
Assistant Chief
Tualatin Valley Fire & Rescue
Aloha, Oregon

> "Seconds later, the entire overhang fell to the ground in a large cloud of dust and debris."

with each other, requires practice and patience on the part of all crew members. The factor to keep in mind is that communication should not focus on who is right, but on what is right.

One effective tactic is to use specific buzzwords, such as "red light" and "red flag," to signal discomfort with a situation. These terms are cues that open the door to inquiry and advocacy. They should be reserved for situations that involve an immediate risk of injury.

Assertive Statement Process

Todd Bishop, from the Error Prevention Institute, teaches a five-step assertive statement process that encompasses the inquiry and advocacy communications steps. The assertive statement runs like this:

- Opening/attention getter: Address the other individual: "Hey, Chief," or "Bill," or whatever appropriate moniker is used to get the individual's attention.
- State your concern: Use an owned emotion: "That smoke is really pushing from those windows. I have a bad feeling about it."
- State the problem as you see it: "It looks like it is going to flash. I think any crews entering that place are going to take a beating."
- State a solution: "Why don't we vent the roof and give the place a minute or two to vent before we send in the attack team?"
- Obtain agreement or buy-in: "Does that sound good to you?"

The inquiry and advocacy process and the assertive statement are essential components of the communication segment of CRM. Of all the components, these two require the most work and attention. They are the toughest to impart and adopt because they often require a wholesale change in interpersonal dynamics. Once mastered, however, inquiry and advocacy can enhance performance, avoid mishaps, and save lives.

Effective listening, covered in Chapter 14, is an important CRM communication skill. One effective listening technique is to purposely refrain from making any response or counterargument until the other individual has drained their emotional bubble.

■ Teamwork

CRM promotes members working together for the common good. This requires developing effective teams through buy-in by all members, leaders, and followers, in the effort to be efficient and safe. Leaders will always be in charge; they must also be open to suggestion and constructive criticism. The National Response Framework and the incident command system, covered in Chapter 15, provide the formal structure for crew resource management at the task, tactical, and strategic levels.

Leadership

Fire officers formally exercise leadership of the team through a combination of rank and authority. In order for officers to become truly effective leaders, they have to earn the trust and respect of their subordinates and demonstrate the skills of effective leadership. These three components comprise the triangle of leadership **Figure 18-3 ▶**.

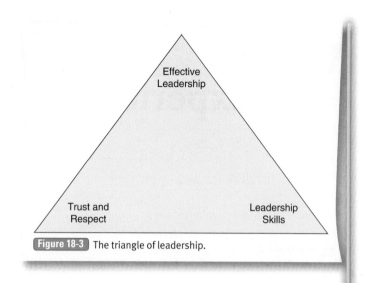

Figure 18-3 The triangle of leadership.

The informal authority to lead is derived through respect. True respect is based on three competencies: personal, technical, and social. Personal competence refers to an individual's own internal strengths, capabilities, and character. Technical competence refers to an individual's ability to perform tasks that require specific knowledge or skills. Social competence refers to the person's ability to interact effectively with other people. All three competencies are essential for leaders to function effectively in a CRM environment.

The fact that a leader must demonstrate effective social skills does not mean that a leader must always be deferential. Social competence requires knowing how to speak to people respectfully, whether praising or chastising them.

Mentoring

Mentors pay special attention to helping others develop their skills. Chapter 7 provides detailed information on training, coaching, and mentoring. CRM provides an effective vehicle for leaders to impart knowledge and skills to their subordinates.

Leading by example is a highly effective technique. Crews notice a supervisor's personal habits, pick up on his or her thought processes, and constantly evaluate the leader's performance. Traits that are admired and respected by the crew members have a major influence on their future behavior. The crew usually sees more than the supervisor may think. Leading by example requires constant effort on the part of the leader, with no opportunities to turn the positive role model on or off at different times.

Human error is a normal, inevitable occurrence. A leader must be willing to admit to making a mistake. Superior leaders readily admit when they have made mistakes, accept responsibility, and focus their attention on moving forward. This trait creates an environment that fosters open communication, promotes safety, and encourages subordinates' belief in the team.

Mentoring also requires sharing knowledge. Knowledge is often closely associated with power. Leaders who are insecure about their positions tend to withhold knowledge and fail to share lessons as they are learned. Mistakes are often repeated in such an environment, and errors that could have been avoided are permitted to occur. A fire officer must maintain technical competence, stay on top of technological advances, and ensure that others have the information that will keep them from being killed or injured.

Technical competency requirements vary from position to position. The street-level fire fighter does not expect the fire chief to pull hose and throw ladders, but the chief must be able to command a fire, mass-casualty incident, or weapons of mass destruction event with the same degree of skill that the fire fighter is expected to use to pull hose and throw ladders.

Handling Conflict

The focal point of CRM in conflict resolution is to focus on what is right, not who is right. This allows the leader to focus on the best, safest outcome. Chapter 11 provides detailed information and recommended skills for handling problems and conflicts.

The fire officer needs to establish an open climate for error prevention. The leader who perceives a subordinate's comment or question as a threat to his or her own authority is part of the problem instead of an essential part of the solution. Leaders are required to keep their own egos in check. A leader who rules by intimidation can often win an argument, but the forces of nature always prevail over words. An officer can order a crew to advance into a dangerous position, but the building will not follow an order to remain standing. An officer who fails to listen or refuses to listen is simply dangerous.

Responsibility

New disciples believe that CRM advocates the creation of "management by committee," in which leadership gives way to consensus. That is not an accurate statement. Leaders need to keep their eyes and ears open, especially for input from subordinates. In order for decision making to be efficient, there has to be someone in charge, someone who has ultimate responsibility for decisions and outcome.

In the cockpit, the captain retains ultimate authority for decision making and ultimate responsibility for getting the plane from airport to airport. Subordinates are encouraged to provide input, but the final decision rests with the recognized authority. Lines of authority are maintained and that ultimate authority rests with the legitimate ranking officer.

Mission analysis requires fire service leaders to look at all situations with a risk-versus-gain mentality. Brunacini's mantra—risk a lot to save a savable life, take a calculated risk to save savable property, and risk nothing to save what is already lost—establishes a concrete foundation on which leaders can manage emergency operations.

Followership

Followership is the appropriate actions of those who are led. All followers should perform a self-assessment of their ability to function as part of a team. The self-assessment should consider four critical areas:

1. Physical condition: People in good physical condition are more aware, alert, and oriented to their surroundings. Maintaining good physical health is critical to the success of any endeavor; however, it is absolutely critical in the fire service.
2. Mental condition: We are constantly pulled in a variety of directions; however, the challenging and dangerous environments where fire fighters have to perform demand our full attention. Fire fighters must ask themselves, "Am I free of distractions that could divert my attention from the task at hand?"

3. Attitude: In order to be an effective team member, a fire fighter must be willing to follow orders and be part of a cohesive team.
4. Understanding human behavior: The effectiveness of CRM is based on understanding human behavior and interpersonal dynamics in a team environment. In order to be effective using CRM, the team members must understand how their individual behavior relates to each other and to the team as a whole.

In order to be effective team players and maximize CRM benefits, each individual must have the following:

- A healthy appreciation for personal safety
- A healthy concern for the safety of the crew
- A respect for authority
- A willingness to accept orders
- A knowledge of the limits of authority
- A desire to help their leader be successful
- Good communication skills
- The ability to provide constructive, pertinent feedback
- The ability to admit errors
- The ability to keep ego in check
- The ability to balance assertiveness and authority
- A learning attitude
- The ability to perform demanding tasks
- Adaptability

These qualities of a good follower are remarkably similar to those required of effective leaders. Followership enhances leadership, which in turn promotes effective teams. Teams that practice CRM make fewer mistakes and are able to recognize and correct errors before they cause tragic outcomes.

■ Task Allocation

Task allocation refers to dividing responsibilities among individuals and teams in a manner that allows for their effective accomplishment. Task overload occurs when the fire officer exceeds capacity to manage all of the simultaneous functions and responsibilities. Safety is compromised with task overload.

Think about the individual who is driving down the interstate at rush hour, cell phone to the ear and attempting to write down a telephone number. He looks up and realizes that traffic has stopped. Cell phone, paper, and pencil all fall to the floor as he slams on the brakes, too late to avoid a collision. Task overload has occurred.

Knowing one's own limits and the capacity of the team is the first step in the CRM task allocation phase. Everyone has a point at which outside stimuli override the ability to process information and perform effectively. If the resources are not available, tasks must be prioritized to ones that can be done safely and effectively.

Fire officers fall into three categories. Some officers are reluctant to admit that they are ever overwhelmed and believe that they become more effective as the situation becomes more hectic. The second group often becomes overwhelmed before the full complexity of the event is even recognized. The third group effectively assesses the incident, calls for additional resources early, and manages to stay ahead of both the incident and the span of control.

A NASA/University of Texas HFRP study noted that pilots believed they were capably handling multiple tasks, when in reality, mistakes began to appear and performance deteriorated. They were so busy that they failed to recognize when the situation was getting out of control.

A fire officer has to evaluate his or her personal capacity to manage complicated situations and identify weak spots. Training and practice can improve performance; however, everyone has limitations. When these are known, we can take action early to compensate.

The fire officer must also know the crew's limits. In some fire departments, the same crew works together as a team for long periods, allowing the fire officer to evaluate strengths and weaknesses and focus on team development. This is difficult to determine when the make-up of a crew changes from day to day or the crew is assembled only at the time of alarm.

The strengths of individuals and crews should be evaluated and built on in multiple nonemergency settings. Performance is enhanced through training classes, live training exercises, table-top modeling, didactic presentations by experts, mentoring, and exchanging Figure 18-4 ▾ .

Figure 18-4 Training exercises help the officer determine the strengths of individuals.

Critical Decision Making

Although CRM promotes the concept of team involvement in all aspects of an operation, emergency scenes often demand rapid decision making by the crew leader. Input is welcome from all team members; however, the final responsibility resides with the crew leader, and experience and training often play pivotal roles in successful outcomes.

Gary Klein, a cognitive psychologist who studies decision making under pressure, found that fire officers and military combat officers often use a very similar decision-making process. During research conducted in the mid-1980s, urban fire commanders told Klein that they made fire-ground decisions based on previous experiences, not on traditional decision-making models. The commanders reported that they often did not even consider options; they simply looked at the incident, recalled previous experiences of a similar nature, and applied the processes that had been used to successfully mitigate the previous incidents. Many of the fire officers of that era had tremendous experience fighting numerous fires in an urban environment; however, much of that experience has been lost through retirements and a declining number of fires.

Klein observed that the fire-ground and combat commanders often have to make decisions in environments that are time compressed and dynamic, involved missing or ambiguous data and multiple players, and lacked real-time feedback. This research identified two decision-making models: **recognition primed decision making (RPD)**, which describes how commanders can recognize a plausible plan of action, and **naturalistic decision making**, which describes how commanders make decisions in their natural environment.

Fire officers trained in enhanced communication skills can provide the missing data and help clarify ambiguous information. Better real-time feedback is also provided through the improved communication skills. Each element of CRM establishes a foundation for better decision making.

Decision making is improved through gaining experience, training constantly, improving communication skills, and pre-planning incidents. Practicing enhanced communication skills and taking advantage of improved crew interaction at all levels provide the following benefits:

- Problem identification is enhanced.
- Supervisors at all command levels maintain better incident control.
- Situational awareness is improved.
- Hazards are more rapidly identified.
- Resource capability is rapidly assessed.
- Potential solutions are more rapidly developed.
- Decision making is improved.
- Surprises and unanticipated problems are reduced.

Situational Awareness

Situational awareness is the accurate perception of what is going on around you. NASA expands the definition to "the awareness of acknowledging and assessing; and is the basis for choosing courses of action of the situation both now and in the future."

Situational awareness affects performance and decision making. When situational awareness is maintained, operations are completed without a hitch. When situational awareness is not maintained, errors occur, performance suffers, and catastrophic events often result. Situational awareness is a human factors element that is somewhat difficult to explain but easy to recognize after an error occurs. Situational awareness also carries over into nonincident operations. Failing to note at shift change that the SCBA cylinders are down to 1900 psi is a situational awareness failure that can have unfortunate consequences.

One of the common human behavior factors that leads to a loss of situational awareness is the tendency to ignore or disregard information that is out of context. If an incident commander is busy thinking about an attack strategy, a brief observation or comment about unusual smoke conditions can be easily brushed aside and forgotten. A critical radio message might be acknowledged without really being understood or placed in context in the situation at hand.

The fire officer needs to constantly check and cross-check the situation and operational performance. This constant review helps ensure that the individual and the entire team are focused on concluding the incident in an efficient and injury-free fashion.

Maintaining Emergency Scene Situational Awareness

The six steps for maintaining emergency scene situational awareness are:

1. Fight the fire. All members have to focus their attention on the details of the incident, while keeping the larger picture in mind. Crash investigations show air crews become distracted by events that took their attention away from flying the plane. These distractions resulted in CFIT (controlled flight into terrain) events.

2. Assess problems in the time available. Emergency scenes do not always permit leisurely, timeless periods for decision making. There has to be a balance between rushing headlong into a burning building and waiting until every possible hazard is evaluated before attacking an incipient fire. Seasoned incident commanders know that taking an extra 30 to 60 seconds to absorb and process as much information as possible often results in better and more confident decisions.

3. Gather information from all sources. Information gathering at the emergency scene has to be a rapid process. One person cannot see everything, know everything, hear everything, or smell everything. Using the rest of the crew, as well as additional arriving command officers and people familiar with the building or terrain, should enhance decision making and help keep situational awareness current.

4. Choose the best option. Once all of the factors have been weighed, choose the option that maximizes results and minimizes risk.

5. Monitor results and alter the plan as necessary. Maintaining situational awareness requires continual evaluation of the effectiveness of decisions. Having a "plan B" or "plan C" ready for implementation may be necessary. An extra set of eyes at a strategic location on the incident scene can often provide valuable information.

6. Beware of situational awareness loss factors. The loss factors include:

   a. Ambiguity: An event, order, or message that has more than one potential meaning.

   b. Distraction: Anything that takes attention away from the larger mission. Distraction can come in the form of outside influences (the slamming of a door down the hall while you are watching television) or inside influences (cultural and personal biases).

   c. Fixation: Tunnel vision.

   d. Overload: More things are happening than one person can process. The first few minutes after arrival at an exceptionally chaotic scene can lead to overload.

   e. Complacency: When humans perform the same process repeatedly, they tend to become bored due to the hypnotic effect of repetitive action. Inattentiveness breeds carelessness. The perception or description of any structure fire as routine is the pinnacle of complacency.

   f. Improper procedure: Advancing a handline into a structure fire before ventilating and positioning apparatus downwind at a hazardous materials release are examples of improper procedures. Immediate action is required to correct these mistakes.

   g. Unresolved discrepancy: You are heading westbound on an interstate and you see a vehicle approaching eastbound in the same lane you are traveling in. There is a brief period where your eyes send the signal to the brain, but you mutter, "I don't believe this." Indications and observations that are in conflict with expectations must not be ignored. Fire fighters must be trained to recognize unresolved discrepancies as needing attention immediately.

Fire Marks

One Meridian Plaza

The 1991 triple fire fighter fatality at the One Meridian Plaza fire in Philadelphia provides a situational awareness example. Engine 11's crew, Captain David Holcombe, Fire Fighter Phyllis McAllister, and Fire Fighter James Chappell, were assigned to open a bulkhead door at the top of a stairway to assist with ventilation at the high-rise fire. The crew became fatigued climbing the stairs and radioed that they had left the stairway and were disoriented in heavy smoke on the 30th floor. Command initiated a search effort, sending additional crews up to the 30th floor. Finding no sign of the missing fire fighters at that level, the crews continued their search by moving to higher floors, eventually becoming victims themselves on the 38th floor. Additional search crews coming down from the roof found the distressed rescue team and safely removed them to the roof for evacuation.

The missing fire fighters were eventually found when a helicopter scanning the exterior of the building observed a lone, broken window on the 28th floor. The missing fire fighters were located and pronounced dead, just inside the broken window.

h. Nobody fighting the fire: An incident commander finds his attention drawn to the actions of one specific company. Emergency scene commanders cannot allow themselves to be drawn away from the big picture.

An aviation practice that can help is to say the checklist out loud when preparing for a low-frequency/high-risk task. Memorizing a checklist and going through it by yourself is not as safe or effective as two people looking at a checklist and speaking the items out loud. This practice is used to reduce paramedic medication errors.

Debriefing

The debriefing component, although not found in all CRM models, is valuable for the fire service. Debriefing offers personnel operating at an incident the opportunity to "replay" the event, extract lessons learned, and evaluate performance.

A formal debriefing should emphasize the elements of communication and teamwork (both leadership and followership), evaluate critical decision making, and reinforce the importance of situational awareness **Figure 18-5 ▾**. One recommendation is to assign an individual who was not involved in the incident to facilitate the debriefing. This individual can maintain a higher level of objectivity about the actions that were taken. The individual should be respected and familiar with departmental standard operating procedures.

Keeping the process simple and direct maximizes the learning environment, minimizes blame, and keeps everyone focused. Rules of decorum need to be established and followed. An environment of candid communication must be provided by the facilitator and respected by the participants.

Although Chapter 15 explains the post-incident review, the following technique is used by the U.S. Navy's precision flying team, the Blue Angels. Each mission is debriefed in an environment known as crossing the "blue line." The debriefing is conducted in an open, honest, candid atmosphere, with a goal of improving performance and ensuring team safety. No member of the team is immune from constructive criticism. Egos and feelings are left outside of the blue line, and whatever is said stays inside the blue line.

Figure 18-5 Debriefing offers personnel the opportunity to "replay" the event, extract lessons learned, and evaluate performance.

Safety Zone

Mental Joggers

A solid strategy for improving situational awareness requires the use of mental joggers for the individual and crew. The crew mental joggers ask:

- "What do we have here?"
- "What's going on here?"
- "How are we doing?"
- "Does this look right?"

The personal mental joggers ask:

- "What do I know that they need to know?"
- "What do they know that I need to know?"
- "What do we all need to know?"

Fire service debriefings should review the incident, capitalize on the positives, identify shortcomings, discuss strategies for improvement, and allow participants an opportunity to see and hear the incident through different sets of eyes and ears. Lubnau and Okray recommend a five-step model that lays out a simple formula for conducting a successful debriefing.

Step 1: Just the facts. Start with the dispatch and walk through the incident from beginning to end. Refer only to what actually took place (e.g., time of dispatch, unit or units that responded, time of arrival, apparatus position, weather, building construction).

Step 2: What did you do? Discuss the actions of the individual companies on the incident. Personnel get an opportunity to describe their actions on the scene. The facilitator needs to maintain order, allowing all participants to describe what they did, without interruption, and maintain a strong instructional tone. Points such as the weather, time of arrival, and building construction are rather concrete. Hose line placement, fire conditions, apparatus placement, and other observations are more argumentative. Recognize that some facts may be difficult to report with precision, and perceptions can vary.

Step 3: What went wrong? Most fire fighters would prefer to admit their own errors than to have their peers point out their shortcomings. The key here is for the facilitator to remind the group that a debriefing is not a kangaroo court. This step often requires a reminder of the rules of decorum. The environment to set here is one of encouraging all to provide an honest self-evaluation rather than be put under the microscope and be dissected by their peers. Avoid discussing the actions of companies or individuals who were at the incident but are not attending the critique.

Step 4: What went right? The concept is to get the poor performance issues out of the way first, and then note the positive actions to rebuild the individual's and team's desire to excel. Asking what went right challenges all team members to analyze the incident and identify as many actions as possible that had either favorable outcomes or resulted in positive results.

Step 5: What are you going to do about it? Most errors on the incident scene can be divided into two categories: poor personal or crew performance and mechanical breakdown.

Near Miss REPORT

Report Number: 06-0000132

Event Description: We were first due at a single story residential fire with heavy fire in the garage and brown, turbulent smoke pushing from the eaves on all four sides. As the officer of the engine, I knew we had fire in the garage and attic. We stretched a 1 3/4" line to the front door and proceeded toward the garage area.

The IC arrived on scene and took command. I gave him a "CAN" report—Conditions—heavy smoke to the floor with no visibility and high heat, Actions—hitting the fire that is rolling into the dining room from the garage and in a "hold" until conditions improve, Needs—second line for attic fire and ventilation both horizontal and vertical.

The IC confirmed my "CAN" report. The second-due engine is required to secure a water source per SOP, but when they arrived on scene they reported that a "mail box was in their way, and cannot get to the hydrant." The second-due engine then proceeded to stretch a second 1 3/4" line. The first-arriving truck arrived on scene and per SOP is required to vent and search.

The truck instead reported "there is a lot of fire, command, do you want us to pull a 2 1/2" line?" The IC confirmed to stretch a 2 1/2" line, which now meant we had two 1 3/4" and one 2 1/2" off our engine without a water supply and nobody performing vent. As I heard this on the radio, conditions were starting to worsen.

Our TIC "whited out" on my crew, the heat was dramatically increasing, and we had water flowing for at least 1 1/2 minutes. We knew our water wasn't going to last much longer. I pulled my crew out while hitting the fire to prevent flashover. Ten seconds after we exited the front door we ran out of water. The third-due engine finally established a water supply and the second-due truck performed vent.

Lessons Learned: There were numerous lessons learned. Command—Must have a strong command presence. When the second engine pulled a second line instead of securing a water supply, command should have said to secure a water supply or give their tank water to the first engine. In addition, when the truck asked to stretch the 2 1/2" line, the IC should have redirected them to vent and search.

Ventilation would have released the heat and smoke allowing us to push to the seat of the fire.

Decision Making—Having made the decision to pull my crew out when my support was not there was a correct decision. The other companies should have put the operation success first instead of "wanting to put water on the fire." Their decisions lead to a defensive attack that caused more property damage. Situational Awareness—As my crew made comments to me about the heat increasing, the conditions not improving, we were aware of our surroundings while taking a beating.

The IC did not recognize the two main factors of why fires don't go out . . . ventilation and water. The truck should have recognized that venting would have dramatically helped the first engine. SOP—The second engine and first truck did NOT follow our Typical Expectations SOP. This lead to a breakdown of priorities on scene.

Training—I trained with my driver never to charge any additional hose lines without a water supply. Reinforce with everyone the roles and responsibilities of their job functions. We need to train our command staff on how to handle situations when things don't go as on their check-off sheets.

A second measure of a debriefing's effectiveness consists of actions taken by the department to correct performance and mechanical issues identified during the debriefing. Debriefings become meaningless if the lessons that are identified are not addressed. Department inaction sets the stage for future debriefings to become unruly character assassinations that further divide crews, stations, and eventually the department itself. The swifter the troops see a response from their officers (including the fire chief), the more value they will attach to the process. Decisive responses from leadership instill confidence and restore morale.

Summary

The fire officer has the best opportunity to dramatically improve the safety and effectiveness of fire operations. The CRM system approach concentrates on the conditions under which individuals work and tries to build defenses to avert errors or mitigate their effects. High reliability organizations, which have less than their fair share of accidents, recognize that human variability is a force to harness in averting errors, but they work hard to focus that variability and are constantly preoccupied with the possibility of failure.

Firefighting shares the features of other high reliability organizations, like nuclear-powered aircraft carriers, air traffic control towers, nuclear power plants, and invasive medicine. They share these factors:

- Complex, internally dynamic, and, intermittently, intensely interactive
- Perform exacting tasks under considerable time pressure
- Carry out these demanding activities with low incident rates and an almost complete absence of catastrophic failures over several years

High reliability organizations that have embraced CRM have seen 70 percent to 80 percent reductions in incidents, accidents, and injuries. Implementing the error management model provides specific practices in six areas: communication skills, teamwork, task allocation, critical decision making, situational awareness, and debriefing.

The formal structure, tradition, and practices of the fire service provide a foundation for any fire officer to make significant improvements in fire fighter safety and incident effectiveness.

You Are the Fire Officer: Conclusion

The Incident Command System provides the structure needed when starting to build the framework for what may become an "Incident of National Importance." The needs greatly exceed the available resources. Clearly describing the scene, calling for more resources, and starting task allocations will start to force order onto chaos.

Much of the first-arriving crew's efforts will be to obtain the data needed to establish situational awareness and avoid task overload. There are far more urgent and life-saving issues that can be handled by the first fire crew with a broken pumper.

The fire officer must clearly set out the initial incident scene objectives and still needs to think 20 minutes ahead of the incident.

Wrap-Up

Chief Concepts

- The aviation industry has reduced its accident rate by 80 percent through the development, refinement, and system-wide delivery of a behavioral modification training system known as crew resource management (CRM).
- National Institute for Occupational Safety and Health reports on fire fighter fatalities routinely indicate that human error plays a role in each fatality.
- The proven principles of CRM are an effective means of reducing the effects of human error and preventing tragedy.
- Gordon Dupont determined that a "dirty dozen" of human factors contribute to tragedy:
  - Lack of communication
  - Complacency
  - Lack of knowledge
  - Distraction
  - Lack of teamwork
  - Fatigue
  - Lack of resources
  - Pressure
  - Lack of assertiveness
  - Stress
  - Lack of awareness
  - Norms
- A six-point model serves the fire service well:
  - Communication skills
  - Teamwork
  - Task allocation
  - Critical decision making
  - Situational awareness
  - Debriefing
- CRM suggests that developing a standard language, maintaining a "sterile cockpit," and teaching appropriate assertive behavior are the keys to reducing errors resulting from miscommunication.
- CRM promotes the concept of working together for the common good.
- Respect, trust, and effectiveness comprise the triangle of leadership.
- Know the crew's limits. Once an officer identifies his or her own limits, the crew needs to be evaluated for the same purpose.
- CRM promotes the concept of team involvement in all aspects of operation. In the area of emergency scene decision making, time, experience, and training play pivotal roles in successful outcomes.
- The loss of situational awareness is frequently the first link in a chain of errors that leads to calamity.
- Situational awareness loss factors include:
  - Ambiguity
  - Distraction
  - Fixation
  - Overload
  - Complacency
  - Improper procedure
  - Unresolved discrepancy
  - "Nobody fighting the fire"
- Debriefing offers personnel operating at an incident the opportunity to "rerun" the event, extract lessons learned, and evaluate performance.

Hot Terms

Active failures Unsafe acts committed by people who are in direct contact with the situation or system.

Crew resource management (CRM) A behavioral modification training system developed by the aviation industry to reduce its accident rate.

Followership The act or condition of following a leader; adherence.

Latent conditions Inevitable "resident pathogens" within the system.

Naturalistic decision making Describes how commanders make decisions in their natural environment.

Recognition primed decision making (RPD) Describes how commanders can recognize a plausible plan of action.

Situational awareness The process of evaluating the severity and consequences of an incident and communicating the results.

Fire Officer *in Action*

The administrative fire officer is working as the incident commander for the gasoline tanker incident. She has sectored the incident into four areas: Hospital, Nursing Home, Tanker, and Underground. Your crew has been reassembled and assigned to perform primary search in the hospital.

Crew resource management provides practices and procedures that allow a crew to operate effectively in a hostile and dangerous environment.

1. There is still fire in the first floor of the hospital. Normally, searches start from the fire area and move outward. One of your fire fighters says, "I am concerned that there is a lot of gasoline still in the basement. We should take a gas monitoring meter." This is an example of:
   A. taskmanship.
   B. an assertive statement.
   C. insubordination.
   D. a quality circle.

2. Error management involves all of the following activities EXCEPT:
   A. entrapment.
   B. avoidance.
   C. establishing expectations.
   D. mitigating consequences.

3. Not wearing a seat belt in a moving vehicle is an example of:
   A. active failure.
   B. incomplete task allocation.
   C. failed followership.
   D. a latent condition.

4. The difference between CRM and traditional fire supervision is:
   A. CRM requires committees and consensus.
   B. effective listening.
   C. increased authority of the informal leader.
   D. application of error management to human behavior.

Chapter 4: Fire Officer I

4.1* General. For qualification at Fire Officer Level I, the candidate shall meet the requirements of Fire Fighter II as defined in NFPA 1001, Fire Instructor I as defined in NFPA 1041, and the job performance requirements defined in Sections 4.2 through 4.7 of this standard.

4.1.1 General Prerequisite Knowledge. The organizational structure of the department; geographical configuration and characteristics of response districts; departmental operating procedures for administration, emergency operations, incident management system and safety; departmental budget process; information management and recordkeeping; the fire prevention and building safety codes and ordinances applicable to the jurisdiction; current trends, technologies, and socioeconomic and political factors that affect the fire service; cultural diversity; methods used by supervisors to obtain cooperation within a group of subordinates; the rights of management and members; agreements in force between the organization and members; generally accepted ethical practices, including a professional code of ethics; and policies and procedures regarding the operation of the department as they involve supervisors and members.

4.1.2 General Prerequisite Skills. The ability to effectively communicate in writing utilizing technology provided by the AHJ; write reports, letters, and memos utilizing word processing and spreadsheet programs; operate in an information management system; and effectively operate at all levels in the incident management system utilized by the AHJ.

4.2 Human Resource Management. This duty involves utilizing human resources to accomplish assignments in accordance with safety plans and in an efficient manner. This duty also involves evaluating member performance and supervising personnel during emergency and nonemergency work periods, according to the following job performance requirements.

4.2.1 Assign tasks or responsibilities to unit members, given an assignment at an emergency incident, so that the instructions are complete, clear, and concise; safety considerations are addressed; and the desired outcomes are conveyed.

(A)   Requisite Knowledge. Verbal communications during emergency incidents, techniques used to make assignments under stressful situations, and methods of confirming understanding.

(B)   Requisite Skills. The ability to condense instructions for frequently assigned unit tasks based on training and standard operating procedures.

4.2.2 Assign tasks or responsibilities to unit members, given an assignment under nonemergency conditions at a station or other work location, so that the instructions are complete, clear, and concise; safety considerations are addressed; and the desired outcomes are conveyed.

(A)   Requisite Knowledge. Verbal communications under nonemergency situations, techniques used to make assignments under routine situations, and methods of confirming understanding.

(B)   Requisite Skills. The ability to issue instructions for frequently assigned unit tasks based on department policy.

4.2.3 Direct unit members during a training evolution, given a company training evolution and training policies and procedures, so that the evolution is performed in accordance with safety plans, efficiently, and as directed.

(A)   Requisite Knowledge. Verbal communication techniques to facilitate learning.

(B)   Requisite Skills. The ability to distribute issue-guided directions to unit members during training evolutions.

4.2.4 Recommend action for member-related problems, given a member with a situation requiring assistance and the member assistance policies and procedures, so that the situation is identified and the actions taken are within the established policies and procedures.

(A)*   Requisite Knowledge. The signs and symptoms of member-related problems, causes of stress in emergency services personnel, adverse effects of stress on the performance of emergency service personnel, and awareness of AHJ member assistance policies and procedures.

(B)   Requisite Skills. The ability to recommend a course of action for a member in need of assistance.

4.2.5* Apply human resource policies and procedures, given an administrative situation requiring action, so that policies and procedures are followed.

(A)   Requisite Knowledge. Human resource policies and procedures.

(B)   Requisite Skills. The ability to communicate orally and in writing and to relate interpersonally.

4.2.6 Coordinate the completion of assigned tasks and projects by members, given a list of projects and tasks and the job requirements of subordinates, so that the assignments are prioritized, a plan for the completion of each assignment is developed, and members are assigned to specific tasks and both supervised during and held accountable for the completion of the assignments.

(A) Requisite Knowledge. Principles of supervision and basic human resource management.

(B) Requisite Skills. The ability to plan and to set priorities.

4.3 Community and Government Relations. This duty involves dealing with inquiries of the community and communicating the role, image, and mission of the department to the public and delivering safety, injury, and fire prevention education programs, according to the following job performance requirements.

4.3.1 Initiate action on a community need, given policies and procedures, so that the need is addressed.

(A) Requisite Knowledge. Community demographics and service organizations, as well as verbal and nonverbal communication, and an understanding of the role and mission of the department.

(B) Requisite Skills. Familiarity with public relations and the ability to communicate verbally.

4.3.2 Initiate action to a citizen's concern, given policies and procedures, so that the concern is answered or referred to the correct individual for action and all policies and procedures are complied with.

(A) Requisite Knowledge. Interpersonal relationships and verbal and nonverbal communication.

(B) Requisite Skills. Familiarity with public relations and the ability to communicate verbally.

4.3.3 Respond to a public inquiry, given policies and procedures, so that the inquiry is answered accurately, courteously, and in accordance with applicable policies and procedures.

(A) Requisite Knowledge. Written and oral communication techniques.

(B) Requisite Skills. The ability to relate interpersonally and to respond to public inquiries.

4.4 Administration. This duty involves general administrative functions and the implementation of departmental policies and procedures at the unit level, according to the following job performance requirements.

4.4.1 Recommend changes to existing departmental policies and/or implement a new departmental policy at the unit level, given a new departmental policy, so that the policy is communicated to and understood by unit members.

(A) Requisite Knowledge. Written and oral communication.

(B) Requisite Skills. The ability to relate interpersonally and to communicate change in a positive manner.

4.4.2 Execute routine unit-level administrative functions, given forms and record-management systems, so that the reports and logs are complete and files are maintained in accordance with policies and procedures.

(A) Requisite Knowledge. Administrative policies and procedures and records management.

(B) Requisite Skills. The ability to communicate orally and in writing.

4.4.3 Prepare a budget request, given a need and budget forms, so that the request is in the proper format and is supported with data.

(A) Requisite Knowledge. Policies and procedures and the revenue sources and budget process.

(B) Requisite Skill. The ability to communicate in writing.

4.4.4 Explain the purpose of each management component of the organization, given an organization chart, so that the explanation is current and accurate and clearly identifies the purpose and mission of the organization.

(A) Requisite Knowledge. Organizational structure of the department and functions of management.

(B) Requisite Skills. The ability to communicate verbally in a clear and concise manner.

4.4.5 Explain the needs and benefits of collecting incident response data, given the goals and mission of the organization, so that incident response reports are timely and accurate.

(A) Requisite Knowledge. The agency's records management system.

(B) Requisite Skills. The ability to communicate both orally and in writing.

4.5* Inspection and Investigation. This duty involves conducting inspections to identify hazards and address violations, performing a fire investigation to determine preliminary cause, securing the incident scene, and preserving evidence, according to the following job performance requirements.

4.5.1 Describe the procedures of the AHJ for conducting fire inspections, given any of the following occupancies, so that all hazards, including hazardous materials, are identified, approved forms are completed, and approved action is initiated:

(1) Assembly
(2) Educational
(3) Health care
(4) Detention and correctional
(5) Residential
(6) Mercantile
(7) Business
(8) Industrial
(9) Storage
(10) Unusual structures
(11) Mixed occupancies

(A)   Requisite Knowledge. Inspection procedures; fire detection, alarm, and protection systems; identification of fire and life safety hazards; and marking and identification systems for hazardous materials.

(B)   Requisite Skills. The ability to communicate in writing and to apply the appropriate codes and standards.

4.5.2 Identify construction, alarm, detection, and suppression features that contribute to or prevent the spread of fire, heat, and smoke throughout the building or from one building to another, given an occupancy, and the policies and forms of the AHJ so that a pre-incident plan for any of the following occupancies is developed:

(1) Public assembly
(2) Educational
(3) Institutional
(4) Residential
(5) Business
(6) Industrial
(7) Manufacturing
(8) Storage
(9) Mercantile
(10) Special properties

(A)   Requisite Knowledge. Fire behavior; building construction; inspection and incident reports; detection, alarm, and suppression systems; and applicable codes, ordinances, and standards.

(B)   Requisite Skills. The ability to use evaluative methods and to communicate orally and in writing.

4.5.3 Secure an incident scene, given rope or barrier tape, so that unauthorized persons can recognize the perimeters of the scene and are kept from restricted areas, and all evidence or potential evidence is protected from damage or destruction.

(A)   Requisite Knowledge. Types of evidence, the importance of fire scene security, and evidence preservation.

(B)   Requisite Skills. The ability to establish perimeters at an incident scene.

4.6* Emergency Service Delivery. This duty involves supervising emergency operations, conducting pre-incident planning, and deploying assigned resources in accordance with the local emergency plan and according to the following job performance requirements.

4.6.1 Develop an initial action plan, given size-up information for an incident and assigned emergency response resources, so that resources are deployed to control the emergency.

(A)   * Requisite Knowledge. Elements of a size-up, standard operating procedures for emergency operations, and fire behavior.

(B)   Requisite Skills. The ability to analyze emergency scene conditions; to activate the local emergency plan, including localized evacuation procedures; to allocate resources; and to communicate orally.

4.6.2* Implement an action plan at an emergency operation, given assigned resources, type of incident, and a preliminary plan, so that resources are deployed to mitigate the situation.

(A)   Requisite Knowledge. Standard operating procedures, resources available for the mitigation of fire and other emergency incidents, an incident management system, scene safety, and a personnel accountability system.

(B)   Requisite Skills. The ability to implement an incident management system, to communicate orally, to manage scene safety, and to supervise and account for assigned personnel under emergency conditions.

4.6.3 Develop and conduct a post-incident analysis, given a single unit incident and post-incident analysis policies, procedures, and forms, so that all required critical elements are identified and communicated, and the approved forms are completed and processed in accordance with policies and procedures.

(A)   Requisite Knowledge. Elements of a post-incident analysis, basic building construction, basic fire protection systems and features, basic water supply, basic fuel loading, fire growth and development, and departmental procedures relating to dispatch response tactics and operations and customer service.

(B)   Requisite Skills. The ability to write reports, to communicate orally, and to evaluate skills.

4.7* Health and Safety. This duty involves integrating health and safety plans, policies, and procedures into daily activities as well as the emergency scene, including the donning of appropriate levels of personal protective equipment to ensure a work environment that is in accordance with health and safety plans for all assigned members, according to the following job performance requirements.

4.7.1 Apply safety regulations at the unit level, given safety policies and procedures, so that required reports are completed, in-service training is conducted, and member responsibilities are conveyed.

(A)   Requisite Knowledge. The most common causes of personal injury and accident to members, safety policies and procedures, basic workplace safety, and the components of an infectious disease control program.

(B)   Requisite Skills. The ability to identify safety hazards and to communicate orally and in writing.

4.7.2 Conduct an initial accident investigation, given an incident and investigation forms, so that the incident is documented and reports are processed in accordance with policies and procedures of the AHJ.

(A)   Requisite Knowledge. Procedures for conducting an accident investigation and safety policies and procedures.

(B)   Requisite Skills. The ability to communicate orally and in writing and to conduct interviews.

4.7.3 Explain the benefits of being physically and medically capable of performing assigned duties and effectively functioning during peak physical demand activities, given current fire service trends and agency policies, so that the need to participate in wellness and fitness programs is explained to members.

(A)   Requisite Knowledge. National death and injury statistics; fire service safety and wellness initiatives; agency policies.

(B)   Requisite Skills. The ability to communicate orally.

Chapter 5: Fire Officer II

5.1 General. For qualification at Level II, the Fire Officer I shall meet the requirements of Fire Instructor I as defined in NFPA 1041 and the job performance requirements defined in Sections 5.2 through 5.7 of this standard.

5.1.1 General Prerequisite Knowledge. The organization of local government; enabling and regulatory legislation and the law-making process at the local, state/provincial, and federal levels; and the functions of other bureaus, divisions, agencies, and organizations and their roles and responsibilities that relate to the fire service.

5.1.2 General Prerequisite Skills. Intergovernmental and interagency cooperation.

5.2 Human Resource Management. This duty involves evaluating member performance, according to the following job performance requirements.

5.2.1 Initiate actions to maximize member performance and/or to correct unacceptable performance, given human resource policies and procedures, so that member and/or unit performance improves or the issue is referred to the next level of supervision.

(A)   Requisite Knowledge. Human resource policies and procedures, problem identification, organizational behavior, group dynamics, leadership styles, types of power, and interpersonal dynamics.

(B)   Requisite Skills. The ability to communicate orally and in writing, to solve problems, to increase team work, and to counsel members.

5.2.2 Evaluate the job performance of assigned members, given personnel records and evaluation forms, so each member's performance is evaluated accurately and reported according to human resource policies and procedures.

(A)   Requisite Knowledge. Human resource policies and procedures, job descriptions, objectives of a member evaluation program, and common errors in evaluating.

(B)   Requisite Skills. The ability to communicate orally and in writing and to plan and conduct evaluations.

5.2.3 Create a professional development plan for a member of the organization, given the requirements for promotion, so that the individual acquires the necessary knowledge, skills, and abilities to be eligible for the examination for the position.

(A)   Required Knowledge. Development of a professional development guide and job shadowing.

(B)   Required Skills. The ability to communicate orally and in writing.

5.3 Community and Government Relations. This duty involves dealing with inquiries of allied organizations in the community and projecting the role, mission, and image of the department to other organizations with similar goals and missions for the purpose of establishing strategic partnerships and delivering safety, injury, and fire prevention education programs, according to the following job performance requirements.

5.3.1 Explain the benefits to the organization of cooperating with allied organizations, given a specific problem or issue in the community, so that the purpose for establishing external agency relationships is clearly explained.

(A) Requisite Knowledge. Agency mission and goals and the types and functions of external agencies in the community.

(B) Requisite Skills. The ability to develop interpersonal relationships and to communicate orally and in writing.

5.4 Administration. This duty involves preparing a project or divisional budget, news releases, and policy changes, according to the following job performance requirements.

5.4.1 Develop a policy or procedure, given an assignment, so that the recommended policy or procedure identifies the problem and proposes a solution.

(A) Requisite Knowledge. Policies and procedures and problem identification.

(B) Requisite Skills. The ability to communicate in writing and to solve problems.

5.4.2 Develop a project or divisional budget, given schedules and guidelines concerning its preparation, so that capital, operating, and personnel costs are determined and justified.

(A) Requisite Knowledge. The supplies and equipment necessary for ongoing or new projects; repairs to existing facilities; new equipment, apparatus maintenance, and personnel costs; and appropriate budgeting system.

(B) Requisite Skill. The ability to allocate finances, to relate interpersonally, and to communicate orally and in writing.

5.4.3 Describe the process of purchasing, including soliciting and awarding bids, given established specifications, in order to ensure competitive bidding.

(A) Requisite Knowledge. Purchasing laws, policies, and procedures.

(B) Requisite Skills. The ability to use evaluative methods and to communicate orally and in writing.

5.4.4 Prepare a news release, given an event or topic, so that the information is accurate and formatted correctly.

(A) Requisite Knowledge. Policies and procedures and the format used for news releases.

(B) Requisite Skills. The ability to communicate orally and in writing.

5.4.5 Prepare a concise report for transmittal to a supervisor, given fire department record(s) and a specific request for details such as trends, variances, or other related topics.

(A) Requisite Knowledge. The data processing system.

(B) Requisite Skills. The ability to communicate in writing and to interpret data.

5.4.6 Develop a plan to accomplish change in the organization, given an agency's change of policy or procedures, so that effective change is implemented in a positive manner.

(A) Requisite Knowledge. Planning and implementing change.

(B) Requisite Skills. The ability to clearly communicate orally and in writing.

5.5 Inspection and Investigation. This duty involves conducting fire investigations to determine origin and preliminary cause, according to the following job performance requirements.

5.5.1 Determine the point of origin and preliminary cause of a fire, given a fire scene, photographs, diagrams, pertinent data, and/or sketches, to determine if arson is suspected.

(A) Requisite Knowledge. Methods used by arsonists, common causes of fire, basic cause and origin determination, fire growth and development, and documentation of preliminary fire investigative procedures.

(B) Requisite Skills. The ability to communicate orally and in writing and to apply knowledge using deductive skills.

5.6 Emergency Service Delivery. This duty involves supervising multi-unit emergency operations, conducting pre-incident planning, and deploying assigned resources, according to the following job requirements.

5.6.1 Produce operational plans, given an emergency incident requiring multi-unit operations, the current edition of NFPA 1600, and AHJ-approved safety procedures, so that required resources and their assignments are obtained and plans are carried out in compliance with NFPA 1600 and approved safety procedures resulting in the mitigation of the incident.

(A) Requisite Knowledge. Standard operating procedures; national, state/provincial, and local information resources available for the mitigation of emergency incidents; an incident management system; and a personnel accountability system.

(B) Requisite Skills. The ability to implement an incident management system, to communicate orally, to supervise and account for assigned personnel under emergency conditions, and to serve in command staff and unit supervision positions within the Incident Management System.

5.6.2 Develop and conduct a post-incident analysis, given multi-unit incident and post-incident analysis policies, procedures, and forms, so that all required critical elements are identified and communicated and the approved forms are completed and processed.

(A) Requisite Knowledge. Elements of a post-incident analysis, basic building construction, basic fire protection systems and features, basic water supply, basic fuel loading, fire growth and development, and departmental procedures relating to dispatch response, strategy tactics and operations, and customer service.

(B) Requisite Skills. The ability to write reports, to communicate orally, and to evaluate skills.

5.6.3 Prepare a written report, given incident reporting data from the jurisdiction, so that the major causes for service demands are identified for various planning areas within the service area of the organization.

(A) Requisite Knowledge. Analyzing data.

(B) Requisite Skills. The ability to write clearly and to interpret response data correctly to identify the reasons for service demands.

5.7 Health and Safety. This duty involves reviewing injury, accident, and health exposure reports, identifying unsafe work environments or behaviors, and taking approved action to prevent reoccurrence, according to the following job requirements.

5.7.1 Analyze a member's accident, injury, or health exposure history, given a case study, so that a report including action taken and recommendations made is prepared for a supervisor.

(A) Requisite Knowledge. The causes of unsafe acts, health exposures, or conditions that result in accidents, injuries, occupational illnesses, or deaths.

(B) Requisite Skills. The ability to communicate in writing and to interpret accidents, injuries, occupational illnesses, or death reports.

NFPA 1021, *Standard for Fire Officer Professional Qualifications*, 2009 Edition	Corresponding Chapter(s)	Corresponding Page(s)
4.1	1	5
4.1.1	1, 3, 4, 5, 13	4–5, 10–12, 15, 17, 45–53, 70–71, 79–82, 84, 87–88, 245–258
4.1.2	1, 14	6, 9, 15, 267–269, 272–281
4.2	4, 8	62–63, 68, 70, 142–144, 147
4.2.1	6, 9, 14	108, 110–111, 162–163, 165, 167–169, 271–272
4.2.2	9, 14	165, 169–170, 267–271
4.2.3	6, 7	99, 102–103, 105–112, 124–129, 132–135
4.2.4	8	142, 148–152, 154–155
4.2.5	3, 4, 5, 8	50–55, 68, 70, 90–92, 142–144, 147–149, 151–152, 154
4.2.6	4	63–72
4.3	10, 13, 14	176–185, 245–253, 279
4.3.1	10, 13, 14	177–185, 245–260, 279
4.3.2	11	203–207
4.3.3	10	180–181
4.4	3, 11	42–45, 201–205
4.4.1	11, 13, 18	206–207, 259–260, 353–362
4.4.2	3	42–44
4.4.3	13	245–258
4.4.4	13	245–258
4.4.5	14	271–281
4.5	12, 17	216–220, 230–236, 332–334, 343–344
4.5.1	12, 17	216–220, 230, 233–236, 341–343
4.5.2	12	220–230, 233
4.5.3	17	343–344
4.6	12, 15, 16	215–237, 287–303, 310–312, 314, 319–321
4.6.1	15, 16	289–300, 310–326
4.6.2	15, 16	287–297, 310–326

NFPA 1021, *Standard for Fire Officer Professional Qualifications*, 2009 Edition	Corresponding Chapter(s)	Corresponding Page(s)
4.6.3	6, 15	115–116, 300, 302–303
4.7	6	99–116
4.7.1	6	99–116
4.7.2	6	110, 114–115
4.7.3	6	99–101, 111–113
5.1	1	5
5.1.1	1, 5, 13	4, 9–11, 15–16, 79–84, 89–90, 246–249, 251–252, 259
5.1.2	1, 13	10, 15–16, 246–249, 251–252, 259
5.2	4, 8	70, 144–147
5.2.1	4, 8, 9	70, 144–147, 151–155, 164–165
5.2.2	8	144–147
5.2.3	2, 7	30, 132, 135–136
5.3	10, 13, 14	185–189, 247–249, 251, 273, 275–279
5.3.1	10, 13	185–189, 247–249, 251
5.4	10, 11, 13	187–188, 206–207, 245–260
5.4.1	11, 13	245–260
5.4.2	13	256–258
5.4.3	13	253, 255–256
5.4.4	10, 14	185–189, 267–281
5.4.5	14	273, 275–280
5.4.6	13, 18	259–260, 352–360
5.5	17	332–343
5.5.1	17	333–343, 345–346
5.6	12, 15, 16	215–237, 290–292, 311–315, 319–324, 326
5.6.1	6, 15, 16	106, 108, 110–111, 116, 288–300, 310–326
5.6.2	15	300, 302–303
5.6.3	14	278–280
5.7	6	99–116
5.7.1	6	99–104, 106, 114–115

Introduction to Fire and Emergency Services Administration (FESHE) Correlation Guide

Introduction to Fire and Emergency Services Administration (FESHE) Course Outcomes	Corresponding Chapter(s)	Corresponding Page(s)
1. Identify career development opportunities and strategies for success.	2, 7	25–36, 128, 135–136
2. Explain the need for effective communication skills both written and verbal.	7, 9, 10, 11, 12, 13, 14, 15, 16, 17, 18	125–128, 169, 178, 180–181, 185–189, 198–199, 203, 205–206, 233, 236–237, 248–249, 255–256, 259–260, 267–279, 281, 291, 296, 298, 300, 302–303, 310, 336–337, 344–345, 354, 356
3. Articulate the concepts of span and control, effective delegation, and division of labor.	1, 4, 16	13, 71–72, 321–322
4. Recognize appropriate appraising and disciplinary actions and the impact on employee behavior.	3, 4, 5, 8, 11, 18	49–50, 70–71, 90–92, 142–154, 206–207, 353
5. Examine the history and development of management and supervision.	1, 4, 5, 9	11–12, 15–16, 63, 79–86, 164–165
6. Evaluate methods of managing available resources.	13, 18	247–253, 357–358
7. Identify roles and responsibilities of leaders in organizations.	1, 3, 4, 5, 6, 7, 8, 9, 11, 12, 13, 16, 17, 18	4–6, 11, 42, 44–47, 49, 52–55, 62, 78–79, 87, 90, 99, 103, 122, 124–125, 128–129, 132, 142–143, 147, 162, 165, 167–170, 196–207, 215, 225, 230–232, 259, 310–311, 332–333, 335, 337, 343–344, 356–357, 362

Introduction to Fire and Emergency Services Administration (FESHE) Course Outcomes	Corresponding Chapter(s)	Corresponding Page(s)
8. Compare and contrast the traits of effective versus ineffective supervision and management styles.	4, 9, 16	62–70, 163–165, 311
9. Identify and assess safety needs for both emergency and nonemergency situations.	6, 7, 9, 10, 12, 13, 15, 16, 17, 18	105, 107–108, 110–116, 129, 133–134, 167, 178, 180–185, 216–218, 225–228, 230, 234–236, 256–257, 291, 300, 311, 314–317, 322–326, 342, 345, 352–354, 360–361
10. Identify the importance of ethics as they apply to supervisors.	1, 3, 11	17, 50, 52, 196, 199
11. Identify the role of a company officer in Incident Command System (ICS).	3, 6, 9, 12, 14, 15, 16, 18	46–47, 99–100, 105–108, 110–111, 163, 165, 167, 169, 216–220, 237, 271–272, 288–300, 311–315, 317–321, 323–324, 326, 358–360
12. Describe the benefits of documentation.	8, 15, 17	154, 300, 302–303, 345–346
13. Identify and analyze the major causes involved in line of duty firefighter deaths related to health, wellness, fitness, and vehicle operations.	6, 8, 18	154–155, 353

Glossary

Accelerant Agent, often ignitable liquid, used to initiate a fire or increase the rate of growth or spread of fire.

Accident An unplanned event that interrupts an activity and sometimes causes injury or damage. A chance occurrence arising from unknown causes; an unexpected happening due to carelessness, ignorance, and the like.

Accreditation A system whereby a certification organization determines that a school or program meets with the requirements of the fire service.

Actionable item Employee behavior that requires an immediate corrective action by the supervisor because dozens of lawsuits have shown that failing to act will create a liability and a loss for the department.

Active failures Unsafe acts committed by people who are in direct contact with the situation or system.

Administrative fire officer IAFC description of a person who has worked as a managing fire officer for 3 to 5 years, is certified at the NFPA Fire Officer III level, and has accomplished formal education equivalent to a bachelor's degree.

Adoption by reference Method of code adoption in which the specific edition of a model code is referred to within the adopting ordinance or regulation.

Adoption by transcription Method of code adoption in which the entire text of the code is published within the adopting ordinance or regulation.

Arbitration Resolution of a dispute by a mediator or a group rather than a court of law. Any civil matter may be settled in this way; some labor–management agreements include a binding arbitration clause.

Arson The crime of maliciously and intentionally, or recklessly, starting a fire or causing an explosion.

Artifacts The remains of the material first ignited, the ignition source, or other items or components in some way related to the fire ignition, development, or spread. An artifact may also be an item on which fire patterns are present, in which case the preservation of the artifact is not for the item itself but for the fire pattern that appears on the item.

Assessment centers A series of simulation exercises to identify a candidate's competency to perform the job that is offered in the promotional examination.

Assistant or division chief Midlevel chief who often has a functional area of responsibility, such as training, and answers directly to the fire chief.

Authority having jurisdiction An organization, office, or individual responsible for enforcing the requirements of a code or standard, or for approving equipment, materials, an installation, or a procedure.

Automatic sprinkler system A system of pipes with water under pressure that allows water to be discharged immediately when a sprinkler head operates.

Base The location at which the primary logistics functions are coordinated and administered. The incident command post may be colocated with the base. There is only one base per incident.

Base budget The level of funding that would be required to maintain all services at the currently authorized levels, including adjustments for inflation, salary increases, and other predictable cost changes.

Battalion chief Usually the first level of fire chief; also called district chief. These chiefs are often in charge of running calls and supervising multiple stations or districts within a city. A battalion chief is usually the officer in charge of a single-alarm working fire.

Binding arbitration The resolution of a dispute by a third and neutral party that is not personally involved in the dispute and may be expected to reach a fair and objective decision based on an informal hearing, at which the disputants may argue their cases and present all relevant evidence. It is usually agreed in advance that such a decision will be binding and not subject to appeal.

Bond A certificate of debt issued by a government or corporation; a bond guarantees payment of the original investment plus interest by a specified future date.

Brainstorming A method of shared problem solving in which all members of a group spontaneously contribute ideas.

Branch A supervisory level established in either the operations or logistics function to provide a span of control.

Branch director A supervisory position in charge of a number of divisions and/or groups. This position reports to a section chief or the incident commander.

Budget An itemized summary of estimated or intended expenditures for a given period, along with proposals for financing them.

Catastrophic theory of reform When fire prevention codes or firefighting procedures are changed in reaction to a fire disaster.

Central tendency An evaluation error that occurs when a fire fighter is rated in the middle of the range for all dimensions of work performance.

Chain of command The superior–subordinate authority relationship that starts at the top of the organization hierarchy and extends to the lowest levels.

Char Carbonaceous material that has been burned and has a blackened appearance.

Chief's trumpet An obsolete amplification device that enabled a chief officer to give orders to fire fighters during an emergency; precursor to a bullhorn and portable radios.

Class specification A technical worksheet that quantifies the knowledge, skills, and abilities (KSAs) by frequency and importance for every classified job within the local civil service agency.

Coaching A method of directing, instructing, and training a person or group of people with the aim to achieve some goal or develop specific skills.

Collective bargaining Method whereby representatives of employees (unions) and employers determine the conditions of employment through direct negotiation, normally resulting in a written contract setting forth the wages, hours, and other conditions to be observed for a stipulated period (e.g., 3 years). Term also applies to union–management dealings during the terms of the agreement.

Command staff Positions that are established to assume responsibility for key activities in the incident management system that are not a part of the line organization; these include safety officer, public information officer, and liaison officer.

Community Emergency Response Team (CERT) A fire department training program to help citizens understand their responsibilities in preparing for disaster and increase their ability to safely help themselves, their families, and their neighbors in the first 72 hours of a catastrophe.

Company journal Log book at the fire station that creates an extemporaneous record of the emergency, routine activities, and special activities that occurred at the fire station. The company journal also records any fire fighter injury, liability-creating event, and special visitors to the fire station.

Complaint Expression of grief, regret, pain, censure, or resentment; lamentation; accusation; or fault finding.

Conflict A state of opposition between two parties. A complaint is a manifestation of a conflict.

Consensus document A code or standard developed through agreement between people representing different organizations and interests. NFPA codes and standards are consensus documents.

Construction type The combination of materials used in the construction of a building or structure, based on the varying degrees of fire resistance and combustibility.

Contrast effect An evaluation error in which a fire fighter is rated on the basis of the performance of another fire fighter and not on the classified job standards.

Controlling Restraining, regulating, governing, counteracting, or overpowering.

Crew resource management (CRM) A behavioral modification training system developed by the aviation industry to reduce its accident rate.

"Data dump" question A promotional question that asks the candidate to write or describe all of the factors or issues covering a technical issue, such as suppression of a basement fire in a commercial property.

Decision making The process of identifying problems and opportunities and resolving them.

Defensive attack When suppression operations are conducted outside the fire structure and feature the use of large-capacity fire streams placed between the fire and the exposures to prevent fire extension.

Demographics The characteristics of human populations and population segments, especially when used to identify consumer markets. Generally includes age, race, sex, income, education, and family status.

Demonstrative evidence Tangible items that can be identified by witnesses, such as incendiary devices and fire scene debris.

Demotion A reduction in rank, with a corresponding reduction in pay.

Dimensions Attributes or qualities that can be described and measured during a promotional examination. Between 5 and 15 dimensions would be measured on a typical promotional examination. The six most common are oral communication, written communication, problem analysis, judgment, organizational sensitivity, and planning/organizing.

Discipline A moral, mental, and physical state in which all ranks respond to the will of the leader. Also, the guidelines that a department sets for fire fighters to work within.

Diversity A fire workforce that reflects differences in terms of age, cultural background, race, religion, sex, and sexual orientation.

Division A supervisory level established to divide an incident into geographical areas of operations.

Division of labor The production process in which each worker repeats one step over and over, achieving greater efficiencies in the use of time and knowledge; also, the formal assignment of authority and responsibility to job holders.

Division supervisor A supervisory position in charge of a geographical operation at the tactical level.

Documentary evidence Evidence in written form, such as reports, records, photographs, sketches, and witness statements.

Education The process of imparting knowledge or skill through systematic instruction.

Employee assistance program (EAP) An employee benefit that covers all or part of the cost for employees to receive counseling, referrals, and advice in dealing with stressful issues in their lives. These may include substance abuse, bereavement, marital problems, weight issues, or general wellness issues.

Environmental noise A physical or sociological condition that interferes with the message in the communication process.

Ethical behavior Decisions and behavior demonstrated by a fire officer that are consistent with the department's core values, mission statement, and value statements.

Evidence The documentary or oral statements and the material objects admissible as testimony in a court of law.

Expanded incident report narrative A report in which all company members submit a narrative on what they observed and activities they performed during an incident.

Expenditures The act of spending money for goods or services.

Failure analysis A logical, systematic examination of an item, component, assembly, or structure and its place and function within a system, conducted in order to identify and analyze the probability, causes, and consequences of potential and real failures.

Fair Labor Standards Act (FLSA) A 1938 act that provides the minimum standards for both wages and overtime entitlement and spells out administrative procedures by which covered work time must be compensated. Public safety workers were added to FLSA coverage in 1986.

Finance/administration section A section of the incident management system that is responsible for the accounting and financial aspects of an incident, as well as any legal issues that may arise.

Fire analysis The process of determining the origin, cause, development, and responsibility, as well as the failure analysis of a fire or explosion.

Fire chief The highest ranking officer in charge of a fire department. The individual assigned the responsibility for management and control of all matters and concerns pertaining to the fire service organization.

Fire mark Historically, an identifying symbol on a building to let fire fighters know that the building was insured by a company that would pay them for extinguishing the fire.

Fire patterns Physical marks left on an object by the fire.

Fire prevention division or hazardous use permit A local government permit renewed annually after the fire prevention division performs a code compliance inspection. A permit is required if the process, storage, or occupancy activity creates a life-safety hazard. Restaurants with more than 50 seats, flammable liquid storage, and printing shops that use ammonia are three examples of occupancies that may require a permit.

Fire scene reconstruction The process of re-creating the physical scene during fire scene analysis through the removal of debris and the replacement of contents or structural elements in their prefire position.

Fire tax district Special service district created to finance the fire protection of a designated district.

Fiscal year Twelve-month period for which an organization plans to use its funds. Local governments' fiscal years are generally from July 1 to June 30.

Floor plan View of a building's interior. Rooms, hallways, cabinets, and the like are drawn in the correct relationship to each other.

Followership The act or condition of following a leader; adherence.

Form of material What the ignited material is being used for; for example, the form of cotton material might be clothing or bales of cotton.

Formal communication Represents an official fire department communication. The letter or report is presented on stationery with the fire department letterhead and generally is signed by a chief officer or headquarters staff member.

Formal written reprimand An official negative supervisory action at the lowest level of the progressive disciplinary process.

Frame of reference An evaluation error in which the fire fighter is evaluated on the basis of the fire officer's personal standards instead of the classified job description standards.

Fuel load The total quantity of all the combustible products that are within a room or space.

General orders Short-term documents signed by the fire chief and lasting for a period of days to 1 year or more.

Geodatabase An object-relational database that uses geometry data types to represent the location of an object in the physical world.

Geographic information system (GIS) A GIS captures, stores, analyzes, manages, and presents data that are linked to a location.

Good faith bargaining A legal requirement of both the union and the employer arising out of Section 8(d) of the National Labor Relations Act. Enforced by the National Labor Relations Board, the parties are required: "To bargain collectively . . . to meet at reasonable times and confer in good faith with respect to wages, hours, and other conditions of employment, or the negotiation of an agreement or any question arising thereunder, and the execution of a written contract incorporating any agreement reached if requested by either party, but such obligation not to compel either party to agree to a proposal or require the making of a concession. . . ."

Governmental Accounting Standards Board (GASB) The mission of the Governmental Accounting Standards Board is to establish and improve the standards of state and local governmental accounting and financial reporting that will result in useful information for users of financial reports and to guide and educate the public, including issuers, auditors, and users of those financial reports.

Grievance A dispute, claim, or complaint that any employee or group of employees may have in relation to the interpretation, application, and/or alleged violation of some provision of the labor agreement or personnel regulations.

Grievance procedure A formal structured process that is employed within an organization to resolve a grievance. In most cases, the grievance procedure is incorporated in the personnel rules or the labor agreement and specifies a series of steps that must be followed in order.

Group A supervisory level established to divide the incident into functional areas of operation.

Group supervisor A supervisory position in charge of a functional operation at the tactical level. This position reports to a branch director, the operations section chief, or the incident commander.

Halo and horn effect An evaluation error in which the fire officer takes one aspect of a fire fighter's job task and applies it to all aspects of work performance.

Hazard Any arrangement of materials and heat sources that presents the potential for harm, such as personal injury or ignition of combustibles.

Health and safety officer The member of the fire department assigned and authorized by the fire chief as the manager of the safety and health program.

High-risk property Structure that contains the potential for a catastrophic property or life loss in the event of a fire.

High-value property Structure that contains equipment, materials, or items that have a high replacement value.

Humanistic management Emphasis on human need and attitude, motivation comes from within the employee and not from authoritarian control. Leads to Maslow's hierarchy of needs.

ICS general staff The group of incident managers composed of the operations section chief, planning section chief, logistics section chief, and finance/administration section chief.

Immediately dangerous to life and health (IDLH) Any condition that would do one or more of the following: (1) pose an immediate or delayed threat to life, (2) cause irreversible adverse health effects, or (3) interfere with an individual's ability to escape unaided from a hazardous environment.

Impasse Occurs when the parties have reached a deadlock in negotiations. Also described as the demarcation line between bargaining and negotiation. A declaration of an impasse brings in a state or federal negotiator that will start a fact-finding process and will lead to a binding arbitration resolution.

In-basket exercise A promotional examination component in which the candidate deals with correspondence and related items accumulated in a fire officer's in-basket.

Incident action plan (IAP) The objectives reflecting the overall incident strategy, tactics, risk management, and member safety that are developed by the incident commander. Incident action plans are updated throughout the incident.

Incident command system (ICS) A system that defines the roles and responsibilities to be assumed by personnel and the operating procedures to be used in the management and direction of emergency operations; the system is also referred to as an incident management system (IMS).

Incident commander The person who is responsible for all decisions relating to the management of the incident and is in charge of the incident site.

Incident safety officer An individual appointed to respond to or assigned at an incident scene by the incident commander to perform the duties and responsibilities specified in NFPA 1521, *Standard for Fire Department Safety Officer*.

Incident safety plan The strategies and tactics developed by the incident safety officer based on the incident commander's incident action plan and the type of incident encountered.

Incident scene rehabilitation The tactical-level management unit that provides for medical evaluation, treatment, monitoring, fluid and food replenishment, mental rest, and relief from climatic conditions of an incident.

Informal communications Internal memos, emails, instant messages, and computer-aided dispatch/mobile data terminal messages. Informal reports have a short life and are not archived as permanent records.

Interrogatory A series of formal written questions sent to the opposing side of a legal argument. The opposition must provide written answers under oath.

Investigation A systematic inquiry or examination.

Involuntary transfer or detail A disciplinary action in which a fire fighter is transferred or assigned to a less desirable or different work location or assignment.

Job description A narrative summary of the scope of a job. Provides examples of the typical tasks.

Job instruction training A systematic four-step approach to training fire fighters in a basic job skill: (1) prepare the fire fighters to learn, (2) demonstrate how the job is done, (3) try them out by letting them do the job, and (4) gradually put them on their own.

Knowledge, skills, and abilities (KSAs) The traits required for every classified position within the municipality. Defined by a narrative job description and a technical class specification.

Latent conditions Inevitable "resident pathogens" within the system.

Leadership A complex process by which a person influences others to accomplish a mission, task, or objective and directs the organization in a way that makes it more cohesive and coherent.

Leading Guiding or directing in a course of action.

Liaison officer The incident commander's representative or a point of contact for representatives from outside agencies.

Line-item budget Budget format in which expenditures are identified in a categorized line-by-line format.

Lobby control officer The lobby control officer controls the entering and exiting of both civilians and fire fighters in the lobby. Also oversees the use of the elevators, operates the local building communication system, and assists in the control of the heating, ventilating, and air conditioning systems.

Logistics section Responsible for providing facilities, services, and materials for the incident. Includes the communications, medical, and food units within the service branch, as well as the supply, facilities, and ground support units within the support branch. The logistics section chief is part of the general staff.

Logistics section chief A supervisory position that is responsible for providing supplies, services, facilities, and materials during the incident. The person in this position reports directly to the incident commander.

Loudermill hearing A predisciplinary conference that occurs before a suspension, demotion, or involuntary termination is issued. The term refers to a Supreme Court decision.

Managing Fire Officer Description from the IAFC *Officer Development Handbook* on the tasks and expectations for a Fire Officer II. Encourages the company officer to acquire the appropriate levels of training, experience, self-development, and education to prepare for the Chief Fire Officer designation.

Masonry wall May consist of brick, stone, concrete block, terra cotta, tile, adobe, precast, or cast-in-place concrete.

Mediation The intervention of a neutral third party in an industrial dispute. The object is to enable the two sides to reach a compromise solution to their differences, which the mediator usually does by seeing representatives of both sides separately and then together.

Mentoring A developmental relationship between a more experienced person, a mentor, and a less experienced person, referred to as a protégé.

Metropolitan (metro-sized) fire department A department with more than 400 fully paid fire fighters. The Metropolitan Fire Chiefs is a special interest group organized by the International Association of Fire Chiefs in 1965 to address the needs of large fire departments. Metro Chiefs is also a section of the National Fire Protection Association.

Mini/max codes Codes developed and adopted at the state level for either mandatory or optional enforcement by local governments; these codes cannot be amended by local governments.

Mistake An error or fault resulting from defective judgment, deficient knowledge, or carelessness. A misconception or misunderstanding.

Mitigation Measures to be taken to limit or control the consequences, extent, or severity of an incident that cannot be reasonably prevented.

Model codes Codes generally developed through the consensus process with the use of technical committees developed by a code-making organization.

National Fire Academy fire flow formula Used to allow a first-arriving fire officer to quickly determine the needed fire flow. Determined on a per-floor calculation, the length is multiplied by the width, and the result is divided by three.

National Fire Incident Reporting System (NFIRS) A nationwide database at the National Fire Data Center under the U.S. Fire Administration that collects fire-related data in order to provide information on the national fire problem.

National Incident Management System (NIMS) Provides a consistent nationwide template to enable federal, state, tribal, and local governments; the private sector; and nongovernmental organizations to work together to prepare for, prevent, respond to, recover from, and mitigate the effects of incidents, regardless of cause, size, location, or complexity, in order to reduce the loss of life, property, and harm to the environment.

National Response Framework (NRF) A comprehensive, national, all-hazards approach to domestic incident response by describing specific authorities and best practices for managing incidents that builds upon the National Incident Management System (NIMS), which provides a consistent template for managing incidents.

Naturalistic decision making Describes how commanders make decisions in their natural environment.

Negotiation Mutual discussion and arrangement of the terms of an agreement.

Occupancy type The purpose for which a building or a portion thereof is used or intended to be used.

Offensive attack An advance into the fire building by fire fighters with hose lines or other extinguishing agents to overpower the fire.

Ongoing compliance inspection Inspection of an existing occupancy to observe the housekeeping and confirm that the built-in fire protection features, such as fire exit doors and sprinkler systems, are in working order.

Operational period A term used with a written incident action plan identifying a period of time during a long-term incident that a specific incident action plan covers. For federally funded incidents, the operational period is 12 hours; local incidents may use operational periods of 8 hours.

Operations section Responsible for all tactical operations at the incident. In the national model, the operations section can be as large as 5 branches, 25 divisions/groups or units, or 125 single resources, task forces, or strike teams.

Operations section chief A supervisory position that is responsible for the management of all actions that are directly related to controlling the incident. This position reports directly to the incident commander.

Oral reprimand, warning, or admonishment The first level of negative discipline. Considered informal, it remains with the fire officer and is not part of the fire fighter's official record.

Ordinance A law of an authorized subdivision of a state, such as a city, county, or town.

Organizing Putting together into an orderly, functional, structured whole.

Performance log Informal record maintained by the fire officer listing fire fighter activities by date and with a brief description. Used to provide documentation for annual evaluations and special recognitions.

Personal bias An evaluation error that occurs when the evaluator's perspective skews the evaluation such that the classified job knowledge, skills, and abilities are not appropriately evaluated.

Personal study journal A personal notebook to aid in scheduling and tracking a candidate's promotional preparation progress.

Personnel accountability system A method of tracking the identity, assignment, and location of fire fighters operating at an incident scene.

Planning Developing a scheme, program, or method that is worked out beforehand to accomplish an objective.

Planning section Responsible for the collection, evaluation, dissemination, and use of information about the development of the incident and the status of resources. Includes the situation status, resource status, and documentation units as well as technical specialists. The planning section chief is part of the general staff.

Planning section chief A supervisory position that is responsible for the collection, evaluation, dissemination, and use of information relevant to the incident. This position reports directly to the incident commander.

Plot plan A representation of the exterior of a structure, identifying doors, utilities access, and any special considerations or hazards.

Point of origin The exact physical location where a heat source and a fuel come in contact with each other and a fire begins.

Policies Formal statements that provide guidelines for present and future actions; policies often require personnel to make judgments.

Political action committee (PAC) An organization formed by corporations, unions, and other interest groups that solicit campaign contributions from private individuals and distribute these funds to political candidates.

Power The capacity of one party to influence another party.

Pre-incident plan A written document resulting from the gathering of general and detailed data to be used by responding personnel for determining the resources and actions necessary to mitigate anticipated emergencies at a specific facility.

Pretermination hearing An initial check to determine if there are reasonable grounds to believe that the charges against the employee are true and support the proposed termination.

Problem A condition in which the desired situation is different from the current situation.

Progressive negative discipline A process for dealing with job-related behavior that does not meet expected and communicated performance standards. Discipline increases from mild to more severe punishments if the problem is not corrected.

Public information officer A command staff position that is responsible for gathering and releasing incident information to the news media and other appropriate agencies. This position reports directly to the incident commander.

Pyrolysis The destructive distillation of organic compounds in an oxygen-free environment that converts the organic matter into gases, liquids, and char.

Rapid intervention crew (RIC) A minimum of two fully equipped personnel on site, in a ready state, for immediate rescue of disoriented, injured, lost, or trapped rescue personnel.

Recency An evaluation error in which the fire fighter is evaluated only on incidents that occurred over the past few weeks rather than on the entire evaluation period.

Recognition primed decision making (RPD) Describes how commanders can recognize a plausible plan of action.

Recommendation report A decision document prepared by a fire officer for the senior staff. The goal is support for a decision or an action.

Regulations Governmental orders written by a governmental agency in accordance with the statute or ordinance authorizing the agency to create the regulation. Regulations are not laws but have the force of law.

Rehabilitation The process of providing rest, rehydration, nourishment, and medical evaluation to members who are involved in extended or extreme incident scene operations.

Restrictive duty A temporary work assignment during an administrative investigation that isolates the fire fighter from the public and usually is an administrative assignment away from the fire station.

Revenues The income of a government from all sources appropriated for the payment of the public expenses.

Right to work A worker cannot be compelled, as a condition of employment, to join or not to join or to pay dues to a labor union.

Risk assessment Involves identifying hazards, monitoring those hazards, determining the likelihood of their occurrence, and assessing the vulnerability of people, property, the environment, and the entity itself to those hazards.

Risk/benefit analysis A decision made by a responder based on a hazard and situation assessment that weighs the risks likely to be taken against the benefits to be gained for taking those risks.

Risk management Identification and analysis of exposure to hazards, selection of appropriate risk management techniques to handle exposures, implementation of chosen techniques, and monitoring of results, with respect to the health and safety of members.

Risk Watch A comprehensive NFPA school-based program focused on injury prevention.

Rules and regulations Developed by various government or government-authorized organizations to implement a law that has been passed by a government body.

Safety officer The person who is responsible for all decisions relating to the management of the incident and is in charge of the incident site as well as safety issues.

Safety unit A member or members assigned to assist the incident safety officer. The tactical-level management unit that can be composed of the incident safety officer alone or with additional assistant safety officers assigned to assist in providing the level of safety supervision appropriate for the magnitude of the incident and the associated hazards.

Scientific management The breakdown of work tasks into constituent elements; the timing of each element is based on repeated stopwatch studies; the fixing of piece rate compensation based on those studies; standardization of work tasks on detailed instruction cards; and generally, the systematic consolidation of the shop floor's brain work.

Service branch A major division within the logistics section of the ICS. Oversees the communications, medical, and food units.

Shop steward A union member appointed or elected to be the first line of labor representation at the workplace. The steward enforces the contract, collective agreement, or memorandum of understanding and represents the union members at that fire station or work location.

Situational awareness The process of evaluating the severity and consequences of an incident and communicating the results.

Size-up A systematic process of gathering and processing information to evaluate the situation and then translating that information into a plan to deal with the situation.

Source of ignition Devices or equipment that, because of their intended modes of use or operation, are capable of providing sufficient thermal energy to ignite flammable gas–air mixtures.

Span of control The maximum number of personnel or activities that can be effectively controlled by one individual (usually three to seven).

Special evaluation period A designated period of time when an employee is provided additional training to resolve a work performance/behavioral issue. The supervisor issues an evaluation at the end of the special evaluation period.

Spoils system Also known as the patronage system. The practice of making appointments to public office based on a personal relationship or affiliation rather than because of merit. The spoils system scandals of the New York City "Tweed Ring" and the Tammany Hall political machine (1865–1871) resulted in Congress passing the Pendleton Civil Service Reform Act of 1883.

Staging A specific function in which resources are assembled in an area at or near the incident scene to await instructions or assignments.

Stairwell support group This group moves equipment and water supply hose lines up and down the stairwells. The stairwell support unit leader reports to the support branch director or the logistics section chief.

Standard operating procedures (SOPs) Written organizational directives that establish or prescribe specific operational or administrative methods to be followed routinely for the performance of designated operations or actions.

Standpipe system An arrangement of piping, valves, hose connections, and allied equipment installed in a building or structure, with the hose connections located in such a manner that water can be discharged in streams or spray patterns through attached hose and nozzles, for the purpose of extinguishing a fire, thereby protecting a building or structure and its contents in addition to protecting the occupants. This is accomplished by means of connections to water supply systems or by means of pumps, tanks, and other equipment necessary to provide an adequate supply of water to the hose connections.

Strategic level Command level that entails the overall direction and goals of the incident.

Strike A concerted act by a group of employees, withholding their labor for the purposes of effecting a change in wages, hours, or working conditions.

Strike team Specified combinations of the same kind and type of resources, with common communications and a leader.

Strike team leader A supervisory position that is in charge of a group of similar resources.

Supervising Fire Officer Description from the IAFC *Officer Development Handbook* on the tasks and expectations for a Fire Officer I. Encourages the company officer to acquire the appropriate levels of training, experience, self-development, and education to prepare for the Chief Fire Officer designation.

Supervisor's report A form that is required by most state workers' compensation agencies that is completed by the immediate supervisor after an injury or property damage accident.

Supplemental budget Proposed increases in spending to provide additional services.

Support branch A major division within the logistics section of the ICS. Oversees the supply, facilities, and ground support units.

Suspension A negative disciplinary action that removes a fire fighter from the work location; he or she is generally not allowed to perform any fire department duties.

T-account A documentation system, similar to an accounting balance sheet, listing credits and debits, in which a single-sheet form is used to list the employee's assets on the left side and liabilities on the right side, appearing like a letter "T."

Tactical level Command level in which objectives must be achieved to meet the strategic goals. The tactical-level supervisor or officer is responsible for completing assigned objectives.

Tactical worksheet A form that allows the incident commander to ensure all tactical issues are addressed and to diagram an incident with the location of resources on the diagram.

Task force Any combination of single resources assembled for a particular tactical need, with common communications and a leader.

Task force leader A supervisory position that is in charge of a group of dissimilar resources.

Task level Command level in which specific tasks are assigned to companies; these tasks are geared toward meeting tactical-level requirements.

Termination The organization ends an individual's employment against his or her will.

Testimonial evidence Witnesses speaking under oath.

Trailers Materials used to spread fire from one area of a structure to another.

Training The process of achieving proficiency through instruction and hands-on practice in the operation of equipment and systems that are expected to be used in the performance of assigned duties.

Transitional attack A situation in which an operation is changing or preparing to change.

Two-in-two-out rule OSHA Respiratory Regulation (29 CFR 1910.134) that requires a two-person team to operate within an environment that is immediately dangerous to life and health (IDLH) and a minimum of a two-person team to be available outside the IDLH atmosphere to remain capable of rapid rescue of the interior team.

Type of material What the ignited material is made of. For example, the type of material might be cotton.

Unconsciously competent The highest level of the Conscious Competence Learning Matrix developed by Dr. Thomas Gordon in the 1970s. At the unconsciously competent level the skill becomes so practiced that it enters the unconscious parts of the brain—it becomes second nature.

Unfair labor practice An employer or union practice forbidden by the National Labor Relations Board or state/local laws, subject to court appeal. It often involves the employer's efforts to avoid bargaining in good faith.

Unit Either a geographical or a functional assignment.

Unity of command The management concept that a subordinate should have only one direct supervisor, and a decision can be traced back through subordinates to the manager who originated it.

Use group Found in the building code classification system whereby buildings and structures are grouped together by their use and by the characteristics of their occupants.

Work improvement plan A written document that is part of a special evaluation period. The plan identifies performance deficiencies and lists the improvements in performance or changes in behavior required to obtain a "satisfactory" evaluation.

Yellow dog contracts Pledges that employers required workers to sign indicating that they would not join a union as long as the company employed them. Declared unenforceable by the Norris-LaGuardia Act of 1932.

Index

Credits

You Are The Fire Officer Courtesy of Captain David Jackson, Saginaw Township Fire Department; **Voices of Experience** Courtesy of Captain David Jackson, Saginaw Township Fire Department; **cloud of smoke** © Greg Henry/ShutterStock, Inc.

Chapter 1
Opener © Glen E. Ellman; **1-1** © Chicago Historical Society ICHi-02808; **1-2** Courtesy of the FASNY Fire Museum of Firefighting, Hudson, New York; **1-3** © fotofacade.com/Alamy Images

Chapter 3
Opener © michael ledray/ShutterStock, Inc.; **3-4** Courtesy of Captain David Jackson, Saginaw Township Fire Department; **3-11** © Joao Virissimo/ShutterStock, Inc.

Chapter 5
5-2 © Matt Rourke/AP Photos; **5-3** © *The Charlotte Observer*/MCT/Landov

Chapter 6
Opener © Glen E. Ellman

Chapter 7
Opener © Mark C. Ide; **7-1B** Courtesy of Captain David Jackson, Saginaw Township Fire Department;

7-7B Courtesy of Captain David Jackson, Saginaw Township Fire Department

Chapter 9
Opener © Glen E. Ellman; **9-1** © Glen E. Ellman

Chapter 10
Opener Courtesy of Captain David Jackson, Saginaw Township Fire Department; **10-3** Courtesy of Captain David Jackson, Saginaw Township Fire Department; **10-4** © National Fire Protection Association; **10-5** © Wilfredo Lee/AP Photos; **10-6** © Patricia Marks/ShutterStock, Inc.; **10-9** © Images-USA/Alamy Images

Chapter 11
Opener © Glen E. Ellman; **11-1** © Damian Dovarganes/AP Photos

Chapter 12
Opener © Dennis Wetherhold, Jr.; **12-1** © Dennis Wetherhold, Jr.; **12-7** © John Foxx/Alamy Images; **12-8** © Dennis Tokarzewski/ShutterStock, Inc.; **12-9** © Ken Hammond/USDA; **12-10** Courtesy of APA - The Engineered Wood Association; **12-13** © Photodisc; **12-14B** © Losevsky Pavel/ShutterStock, Inc.

Chapter 14
Opener © Glen Ellman; **14-3** © Glen E. Ellman

Chapter 15
Opener © Glen E. Ellman; **15-7** © Keith D. Cullom

Chapter 16
Opener © Glen E. Ellman; **16-1** Photographed by Mike Legeros; **16-3** Courtesy of Dennis Walus; **16-4A-D** Courtesy of Allen Sklar <www.allensklar.smugmug.com>

Chapter 17
17-1 © Rhode Island Attorney General's Office/AP Photos; **17-2, 17-3** Courtesy of Eddie D. Smith/Unified Investigations & Sciences, Inc.; **17-5** Courtesy of Robert A. Corry/SceneInvestigator.com; **17-6** © Frances Roberts/Alamy Images; **17-7** Courtesy of Captain David Jackson, Saginaw Township Fire Department

Chapter 18
Opener © Mark C. Ide; **18-4** © 2003, Berta A. Daniels; **18-5** Photographed by Mike Legeros

Unless otherwise indicated, all photographs and illustrations are under copyright of Jones and Bartlett Publishers, LLC, courtesy of Maryland Institute for Emergency Medical Services Systems, or have been provided by the authors.